John M. Doyle
Consultant, Engineering Publications

PULSE

FUNDAMENTALS

Under the editorship of Dr. Irving L. Kosow

PRENTICE-HALL, INC.
Englewood Cliffs, N. J.

PRENTICE-HALL INTERNATIONAL, INC., *London*
PRENTICE-HALL OF AUSTRALIA, PTY., LTD., *Sydney*
PRENTICE-HALL OF CANADA, LTD., *Toronto*
PRENTICE-HALL OF JAPAN, INC., *Tokyo*

TO MY SON *John*

Current printing (last digit):
12   11   10   9   8   7   6   5   4   3

© 1963
by *PRENTICE-HALL, Inc.,*
Englewood Cliffs, New Jersey

All rights reserved. No part of this book may be reproduced in any form, by mimeograph or any other means, without permission in writing from the publisher.

Printed in the United States of America.

Library of Congress Catalog Card Number   63-15290

C-74110

PREFACE

This book presents a comprehensive course in pulse circuit techniques at the technical institute level. A working knowledge of d-c and a-c circuit theory is presupposed. A working knowledge of basic algebra is the only mathematical requirement.

The introductory chapter describes the essential differences between pulse and communication-type networks and shows typical applications of the former. Various nonsinusoidal waveshapes and the parameters by which they are described are also included.

Chapter 2 discusses linear networks (resistors, inductors, and capacitors), and their response to the nonlinear waveshapes studied in Chapter 1.

In Chapter 3, the operation of resistance-capacitance coupled vacuum-tube amplifiers is reviewed. Equivalent circuits are derived and their importance as an analytical tool is demonstrated.

Chapters 4 and 5 are devoted to the transistor. No attempt is made to discuss the physics of these devices, but problems of such practical importance as stability of the operating point, load line analysis, and equivalent circuit parameters and their application are covered thoroughly. From this point forward, vacuum-tube and semiconductor devices are treated in a unified, integrated manner.

Methods of amplifier compensation for wideband application are investigated next. High frequency shunt, series, shunt-series, four-terminal compensating networks, and low-frequency compensation networks are described. Important factors in selecting vacuum tubes and transistors for pulse work are also studied.

Cathode and emitter followers are discussed in Chapter 8. Equivalent circuits are derived for both type "followers" and gain, and impedance calculations are again demonstrated.

Chapter 9 discusses characteristics of vacuum tube and transistors as switching circuit components. The principles discussed in this chapter are most important for a thorough understanding of the remainder of the book.

Following this are chapters on nonlinear waveshaping circuits (those using active elements — tubes and transistors), clamping circuits, the various types of multivibrators, blocking oscillators, time base generators, counting circuits, transmission gates, pulse modulation systems, and delay lines. Typical applications for each type of circuit are described. Free use is made of circuit diagrams and waveform illustrations to clarify the explanations.

In the final chapter, all circuits previously described are shown working together in a laboratory type oscilloscope.

Numerous references are given throughout the book. Wherever possible, the selected references are also at the technical institute level. Of necessity, however, some are of a more difficult nature, but the reader should still find them useful.

Questions are given at the end of each chapter. They are intended primarily as a stimulant to group discussion, but may also be used as objective "homework" assignments to classroom instructors. Full answers may be obtained by qualified instructors upon written request to the publisher.

It is my pleasure to acknowledge assistance from the following persons and organizations in the preparation of this book: Dr. Irving L. Kosow, Professor and Head, Electronic Technology, Staten Island Community College, Staten Island, N. Y., whose contributions to the organization and scope of the text were invaluable; Dr. Charles M. Thompson, Wentworth Institute, Boston, Mass., for reading the complete manuscript and making suggestions for improvement; Darrell L. Geiger and G. O. Allen, Chief Instructor and President, respectively, Cleveland Institute of Electronics, Cleveland, Ohio, for initial prompting to write the book, constant encouragement, and permission to use portions of texts originally written for the C.I.E.; Jake Davis Jr., Educational Director, National Radio Television School, Cleveland, Ohio, for permission to use portions of this text in resident study courses and to my students for their many worthwhile comments; my editorial associates at the National Radio Institute, Washington, D. C., particularly, John F. Zung, for manuscript reading and comment; Lindbergh H. Trent, independent service technician, for his most observant comments on the practical aspects of circuit operation; Nolan G. Thorsteinson, free lance technical writer, and his good wife Patty,

Preface

for assistance in manuscript planning and constant encouragement; Carl W. Anderson, Hewlett-Packard Co., for supplying an abundance of technical literature on counters and oscilloscopes; Electronics Digest, Cleveland, Ohio, for permission to use portions of material originally written for that publication; Mrs. Emma E. Grist, head of the correspondence section at National Radio Institute for complete manuscript typing; and finally, my wife Joanne for her unfailing confidence and consideration.

An effort has been made to acknowledge all sources of information; however, original authorship could not be determined in some instances. Any omissions are unintentional and deeply regretted.

CONTENTS

1 Introduction   1

1.1 Historical developments, 1; 1.2 Typical pulse applications, 2; 1.3 Definitions, 7; 1.4 General considerations, 10.

2 Linear Wave Shaping   12

2.1 The high-pass R-C filter, 12; 2.2 The R-C time constant, 15; 2.3 The universal time-constant chart, 16; 2.4 The capacitor discharge period, 17; 2.5 Network response, 19; 2.6 Response to a square wave input, 23; 2.7 Effect of rise time on amplitude of output waveform, 24; 2.8 Response of the high-pass R-C filter to a ramp voltage, 24; 2.9 The R-C differentiator, 26; 2.10 The low-pass R-C filter, 28; 2.11 Response of the low-pass R-C filter to a pulse input, 29; 2.12 Response of low-pass filter to square wave input, 31; 2.13 Response of low-pass filter to a ramp input, 31; 2.14 The R-C integrator, 32; 2.15 R-L circuits, 33; 2.16 R-L-C circuits, 34; 2.17 Ringing oscillator, 37; 2.18 R-L-C peaker, 38.

3 Vacuum Tube R-C Amplifiers — 41

3.1 Static, quiescent, and dynamic operation, 41; 3.2 Vacuum tube parameters, 43; 3.3 The constant-voltage equivalent circuit, 45; 3.4 How frequency affects the constant voltage equivalent circuit, 47; 3.5 The constant-current equivalent circuit, 50; 3.6 A universal response curve for R-C coupled amplifiers, 51; 3.7 Distortion, 52.

4 Introduction to Transistors — 55

4.1 Basic circuits and circuit notation, 56; 4.2 Leakage currents, 60; 4.3 Graphs, 61; 4.4 The operating point, 64; 4.5 Important variations in transistor replacement, 66; 4.6 Bias circuits, 67; 4.7 Thermal runway, 71; 4.8 Small-signal graphical analysis, 74.

5 Transistor Equivalent Circuits and Parameters — 79

5.1 The r parameters, 80; 5.2 The h (hybrid) parameters, 89; 5.3 Relating the h and r parameters, 96; 5.4 Transistor frequency response, 97; 5.5 Transistor R-C amplifier, 99.

6 Wideband Amplifiers — 104

6.1 Defects in the reproduced waveform and the meaning of transient response, 105; 6.2 Vacuum tube gain-bandwidth product, 106; 6.3 Transistor gain-bandwidth product, 107; 6.4 Vacuum tube figure-of-merit, 107; 6.5 Transistor figure-of-merit, 108; 6.6 Need for compensation, 109; 6.7 Methods of high frequency compensation, 109; 6.8 Low-frequency response, 112; 6.9 Direct-coupled vacuum tube amplifiers, 115; 6.10 Direct-coupled transistor amplifiers, 117.

7 Feedback Amplifiers — 120

7.1 Understanding feedback, 121; 7.2 Improved stability, 122; 7.3 Using decibels (db), 122; 7.4 Reduced nonlinear distortion, 123; 7.5 Maintaining stability over a wide frequency range, 125; 7.6 Voltage and current feedback, 128; 7.7 Two-stage feedback, 128; 7.8 Three-stage feedback, 130; 7.9 Effect of voltage feedback on output impedance, 131; 7.10 Effect of voltage feedback on input impedance, 132; 7.11 Effects of negative current feedback on output and input impedances, 133; 7.12 Illustrating improved frequency response, 135.

8 Cathode-Follower Circuits, Emitter Followers, and Phase Inverters — 139

8.1 The cathode-follower circuit, 140; 8.2 Voltage gain, 141; 8.3 Power gain, 142; 8.4 Input resistance and capacitance, 142; 8.5 Output impedance, 143; 8.6 Frequency response, 144; 8.7 The common-collector amplifier-general, 144; 8.8 Common-collector r parameters, 145; 8.9 Common-collector h parameters, 147; 8.10 Relating the r and h parameters, 148; 8.11 Phase inverters — general, 149; 8.12 Single-tube phase inverters, 149; 8.13 Two-tube phase inverters, 152; 8.14 Transistor split-load inverters, 155; 8.15 Two-stage transistors phase inverters, 156.

9 The Vacuum Tube and the Transistor as Switches — 160

9.1 Ideal switches, 160; 9.2 Practical forms of electronic switching devices, 161; 9.3 The high-vacuum thermionic diode, 162; 9.4 The semiconductor junction diode and breakdown, 162; 9.5 Transient response of junction diodes, 164; 9.6 The high-vacuum thermionic triode, 167; 9.7 The junction transistor, 170; 9.8 Pentodes, 173.

10 Nonlinear Waveshaping — 176

10.1 Series-diode clipper, 177; 10.2 Parallel-diode clipper, 178; 10.3 Biased-diode clipper, 179; 10.4 Grid limiting, 183; 10.5 Saturation limiting, 185; 10.6 Cutoff limiting, 186; 10.7 Overdriven amplifiers, 188; 10.8 Cathode-coupled clipper, 188; 10.9 Transistor limiters, 190; 10.10 Limiter applications, 191.

11 Clamping Circuits — 196

11.1 General, 196; 11.2 Action of an R-C coupling network, 196; 11.3 Diode clampers, 199; 11.4 Biased diode clamps, 207; 11.5 Notes on using semiconductor diodes in clamping circuits, 209; 11.6 Grid clamps, 210; 11.7

Disadvantages of clamping circuits, 212; 11.8 Keyed (synchronized) clamping circuits, 214.

12 Astable Multivibrators 221

12.1 Definitions, 221; 12.2 Astable MV, 222; 12.3 Equivalent circuit analysis, 225; 12.4 Symmetrical wave forms, 227; 12.5 Asymmetrical operation, 232; 12.6 Astable cathode-coupled multivibrator, 232; 12.7 Equivalent circuit analysis, 236; 12.8 Electron-coupled astable multivibrator, 237; 12.9 Frequency of multivibrator, 238; 12.10 Synchronization of the MV, 240; 12.11 Typical applications, 244.

13 Bistable Multivibrators 251

13.1 Detailed operation, 252; 13.2 Self-biasing, 253; 13.3 Triggering vacuum-tube bistable multivibrators, 255; 13.4 Triggering transistor bistable multivibrators, 257; 13.5 Nonsaturated flip-flops, 260; 13.6 Cathode-coupled binary, 262; 13.7 Direct-coupled bistable multivibrator, 264; 13.8 Bistable MV applications, 265.

14 Monostable Multivibrators 269

14.1 Operation of the vacuum-tube monostable MV circuit, 269; 14.2 Vacuum-tube monostable cathode-coupled multivibrator, 273; 14.3 Other vacuum-tube circuit modifications, 275; 14.4 Transistor monostable MV, 278; 14.5 Modifications of the basic circuit, 279; 14.6 Applications, 283.

15 Blocking Oscillators 286

15.1 General, 287; 15.2 Pulse transformers, 288; 15.3 The vacuum-tube type of blocking oscillator, 295; 15.4 Shape of the output pulse, 298; 15.5 Frequency and duration of output, 300; 15.6 The vacuum-tube monostable blocking oscillator, 301; 15.7 Transistor blocking oscillator, 302; 15.8 Triggering methods, 303; 15.9 Applications, 308.

Contents   xi

16   Time-base Generators   314

16.1 A neon sawtooth generator (relaxation oscillator), 316; 16.2 Thyratron sweep generator, 318; 16.3 Improving linearity of the thyratron time-base generator, 319; 16.4 Vacuum-tube and transistor sweep circuits, 321; 16.5 Improving linearity of the vacuum-tube sweep circuit, 321; 16.6 The Miller sweep circuit, 325; 16.7 The monostable screen-coupled phantastron, 328; 16.8 Current time-base generators, 330.

17   Transmission Gates   334

17.1 Unidirectional gate, 334; 17.2 The coincidence gate, 336; 17.3 Bidirectional gates, 338; 17.4 Four-diode gate, 340; 17.5 Six-diode gate, 341; 17.6 A simple triode gate, 342; 17.7 A two-triode gate, 343; 17.8 Removal of pedestal, 344; 17.9 Pentode gates, 345; 17.10 Multicoincidence pentode gate, 345.

18   Counters   347

18.1 Counting to the base 2, 348; 18.2 Counting to a base other than 2, 354; 18.3 Counting to a scale of 3, 354; 18.4 Decade counters, 361; 18.5 Preset counters, 365; 18.6 Gate and count types of instruments, 368; 18.7 Readout indicators: the Nixie and the Pixie, 384; 18.8 The Beam-X switch, 385; 18.9 Transistor counters, 403.

19   Pulse Modulation   412

19.1 Pulse-amplitude modulation (PAM), 413; 19.2 Pulse-duration modulation (PDM), 414; 19.3 Pulse-position modulation (PPM), 417; 19.4 Pulse-code modulation (PCM), 421.

20   Electromagnetic Delay Lines   429

20.1 Transmission lines — general, 430; 20.2 Distributed-parameter delay lines, 437; 20.3 Lumped-parameter-delay lines, 441; 20.4 Delay-line applications, 441.

21 Conclusion 448

21.1 The oscilloscope — general, 448; 21.2 Main vertical amplifier, 450; 21.3 Sweep generator, 455; 21.4 Horizontal amplifier, 459; 21.5 High-voltage power supply, 460; 21.6 Calibrator, 461; 21.7 Dual-trace plug-in amplifier Model 162A, 463; 21.8 High-gain vertical plug-in amplifiers Model 162D, 471; 21.9 Wide-band vertical plug-in amplifiers Model 162F, 474; 21.10 Time mark generator Model 166B, 475; 21.11 Delay generator Model 166D, 476.

Index 483

1
INTRODUCTION

Pulse circuits are defined as those circuits required for the generation and control of precisely timed waveforms. They differ widely from those used in radio communications equipment, in which the waveform of the operating voltage is usually sinusoidal or a simple combination of sinusoidal waves. Some pulse circuits are used to develop square, sawtooth, trapezoidal, or peaked waves of voltage that are required in indicating, timing, and modulating circuits of television, radar equipment, and so forth. The circuit operating conditions in most cases range from *full on* to *full off* and do not fall into the simple classifications of Class A, B, and C operation. The circuits, therefore, are named for the function they perform rather than their type of operation.

1.1 Historical developments

Pulses, in the form of dots and dashes, were used to convey intelligence by wire in the earliest days of electrical communication. Marconi used short and long pulses in his "wireless" to form the modulation envelope for radio-frequency energy.

The advent of the triode vacuum tube made possible modern radio communication and the transmission of intelligence by means of an amplitude-modulated carrier. In this scheme, the prime-signal source is a sinusoidal-signal generator. Recently, many electronic systems have been developed that require the use of pulses as the prime-signal source.

1.2 Typical pulse applications

Pulses are used in radar, which is an electronic system employed to detect the presence, range, and direction of objects. The block diagram of a typical radar system is shown in Fig. 1-1. It consists

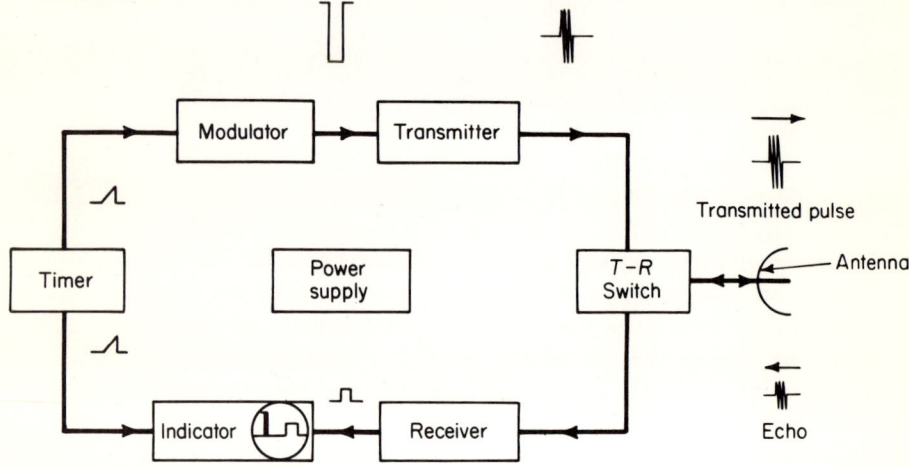

Figure 1-1 Block diagram of a radar set.

of a timer, modulator, transmitter, transmit-receive (T-R) switch, antenna, receiver, indicator, and power supply.

The timer synchronizes the indicator and transmitter circuits. At regular intervals it produces a pulse, that causes the sweep (horizontal movement of the scanning spot) to start in the indicator. At the same instant, or after a precise predetermined time, the timer produces a signal, which is also applied to the modulator.

When the modulator receives the synchronizing timing pulse it develops a high-voltage, high-power pulse, which turns the transmitter on for a short time interval.

The transmitter is a very high frequency (vhf) high-power generator of radio-frequency (r-f) energy. It produces a radio wave

Sec. 1.2   Typical pulse applications   3

of constant frequency and amplitude for the short time during which it is turned on by the modulator.

The T-R switch is an electrically operated switch that effectively disconnects the antenna from the receiver during the production of the transmitter pulse and connects it to the transmitter. During the remainder of the operating period, the T-R switch connects the antenna to the receiver.

The antenna acts as a radiator of the transmitter-produced energy when the transmitter is pulsed. A small portion of the radio wave transmitted from the radar set travels to the object (target) and is reflected as shown in Fig. 1-1. The reflected waves, called *echoes*, are picked up by the antenna and conveyed to the receiver. The antenna is designed to be directive for both transmission and reception so that the *bearing* (direction) as well as the *range* (distance) of the target may be determined.

The receiver amplifies the echoes and provides *video* pulses, called *pips*, of sufficient amplitude to produce visual indications on the indicator.

The indicator can be thought of as an electrical stop watch that measures precisely the small time interval required for the transmitter pulses to travel to the target and for the echoes to return. Because the speed of propagation of radio waves is known with great accuracy, the range can be determined as accurately as the time interval can be measured.

Pulses are also used in *radio telemetering*. In this application, certain variables, such as changes in temperature of the outer casing of a rocket, are sensed and measured in flight. This information is then transmitted to a ground station where permanent records are made for analysis by engineers.

A simplified explanation of telemetering can be given with the help of Fig. 1-2. The output signal of a pickup device, termed a *trans-*

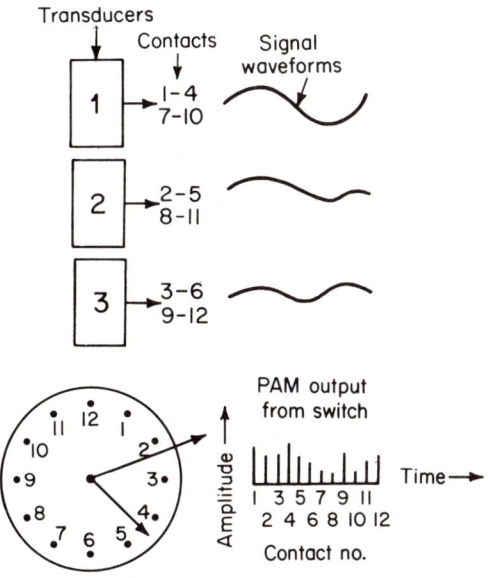

Figure 1-2 Simplified illustration of how the output of a transducer is "sampled."

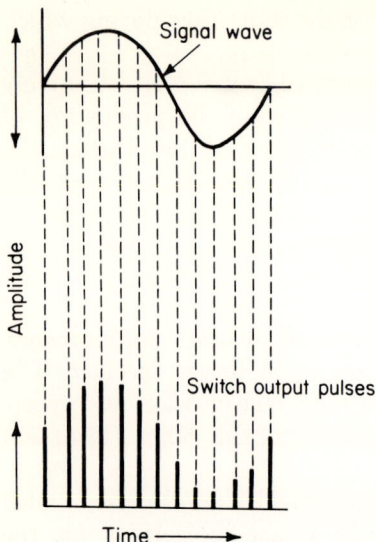

Figure 1-3 Illustration of how switch output is a series of pulses having uniform width and amplitudes proportional to the corresponding amplitude of the signal wave.

ducer, is shown as a sinusoidal waveform. This signal is applied to several contacts on the stationary plate of a mechanical switch. A revolving brush then passes over each contact and samples the transducer output. Assuming that each contact is of uniform area and that the brush is revolving at a constant speed, the switch output is a series of pulses having uniform width, and the amplitude of each successive pulse is proportional to the corresponding amplitude of the signal wave, as shown in Fig. 1-3. The output signal of the switch is now *pulse-amplitude modulated*, abbreviated PAM. It is apparent that the successive pulses reproduce the signal wave rather faithfully. If the number of samplers per second exceeds twice the highest frequency contained in the signal wave, the original signal can be reconstructed from the succession of pulses.

The PAM signal is next used to modulate an FM transmitter whose output is radiated by means of a suitable antenna.

At the ground station the signal is picked up by an antenna of a special type and applied to the input terminals of a highly sensitive receiver. The receiver amplifies the weak signal and separates the FM carrier signal from the PAM signal. The PAM output signal of the receiver is then applied to a detector, whose instantaneous output voltage is proportional to the instantaneous amplitude of the input voltage. This voltage is then used to operate a mechanical reproducer that reconstructs an essentially faithful reproduction of the signal originally produced by the pick-up transducer.

A re-examination of Fig. 1-3 shows considerable unallocated time between successive pulses. For example, the time required for each pulse may be 10 μsec and the elapsed time between pulses may be 100 μsec.

Sec. 1.2   Typical pulse applications

Use of this unallocated time introduces the possibility of *time-division multiplexing*, in which successive intervals of time are assigned to different signals or information channels. In practice, this is exactly what is done in radio telemetering. One transducer may be used to record, say, temperature variations; separate transducers may measure other variables, such as atmospheric pressure and cosmic radiation. The output signal of each transducer is applied to successive contacts on the mechanical switch. This operation is illustrated in Fig. 1-4. The amplitudes of the successive output pulses

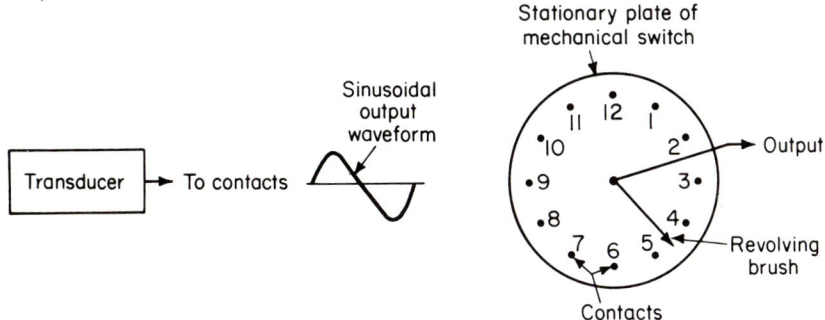

Figure 1-4 Simple multiplex system.

from the switch are now proportional to the corresponding amplitudes of each signal wave.

At the output of the ground-station receiver, suitable filters are required to separate the various transducer signals from the composite PAM signal, but this filtering represents no great technical difficulty. Once separated, the individual signals are handled in the manner previously described.

A television system also serves to illustrate another practical application of pulses. At the television studio, a camera is trained on the scene to be televised. A *horizontal line pulse*, occurring at a frequency of 15,750 cps, causes the electron scanning beam to begin in the upper left-hand corner of the camera tube and move at a uniform rate from left to right along lines that lie at a constant distance from each other. This is illustrated by the solid lines in Fig. 1-5. At the end of each horizontal line, the scanning beam returns to the left and starts a new line. During the return interval, a *horizontal blanking pulse* is applied to the camera tube, and the spot does not appear in the reproduced picture. When the scanning

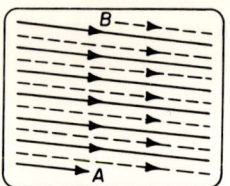

Figure 1-5 Scanning sequence for interlaced scanning.

spot reaches point *A* at the bottom of the picture, another pulse, called the *vertical blanking pulse*, is applied to the camera tube, and again, the spot does not appear in the reproduced picture. When the vertical blanking pulse ends, the scanning spot begins at point *B* at the top of the figure, and again moves downward as indicated by the dashed lines. The lines now scanned ideally fall equidistant between those previously scanned. This process is known as *interlacing*. Because the scanning spot moves from top to bottom 60 times each second, the frequency of the vertical blanking pulse is 60 cps.

Irregularities in the interlace process sometimes cause the scanned lines to be closer together than desired. To correct this condition it is necessary to have some means of compelling the scanning spot to begin each downward movement from a particular desired position. This is accomplished by inserting six *equalizing pulses* immediately before and after each vertical blanking pulse. Details of the blanking,

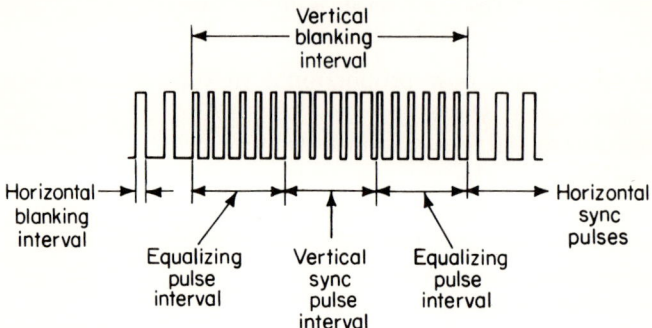

Figure 1-6 Details of blanking, synchronizing, and equalizing pulses used in television.

synchronizing, and equalizing pulses are shown in Fig. 1-6. Notice that horizontal synchronizing is maintained during the presence of the vertical synchronizing pulses by serrations that break up the vertical synchronizing pulses into blocks. These serrations, occurring at a frequency of 31,500 cps, are so timed that the rise of every other serration occurs at the instant a horizontal synchronizing pulse would have risen in amplitude if it had been present.

The foregoing examples, presented as typical applications of pulse

Sec. 1.3   Definitions   7

circuitry, have been advanced without defining certain essential terms. The required definitions are given in the next section.

1.3 Definitions

A *pulse* may be defined as a brief surge of voltage or current. It may be recurrent, but it is not cyclic except in the sense of being a highly distorted waveform.

An *ideal rectangular pulse* is shown in Fig. 1-7(a). At time t_1,

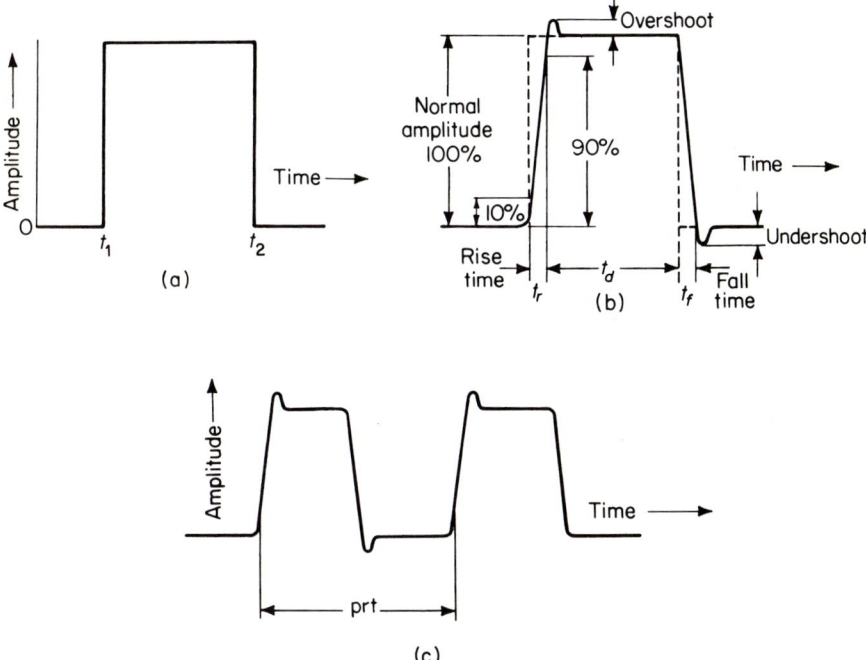

Figure 1-7 An (a) *ideal rectangular pulse,* (b) *modified pulse,* (c) *pulse train.*

the voltage or current rises instantly from zero to some maximum value. It remains at this value until time t_2, and then instantly returns to zero.

In practical circuits, lumped or distributed inductance, capacitance, and resistance always exist. As a result, such instantaneous changes as shown in Fig. 1-7(a) cannot occur. A finite period of time must elapse before the voltage rises to a normal value or drops to zero.

This causes a modification of the ideal rectangular pulse, and it may assume the form shown in Fig. 1-7(b).

The time required for the pulse to increase from 10 to 90 per cent of normal amplitude is called the *rise time*, and is indicated by the letter symbol t_r.

The time required for the pulse to decrease from 90 to 10 per cent of normal amplitude is called the *fall time*, and is indicated by the letter symbol t_f.

The time interval between the end of rise time and the beginning of fall time is called the *duration*, and is indicated by the letter symbol t_d.

In some pulse generation, the initial amplitude rise exceeds the correct value and, as shown in Fig. 1-7(b), a pip called an *overshoot* is produced on the waveform. There may also be a corresponding *undershoot* when the amplitude suddenly falls.

When pulses occur at regular intervals, as shown in Fig. 1-7(c), the time between a point on one pulse and the corresponding point on an adjacent pulse is called the *pulse repetition time*, abbreviated prt.

A series of successive pulses is called a *pulse train*. The number of pulses that occur per second in a pulse train is called the *pulse repetition frequency*, abbreviated prf, or *pulse repetition rate*, abbreviated prr.

A *step* voltage waveform is defined as one which maintains a value of zero for all times before t_0, and then rises instantaneously to a value of E after t_0. Such a waveform is shown in Fig. 1-8. An ideal rectangular pulse, such as that shown in Fig. 1-7(a), is actually the sum of two step voltages. This may be demonstrated by Fig. 1-9. The first step voltages, termed $+E$, occurs at time t_0 and

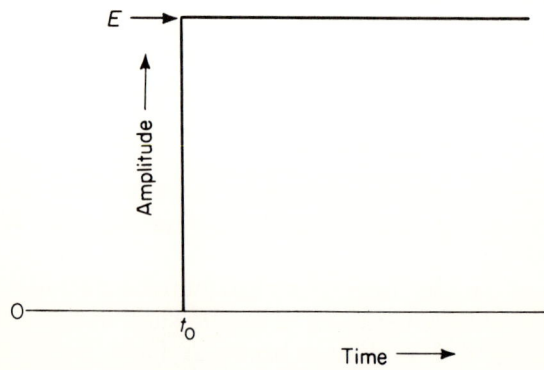

Figure 1-8 Step voltage.

Sec. 1.3   Definitions   9

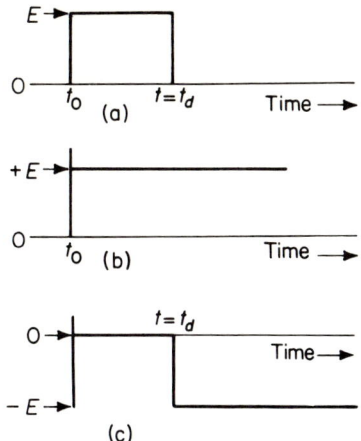

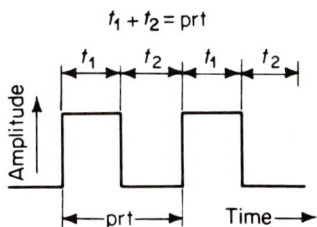

Figure 1-9 (a) A pulse, (b) and (c) the step voltages, +E and −E, that make up the pulse.

Figure 1-10 Square wave.

continues for the duration of the pulse. The second step voltage, termed $-E$, begins at the end of the pulse duration, time t_d, and continues until another pulse is applied. The resulting ideal pulse and the two step voltages are shown in Figs. 1-9(a), (b), and (c), respectively.

A *square* wave, shown in Fig. 1-10, is a waveform which has a constant level for a time t_1 and another constant level for a time t_2. The repetition time of a square wave is the sum of $t_1 + t_2$.

A *ramp* waveform shown in Fig. 1-11, has zero amplitude for all times before t_0, and then increases linearly after t_0.

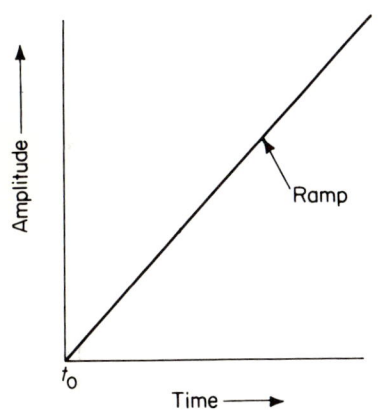

Figure 1-11 Ramp waveshape.

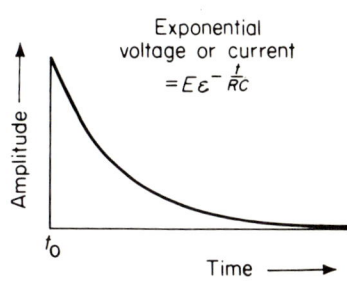

Figure 1-12 Exponential curve. Variable t/RC is used as the exponent of the equation from which the graph is derived.

An *exponential* waveform, shown in Fig. 1-12, is a graph of an equation in which the independent variable, the ratio of t/RC, appears as an exponent. The rate of change in such a curve is rapid at first, and then becomes increasingly slower.

1.4 General considerations

In some applications, certain pulse characteristics (parameters) can be changed without seriously affecting the transmission of intelligence. In a telegraph system, for example, rise and fall time may be altered within certain limitations without seriously affecting the transmitted intelligence. The only requirement is the operator's ability to distinguish between dots and dashes.

The situation is more critical, on the other hand, in a television receiver. If the video amplifiers are unable to reproduce the complicated waveforms with reasonable fidelity, the quality of the picture deteriorates.

The capability with which a circuit passes a given waveform depends on its *response characteristics*. Familiarity with the response characteristics of various networks is the required foundation for understanding pulse-type circuits, and this is the general subject of the next few chapters.

EXERCISES

1-1 How may a pulse be defined?

1-2 Why is an ideal rectangular pulse not encountered in practice?

1-3 What portion of a pulse is termed the rise time?

1-4 What portion of a pulse is termed the fall time?

1-5 The pulse duration extends between the _____ and the _____. It is indicated by the letter symbol _____.

1-6 When does overshoot occur?

1-7 Define "prt" and "prf."

1-8 The repetition time of a square wave is:
    ( ) The product of $t_1 \times t_2$.
    ( ) The difference between t_1 and t_2.
    ( ) The sum of $t_1 + t_2$.
    ( ) The time for which the waveform has a constant level.

1-9 In what manner does the amplitude of a ramp waveform increase for all times greater than t_0?

Exercises

1-10 Why are certain waveforms termed "exponential"?

1-11 Draw the following waveforms and label the significant parameters: (a) A modified rectangular pulse. (b) A pulse train. (c) A step voltage. (d) A pulse, and the step voltages that make up the pulse. (e) A square wave. (f) A ramp waveform. (g) An exponential curve.

1-12 A rectangular pulse has a repetition rate of 20,000 pps and a pulse duration of 5 μsec.
    (a) What is the pulse repetition time?
    (b) Draw the waveform, labeling prt and t_d.
    (c) For what period of time is the pulse zero?

2

LINEAR WAVE SHAPING

Simple networks composed of resistors, capacitors, and inductors are called *linear* networks because they do not distort a *sine wave* when it is transmitted through them. The output waveform from such a network is a faithful reproduction of the input waveform, although its phase and amplitude may be changed.

When *nonsinusoidal* waveforms are applied to *linear* networks, the output waveform may bear little resemblance to the input. The process whereby *non-sinusoidal* waveforms are altered in passing through linear networks is called *linear wave shaping.* Some of the more common linear-wave-shaping networks are described in this chapter.

2.1 The high-pass R-C filter

A high-pass R-C filter circuit (low frequency discriminator) consists of a resistor and capacitor connected in series, with the output taken across the resistor as shown in Fig. 2-1(a). The term *high pass* is used because the output voltage increases as the *frequency* of the input voltage increases. This

Sec. 2.1   The high-pass R-C filter   13

effect results from the decreasing reactance of the capacitor with an increase in frequency.

A pass-pass R-C filter may be connected to a d-c source as shown in the circuit of Fig. 2-1(b) which also includes a switch, S_1 that offers zero resistance to current, and a battery whose voltage

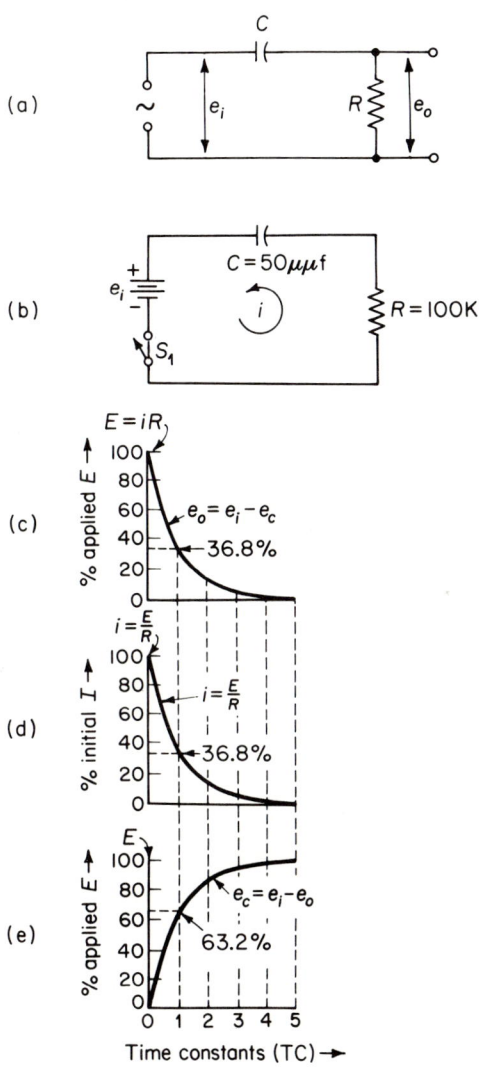

Figure 2-1 High-pass R-C filter and waveforms e_c, e_o, and i during capacitor charge.

is E. The battery applies an input voltage, termed e_i, to the R-C circuit whenever switch S_1 is closed. The internal impedance of the battery may be considered negligible, and the rise time of the battery voltage to amplitude E is considered to be zero. Thus, e_i is zero at all times before the switch is closed and rises to amplitude E immediately after it is closed. The input voltage waveform is, therefore, a step voltage.

Switch S_1 is closed at the time interval marked 0 on the horizontal axis of the three graphs shown in Fig. 2-1 (c), (d), and (e). Voltage e_i rises instantaneously to amplitude E, producing a current that charges capacitor C. At every instant, the voltage across the capacitor is proportional to the charge on it. This is expressed by the familiar formula

$$q = Ce \qquad (2\text{-}1)$$

where q = instantaneous charge on capacitor C in coulombs
e = instantaneous voltage across capacitor C in volts
C = capacitance of capacitor C in farads

Thus, any change in the charge on the capacitor causes a corresponding change in the voltage across its terminals.

Assuming no charge exists on the capacitor when switch S_1 is closed, the voltage across the capacitor terminals is zero. An instantaneous output voltage, termed e_o, which is equal to E at the instant of closing the switch, therefore, appears across resistor R. The instantaneous circuit current, termed i, is equal to the quotient E/R. As the charge builds up a voltage across the capacitor in opposition to the input voltage, current decreases rapidly at first, then gradually at a slower rate. Waveforms indicating the changes in e_o, i, and the voltage across the capacitor, termed e_c, at various time intervals are shown in Fig. 2-1(c), (d), and (e) respectively.

All of the curves shown are called *exponential* curves because each is the graph of an equation in which the independent variable appears as an exponential function.

The instantaneous resistor voltage curve, e_o, shown in Fig. 2-1(c), can be determined from the instantaneous current values by using a form of Ohm's law, $e_o = iR$. Because R is constant, e_o is directly proportional to i. Its curve, therefore, has the same exponential slope as curve i, shown in Fig. 2-1(d). Initially, there is no voltage across the capacitor, as shown in Fig. 2-1(e), and the output voltage equals the battery voltage, E. As the capacitor charge builds up and opposes the applied voltage, however e_o decreases. When e_o is zero, notice that e_c equals E.

2.2 The R-C time constant

The rate of charge for the capacitor with a given voltage depends on the total resistance in series with the capacitor and the capacitance of the capacitor. If the resistance is increased, the current is decreased and the amount of charge on the capacitor in a given time is also decreased. Similarly, if the capacitance is increased, a greater charge is needed to charge the larger capacitor to the same voltage, and a longer time is, therefore, required as well.

The time taken for a capacitor to charge completely and for the charging current to become zero is theoretically infinite. The time taken for the capacitor to reach a certain percentage of its final charge varies directly as the resistance and capacitance if the charging voltage remains the same. In order to compare the rate of charge of different R-C circuits the *time constant*, termed T, of the circuit is used. The time constant is defined as

$$T = RC \qquad (2\text{-}2)$$

where T = time in seconds
 R = total resistance in ohms
 C = capacitance of capacitor in farads

For the values used in the circuit of Fig. 2-1(b), where $R = 100 \text{ K}$ and $C = 50 \,\mu\mu\text{f}$

$$T = 1 \times 10^5_\Omega \times 5 \times 10_f^{-11} = 5 \times 10^{-6}_{\text{secs}} = 5_{\mu\text{secs}}$$

The value computed above is called the time constant of the circuit. It may seem as though 5 μsec is a short time, but any distinction between long, short, or medium circuit time constants is purely arbitrary. Generally, however, a circuit is considered to have a *long* time constant when the RC product is equal to ten times the duration (t_d) of the applied waveform. When the RC product is one-tenth of the duration of the applied waveform, or less, the time constant is considered *short*. When the RC product lies between these two extremes, the time constant is considered *medium*.

Each division on the common horizontal scale of the waveform graphs in Fig. 2-1(c), (d), and (e), respectively, represents one time constant. Notice that voltage e_c in Fig. 2-1(e) has reached approximately 63.2 per cent of its final value at the end of one time constant. This is true of all R-C circuits, and is sometimes used as the basis for defining the time constant, i.e., the time in which the voltage across the capacitor rises to 63.2 per cent of the input voltage.

Notice also that on the waveform showing current variation, Fig. 2-1(d), the initial current has decreased in value approximately 63.2 per cent at the end of one time constant. Further, because $e_o = iR$, voltage e_o has also decreased 63.2 per cent to $0.368E$ after one time constant. At the end of five time constants, voltage e_c is about 99.4 percent of its final value and for practical purposes may be considered to be fully charged. The network output voltage is then $E - e_c = 0$, so that i and e_o are also zero.

2.3 The universal time-constant chart

A *universal time-constant chart* is shown in Fig. 2-2. It is used to determine any and all instantaneous values of e_o, e_c and i when dealing with *R-C* networks. Curve A can be used to determine the instantaneous capacitor voltage, e_c, during the charging period and, with certain assumptions discussed later, the instantaneous output voltage, e_o, across the resistor during the discharge period. Curve B can be used to determine e_c during the *discharge* period and e_o during the *charge* period.

In learning how to use this chart, with the above circuit parameters, let us first determine e_o in Fig. 2-1 after e_i has been applied

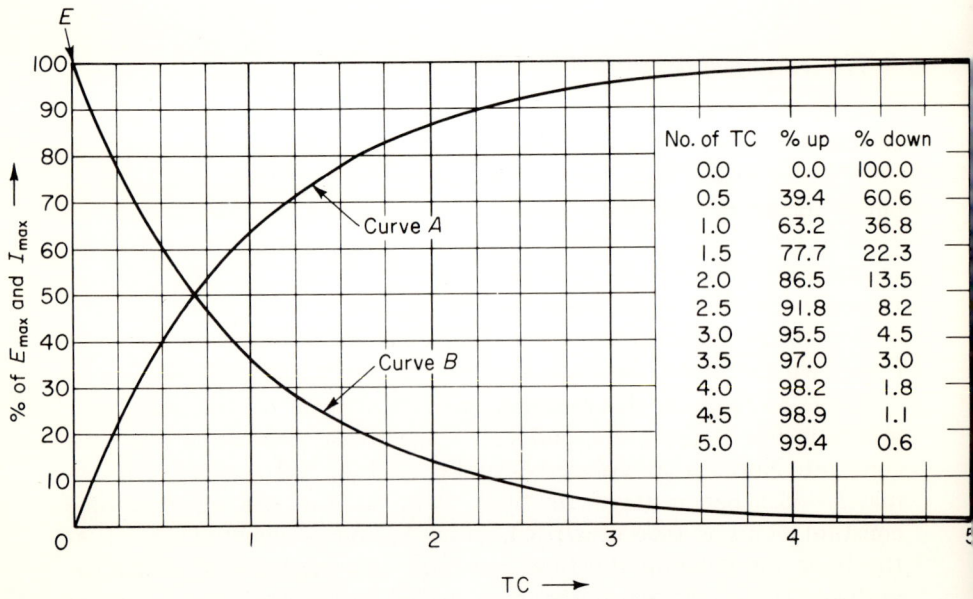

Figure 2-2 Universal time-constant chart.

Sec. 2.4   *The capacitor discharge period*   17

to the input terminals for 15 μsec. The time constant of this circuit was calculated previously as 5 μsec. Thus, in 15 μsec, three time constants have elapsed.

First, find the third time constant division in Fig. 2-2 on the horizontal scale, and follow the vertical line to curve B. From here, follow the horizontal line to the left-hand vertical scale. The point last determined shows that, after an elapsed time equal to three time constants, voltage e_o is approximately 4.5 per cent of the applied voltage, i.e., $0.045E$. Thus, if the battery voltage was 10 v, voltage e_o, across the resistor is 0.45 v after 15 μsec.

As a second example, let us find voltage e_o after 2.5 μsec, which is equal to 0.5 time constant. From 0.5 on the time constant scale, follow the vertical line to curve A. From the point of intersection, follow the horizontal line to the left-hand vertical scale. This last determined point shows that, after 0.5 time constant, voltage e_c is equal to approximately 39.4 per cent of the applied voltage, i.e., $0.394E$. Again, using a supply voltage of 10 v, the voltage across the capacitor is 3.94 v after 2.5 μsec.

2.4 The capacitor discharge period

Figure 2-3(a) is a modification of Fig. 2-1(b). Another switch, S_2 which also offers zero resistance to current, has been placed across the supply in parallel with the series R-C capacitor-and-resistor combination. The two switches are assumed to be ganged in such a manner that S_2 is closed the instant S_1 is opened, and vice versa.

When the ganged switch is thrown as shown in Fig. 2-3(a), the battery is effectively removed from the circuit. Capacitor C,

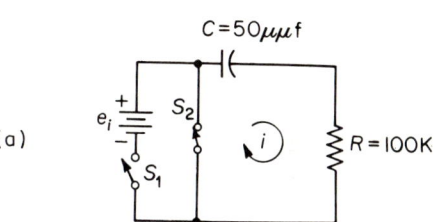

(a)

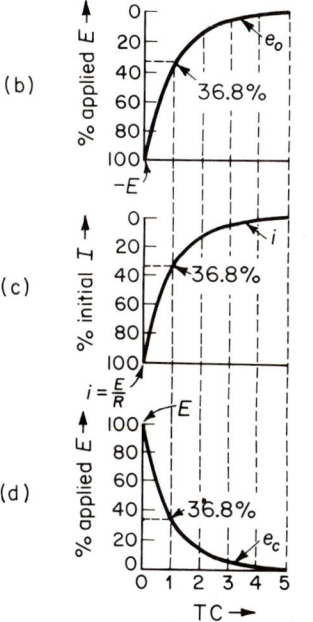

(b)

(c)

(d)

Figure 2-3 High-pass R-C filter and waveforms for e_c, e_o, and i during capacitor discharge.

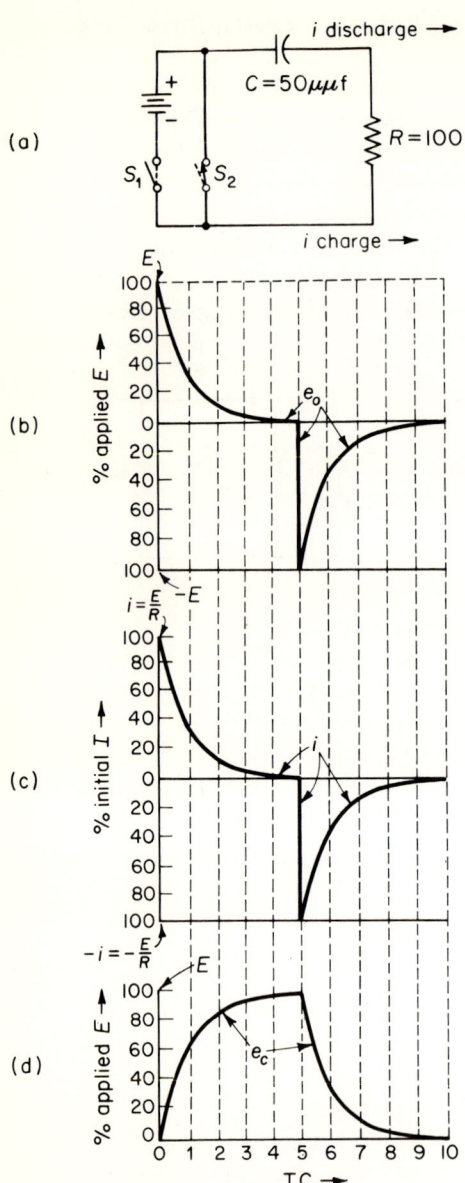

Figure 2-4 Response of a high-pass R-C filter to an input pulse waveshape.

which was originally charged to a voltage of amplitude E, now discharges through resistor R. After an elapsed time equal to one time constant, e_o drops to 36.8 per cent of its initial value, E. After five time constants, capacitor C is considered to be discharged, and current and output voltage drop to zero. Notice that again all waveforms are exponential.

During the discussion of the universal time-constant chart, it was stated that curve A represented e_o during the discharge period if a certain assumption was made. This assumption is that the bottom of the horizontal scale marked 0 in Fig. 2-2 is changed to $-E$, and the top of the scale marked E is changed to 0. The reason for this required assumption is made apparent from a study of the curves associated with Fig. 2-3.

The direction of discharge current in Fig. 2-3(a) is seen to be opposite to the charging current in Fig. 2-1(b), and the curves for e_o and i are, therefore, drawn in the opposite polarity with respect to the zero line. Because the capacitor was charged to amplitude E, the initial amplitude of e_o during the discharge period is labeled $-E$, and the intial amplitude of i is labeled $i = E/R$ as shown in Fig. 2-3(c). Because the voltage across capacitor C is of the same polarity with respect to the zero line during both charge and discharge, waveform e_c lies entirely above the zero line.

Sec. 2.5 Network response   19

In summary, therefore, curve A of Fig. 2-2 represents the current (and output voltage) waveform during discharge as well as the voltage across the capacitor during the charging period. Similarly, curve B represents the voltage across the capacitor during discharge and the current (and output voltage) during the charging period.

2.5 Network response

In Fig. 2-4 the waveforms from Fig. 2-1 are combined with those of Fig. 2-3. Because the closing of switch S_1 in Fig. 2-1 is equivalent to the application of a step voltage or rectangular pulse, the switching action in Fig. 2-3 is equivalent to applying a second step voltage of opposite polarity. Two such step voltages when combined form a rectangular pulse having some positive d-c average value. The waveforms of Fig. 2-4, therefore, represent the current and voltage response of a high-pass R-C filter to a rectangular input pulse waveform of some positive d-c value.

The input pulse waveform of Fig. 2-4 is in the form of an ideal pulse having a duration equal to five time constants of the R-C network so that the capacitor charges completely in this time. The network may be considered as having a *medium* time constant with respect to the pulse duration. The output voltage waveform as shown in Fig. 2-4(b) first rises immediately to maximum amplitude E, then declines exponentially to zero after a time interval equal to five time constants. It then rises instantaneously to amplitude $-E$, when the input pulse drops to zero, and again declines exponentially to zero in a time interval equal to five time constants. The current waveform is shown in Fig. 2-4(c) and the waveform of voltage across the capacitor is shown in Fig. 2-4(d).

It is important to note that this particular output waveform is obtained only in the special case where the duration of the ideal rectangular pulse is equal to five circuit time constants. The capacitor is permitted to charge and discharge fully at the instant that the circuits are switched.

Now let us see how the same circuit responds when the time constant is shorter, say one-tenth the duration of a rectangular pulse. In Fig. 2-5 the R-C circuit time constant is $5\,\mu\text{sec}$, and the duration of the applied pulse is $50\,\mu\text{sec}$. The R-C network may be considered as having a *short* time constant, with respect to the pulse duration. The voltage and current waveforms from time $t = 0$ to time $t = 25\,\mu\text{sec}$ are exponential, as previously shown. At time $t = 25\,\mu\text{sec}$ (5 RC units), however, as shown in Fig. 2-5, voltage e_c equals the

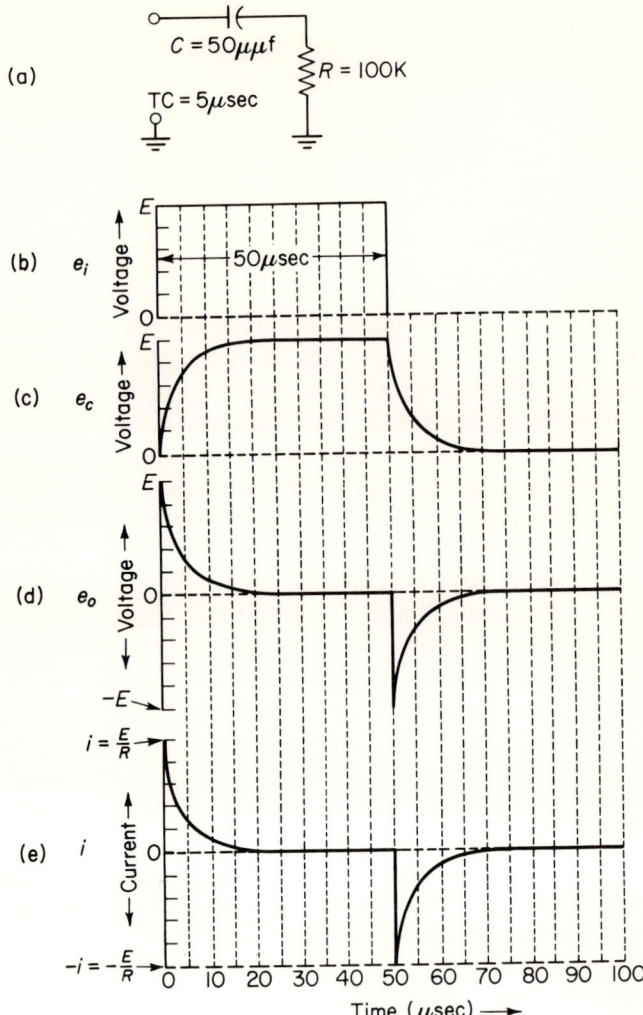

Figure 2-5 *Response of a high-pass R-C filter circuit, time constant $\frac{1}{10}$ duration of input pulse.*

applied voltage, and the circuit current and output voltage drop to zero. As long as the applied voltage remains the same, therefore, there is no output voltage developed across resistor R. At time $t = 50\,\mu\text{sec}$, the input voltage falls abruptly by the amount E, and the output voltage becomes $-E$. From $t = 50$ to $t = 75\,\mu\text{sec}$ (5 RC units), the output voltage decays exponentially to zero. From $t = 75$ to $t = 100\,\mu\text{sec}$, the output voltage is again zero.

Sec 2.5   Network response   21

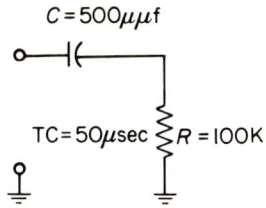

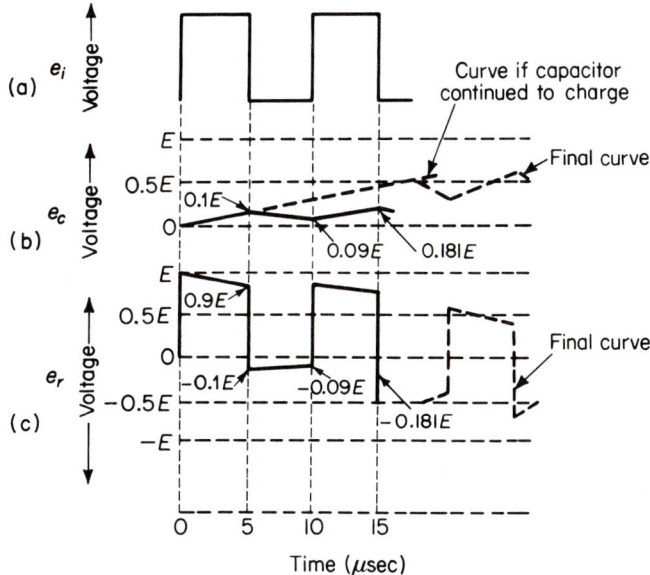

Figure 2-6 Response of a high-pass R-C filter circuit, time constant 10 times duration of input pulse.

In Fig. 2-6, the R-C network time constant is 50 μsec and the input pulse duration is 5 μsec. The R-C network may be considered as having a *long* time constant with respect to the pulse duration. Referring to the universal time-constant chart, Fig. 2-2, it is seen that the capacitor charges to a value of approximately $0.1E$ in 5 μsec (one-tenth of the time constant), and that the curve is very nearly a straight line. Voltage e_c as shown in Fig. 2-6(c) is, therefore, essentially linear from $t = 0$ to $t = 5$ μsec. The output voltage, e_r or e_o, as shown in Fig. 2-6(d), declines to $0.9E$ at $t = 5$ μsec. The input voltage, as shown in Fig. 2-6(b), then drops abruptly to zero.

When the input voltage is removed, voltage e_c becomes the network voltage source and has an initial value of $0.1E$. From $t = 5$ to $t = 10\,\mu$sec, this voltage declines by 90 per cent of its initial value, i.e., to $0.09E$. Because voltage e_c cannot change instantaneously, the output voltage drops to $-0.2E$ at $t = 5\,\mu$sec and decays exponentially to $0.09E$ at $t = 10\,\mu$sec.

When the second pulse is applied to the network at $t = 10\,\mu$sec, the voltage across the capacitor does not increase to $0.1E$ within one-tenth time constant as before, because its initial value is not zero. At $t = 10\,\mu$sec, $e_c = 0.09E$. This voltage opposes the input voltage e_i and effectively reduces its value to $0.91E$. Thus, in the next one-tenth time constant (from $t = 10$ to $t = 15\,\mu$sec), capacitor C charges to 10 per cent of the effective input voltage plus its beginning voltage, i.e., $0.09E = 0.181E$.

At $t = 10\,\mu$sec, the output voltage changes instantaneous from $-0.09E$ to $+0.91E$ and decays exponentially to $0.819E$ by $t = 15\,\mu$sec. It then drops instantaneously by the amount E and becomes $-0.181E$.

The gradual increase in e_c and decrease in e_R with time continues until the average (d-c) value of each waveform is zero. This is also indicated in Figs. 2-6(c) and 2-6(d), respectively. Because a capacitor cannot pass a d-c voltage, e_R becomes pure a-c (zero d-c) as soon as the capacitor is charged. The d-c component of the input signal has been removed by (and appears across) the capacitor. The output waveform then varies an equal amount on the zero axis.

Thus, if the distortion resulting from the passage of a pulse through a linear network is to be minimized, the RC time constant must be very large as compared to the input pulse duration.

It should be noted, therefore, that if the circuit time constant is ten times the pulse duration, the amplitude of the output waveform declines only to about $0.9E$, and the undershoot (negative portion of the output waveform) is only $-0.1E$ after one half the pulse duration, as shown in Fig. 2-6(d). On the other hand, if the time constant is one-tenth the pulse duration, the amplitude of the output voltage is zero after one-half the pulse duration, as shown in Fig. 2-5(d). The undershoot occurring at the end of the pulse has an amplitude of $-E$. Thus, for a square wave input having a d-c level, positive and negative *spikes* are produced in the output. It should also be noted that the output voltage, e_r, has a zero average value although the input voltage does not.

2.6 Response to a square wave input

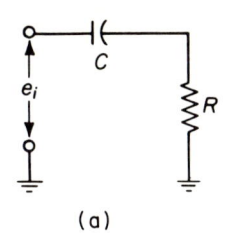

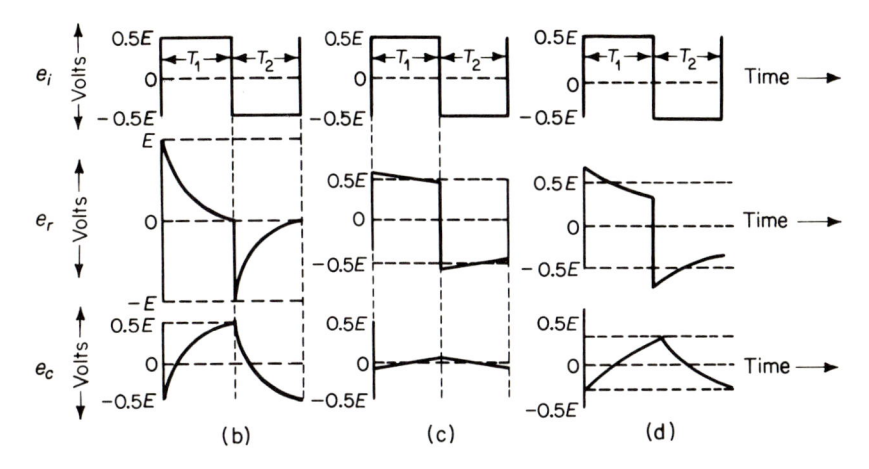

Figure 2-7 Illustration of the effect of time constant when square wave is applied to input terminals of a high-pass R-C filter.

In Fig. 2-7(b), a square wave, with an average d-c value of zero, is applied to the input of a high-pass filter (a). Voltage e_i charges the capacitor alternately positively and negatively by an amount determined by the circuit time constant. At each instant the vector sum of $e_R + e_c = e_i$.

Figure 2-7(b) illustrates the effect of a time constant that is one-tenth as long as the input pulse period $(T_1 + T_2)$. The output e_R consists of alternate positive and negative spikes that have an amplitude equal to twice the input. In Fig. 2-7(c), the time constant is ten times as long as the pulse period and e_R closely resembles the input. When the RC time constant lies somewhere between one-tenth and ten times the pulse period, the output waveform resembles that shown in Fig. 2-7(d).

Notice that times T_1 and T_2 in all parts of Fig. 2-7 are equal and the waveforms are *symmetrical*. When times T_1 and T_2 are not equal, the waveforms are *unsymmetrical*. Note also that the average values at e_i, e_r, and e_c are all zero in Fig. 2-7.

2.7 Effect of rise time on amplitude of output waveform

As the R-C circuit time constant is made increasingly shorter in comparison to the duration of the input waveform, the width of the output pulses becomes narrower. Narrow pulses are often required in pulse-type circuits, but they are acquired only at the cost of a smaller output peak, because the rise time of a practical waveform is not zero as assumed thus far, but is finite and must, therefore, be taken into consideration.

With the ideal input waveforms considered thus far, voltage rises instantaneously from 0 to amplitude E at time $t = 0$. Because the capacitor charge cannot change in an identical manner, and because the sum of the voltage drops around the circuit must equal e_i, the full applied voltage appears across resistor R as output voltage e_o.

In practice, e_i rises to amplitude E, but not instantaneously; and, in a short-time-constant circuit, the capacitor is capable of charging during the time before e_i reaches amplitude E. Because $e_i = e_c + e_o$, some value of voltage less than E appears at the output. The shorter the time constant, the more capacitor C can charge during the rise time of e_i and the lower e_o becomes. By the same reasoning, the capacitor discharges very rapidly, and the output spikes are very sharp, when the time constant is small.

2.8 Response of the high-pass R-C filter to a ramp voltage

When a waveform such as a ramp voltage is applied to the high-pass R-C filter circuit the output waveform is also a function of the time constant. A ramp voltage starts at zero and increases linearly at a constant rate. However, unlike the case of the square wave input, where the capacitor charges exponentially to the final value E, the ramp voltage continues to increase as the capacitor charges toward it, and the capacitor must continually charge to a new higher value. The input waveform, and voltages across the capacitor and output, are shown in Figs. 2-8(a), (b), and (c), respectively.

Assuming the capacitor is uncharged at $t = 0$, current rises with the ramp voltage when it is initially applied. The resulting charge on the capacitor opposes the input voltage, and the current should

Sec. 2.8 *Response of the high-pass R-C filter to a ramp voltage*

seemingly decrease. The input voltage, however, is continually increasing and more than overcomes the opposing capacitor voltage, so that current actually increases, although not as rapidly as before. Because voltage across a resistor is in phase with current, the output voltage represents the current waveform. The capacitor becomes charged to the higher voltage as the current increase continues and the rate of current increase falls off. Finally, the current is high enough to raise the capacitor voltage to the same rate of increase as the input voltage waveform. At this point, current becomes constant, as shown in Fig. 2-8(c), and the capacitor voltage slope is essentially the same as the ramp voltage slope, as in Fig. 2-8(b). Because the current is constant, the output voltage across R is also constant.

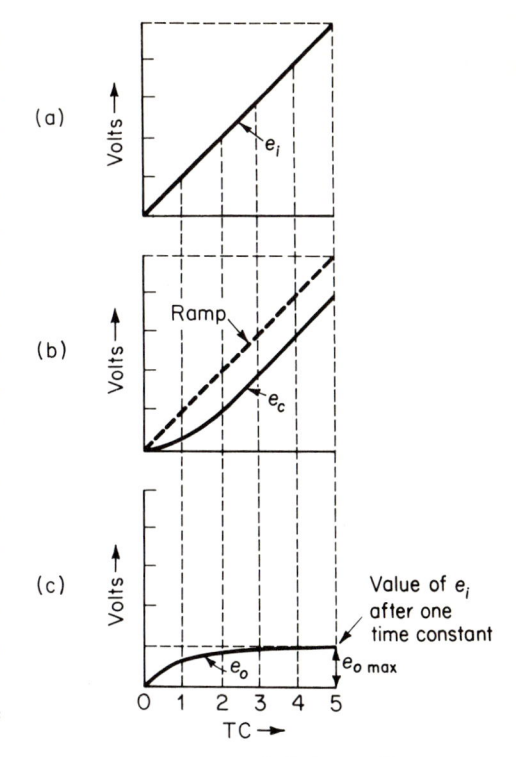

Figure 2-8 Response of high-pass filter to ramp voltage input.

Because the curve representing e_o and the current is exponential, after one time constant the value of e_o is approximately 63.2 per cent of the voltage applied at that time. At the end of five time constants, e_o is practically equal to the applied voltage after one time constant.

The maximum value of e_o is a function of the time constant and rate of rise. Mathematically

$$e_{o_{max}} \frac{de_i}{dt} \cdot RC \qquad (2\text{-}3)$$

where $e_{o_{max}}$ = maximum output voltage across resistor R
R = resistance of circuit in ohms
C = capacitance of circuit in farads
de_i/dt = rate of change of input volts in volts per second

The symbol d, in the above ratio, is used in the differential calculus to express "*a little bit of.*" It is not a multiplying factor, and both de_i and dt are single quantities. Any such quantity is called a *differential* and is considered infinitesimally small. The ratio of two differentials is known as a *derivative:* in this case, the derivative of input voltage with respect to time. A derivative expresses the exact (instantaneous) rate of change.

Sometimes, you will see $e_{o_\max}$ expressed as $(\Delta e_i/\Delta t)\ RC$. In this form, *incremental* values are used in the ratio and express the *average* rate of change. Δe_i then equals $e_{i2} - e_{i1}$, where these two values of e_i are arbitrarily selected points on the input voltage curve. $\Delta t = t_2 - t_1$, where the two values of time correspond to the two values of input voltage.

If the time intervals are made increasingly smaller, t approaches but never reaches zero. When t is infinitesimally small, we can say

$$\lim_{\Delta t \to 0} \frac{\Delta e_i}{\Delta t} = \frac{de_i}{dt}$$

where the notation $\lim \to 0$ indicates that Δt approaches zero as a limit.

2.9 The R-C differentiator

A high-pass R-C filter circuit is usually known as a *differentiator* when the time constant is short. The circuit is shown in Fig. 2-9. The RC time constant is most often made small in comparison to the pulse duration by reducing the value of the capacitor. Such circuits are used to develop a series of sharp positive and negative spikes of output voltage from a square wave of voltage applied to the input terminals. This process is known as differentiation because the output voltage represents the rate of change of input voltage.

The effect of rise time on the output voltage amplitude has already been discussed. It was shown that the leading edge of the output

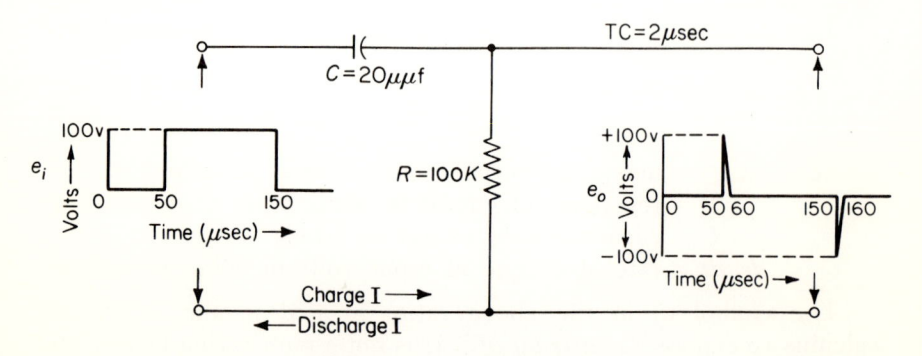

Figure 2-9 R-C differentiator.

Sec. 2.9   The R-C differentiator   27

pulse could not be exactly vertical and equal to amplitude E because the voltage across R cannot build up faster than the input voltage changes. This should be kept in mind during the following discussion as an ideal waveform is again used as the input waveform. This simplification is used to permit an easier understanding of circuit action.

Returning to Fig. 2-9, the input voltage remains constant at 0 v from $t = 0$ to $t = 50\,\mu\text{sec}$. Because the capacitor is neither charging or discharging, there is no current through R and the output is zero. At $t = 50\,\mu\text{sec}$, the input voltage rises instantaneously from 0 to 100 v (previously designated as amplitude E). Because capacitor C cannot charge instantaneously, its voltage cannot immediately change and the input voltage increase appears across R. As the capacitor starts to charge, a voltage drop is produced across resistor R with a negative polarity at the lower end. The rate of change is determined by the RC time constant of the circuit, which is very short ($2\,\mu\text{sec}$) compared to the duration of the input waveform ($100\,\mu\text{sec}$). As C charges, the voltage across R decreases exponentially and becomes zero when C is charged to 100 v at the end of five time constants ($10\,\mu\text{sec}$). As shown in Fig. 2-9, from $t = 60$ to $t = 150\,\mu\text{sec}$, the output voltage remains at zero because the charge across C equals but opposes the input voltage.

At $t = 150\,\mu\text{sec}$, the input voltage drops 100 v and the capacitor discharges. This discharge results in a reversal of current and appears as a negative voltage across resistor R. As the capacitor discharges through R and the input circuit, the voltage across R decreases exponentially, becoming zero when the capacitor voltage is 0 in $10\,\mu\text{secs}$.

Pulse-forming by means of a differentiating circuit, therefore, requires that the input voltage waveform have reasonably steep edges and that the input circuit be capable of maintaining the square waveform in spite of the loading effect of the differentiating circuit. The

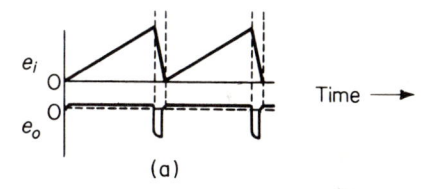

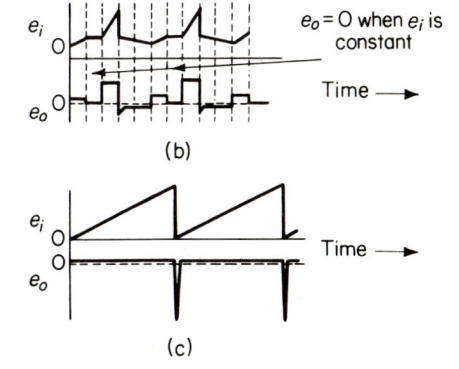

Figure 2-10 Differentiator outputs for different input voltage waveshapes.

shorter the time constant of the differentiating circuit, the shorter the duration of the spikes or "pips" produced.

Differentiator outputs for various types of input waveforms are shown in Fig. 2-10. In all of these diagrams it is important to note that the output, at each instant, represents the *rate-of-change* or the *differentiation* of input. When the input voltage is rising in a positive (+) direction, the rate of change is positive; and when the input voltage is rising in a negative (−) direction, the rate of change is negative. When the input voltage is constant (but not zero) the rate of change is zero.

In Fig. 2-10(a), the slope of the trailing edge of e_i exceeds the slope of the leading edge, and the output voltage is greater during the trailing edge because the rate-of-change is greater. The waveform shown in Fig. 2-10(b), in addition, illustrates zero output whenever there is a constant input voltage. In Fig. 2-10(c), the relationship of the amplitude of the output voltage and the rate-of-change of input voltage is again illustrated. In each of these examples, it is assumed that the circuit time constants are short in comparison to the input duration to produce a true differentiator output.

2.10 The low-pass R-C filter

If the positions of the capacitor and resistor in the high-pass R-C filter are interchanged, as shown in Fig. 2-11, the circuit becomes a low-pass filter (or high-frequency discriminator). As the frequency increases, the reactance of the capacitor decreases. The higher frequencies are, therefore, shunted to ground and the lower frequencies develop an appreciable voltage across the capacitor that is passed to the following circuit.

A step voltage is applied to the input terminals of the low-pass filter as shown in Fig. 2-11. At $t = 0$, this voltage rises instan-

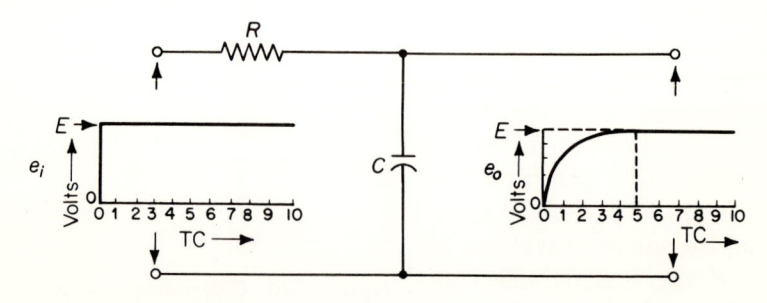

Figure 2-11 Response of a low-pass filter to a step input voltage.

Sec. 2.11   Response of the low-pass R-C filter to a pulse input   29

taneously to amplitude E. Because the capacitor cannot charge instantaneously, the output voltage taken across C at time $t = 0$ is zero. The capacitor then charges exponentially through the resistor, and at the end of five time constants is considered to be charged to E. It remains at this value (E) as long as the input waveform is applied.

2.11 Response of the low-pass R-C filter to a pulse input

As shown in Fig. 2-12, a pulse waveform of 25 μsec duration is applied to the low-pass R-C filter. The time constant of the R-C filter is 5 μsec, ($RC = 1 \times 10^6_\Omega \times 5 \times 10^{-12}_f$). In this particular case,

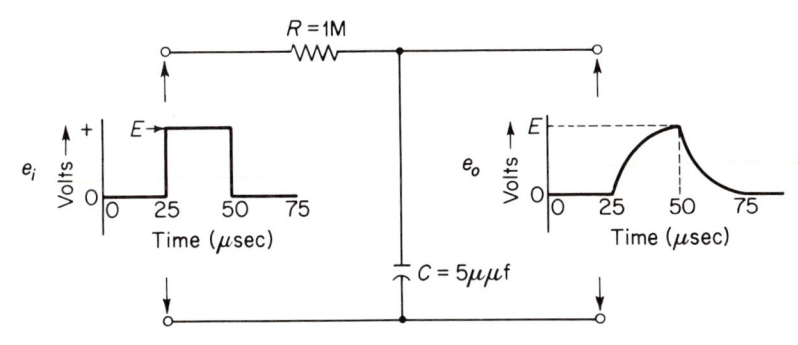

Figure 2-12 Response of a low-pass filter having a time constant of 5 microseconds to an input pulse having a duration of 25 microseconds (5 TC).

the pulse duration is exactly equal to five time constants, and the capacitor can charge fully (practically) in this time to amplitude E, as shown. The output waveform from $t = 0$ to $t = 25$ μsec is the same as that obtained for the step-voltage input. At $t = 25$ μsec, the input voltage drops instantaneously to zero. Capacitor C then acts as the network voltage source and discharges exponentially through resistor R. After five time constants (at $t = 50$ μsec), the output voltage is considered to be zero, as shown.

In Fig. 2-13, a pulse of 5 μsec duration is applied to the same low-pass filter, having a time constant of 5 μsec. At $t = 0$, the input pulse rises instantly to amplitude E. Again the output voltage is zero at $t = 0$, but it begins to rise exponentially immediately thereafter. At the end of one time constant (5 μsec), the output voltage is 0.632E. At this instant, e_i drops to zero. With e_i removed, capacitor C can charge no further. Thus, the maximum possible

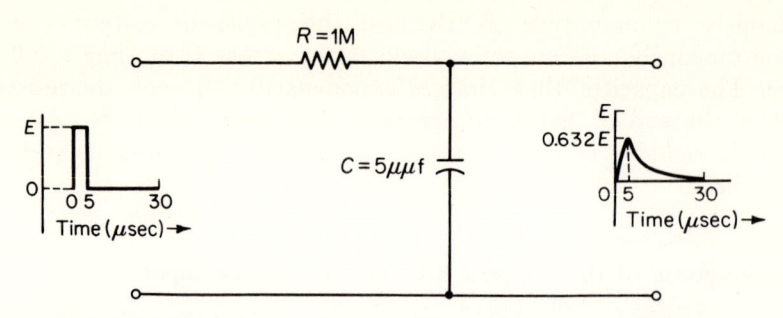

Figure 2-13 Response of a low-pass filter having a time constant of 5 microseconds to an input pulse having a duration of 5 microseconds.

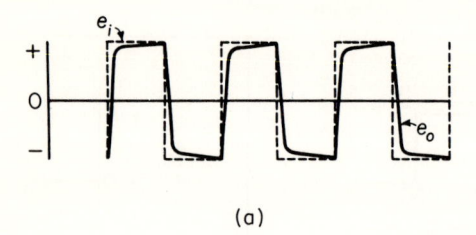

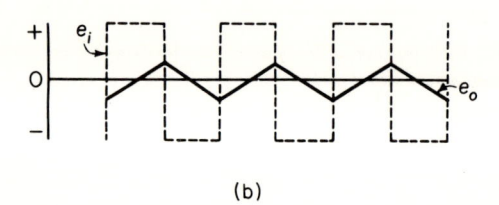

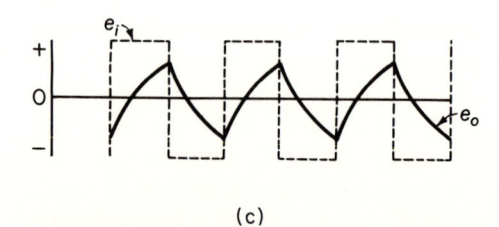

Figure 2-14 Response of a low-pass filter to a square-wave input signal when (a) time constant is short compared to square-wave duration, (b) time constant is long compared to duration, (c) time constant is medium compared to duration.

Sec. 2.13   *Response of low-pass filter to a ramp input*   31

output voltage is $0.632E$. From $t = 5$ to $t = 30\,\mu$sec, the capacitor discharges exponentially through R to the zero voltage level.

It should be noted that the maximum value as well as the shape of the output waveform is determined by the time constant of the circuit.

2.12 Response of low-pass filter to square wave input

As in the case of the high-pass filter, distortion of voltage waveforms across the capacitor is minimized if the time constant of the network is short compared to the duration of the input waveform. Since in a low-pass filter the output is taken across the capacitor, the output waveform is then essentially a reproduction of the input for a short time constant, as shown in Fig. 2-14(a). On the other hand, if the time constant of the low-pass filter is long compared to the duration of the input pulse, the output is a triangular wave as shown in Fig. 2-14(b). For a medium time constant, the output waveform shown in Fig. 2-14(c) is obtained.

2.13 Response of low-pass filter to a ramp input

If the output waveform is to resemble the input closely, the time constant of the low-pass filter must be small compared to the total ramp time, as shown in Fig. 2-14(a) and in Fig. 2-15. Some distortion still occurs near the origin (in the case of the short-time-

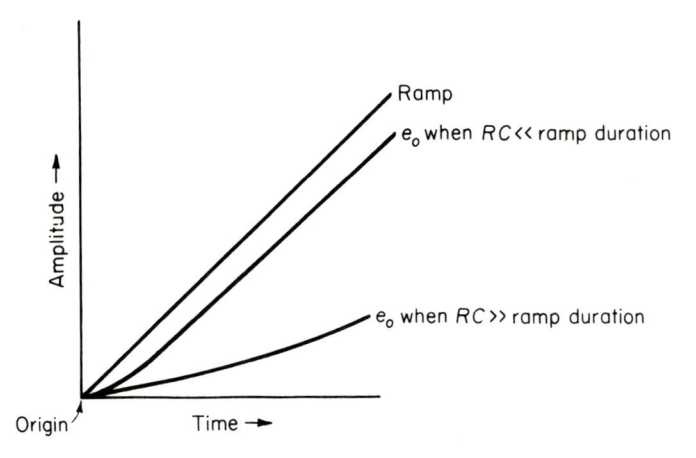

Figure 2-15 Response of a low-pass filter to a ramp input signal.

constant circuit), and the output is displaced in time from the input; however, this distortion is usually insignificant. If the time constant of the filter is large compared to the ramp duration, however, the output bears little resemblance to the input as shown by the lower waveform.

2.14 The R-C integrator

A low-pass filter is usually known as an *integrator* when the circuit time constant is long compared to the duration of the pulse. Such a circuit is shown in Fig. 2-16. The time constant of the circuit

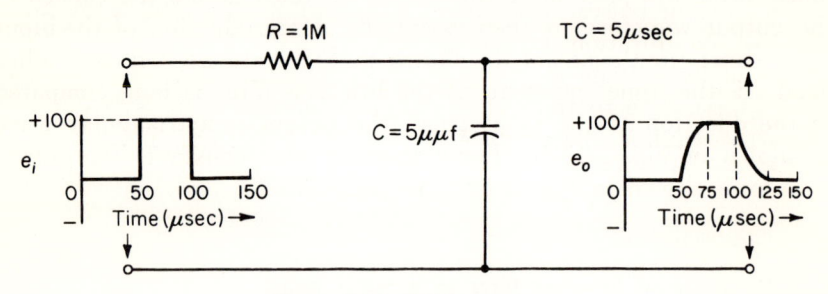

Figure 2-16 R-C integrator.

shown is 5 μsecs compared to a pulse of 50 μsec duration. If a rectangular pulse is applied across the input terminals, the output voltage across the capacitor at every instant is the difference between the input voltage and the IR drop across the resistor. From $t = 0$ to $t = 50$ μsec, there is no voltage drop across resistor R since the input voltage is zero. When the input voltage rises to 100 v at $t = 50$ μsec, the capacitor charges exponentially at a rate determined by the time constant of the circuit. For the values shown, the time constant is 5 μsec. Capacitor C would be charged to 100 v after a period of time equal to five time constants (25 μsec) at $t = 75$ μsec. The voltage across R decreases from its maximum to its minimum value during the time C is charging to 100 v. The voltage across the capacitor remains at 100 v from $t = 75$ to $t = 100$ μsec. At that time, the applied voltage drops suddenly to zero, and the capacitor discharges exponentially. The output waveform tends to resemble the input waveform.

The practical use of such an integrator in TV receivers is illustrated in Fig. 2-17. The input signal is made up of six equalizing pulses, the serrated vertical sync pulse, and then another series of

Sec. 2.15   R-L circuits   33

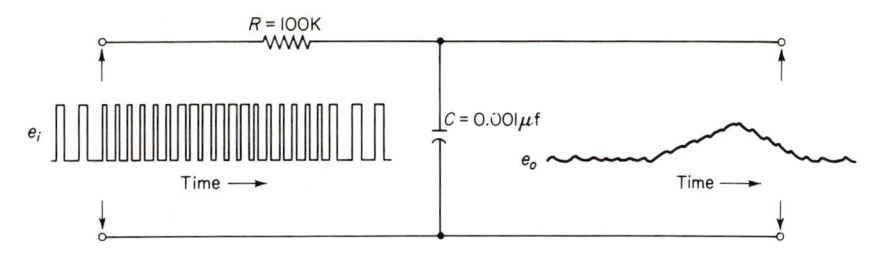

Figure 2-17 Separation of vertical sync signal from total sync by means of an R-C integrator.

six equalizing pulses. The circuit time constant is long in comparison to the duration of any of the pulses shown. The equalizing pulses apply voltage at half-line intervals (a pulse-repetition rate of 31,500 cps). Due to their short duration, as shown in Fig. 2-17, the capacitor cannot charge to any appreciable voltage. When the serrated vertical sync pulse is applied, however, the integrated voltage across the capacitor rises to an amplitude sufficient to trigger the vertical oscillator.

In the circuit shown, the integrator time constant is 100 μsec. Each pulse of the serrated vertical pulse has a duration of 27 μsec and the serrations have a duration of 4.4 μsec. Referring to a universal time-constant chart, it can be seen that the capacitor charges to approximately 27 per cent of the applied voltage for the first vertical pulse. Because the duration of a serrated pulse represents only 0.044 time constant, the capacitor loses little voltage before the next pulse provides voltage to recharge it. Thus, the integrated voltage across C reaches its maximum value at the end of the vertical sync pulse and then rapidly declines almost to zero for the equalizing pulses that follow thereby producing the output waveform shown for the complete vertical synchronizing pulse. This process is repeated at the field frequency, 60 times per second, the rate at which the vertical oscillator is triggered.

2.15 R-L circuits

Linear wave-shaping circuits using an inductor in place of a capacitor find rather limited application. The principal disadvantages of this type of circuit compared to the R-C circuit are: (1) any practical inductor has distributed capacitance and series resistance which makes the prediction of performance from theoretical calculation difficult, particularly over a wide frequency; and (2) the cost of an inductor, in general, exceeds that of a capacitor which can perform the same function with a high degree of predictability.

The analysis of an R-L circuit is similar to that of an R-C circuit except that L/R is substituted in place of RC for the time constant.

Inductors most frequently appear as part of R-L-C circuits rather than the simpler R-L circuits, because even if the actual capacitor is not placed in the circuit, the distributed capacitance of the inductor is large enough to be important, particularly at the higher frequencies.

2.16 R-L-C circuits

Two constants, resonant frequency and Q (quality factor), are associated with parallel R-L-C circuits. The higher the Q of such a circuit, the sharper its response at a given frequency.

At the resonant frequency, f_o, the 90° leading current in the capacitor is exactly equal in magnitude to the 90° lagging current of a theoretically pure inductor. Under these conditions, the resistance of the circuit represents the total impedance, and the net current drawn from the source is that due to the resistance. The current drawn is a minimum and is in phase with the supply voltage. At frequencies below f_o, current drawn by the pure inductor is greater than that through the capacitor, so that total current from the source increases and the total impedance is less than the resistance at resonance. The impedance also decreases to a value less than R at frequencies above resonance, because current through the capacitor exceeds that through the inductor. Maximum impedance, therefore, occurs at the resonant frequency.

Because $X_L = X_C$ at frequency f_o

$$\omega_o L = \frac{1}{\omega_o C} \tag{2-4}$$

Multiplying both sides of the equation by $\omega_o C$

$$\omega_o^2 LC = 1 \tag{2-5}$$

Rewriting ω_o^2 as $4\pi^2 f_o^2$, and dividing both sides of the equation by $4\pi^2 LC$

$$f_o^2 = \frac{1}{4\pi^2 LC} \tag{2-6}$$

Taking the square root of both sides of the equation

$$f_o = \frac{1}{2\pi\sqrt{LC}} \tag{2-7}$$

Sec. 2.16 R-L-C circuits

or
$$T_o = 2\pi\sqrt{LC} \tag{2-8}$$

where T_o is the *period* of a resonant-frequency wave.

The factor Q (figure-of-merit) also specifies the *suddenness* of a decrease of impedance above and below the resonant frequency f_o. Thus, Q can be reduced mathematically to

$$Q = \omega_o CR \tag{2-9}$$

$$= \frac{R}{\omega_o L} \tag{2-10}$$

$$= \frac{R}{\sqrt{L/C}} \tag{2-11}$$

for a parallel resonant circuit.

Each of the above expressions is the ratio of resistance to reactance (R/X) at resonance, of either the inductor or capacitor. Because the voltage is applied to both R and X, an equivalent expression for Q at resonance is

$$Q_{f_o} = \frac{I_L}{I_R} \quad \text{or} \quad \frac{I_C}{I_R} \tag{2-12}$$

where I_L, I_R and I_C, represent the current through the inductor, resistor, and capacitor, respectively.

If, at resonance, the balanced currents through L and C are large compared to the current in R, a small shift from f_o causes an appreciable increase in current drawn from the source and a correspondingly large decrease in impedance. If, on the other hand, the balanced currents through L and C are small compared with the current in R, a shift from f_o cause a smaller increase in current and a smaller decrease in impedance. Thus, a high Q for a circuit corresponds to a sharp resonance curve, whereas a low Q indicates a broad response curve, at f_o.

The Q of a *series* R-L-C circuit is frequently expressed as

$$Q = \frac{\omega_o L}{R} \tag{2-13}$$

$$= \frac{1}{\omega_o CR} \tag{2-14}$$

$$= \frac{\sqrt{L/C}}{R} \tag{2-15}$$

Notice that the above definitions are the reciprocals of those given for a parallel resonant circuit. These last expressions may also be applied to a circuit where C is in parallel with a series combination of R and L.

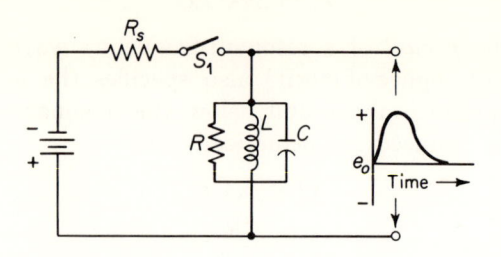

Figure 2-18 Shape of an output voltage when $R = 0.5\sqrt{L/C}$; i.e., $Q = 0.5$.

The shape of output voltage e_o for the series-parallel circuit of Fig. 2-18 depends on the Q of the circuit. Note that the circuit is connected to a d-c supply of zero frequency. It is assumed that an initial steady current is built up in the inductance, and switch S_1 is then opened. If R were removed, a sine wave of constant amplitude would result, as the energy oscillates between L and C. If a value of R greater than $0.5\sqrt{L/C}$ was then inserted, e_o would become a damped wave, (one of rapidly decreasing amplitude) and would soon decrease to zero. Under these conditions, Q is greater than 0.5. If the circuit Q is now made less than 0.5 by making R less than $0.5\sqrt{L/C}$, then e_o would first rise to some relatively low value and then drop slowly until, at the end of time t, it would become zero. This is *overdamping*. Finally, if the value of $R = 0.5\sqrt{L/C}$, so that $Q = 0.5$, a *critically damped* output waveform occurs in which a positive swing takes place, but no negative swing. This is the output voltage waveform shown in Fig. 2-18. The elimination of the negative portion of the waveform is due to the rapid damping action of R. The peak value of e_o is appreciably lower than it would be without damping because much of the energy in the inductance is lost initially to the resistor before e_o reaches its peak.

In actual circuits, the capacitor may be a physical component or merely the distributed capacitance between windings of the inductor and the shunt capacitance of the circuit. If a critically damped or overdamped response is desired, a resistor can be placed in parallel with the tank circuit. The resistance of the inductor by itself is sufficient to produce a damped wave output if this is desired. Further, the battery shown in Fig. 2-18 may be replaced by a rectangular pulse.

2.17 Ringing oscillator

The *ringing oscillator*, shown in Fig. 2-19, is typically used for calibration of the time axis of a cathode-ray oscilloscope (CRO). This and similar applications can be understood from the following explanation of circuit operation.

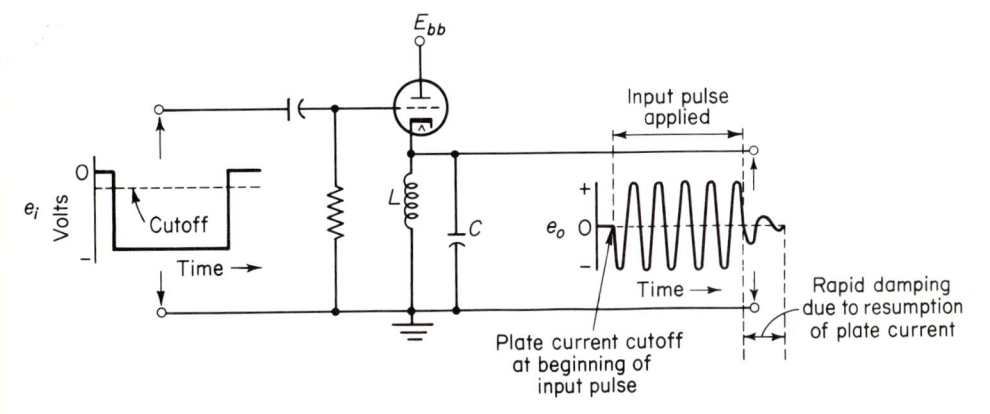

Figure 2-19 *Ringing circuit.*

In the absence of an input signal (zero grid bias), the circuit is adjusted so that a steady cathode-to-plate current passes through the tube and a high Q inductor, L. A steady magnetic field then exists around the inductor.

When a negative-going pulse, of sufficient amplitude to cause plate-current cutoff, is applied to the input circuit, the steady magnetic field around the inductor collapses and the L-C circuit is left free to oscillate at its natural frequency. The voltage induced in the inductor by the collapsing field tends to maintain current. This current charges capacitor C, and natural oscillations are set up in the L-C circuit. Although some damping occurs, due to the resistance of the coil, equally spaced oscillations continue until the end of the input pulse at which time the grid voltage is reduced to a negative value beyond cutoff. The ensuing cathode-to-plate current then rapidly damps out oscillations in the L-C circuit.

In practice, the input pulse which initiates the oscillatory circuit action also generates the CRO time base. The ringing circuit output is displayed on the CRO screen and, because the ringing frequency is known, serves as an accurate indicator of sweep speed.

2.18 R-L-C peaker

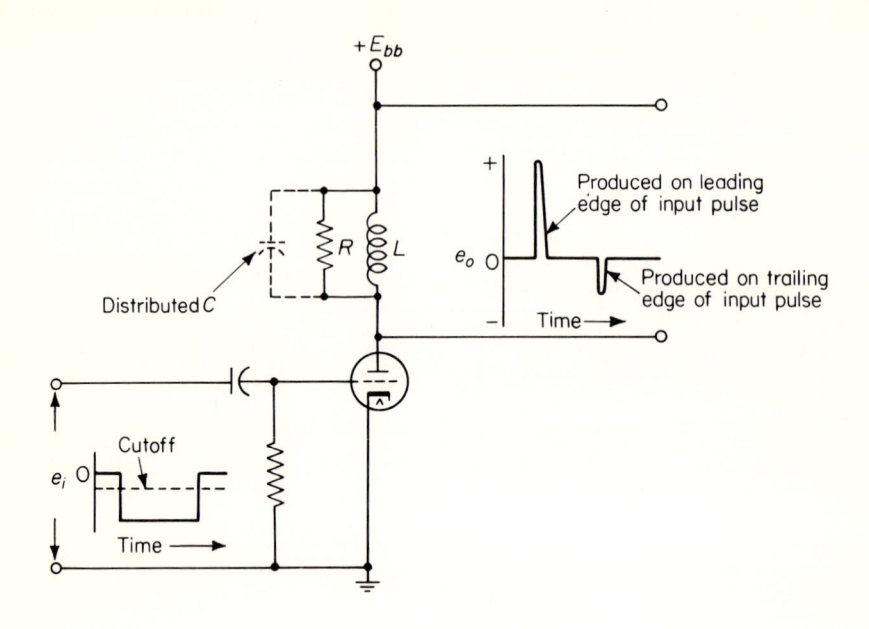

Figure 2-20 R-L-C peaker.

The R-L-C *peaker*, shown in Fig. 2-20, is used to produce narrow pulses from rectangular waves. The pulses produced are suitable for triggering gas tubes and waveform generators such as *multivibrators* and *blocking oscillators*, as described in later chapters.

The circuit resembles the ringing oscillator, but there are significant differences. Instead of an actual capacitor in the oscillating circuit, only the capacitance between turns of the coil and the unavoidable disturbed circuit capacitance are used. This procedure ensures a high resonant frequency since the capacitance, as such, is small. In addition, a resistor is placed across the L-C circuit of such value as to produce nearly critical damping when the tube is cut off. Instead of an oscillation, therefore, a single sharp positive *peak* is produced when the leading edge of the negative-going input pulse is applied, and a single sharp negative peak is produced when the trailing edge of the input pulse again raises the grid positively above cutoff. Finally, the inductor is connected to the plate rather than to the cathode of the tube, so that an output pulse of greater amplitude than the input pulse can be obtained.

EXERCISES

2-1 Why are simple networks composed of resistors, capacitors, and inductors called "linear"?

2-2 In a high-pass filter the output is taken across the _____.

2-3 Define "time constant."

2-4 In an R-C low-pass filter circuit the voltage across the capacitor rises to an amplitude of _____ when an ideal rectangular waveshape of 50 v amplitude is applied to the input terminals for a time interval equal to one time constant.

2-5 The charge or discharge process in an R-C network is considered to be complete after _____.

2-6 What is meant by "differentiation"; "integration"?

2-7 Assuming a square-wave input, draw the output voltage waveforms obtained from a differentiating network; from an integrating network.

2-8 What is meant by the terms "symmetrical" and "unsymmetrical" when referring to a waveform?

2-9 When a ramp voltage is applied to a high-pass filter, what is the amplitude of the output voltage after an elapsed time equal to five time constants?

2-10 In a low-pass R-C network, the output is taken across the _____.

2-11 If the time constant of the low-pass R-C filter is large compared to the duration of a ramp input voltage, how does the output waveform compare to the input waveform?

2-12 In a high-pass R-C filter, how does the rise time of the input waveform affect the output?

2-13 What are the disadvantages of an R-L linear waveshaping circuit compared to an R-C circuit?

2-14 What would be the shape of the output voltage waveform for the circuit of Fig. 2-18, when: (a) $R > 0.5\sqrt{L/C}$; (b) $R < 0.5\sqrt{L/C}$; (c) $R = 0.5\sqrt{L/C}$?

2-15 How does the Q of an R-L-C circuit affect the shape of the response curve at resonance?

2-16 If the R-C time constant of a high-pass filter is made increasingly smaller in comparison to the duration of the input waveform, is the width of the output pulse increased or decreased? Explain your answer.

2-17 What distinction is generally made in saying that a time constant is long, short, or medium?

2-18 Draw the differentiator (short-time constant) and integrator (long-time constant) outputs for the following input waveforms.

(a)

(b)

(c)

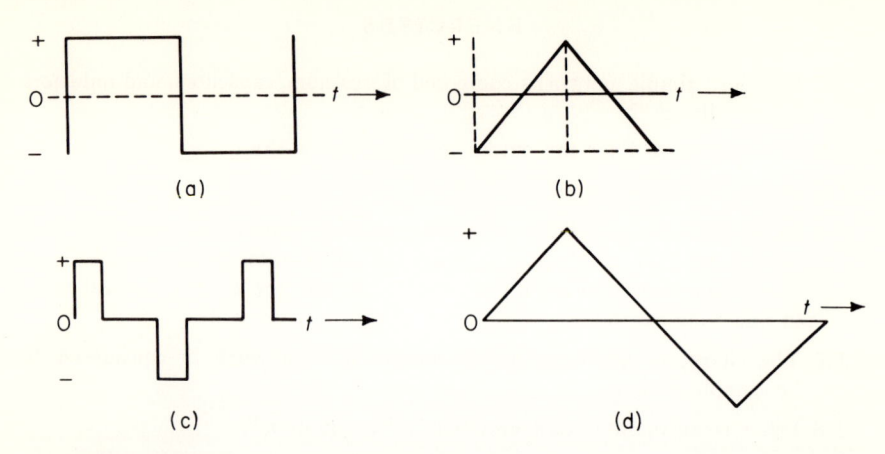

(d)

2-19 Given an R-C circuit in which $R = 100\,\text{K}$ and $C = 0.02\,\mu\text{f}$; the input pulse duration is $200\,\mu\text{sec}$. What is the circuit time constant; and it is considered to be long, short, or medium?

2-20 A triangular waveform is applied to the input of a high-pass R-C filter as shown below. What is the value of $e_{o_{\max}}$?

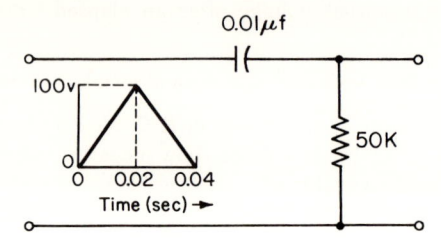

2-21 In an L-R circuit, $L = 10\,\text{mh}$ and $R = 100\,\text{ohms}$. Calculate the circuit time constant.

2-22 In the circuit of Fig. 2–12, what is the value of the output voltage at $t = 35\,\mu\text{sec}$?

2-23 In an R-L-C circuit, what is the period of a resonant frequency wave if $L = 10\,\text{mh}$, $R = 25\,\text{ohms}$ and $C = 50\,\mu\text{f}$?

2-24 What is the Q of a series R-L-C circuit if $L = 10\,\text{mh}$, $R = 5\,\text{ohms}$ and $C = 100\,\mu\text{f}$?

2-25 In an R-C differentiator, the input voltage has a constant value of $+10\,\text{v}$. What is the output?

3

VACUUM TUBE
R-C AMPLIFIERS

To ensure an understanding of material covered in succeeding chapters, a review of the basic theory of vacuum-tube R-C amplifiers is presented in this chapter. Equivalent circuits are also derived and their importance as an analytical tool is demonstrated.

3.1 Static, quiescent, and dynamic operation

A typical resistance-capacitance (R-C) coupled vacuum-tube amplifier is shown in Fig. 3-1. A plate supply voltage, E_{bb}, is connected in series with load resistor R_L between cathode K and plate P of the tube. When the plate of the tube is positive with respect to its cathode, conduction occurs. A d-c current, i_b, passes through the tube, the load resistor, and the self-biasing cathode resistor R_k, in the direction indicated by the arrow.

In passing through R_k, the current develops a d-c voltage of such polarity as to make the grid G negative in respect to the cathode. As long as this condition exists, there is no grid current.

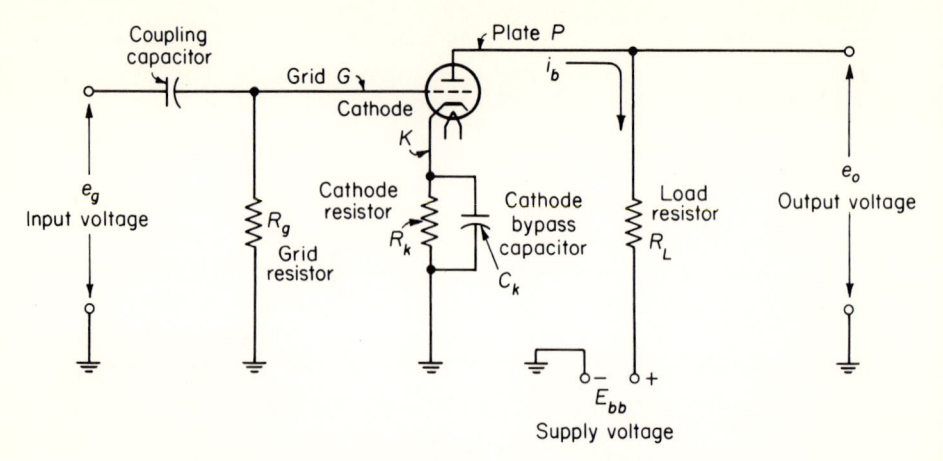

Figure 3-1 *Resistance-capacitance (R-C) coupled amplifier.*

When an alternating-signal voltage, e_g, is applied across the input terminals, plate current variations occur; these variations consist of a d-c component caused by the supply voltage and an a-c component caused by the input signal. As a result of tube amplification, the a-c output voltage developed across R_L is of greater amplitude than the a-c input voltage.

If the load resistor is removed from the circuit, no useful output voltage is obtained, and the plate voltage at every instant equals the supply voltage, even though plate current varies with changes in grid voltage. When a tube is operated in this manner, it is said to be operating under *static* conditions. Plate characteristic curves, published in tube manuals, are obtained by static operation.

Static operation should not be confused, however, with *quiescent* operation. If the input-signal voltage to the amplifier is removed, a steady voltage, E_{cc}, appears as grid bias from grid to cathode. If the grid bias is not of sufficient amplitude to cause plate-current cutoff, a direct current, I_{bo}, produces a direct-voltage drop, $I_{bo}R_L$, across the load, and places a direct voltage, E_{bo}, across the tube. The values E_{cc}, I_{bo}, and E_{bo} are known as quiescent values of grid voltage, plate current, and plate voltage, respectively. When a tube is operated in this manner, (without input signal) it is said to be operating under quiescent conditions.

When a varying signal voltage, e_g, is applied to the input terminals of the amplifier, the current and all voltages change from instant to instants; then the tube is said to be operating under *dynamic* conditions. The instantaneously changing quantities are designated i_b, e_c, i_bR_L, and e_b, for total plate current, total grid cathode voltage, varying voltage drop across the load, and total plate voltage, respectively. The dynamic values vary above and below the quiescent values. The excursions of instantaneous plate voltage and current about I_{bo} and E_{bo} are the a-c components of these quantities, and are designated e_p and i_p, respectively.

When an R-C amplifier is operated under dynamic conditions, the input and output-signal voltages are 180° out of phase. This condition may be readily understood by noting that R_L, R_k, and the tube act as a voltage divider for supply voltage E_{bb}. During the positive alternation of input signal voltage e_g, the instantaneous value of the a-c component of grid voltage e_c ($E_{cc} + e_g$) increases (becomes less negative). This causes an increase in plate current and greater voltage drops across R_k and R_L. Because the sum of the individual voltage drops across the complete series circuit must, at all times, equal E_{bb}, it is apparent that less tube voltage e_b is available. In brief, as e_c increases, e_b decreases. The opposite of this action occurs during the negative alternation of e_g during which e_b increases. At cutoff (and beyond) e_b equals E_{bb}, since i_b is zero. Thus, at each instant, the input signal voltage and the tube plate voltage are 180° out of phase.

3.2 Vacuum tube parameters

Under static conditions of operation, a small increase in plate voltage causes a corresponding change in plate current. To reduce plate current to its original value, the negative grid bias must be increased by a slight amount. Generally, the change in grid voltage required to accomplish this reduction is much smaller than the change in plate voltage causing the increase. The ratio of a small change in plate voltage to a small change in grid voltage required to hold the plate current constant is called the amplification factor (μ) of a tube. The equation for this ratio is

$$\mu = \frac{\Delta E_p}{\Delta E_g} \text{ with } I_p \text{ constant} \qquad (3\text{-}1)$$

where the triangle (Greek letter delta) is the symbol for "a small change." The reason why delta is used in the equation will be explained shortly.

Because an a-c voltage is varying continually, the amplification factor indicates how much greater a fictitious a-c plate voltage would have to be, as compared with the a-c signal applied to the grid, to produce the same a-c component of plate current.

Electrons pass from the cathode, through the space charge, and past a charged control grid to the plate of a tube, where their energy is changed to heat. The loss of energy within the tube is, therefore, analogous to the loss due to resistance, and the plate-to-cathode path of the tube is considered as offering resistance to the flow of electrons. It is also obvious that, because a d-c voltage drop occurs across the tube when it is carrying current, it also has resistance. In operating circuits, the opposition this resistance offers to alternating current is of primary interest. This effective resistance is called the *dynamic plate resistance* (r_p) of the tube, and is expressed in ohms. In equation form

$$r_p = \frac{\Delta E_p}{\Delta I_p} \text{ with } E_g \text{ constant} \tag{3-2}$$

As noted before, a small change in plate voltage causes a corresponding change in plate current under static conditions of operation. This same small change in plate current can also be produced by a small change in grid voltage. The small plate current change resulting from a small grid voltage change, under conditions of constant plate voltage, expresses in mhos the *transconductance* (g_m) of a tube. In equation form

$$g_m = \frac{\Delta I_p}{\Delta E_g} \text{ with } E_p \text{ constant} \tag{3-3}$$

Amplification factor, dynamic plate resistance, and transconductance are interrelated. The amplification factor is a product of the other two. This relationship is expressed in equation form as

$$\mu = g_m r_p \tag{3-4}$$

By transposition, it can also be seen that

$$r_p = \frac{\mu}{g_m} \tag{3-5}$$

and that

$$g_m = \frac{\mu}{r_p} \tag{3-6}$$

A family of curves (plate characteristics) for the 12BZ7 twin triode is shown in Fig. 3-2. They represent the average curves for many tubes of this type. Relatively long portions of the curves are represented as essentially straight lines. These curves may not hold true for an individual tube of this type. (The same statements apply to any type of tube. The 12BZ7 merely represents a specific case.)

The equations given for μ, r_p, and g_m are accurate only when the characteristic is linear (straight). This is not the condition that

Sec. 3.3 The constant-voltage equivalent circuit   45

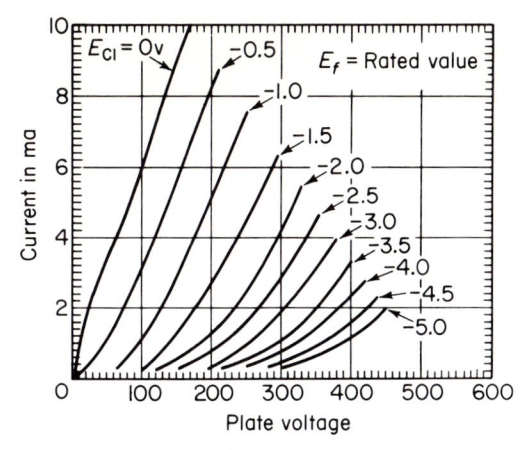

Figure 3-2 Family of curves for the 12BZ7 twin triode.

exists in practice, because the operating point is being moved up and down over a considerable portion of the characteristic. To prevent significant error, the value of voltage and current variations used in the equations must, therefore, be small, which explains the use of delta in the first three equations given.

3.3 The constant-voltage equivalent circuit

If the voltage and current variations are small enough so that μ, g_m, and r_p may be considered constant over the range of operation, a vacuum tube can be represented by an equivalent circuit containing only linear (resistive) elements. This greatly simplifies the prediction of gain.

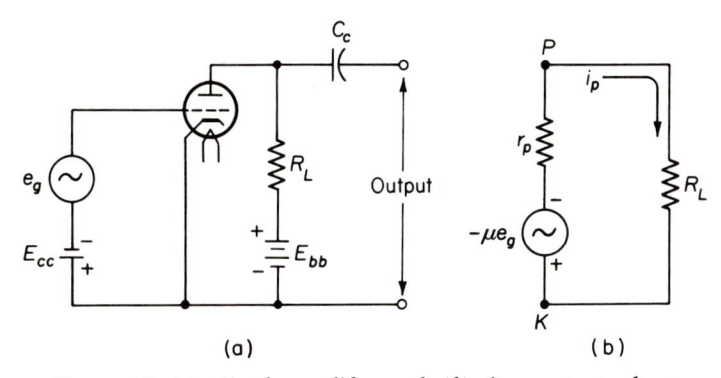

Figure 3-3 (a) Simple amplifier and (b) its constant-voltage equivalent circuit.

In Fig. 3-3(a), an a-c voltage, e_g, in the grid circuit, has the same effect as if, due to amplification, an a-c voltage of value μe_g were inserted directly in the plate circuit. As a result, the tube may be replaced by an equivalent alternator and effective resistance, as shown in Fig. 3-3(b), $-\mu e_g$ and r_p, respectively.

It should be understood that this is not an actual circuit, but a fictitious *equivalent* circuit carrying the same alternating current as the actual circuit and having the same value of R_L. Note also that the a-c alternator in the equivalent circuit has an instantaneous voltage of $-\mu e_g$. The minus sign indicates the polarity reversal between grid and plate voltages in the actual circuit. This reversal is represented by assigning polarities to the generator terminals just opposite from their actual polarities; i.e., plate terminal P negative in respect to cathode terminal K. This indicates that during the positive half-cycle of plate current and grid voltage, electrons move toward the positive generator terminal.

Figure 3-3(b) is a simple series circuit. The methods used to calculate the a-c plate current and voltage are determined by Ohm's law, as follows

$$\text{Plate current} = i_p = \frac{-\mu e_g}{r_p + R_L} \tag{3-7}$$

and

$$\text{Output voltage} = i_p R_L = -\mu e_g \left(\frac{R_L}{r_p + R_L}\right) \tag{3-8}$$

Equation 3-8 points out a most important fact. It shows that the output voltage ($i_p R_L$) is not equal to $-\mu e_g$, but is less than this value by the ratio of resistance R_L to the total resistance in the circuit. As the value of R_L is increased in comparison to r_p, the output voltage also increases. This statement seemingly implies that the output voltage can be made approximately equal to $-\mu e_g$ simply by making the value of R_L very high in comparison to r_p. In practice, however, R_L can be increased to a certain value only if the size of the d-c supply voltage is to be kept within reason. Under quiescent conditions of operation, $E_{bo} = E_{bb} - I_{bo}R_L$. If R_L is increased excessively, the I-R drop across it is so great that only a very small voltage is applied to the plate.

Two assumptions must hold true if the results obtained by using the constant-voltage equivalent circuit are to be of practical use: (1) the dynamic plate resistance and amplification factor of the tube must remain constant under dynamic conditions of operation; and (2) negligible distortion is generated by the tube.

Nothing can be done to eliminate the slight variations in r_p and

Sec. 3.4   How frequency affects constant voltage equivalent circuit   47

μ that always occur under dynamic conditions of operation. The parameters given in a manufacturer's tube manual are, however, satisfactory in all but extreme cases. Distortion is minimized by operating the amplifier as Class A, i.e., along the straight portion of the characteristic curve.

3.4 How frequency affects the constant voltage equivalent circuit

A typical resistance-capacitance (R-C) coupled amplifier is shown in Fig. 3-4. The arrangement of triode V_1 and its associated components

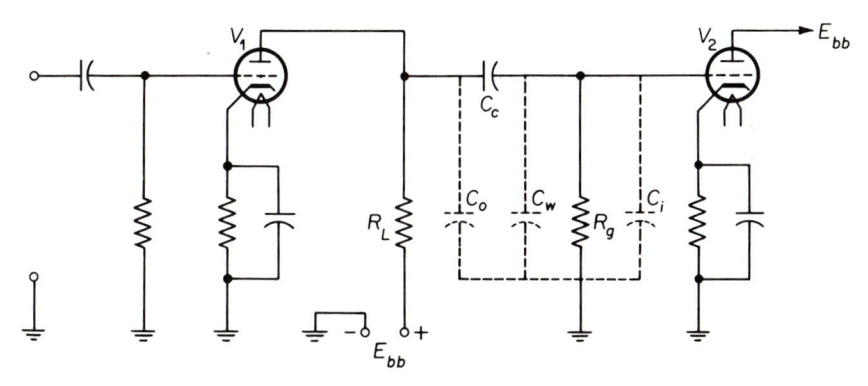

Figure 3-4 R-C coupled amplifier. Capacitor C_c passes the a-c component of voltage developed across R_L and blocks the d-c component. Resistor R_g provides a d-c return path for V_2.

is identical to that shown in Fig. 3-1. Capacitor C_c transfers the a-c component of plate voltage developed across R_L to the input circuit of V_2, and at the same time blocks the d-c component. Resistor R_g provides a d-c return path to ground for any electrons that accumulate on the grid of V_2. Even when V_2 is operated in Class A, a minute electron current occurs in the grid circuit, primarily as a result of residual gas remaining after the evacuation process. If a return path to ground is not provided, electrons accumulate on the grid side of C_c and cause plate-current cutoff in tube V_2. Resistor R_g also acts as a path for the application of bias voltage to the grid of V_2.

The frequency response of any amplifier is largely determined by the nature of its load impedance. In Fig. 3-4, it might be assumed that the load impedance for V_1 consists entirely of resistor R_L. This assumption is not accurate. The load impedance is actually a complex

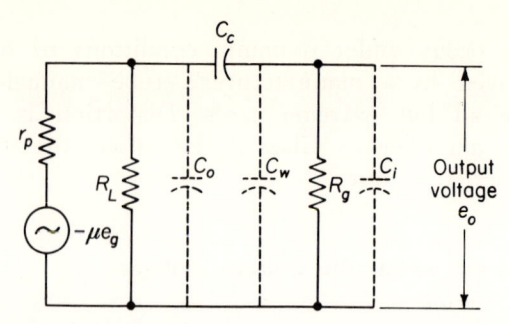

Figure 3-5 Equivalent circuit of an R-C coupled amplifier used to show that load for the input stage is a complex impedance.

network that includes resistor R_L, grid-leak resistor R_g, coupling capacitor C_c, the distributed capacitance C_w of the coupling network, the plate-to-cathode (output) capacitance C_o of the tube V_1, and the input capacitance C_i of V_2.

In the equivalent circuit of Fig. 3-5, that part which includes $-\mu e_g$, r_p, and R_L make up the simple constant voltage-equivalent circuit developed in Section 3.3. Capacitors C_o, C_w, and C_i are shown by dashed lines to indicate that they are not physical (lumped) components such as capacitor C_c.

At low audio frequencies, up to about 300 cycles, the *combined* value of capacitances C_o, C_w, and C_i might be 10 $\mu\mu f$; thus their shunting effect on resistors R_g and R_L is completely negligible. At a frequency of 100 cycles, their combined capacitive reactance is nearly 160 megohms, and at 300 cycles their reactance is 53 megohms.

Capacitor C_c and resistor R_g form in effect an a-c voltage divider, a high-pass filter. The only useful voltage is that developed across R_g and applied to the grid-cathode circuit of V_2. Any voltage developed across the reactance of C_c represents a decrease in output voltage. At low frequencies, the reactance of C_c is high compared to resistance R_g, and this must be considered in analyzing operation.

Knowing the effect at low frequencies of the various negligible shunting capacitances (C_o, C_w, and C_i) the equivalent circuit of Fig. 3-5 may be redrawn as shown in Fig. 3-6, with these capacitances omitted.

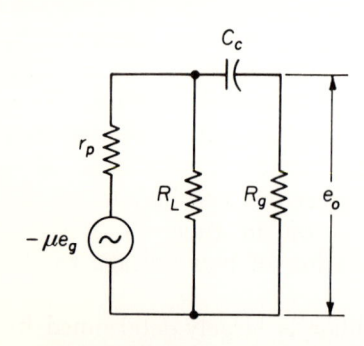

Figure 3-6 Equivalent circuit of an R-C coupled amplifier at low frequencies.

Sec. 3.4   How frequency affects constant voltage equivalent circuit   49

At middle-range audio frequencies, approximately 300 to 3000 cycles, the reactance of capacitor C_c is relatively small compared to resistor R_g, and the reactances of C_o, C_w, and C_i are still high compared to R_L and R_g. Thus, at middle-range audio frequencies, the effects of all the various capacitances may be neglected and the equivalent circuit may be represented by Fig. 3-7.

At frequencies above 3000 cycles, the reactance of the coupling capacitor C_c is equivalent to a short circuit and it can be neglected. Capacitances C_o, C_w, and C_i are in parallel and can be represented by a single shunting capacitor termed C_t. The high-frequency equivalent circuit is shown in Fig. 3-8. The low reactance of C_t lowers the total load impedance, resulting in a lower output voltage and decreased frequency response.

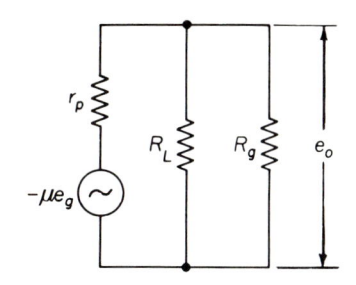

Figure 3-7 Equivalent circuit of an R-C coupled amplifier at middle-range audio frequencies.

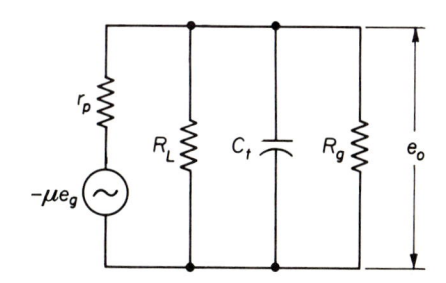

Figure 3-8 Equivalent circuit of an R-C coupled amplifier at high frequencies.

A careful study of the low-, middle-, and high-frequency equivalent circuits leads to the conclusion that output impedance and gain are highest at the middle frequencies. The reduced response at high and low frequencies is expressed in terms of the mid-frequency gain, termed A.

The mid-frequency voltage gain of an R-C coupled amplifier is derived from the equivalent circuit of Fig. 3-7. If Z_L is used to represent R_L and R_g in parallel,

$$A = -\mu \left(\frac{Z_L}{r_p + Z_L} \right) \qquad (3\text{-}9)$$

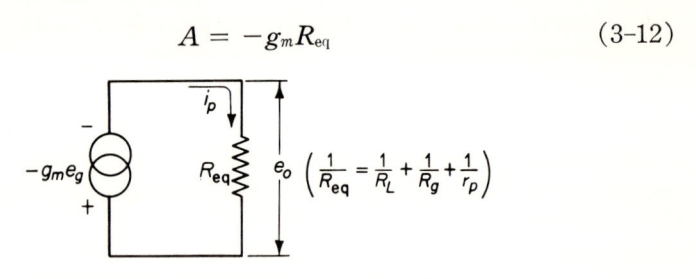

3.5 The constant-current equivalent circuit

A second form of equivalent circuit may be obtained from the first. Equation 3-8 states that the output voltage of the constant-voltage equivalent circuit is the ratio $-\mu e_g R_L/(r_p + R_L)$. If $-g_m/r_p$ is substituted for $-\mu$, the equation becomes

$$e_o = \frac{-r_p g_m e_g R_L}{r_p + R_L} \qquad (3\text{-}10)$$

which can be rearranged as

$$e_o = -g_m e_g \left(\frac{r_p R_L}{r_p + R_L}\right) \qquad (3\text{-}11)$$

The units of $g_m e_g$ are mhos × volts. To multiply by a value is the same as to divide by its reciprocal; therefore, mhos × volts is the same as volts/ohms which, by Ohm's law, equals current. The expression $r_p R_L/(r_p + R_L)$ is the parallel combination of the two resistors, and can be represented by the term R_{eq}, where the subscript "eq" designates "equivalent." When current $g_m e_g$ passes through R_{eq}, it produces a voltage drop which is the output voltage of the network. These relations are shown in the constant-current equivalent circuit of Fig. 3-9. Using this equivalent circuit, the output voltage amplification at mid-frequencies is

$$A = -g_m R_{eq} \qquad (3\text{-}12)$$

Figure 3-9 Constant-current equivalent circuit.

Two equivalent circuits have been developed. Either version can be used for the same amplifier because they are equivalent to each other. In general, however, the constant-voltage equivalent circuit is used to analyze triodes, where the value of r_p is low compared to R_L and cannot be neglected. The constant-current equivalent circuit is useful for analyzing tetrodes and pentodes, where the value of r_p is very high compared to R_L and can be neglected. If r_p is at least ten times R_L, Eq. 3-12 can be used in the simple form

$$A = g_m R_L \qquad (3\text{-}13)$$

Sec. 3.6 A universal response curve for R-C coupled amplifiers   51

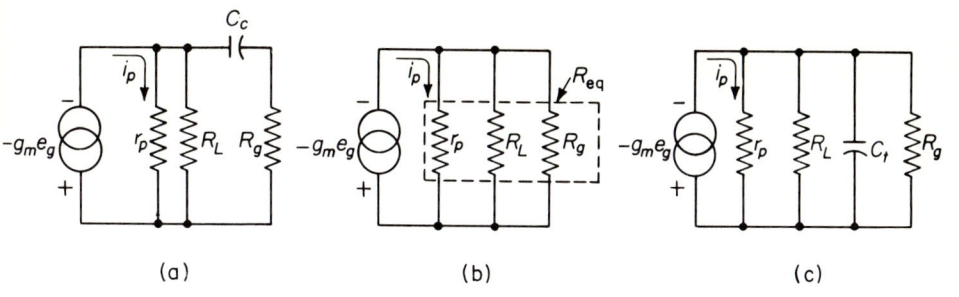

Figure 3-10 Constant-current equivalent circuits (a) low frequencies, (b) mid-frequencies, (c) high frequencies.

Because frequency affects both constant-voltage and constant-current equivalent circuits in the same manner, the three arrangements shown in Fig. 3-10 are used in drawing the constant-current equivalent circuit at different frequencies. The constant-current method is used extensively in tetrode and pentode amplifier design because of its simplicity, as well as in transistor design.

3.6 A universal response curve for R-C coupled amplifiers

The universal-response curve shown in Fig. 3-11 is representative of R-C coupled amplifiers. The distinguishing feature of this charac-

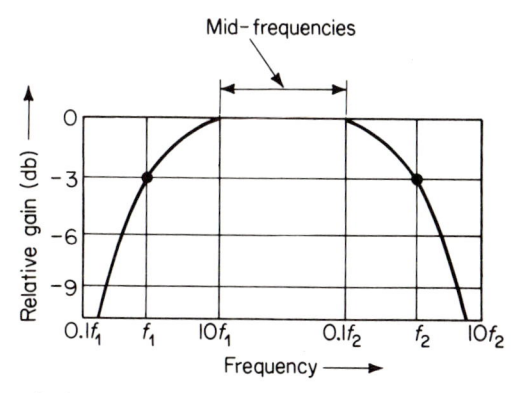

Scales: Vertical — logarithmic; horizontal — linear

• Indicates lower-and-upper frequency half-power points (−3db A)

Figure 3-11 Universal response curve of an R-C coupled amplifier.

teristic is an amplification that is substantially constant over a wide range of frequencies but drops off at very high and very low frequencies.

At some high frequency, termed f_2, the reactance of shunt capacitance C_t equals the resistance of R_{eq} and the voltage gain drops to 70.7 per cent of A, the mid-frequency gain. Amplification can also be expressed by the decibel equivalent of the voltage ratio. Thus, when the voltage drops to 70.7 per cent, power is reduced to 3 db ($0.707^2 = 0.5$), and gain at frequency f_2 is expressed as -3 db. Because -3 db represents a 50 per cent decrease in power, frequency f_2 is referred to as the *high-frequency half-power point*.

At low frequencies, C_t represents an open circuit. At some low frequency, termed f_1, however, the reactance of the coupling capacitor equals R_{eq}, and amplification again falls to 70.7 per cent (-3 db) of its mid-frequency value. For this reason, frequency f_1 is referred to as the *low-frequency half-power point*.

The flat portion of the universal response curve between $10 f_1$ and $0.1 f_2$ represents the frequency range over which amplification never falls below 99.5 per cent of its mid-frequency value. The frequency-response characteristics of an amplifier are usually expressed in terms of the frequency band lying between the half-power points of the upper and lower frequencies.

3.7 Distortion

Ideally, the output of an amplifier should be an exact duplicate of the input in all respects except amplitude. This ideal is not achieved in practice. Three types of distortion occur: (1) variations in the relative amplitudes of the frequency components of the output signal compared to the relative amplitudes of the frequency components of the input signal; (2) the existence of frequency components in the output which are not present in the input; and (3) variations in the relative phases of the frequency components of the output waveform compared to the relative phases of the frequency components of the input waveform.

The first type of distortion noted above is called *frequency distortion*. It is caused by variations in the impedance of components and circuits associated with the amplifier. The manner in which it limits the range of uniform frequency response has been described. The curve of Fig. 3-11 results from frequency distortion.

The second type of distortion is called *amplitude* (nonlinear) distortion, and results from the nonlinear relations between voltage and current in the amplifier tube. Amplitude distortion arises when large signal voltages are applied and the operating point is shifted to the highly nonlinear portion of the characteristic curve. Amplitude distortion produces two effects: the production of *harmonics* of the frequencies present in the input signal (harmonic distortion) and the production of frequencies that are *sums and differences* of those frequencies present in the input signal. Thus, amplitude distortion may be measured as the arithmetic or rms sum of harmonic voltages produced when a fundamental (single) frequency is used as the input signal; this sum is expressed as a percentage of the fundamental voltage produced at the output. An alternative method, based on the second effect, is commonly called *intermodulation* distortion. This method evaluates amplitude distortion in terms of the amplitude of the sum and difference frequencies produced using a complex wave as the input signal. Thus, the same nonlinearity which leads to production of harmonics of a single frequency also leads to production of sum-and-difference frequencies of a complex wave. Intermodulation distortion, therefore, should not be considered as a separate form the distortion. It is merely another means of determining amplitude distortion. In summary, *nonlinear distortion*, *harmonic distortion*, and *intermodulation distortion* are all forms of *amplitude distortion*.

The third type of distortion is called *phase* (time delay) distortion. The output voltage of an amplifier with a resistance load is 180° out of phase with the input voltage. An R-C or R-L circuit, in addition to frequency discrimination, is capable of producing phase shift. If the load is reactive, this phase difference between output and input voltage is not 180°. It also varies with changes in frequency, since the amount of phase shift is caused by the magnitude of reactance. Thus, if the input voltage contains components of various frequencies, the phase relationships of the components are changed in the output. As a result, the shape of the output waveform differs from that of the input, and distortion has been introduced. Phase distortion usually occurs in the circuits having inductive or capacitive components such as coupling circuits, cathode-bias circuits, and screen-grid decoupling circuits.

Compensation for frequency and phase distortion is discussed in a later chapter. Controlled amplitude distortion is made use of in pulse-type circuits, as will be shown later.

EXERCISES

3-1 Define static, quiescent, and dynamic operation.

3-2 If, in a given tube, a 2-v increase in plate voltage produces a given change in plate current, and an increase in negative bias voltage of 0.1 v returns plate current to its originally value, what is the amplification factor of the tube?

3-3 If the dynamic plate resistance of the tube used in Problem 3-2 is 6000 ohms, what is its transconductance?

3-4 Draw the constant-voltage equivalent circuits of an R-C coupled amplifier at low-, high-, and mid-frequencies. Explain why the circuits differ.

3-5 Draw the constant-current equivalent circuit of an R-C coupled amplifier at low-, high-, and mid-frequencies. How can the quantity $g_m e_g$ represent a current?

3-6 Calculate the mid-frequency voltage gain of an R-C amplifier using the constant-voltage equivalent circuit when $\mu = 70$, $r_p = 61{,}000$ ohms, $R_L = 470{,}000$ ohms, and $R_g = 470{,}000$ ohms.

3-7 Define the upper and lower frequency half-power points.

3-8 What types of distortion occur in an amplifier? Define each.

3-9 A 100-cps signal having a peak-to-peak amplitude of 2 v is applied across the coupling network of an R-C amplifier. Capacitor $C_c = 0.01\ \mu\mu\text{f}$ and $R_g = 470{,}000$ ohms. What portion of the signal (approximately) is lost due to the reactance of C_c?

3-10 Two triodes are connected as shown in Fig. 3-4. $C_o = 1.0\ \mu\mu\text{f}$, $C_i = 5.0\ \mu\mu\text{f}$, $C_w = 4.0\ \mu\mu\text{f}$, $\mu = 55$, $g_m = 1200\ \mu$ mhos, $C_c = 0.02\ \mu\text{f}$, and R_g and R_L each $= 447{,}000$ ohms. Using the constant-voltage equivalent circuit, compute the stage gain at (a) 200 cycles; (b) 1000 cycles; and (c) 5000 cycles. Also determine the high-frequency and low-frequency half-power points.

4
INTRODUCTION TO TRANSISTORS

Semiconductors are assuming an increasingly important role in electronics, particularly in the areas of application with which this book is concerned. Foremost among these devices is the *transistor*. Its advantages over the vacuum tube include: (1) a transistor has no filament and consequently uses no standby power; (2) it can be operated from low-voltage B batteries; (3) it has low heat dissipation; (4) it has generally longer life; and (5) it has greater mechanical ruggedness.

By no means should the reader infer from the above that the vacuum tube is doomed to obsolescence. It is still superior to the transistor in large-power-handling capabilities, frequency response, and noise figure. The vacuum tube is also less susceptible to damage due to electrical overloads and voltage ripple, and it has characteristics that are practically independent of ambient temperature.

Thus, it is advantageous when studying pulse-type circuits to consider both vacuum-tube and semiconductor arrangements. We shall do so in subsequent chapters.

Because most technicians are less familiar with the transistor than with the vacuum tube, the discussion in this and the following chapter is quite thorough. No attempt is made, however, to study transistor physics. Previous knowledge is assumed, as noted in the introduction. Only the *common-base* and *common-emitter* configurations are discussed; the *common-collector* amplifier is studied later with *cathode followers*. In Chap. 5 equivalent circuits are developed and their use as an analytical tool is demonstrated. A discussion of semiconductor devices other than the transistor is left to a later chapter.

4.1 Basic circuits and circuit notation

An NPN junction transistor amplifier using the common-base configuration is shown in Fig. 4-1. A collector-supply voltage, V_{CC}, is connected in series with load resistor R_L between the base and

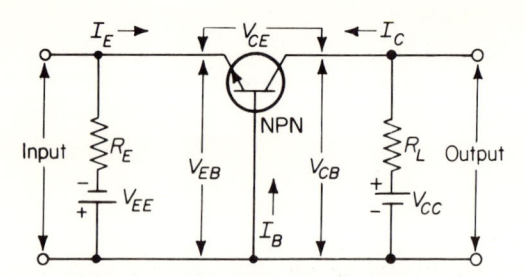

Figure 4-1 Common-base amplifier using an NPN junction transistor.

collector. This battery *reverse-biases* the collector, with respect to the base. An emitter supply, V_{EE}, is connected in series with emitter resistor R_E between the emitter and base. This battery *forward-biases* the emitter, with respect to the base.

When an alternating signal voltage is applied to the input, collector current variations occur; these variations consist of a d-c component caused by the supply voltage and an a-c component caused by the input signal. As a result of transistor amplification, the a-c output voltage developed across R_L is of greater amplitude than the a-c input voltage.

If the load resistor is removed from the circuit, no useful output voltage is obtained, and the collector-to-base voltage at every instant equals the supply voltage, even though collector current varies with changes in emitter current resulting from the applied a-c signal. When a transistor is operated in this manner, it is said to be operating under static conditions, exactly as in the case of the vacuum

Sec. 4.1   Basic circuits and circuit notation

tube. Characteristic curves, published in transistor manuals, are obtained by static operation.

Static operation should not be confused however, with quiescent operation. In the absence of an input signal voltage, battery voltage V_{EE} forward-biases the emitter and voltage V_{EB} exists between emitter and base. This forward bias lowers the emitter-base potential barrier, and the emitter injects electrons into the base. The vast majority of these electrons (95 to 99 per cent) diffuse across the base and are collected by the collector. The remaining electrons (1 to 5 per cent) recombine with *holes* in the base region. The ratio of a change in collector current to a change in emitter current with V_{CB} constant is termed *alpha* and is designated α. In equation form

$$\alpha = \frac{I_c}{I_E}\bigg|\text{const. } V_{CB} \qquad (4\text{-}1)$$

In a junction transistor, α is always less than unity (1).

Although α does not exceed unity in the common-base configuration, power gain occurs because of different impedance levels. The input resistance, R_i, generally is very low, about 50 ohms, when the device is connected to a load, R_L, of say 50 K ohms. Because power gain, G, is the ratio of signal power delivered to the load in relation to signal power supplied to the input terminals, for the values given

$$G = \frac{P_o}{P_i} = \frac{I_c^2 R_L}{I_E^2 R_i} \simeq \frac{\alpha^2 R_L}{R_i} \simeq 1000$$

The d-c collector-to-emitter voltage is V_{CE}. Collector current I_C in passing through the load resistor produces a direct voltage drop, $I_C R_L$, and voltage V_{CB} appears between collector and base. Notice that $V_{CB} = V_{CC} - V_{RL}$. The values V_{EB}, V_{CE}, V_{CB}, I_E, I_B, and I_C, are *quiescent* values of emitter-to-base voltage, collector-to-emitter voltage, collector-to-base voltage, emitter current, base current, and collector current, respectively. When the transistor is operated in this manner (without signal input) it is operating under *quiescent* conditions.

When a signal voltage is applied to the input terminals of the amplifier, the current and all voltages change from instant to instant, and the transistor is operating under *dynamic* conditions. The instantaneously changing quantities are designated by using lower-case letters corresponding to those shown above for the quiescent quantities. The dynamic values vary above and below the quiescent values.

When the common-base amplifier is operated under dynamic conditions, the input and output signal voltages are *in phase*. In this respect common-base transistor amplifier circuits are similar to grounded-grid amplifier circuits. With the NPN transistor amplifier shown in Fig. 4-1, a negative-going input signal increases the forward bias voltage v_{eb}, which causes an increase in i_e and a corresponding increase in i_c. The increased collector current through R_L increases $i_c R_L$ and, therefore, decreases v_{cb}. Thus, a negative-going input signal produces a negative-going output signal. The opposite of this action occurs during the positive alternation of the input signal. Under these conditions, v_{eb}, i_e, i_c, and $i_c R_L$ decrease, and v_{cb} increases.

A common-emitter amplifier is shown in Fig. 4-2. The emitter is now the common element between input and output. When a

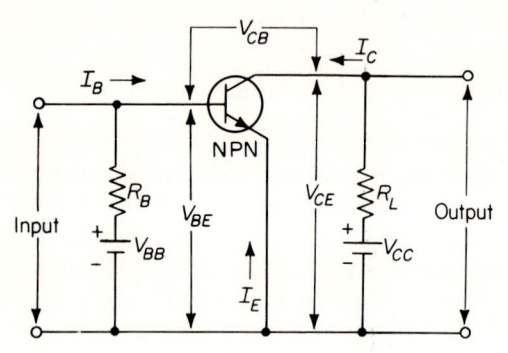

Figure 4-2 Common-emitter amplifier using an NPN junction transistor.

positive-going signal is applied to the input, the increase in base-to-emitter voltage v_{be} causes an increase in base current i_b and an increase in collector current i_c. The increase in the i_c reduces the collector-to-emitter voltage v_{ce}. Thus, a positive-going input signal produces a negative-going output signal, which indicates that the common emitter circuit produces *signal inversion*, similar to that of a basic grounded-cathode vacuum-tube amplifier circuit.

A second current amplification factor is used in the common-emitter circuit. This factor, called *beta*, is indicated by the symbol β. It is defined in terms of the ratio of a change in collector current to a change in base current with the collector-to-emitter voltage constant; that is

$$\beta = \frac{\Delta I_C}{\Delta I_B}\bigg|\text{const. } V_{CE} \tag{4-2}$$

Because base current results from recombination and amounts to only a small percentage (1 to 5 per cent) of the total electrons (or holes) injected into the base region, a small change in base

Sec. 4.1 Basic circuits and circuit notation

current produces a much greater change in collector current than did a change in emitter current in the grounded-base configuration. Thus β can, and does, exceed unity, with common values up to about 200.

Since $I_B = I_E - I_C$, we can substitute this value in Eq. 4-2 and it becomes

$$\beta = \frac{\Delta I_C}{\Delta I_E - \Delta I_C}$$

Now, if we divide both the numerator and denominator of the right member by ΔI_E we obtain

$$\beta = \frac{\Delta I_C/\Delta I_E}{1 - \Delta I_C/\Delta I_E}$$

But by Eq. 4-1, the ratio I_C/I_E is defined as α. Thus we can write

$$\beta = \frac{\alpha}{1 - \alpha} \qquad (4\text{-}3)$$

It is often useful to solve for α in terms of β. To do so, we first multiply both sides of Eq. 4-3 by $(1 - \alpha)$, which gives

$$\beta(1 - \alpha) = \alpha$$

and performing the indicated multiplication in the left member of the equation

$$\beta - \beta\alpha = \alpha$$

Then, by adding $\beta\alpha$ to both members

$$\beta = \alpha + \beta\alpha$$

Factoring α in the left members

$$\beta = \alpha(1 + \beta)$$

and, dividing both members by $1 + \beta$

$$\alpha = \frac{\beta}{1 + \beta} \qquad (4\text{-}4)$$

Equations 4-3 and 4-4 show the relationship of the two basic current gains of a transistor, and they should be memorized.

The common emitter configuration has a higher input resistance but a lower output resistance than the common-base configuration. Due to the greater value of current gain, however, power gain is high. Suppose, for example, $\beta = 50$, $R_i = 1000$ ohms, and $R_L = 2000$ ohms. Then

$$G = \frac{I_C^2 R_L}{I_B^2 R_i} \simeq \frac{\beta^2 R_L}{R_i} = 5000$$

where $R_i =$ input resistance.

It should be noted that the power gain calculations for both the common-base and common-emitter circuits, as given thus far, are approximations only, used to show basic characteristics of the two configurations. The figures are approximate because α and β are both *short-circuit current-amplification* factors, and in both examples given R_L was not zero and, therefore, v_c varied. More exact power gain expressions are given later.

4.2 Leakage currents[1,2,3,4]

In a common-base amplifier, if the emitter is d-c open-circuited, and a voltage is applied to the collector that reverse-biases the collector-to-base junction, a small collector current exists. This current is called the *reverse (leakage) current* of the collector-to-base junction, and is denoted either as I_{CO} or as I_{CBO}. This leakage current exhibits wide variance in transistors of similar types, and is extremely sensitive to temperature. At a specified temperature, however, I_{CO} is considered a constant for a particular transistor. For the time being, we will simply note that variations of I_{CO} have a very marked effect upon the operating point. More will be said about this matter later.

From what has been said, it is quite obvious that

$$I_C \cong \alpha I_E + I_{CO} \quad (4\text{-}5)$$

at any particular collector potential.

In a common-emitter amplifier, if the collector is d-c open-circuited and a voltage is applied from emitter to base that reverse-biases the emitter-base junction, a leakage current, designated either as I_{EO} or as I_{EBO}, occurs. This leakage current is called the reverse (leakage) current of the emitter-to-base junction, and it is also sensitive to temperature.

Again, in a common-emitter amplifier, if the base is d-c open-circuited, and a voltage is applied from collector to emitter that reverse-biases the collector-to-base junction, a collector current exists. This current is termed I_{CEO}. The relative magnitude of I_{CEO} is indicated by recalling that $I_E = I_C + I_B$; and substituting for I_E in Eq. 4-5

$$I_C \cong \alpha(I_C + I_B) + I_{CO} \quad (4\text{-}6)$$

Rearranging

$$I_C \cong \frac{\alpha I_B}{1 - \alpha} + \frac{I_{CO}}{1 - \alpha} \quad (4\text{-}7)$$

If Eq. 4-3 is substituted into Eq. 4-7

$$I_C \cong \beta I_B + (\beta + 1) I_{CO} \quad (4\text{-}8)$$

Sec. 4.3 Graphs

As noted above, when the base is d-c open-circuited, $I_C = I_{CEO}$. Therefore, for any particular collector potential

$$I_{CEO} \cong (\beta + 1) I_{CO} \qquad (4\text{-}9)$$

Then Eq. 4-8 becomes

$$I_C \cong \beta I_B + I_{CEO} \qquad (4\text{-}10)$$

In summary, I_{CO} is the collector-to-base leakage current, when a normal collector potential is applied and the input circuit (emitter) is open for the common-base configuration. For the common-emitter circuit, collector current is I_{CEO} when the input circuit (base) is open; I_{CEO} is many times greater than I_{CO}.

4.3 Graphs[1]

Various volt-ampere plots are used in studying transistors. A complete forward and inverse collector characteristic for an n-p-n transistor is shown in Fig. 4-3(a), and for a PNP transistor in Fig. 4-3(b). In both illustrations, emitter current is the parameter represented. For the NPN transistor, both collector current and collector voltage are positive; for the PNP transistor, they are negative. For the PNP transistor, the results are often plotted in the first quadrant in spite of the negative values of I_C and V_{CB}.

The forward volt-ampere characteristic of the collector junction, effectively a diode, is plotted in the third quadrant for the NPN transistor and in the first quadrant for the PNP transistor. Because the transistor collector junction is reversed-biased, however, this region is of no interest and is commonly omitted.

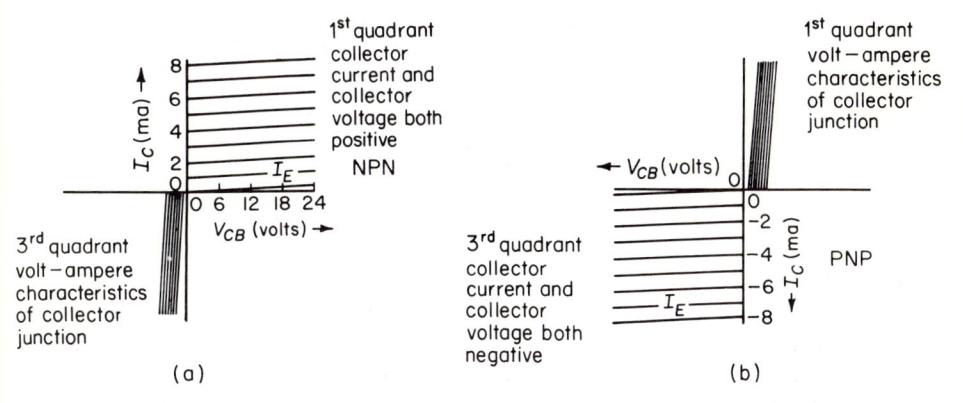

Figure 4-3 Complete forward and inverse collector characteristics of (a) NPN transistor and (b) PNP transistor.

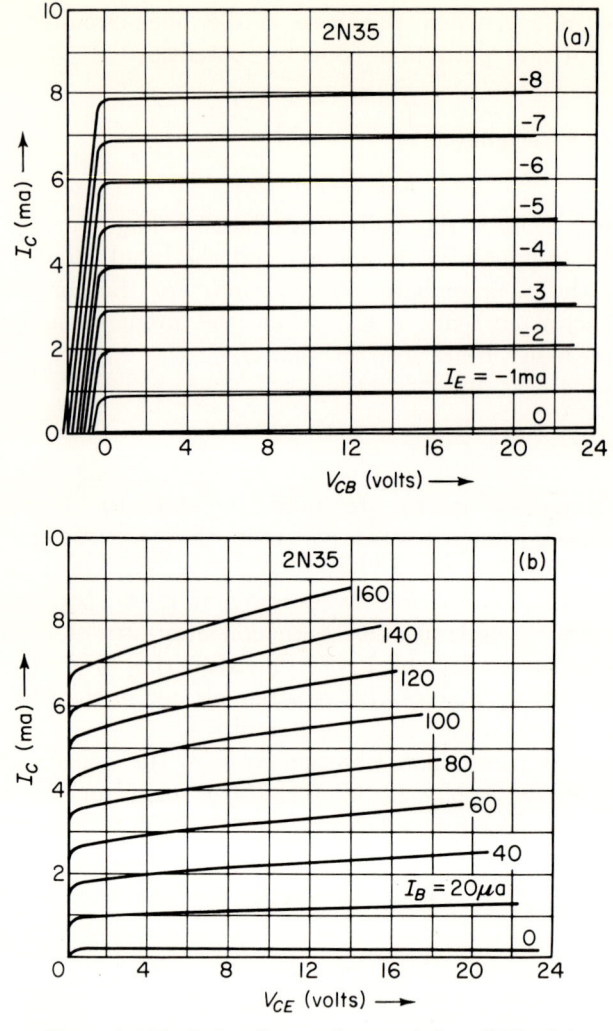

Figure 4-4 Typical collector characteristics—NPN—(a) common base and (b) common emitter (Courtesy of Sylvania Electric Products, Inc.).

Typical common-base and common-emitter collector characteristics are shown in Figs. 4-4(a) and (b), respectively. In both, I_C is the independent variable. The parameter in (a) is I_E. The parameter in (b) is I_B.

Notice the degree of linearity in the common-base characteristic. An increase in I_E of 2 ma produces a nearly identical change in I_C, which is to be expected from what has been said regarding α. In

Sec. 4.3 Graphs

the common-emitter characteristic shown in (b), however, this linearity no longer holds true. An increase of 5 μa in I_E produces a change in I_C of approximately 200 μa. This statement indicates that for this transistor, β is approximately 40, and is also in agreement with what was said earlier.

The plots shown in Fig. 4-3 are the most commonly used and are analogous to the family of curves for the vacuum-tube plate. The slope of the curves represent the parameter I/R_0 (where R_0 is the output resistance). The nearly horizontal slope of the curves indicates that R_0 is large.

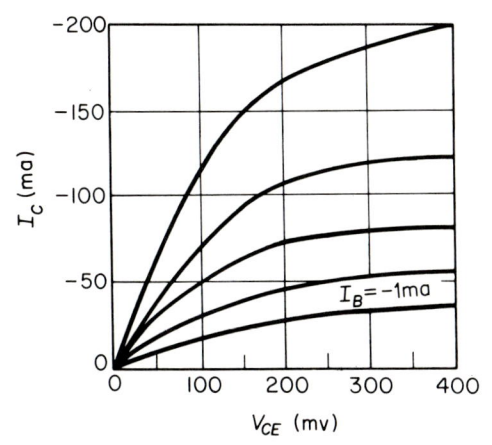

Figure 4-5 Saturation (bottomed) region—common-emitter configuration (from Fitchen, Transistor Circuit Analysis and Design, 1960, D. Van Nostrand Co., Inc., Princeton, N.J.).

If we were to expand the scale of Fig. 4-4(b) in the vicinity of the origin, it would appear somewhat as shown in Fig. 4-5. In this low-voltage region, notice that incremental changes in base current do not cause the correspondingly large collector-current changes found at higher collector voltages. This region is called either the *saturation region* or the *bottomed region*.

In the saturation region, both junctions are forward-biased. The existence of the saturation region is specified by a *saturation resistance*, R_{CS}, which is determined from the slope of the I_B lines in Fig. 4-5. In the majority of applications where germanium transistors are used, R_{CS} usually has a value less than 20 ohms and is neglected; in low-power silicon transistors, however, R_{CS} may have a value of several hundred ohms; such a difference imposes a serious limit on the allowable operating portion of the characteristic. Obviously, the magnitude of R_{CS} depends upon I_B and, therefore, a specified value of R_{CS} exists at the specified values of I_C and I_B.

It has been shown that $I_{CEO} = I_C$ when $I_B = 0$, and our curves indicate the relations between I_C and V_{CE} when I_B is leaving the base terminal of a PNP transistor. If we inspect typical common-emitter collector characteristics, there is no indication that operation is possible below I_{CEO}; i.e., there is no indication that the transistor can be operated with reverse base current, because lines of constant current are not drawn for that region. However, recall that in the common-base configuration, I_{CO} flows into the base of a PNP transistor. Consequently, operation is possible as low as I_{CO} in the common-emitter configuration.

Figure 4-6 clarifies the relationships among the quantities of interest in the *cutoff* region. Base current cannot assume values more positive than I_{CO}, since this quantity is a leakage current that is generally independent of applied potential.

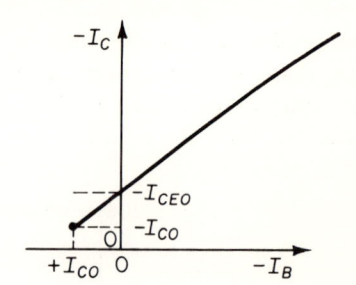

Figure 4-6 Clarification of the relationships among the quantities of interest in the cutoff region (from Fitchen, Transistor Circuit Analysis and Design, *1960*, D. Van Nostrand Co., Inc., Princeton, N.J.).

4.4 The operating point[1-4]

Variations occur in transistors of a given type as a result of manufacturing tolerance, aging, and temperature changes. To establish the correct operating point and maintain it is not an easy task. Only when the operating point is established correctly is the output waveform an essentially undistorted amplified reproduction of the input waveform.

This situation is illustrated in Fig. 4-7, where the common-emitter stage (a) is being operated Class A as indicated in (b). The operating point is defined by I_B and V_{CE}, with I_B determined by the value of resistor R_B. The product $I_B R_B \cong V_{BB}$, since V_{BE} is very small, (0.2 to 0.8 v). As indicated by Eq. 4-10, I_C is directly dependent upon I_B. Also, V_{CE} is dependent upon I_C, R_L, and V_{CC}, since

$$V_{CE} = V_{CC} - I_C R_L \tag{4-11}$$

Sec. 4.4   The operating point

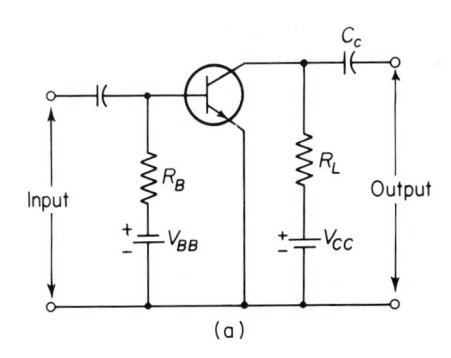

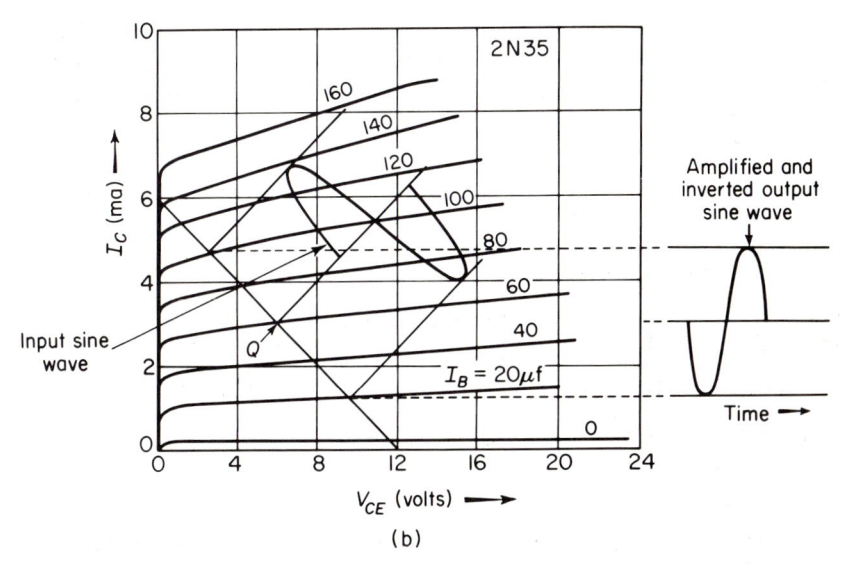

Figure 4-7 (a) Class A amplifier produces (b) an essentially undistorted reproduction of the input (Graph of (b) courtesy of Sylvania Electric Products, Inc.).

Under these conditions, a variation in I_B causes an amplified variation in I_C that is a faithful reproduction of the signal.

Now let us suppose that the transistor used in the circuit of Fig. 4-7(a) became defective for some reason and had to be replaced by another unit of the same type. If the leakage, gain, or saturation of the new transistor is different, or if the temperature is varied, the amplified output signal is distorted even though the Q point is unchanged. If, for example, the gain of the new transistor is higher than that of the unit it replaced, the stage is driven into the saturation region of the characteristics, and the positive portion of the output waveform is clipped.

Thus, maintaining a constant value of I_B does not provide an operating point that minimizes the effects of aging, etc. However, if we keep the Q point at a constant value of I_C, Class A operation is maintained for the replacement transistor.

4.5 Important variations in transistor replacement

For the purpose of d-c circuit analysis, the junction transistor may be considered as two back-to-back junction diodes. It has also been shown that I_C consists of an amplified input signal portion and a leakage current. Using this information, we can draw simplified d-c circuits for the common-base and common-emitter configurations, as shown in Figs. 4-8(a) and (b), respectively. The emitter-base diode is

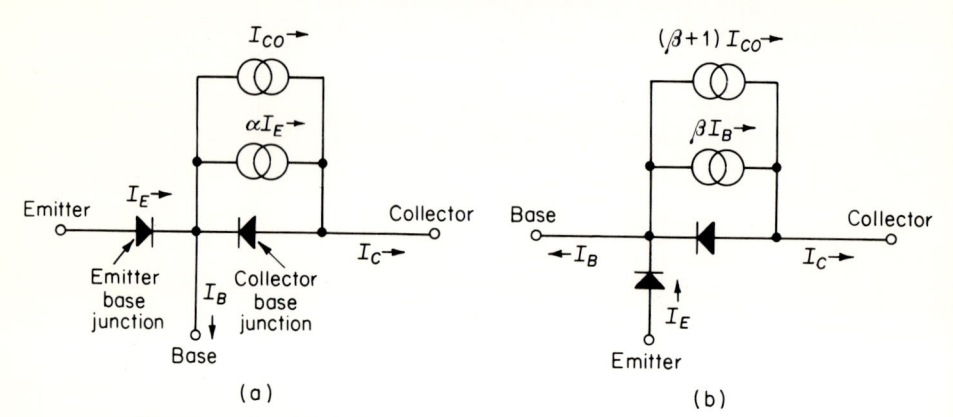

Figure 4-8 Simplified d-c equivalent circuits for the (a) common-base and (b) common-emitter configurations.

forward-biased. Large variations in the forward properties can be minimized by use of an external resistance. The collector-base diode is reverse-biased, and thus presents a high-value resistance. Any variations in its properties can be neglected. In the current generators, α does not vary sufficiently between transistors of the same type to have any important effect on bias stability, but the term $(1 - \alpha)$, as found in β, is extremely important, and it is primarily this effect which must be minimized. To illustrate this variation, if α is 0.99, $(1 - \alpha) = 0.01$; but if α is 0.96, $(1 - \alpha)$ is 0.04. Thus, for a change in α of 3 per cent, we have a change in $(1 - \alpha)$ of 75 per cent.

In germanium transistors, I_{CO} is approximately doubled for every

Sec. 4.6　Bias circuits

11°C rise in temperature. In silicon transistors, I_{co} doubles for about every 18°C rise in temperature. It was shown in Eq. 4-9 that leakage current I_{CEO} in the grounded-emitter configuration varies as I_{co} and is greater than I_{co} by the factor $(\beta + 1)$.

Our discussion indicates that gain and leakage current are the most important variations to be considered in the replacement of transistors.

4.6 Bias circuits[1-7]

A typical common-emitter amplifier using *fixed bias* is shown in Fig. 4-9. This is the simplest method of establishing the proper Q point, but has the disadvantage of relatively poor stability in maintaining the Q point and requries two separate supplies for the base and collector.

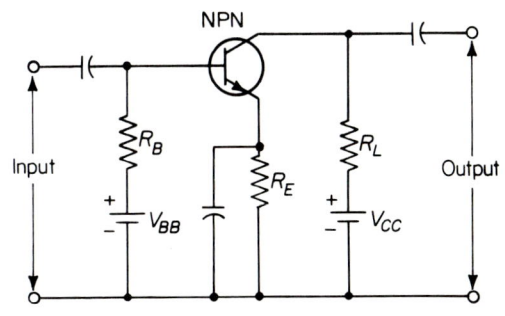

Figure 4-9 Common-emitter amplifier using fixed bias.

In the circuit of Fig. 4-9, I_B is maintained constant. The emitter resistor R_E and base resistor R_B are used primarily to minimize variations in the transistor input resistance.

The necessary equations for the circuit of Fig. 4-9 are

$$I_C = \beta I_B + (\beta + 1) I_{co}$$

$$I_C = I_E - I_B$$

and

$$V_{BB} = I_E R_E + I_B R_B$$

Solving these three equations for I_C

$$I_C = \frac{\alpha V_{BB} + I_{co}(R_E + R_B)}{R_E + R_B\,(1 - \alpha)} \qquad (4\text{-}12)$$

To obtain a measure of the bias stability, we can rewrite Eq. 4-12 in the form

$$S = \frac{R_E + R_B}{R_E + R_B\,(1 - \alpha)} \qquad (4\text{-}13)$$

where $$S = \frac{dI_C}{dI_{CO}} = \text{current stability factor}$$

Recall that dI_C and dI_{CO} are differentials and that their ratio is a derivative: in this case, the derivative of I_C with respect to I_{CO}. The derivative expresses the exact rate of change.

The stability factor S permits a measure of comparison among circuits. To illustrate, let us assume that we have an unstabilized circuit where $S = 50$, so that a ΔI_{CO} of 20 μa becomes a ΔI_C of 1000 μa. Further, let us suppose that in Fig. 4-9, $R_E = 1000$ ohms, $R_B = 10,000$ ohms, and $\alpha = 0.98$. Substituting these values in Eq. 4-13

$$S = \frac{1000 + 10,000}{1000 + 10,000\ (0.02)} = \frac{1.1 \times 10^4}{1.2 \times 10^3} \cong 9.2$$

Now, for $\Delta I_{CO} = 20\,\mu$a, the ΔI_C value is only 184 μa, and this is a very marked improvement. It can be seen that the closer S approaches unity, the less the operating point shifts. Once we know what to expect in the way of variation in I_C, we can determine whether or not this variation is permissible to a given application.

Notice that in Eq. 4-13, we can minimize the effect of the term $(1 - \alpha)$ in the denominator by making R_E as large as possible with respect to $R_B\,(1 - \alpha)$. Seemingly, we could eliminate the effect of $(1 - \alpha)$ altogether by making R_E sufficiently large; but in practice, the value of R_E is limited by the type of biasing supply used. If a separate low-voltage source is used for V_{BB}, the value of R_E can be appreciably greater than in those applications where a single supply is used for both the base and collector.

Single-battery bias is shown in Fig. 4-10. The addition of resistor

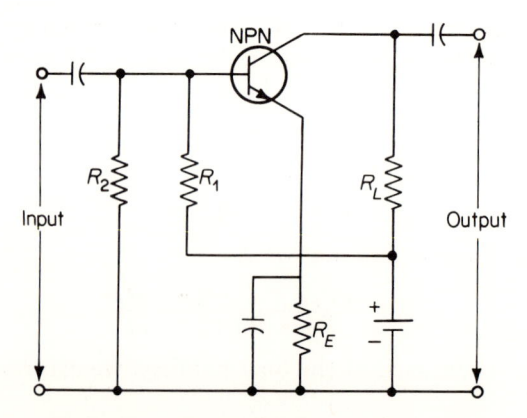

Figure 4-10 Common-emitter amplifier using single-battery bias.

Sec. 4.6   Bias circuits

R_2 results in considerable improvement in Q point stability. The equations necessary to describe circuit action are

$$I_C = I_E - I_B$$
$$I_C = \beta I_B + (\beta + 1) I_{CO}$$
$$I_B = I_1 - I_2$$
$$V_{BB} = I_1 R_1 + I_2 R_2$$

and $I_2 R_2 - I_E R_E = 0$

where I_1 is the current through resistor R_1, and I_2 is the current through resistor R_2. When these equations are solved for I_C we obtain

$$I_C = \frac{\alpha R_2 V_{BB} + I_{CO}[R_E(R_1 + R_2) + R_1 R_2]}{R_E(R_1 + R_2) + (1 - \alpha) R_1 R_2} \quad (4\text{-}14)$$

which can be rewritten as

$$S = \frac{R_E(R_1 + R_2) + R_1 R_2}{R_E(R_1 + R_2) + (1 - \alpha) R_1 R_2} \quad (4\text{-}15)$$

With the assumption that $R_1 \gg R_2$, Eq. 4-15 becomes

$$S = \frac{R_E + R_2}{R_E + R_2 (1 - \alpha)} \quad (4\text{-}16)$$

In using Eq. 4-16, if $R_1 = 5 R_2$ or greater, an error in S is unimportant.

To illustrate the improvement in S, again let $R_E = 1000$ ohms, $\alpha = 0.98$, $R_1 = 10{,}000$ ohms (the same as R_B in the previous S calculation), and R_2 (the added resistor) $= 1000$ ohms. By Eq. 4-15

$$S = \frac{2.1 \times 10^7}{1.12 \times 10^7} \simeq 1.8$$

and by Eq. 4-16

$$S = \frac{2 \times 10^3}{1.02 \times 10^3} \simeq 1.9$$

Using either equation, it is apparent that stability is greatly improved. Now the current $I_{CO} = 20\ \mu\text{a}$ becomes an I_C of $36\ \mu\text{a}$ by Eqs. 4-14 and 4-15.

Very good stability of the Q point is achieved with *emitter biasing*, shown in Fig. 4-11.

The equations necessary for this circuit are

$$I_C = I_E - I_B$$
$$I_C = \beta I_B + (\beta + 1) I_{CO}$$
$$V_{EE} = I_E R_E + I_B R_B$$

Again solving for I_C

$$I_C = \frac{\alpha V_{EE} + I_{co}(R_E + R_B)}{R_E + R_B(1-\alpha)} \qquad (4\text{-}17)$$

which is rewritten as

$$S = \frac{R_E + R_B}{R_E + R_B(1-\alpha)} \qquad (4\text{-}18)$$

Although Eq. 4-18 is identical to Eq. 4-13, and Eq. 4-17 differs from Eq. 4-12 only in the use of V_{EE} in place of V_{BB}, the numerical values of the components are different.

In the circuit of Fig. 4-11, I_E is set instead of I_B. Resistor R_B serves only as a return for I_B and has a small value. Thus, a very small voltage drop $I_B R_B$ occurs across resistor R_B, and

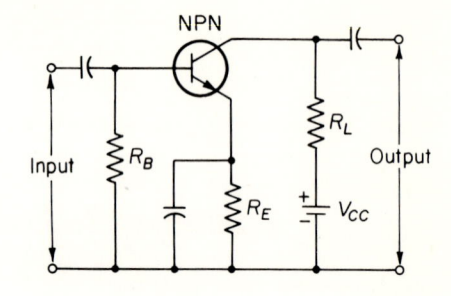

Figure 4-11 Common-emitter amplifier using emitter biasing.

the base is practically at ground potential. Also, because V_{BE} is very small (a few tenths of a volt), the emitter terminal is nearly at ground. Under these conditions

$$I_E = \frac{V_{EE}}{R_E} \qquad (4\text{-}19)$$

If $V_{EE} = 5$ v and $I_E = 1$ ma, then by Eq. 4-19, $R_E = 5000$ ohms. Let $R_B = 6800$ ohms. Then

$$S = \frac{5000 + 6800}{5000 + 6800(0.02)} \simeq 2.3$$

If we reduced R_B, S would be further improved; but this reduction has practical limitations. Although R_B could even be omitted for biasing purposes, it is required for a-c amplification.

Self-biasing is illustrated in Fig. 4-12. The collector is connected to the base through resistor R_F, and an operating point is set at a quiescent value of I_B. When an input signal causes an increase in collector current, the voltage drop across R_L increases and the voltage across R_F decreases, indicating

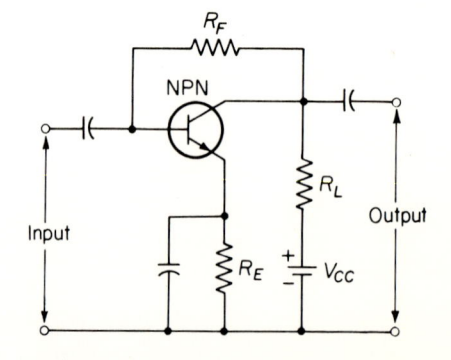

Figure 4-12 Common-emitter amplifier using self bias.

a decrease in I_B. Thus, I_B varies inversely with I_C and provides a compensating effect.

An undesirable feature of this circuit is the loss in signal gain due to *feedback* through R_F. However, if two resistors, each equal to 0.5 R_F, are used in place of R_F, and if their center point is connected to ground through a low-reactance capacitor, the undesired *degenerative effect* is eliminated.

The d-c equations for this circuit are

$$I_C \cong \beta I_B + (\beta + 1)I_{co}$$
$$I_C = I_E - I_B$$
$$V_{CC} = I_E R_E - I_B R_F + I_E R_L$$

Solving for I_C

$$I_C = \frac{\alpha V_{CC} + I_{co}(R_F + R_E + R_L)}{R_E + R_L + (1 - \alpha) R_2} \tag{4-20}$$

which is rewritten as

$$S = \frac{R_E + R_F + R_L}{R_E + R_F + R_L(1 - \alpha)} \tag{4-21}$$

Using the values $R_E = 1$ K, $R_L = 22$ K, $R_F = 100$ K, and $\alpha = 0.98$

$$S = \frac{123{,}000}{101{,}000 + 22{,}000(0.02)} = 1.2$$

a very satisfactory value.

In the preceding discussion of biasing methods the calculation of S using the component values given is intended only to indicate the relative stability of several biasing arrangements often encountered in practice. For a specific application, it should be noted that S may be higher or lower than the values computed here. Remember, we have studied only the general case, not the specific.

4.7 Thermal runaway[2]

The junction temperature T_J of a transistor is determined by the total power dissipation P, the ambient temperature T_A, and the thermal resistance R_T, and

$$T_J = T_A + PR_T \tag{4-22}$$

When a transistor is used at high junction temperatures, regenerative heating may occur and produce a condition known as *thermal runaway*, which generally results in destruction of the transistor.

If T_A increases, T_J increases an equal amount, provided P remains constant. Since β and I_{co} both increase with temperature, however,

I_C also increases with temperature, and this increases P. Thermal runaway occurs when the rate of increase of T_J with respect to P is greater than thermal resistance R_T.

Thermal runaway can be prevented, but the method of attack depends on whether the transistor is used as a linear amplifier or as a switch. (Switching circuits are fully described in later chapters.)

In switching circuits, transistors are commonly operated either in saturation (low V_{CE}) or in cutoff, where the base-to-emitter circuit is reverse-biased. In saturation, P does not change appreciably with temperature and this relative stability acts to prevent thermal runaway. In cutoff, however, P depends on I_{CO} and increases rapidly as temperature increases.

A simplified switching circuit arrangement that insures reverse bias of the emitter-to-base junction at all temperatures is shown in Fig. 4-13. In this circuit $P = I_{CO}V_{CE}$ which can be rewritten as

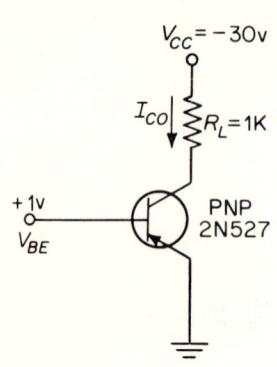

Figure 4-13 Simplified switching circuit arrangement that ensures reverse bias of the emitter-to-base junction at all temperatures (from General Electric Transistor Manual, 6th edition, 1960, General Electric Semiconductor Products Div., Liverpool, N.Y.).

$$P = I_{co}(V_{cc} - I_{co}R_L) \qquad (4\text{-}23)$$

The rate of change of P with respect to temperature is

$$\frac{dP}{dT} = (V_{cc} - 2I_{co}R_L)0.08\ I_{co} \qquad (4\text{-}24)$$

where 0.08 is the approximate fractional increase in I_{co} with temperature.

The condition for thermal runaway occurs when $dP/dT = 1/R_T$; that is, when

$$\frac{1}{R_T} = (V_{cc} - 2I_{coT}R_L)0.08\ I_{coT} \qquad (4\text{-}25)$$

In Eq. 4-25, I_{coT} is the value of I_{co} at the thermal runaway point. Solving this equation for I_{coT}

$$I_{coT} = \frac{V_{cc} \pm \sqrt{(V_{cc})^2 - 8R_L/0.08\ R_T}}{4R_L} \qquad (4\text{-}26)$$

If the negative sign is used to solve Eq. 4-26, the value of I_{coT} is obtained; if the positive sign is used, the value of I_{co} *after* runa-

Sec. 4.7   Thermal runaway

way is obtained. The equation also shows that the value of I_{co} after runaway cannot exceed $V_{cc}/2R_L$, which means that V_{CE} after runaway can never be less than $V_{cc}/2$. If the term under the square root is zero or negative, thermal runaway *cannot* occur. Also, if runaway does occur, it is only when $V_{CE} > 0.75 V_{cc}$. The power dissipated due to I_{COT} is found by substituting I_{COT} for I_{co} in Eq. 4-23. The ambient temperature at which runaway occurs is calculated from Eq. 4-22.

In Fig. 4-14, there is appreciable resistance in the base circuit, and the base-to-emitter junction is reverse-biased only over a limited temperature range. Current results when the temperature increases to the point where this junction is no longer reverse-biased and dissipation increases rapidly. In this case

Figure 4-14 Base-to-emitter junction biased only over a limited temperature range due to appreciable resistance in the base circuit (from General Electric Transistor Manual, 6th edition, 1960, General Electric Semiconductor Products Div., Liverpool, N. Y.).

$$I_{COT} = \frac{V_{cc} - 2R_L\beta I_x \pm \sqrt{(V_{cc} - 2R_L\beta I_x)^2 - 8R_L/0.08R_T}}{4R_L\beta} \quad (4\text{-}27)$$

where $I_x = V_B/R_B$.

For the purpose of discussing thermal runaway, it is best to divide linear amplifiers into two groups; preamplifiers and power amplifiers. Preamplifiers operate at low signal levels so that bias voltage and current are very low. In R-C coupled stages, a large collector-load resistor is used for increased gain, and a large emitter resistor is used to improve stability. As a result, thermal runaway seldom occurs.

In power amplifiers, however, operating power levels are near the runaway condition. Also, for the sake of efficiency, the load is

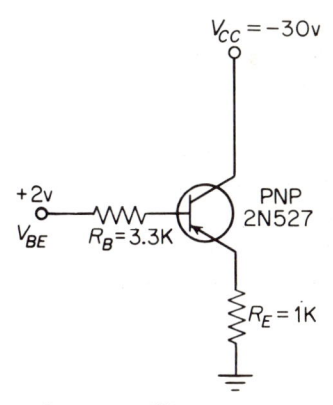

Figure 4-15 Worst-case conditions applied to a transistor in a general bias circuit (from General Electric Transistor Manual, 6th edition, 1960, General Electric Semiconductor Products Div., Liverpool, N. Y.).

transformer-coupled, which reduces the effective collector series resistance, and biasing networks of marginal stability are used. Because runaway in a power stage is likely to destroy the transistor, worst-case design principles are used for its prevention. The worst-case conditions exist when β approaches infinity ($\beta \to \infty$), $V_{BE} = 0$, $R_L = 0$, and I_{CO} is maximum ($I_{CO\,max}$).

If these conditions are applied to a transistor in the general bias circuit of Fig. 4-15, the total transistor dissipation is

$$P = (V_{CC} - V_{BE} - I_{CO}R_B)\left(I_{CO} + \frac{V_{BE} + I_{CO}R_B}{R_E}\right) \quad (4\text{-}28)$$

Equating dP/dT with $1/R_T$, and solving for I_{COT} as before

$$I_{COT} = \frac{(V_{CC} - R_1 V_{BE}) \pm \sqrt{(V_{CC} - R_1 V_{BE})^2 - R_2/0.08 R_T}}{4R_B} \quad (4\text{-}29)$$

where
$$R_1 = \frac{R_E + 2R_B}{R_E + R_B}$$

$$R_2 = \frac{8R_E R_B}{R_E + R_B}$$

As before, if Eq. 4-29 is solved using the minus sign, the value of I_{COT} is obtained; and if it is solved using the plus sign, the final value of I_C after runaway occurs is obtained. If the quantity under the square root is zero or negative, runaway *cannot* occur.

In Class B power amplifiers, the maximum transistor power dissipation occurs when the power output is at 40 per cent of its maximum value, at which point the power dissipation in each transistor is 20 per cent of the maximum power output. In Class A amplifiers, maximum transistor dissipation occurs under quiescent conditions. The maximum power dissipation is obtained by substituting I_{COT} in Eq. 4-28, and the maximum junction temperature is obtained from Eq. 4-22.

In power amplifiers, the circuit is designed to meet the requirements for gain, power output, distortion, and bias stability, and is then analyzed to determine the conditions under which runaway can occur to determine if these conditions meet the operating requirements.

4.8 Small-signal graphical analysis

A common-base amplifier is shown in Fig. 4-16(a). The load line for $R_L = 2000$ ohms is shown in Fig. 4-16(b). The load line is obtained from the equation

Sec. 4.8   Small-signal graphical analysis   75

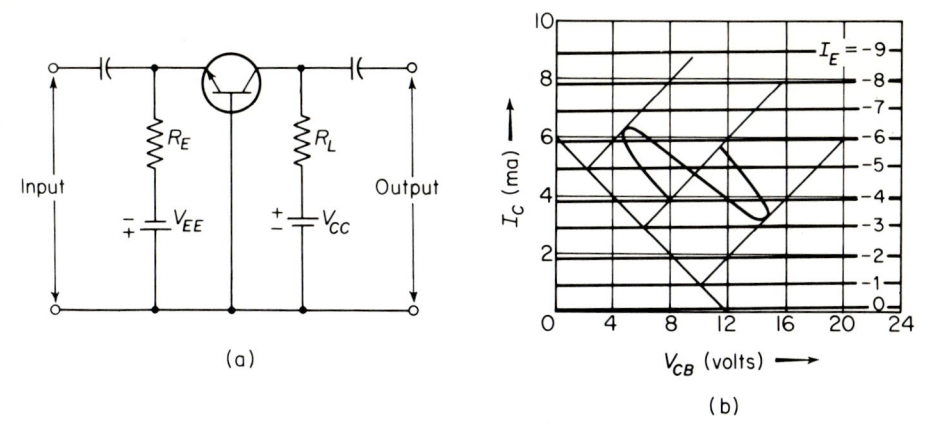

Figure 4-16 Small signal graphical analysis of a common-base amplifier.

$$V_{cc} = v_{cb} + i_c R_L \qquad (4\text{-}30)$$

In this equation, if i_c is zero, v_{cb} equals V_{cc}; and if v_{cb} is zero, i_c equals V_{cc}/R_L. Thus, the ends of the load line in Fig. 4-16(b) are $V_{cc} = 12$ v and $V_{cc}/R_L = 6$ ma.

The Q point is set by d-c bias conditions and, in this case, is 3 ma. If a sinusoidal input current, i.e., a current of 1.414 ma rms, is applied to the input terminals, the emitter current swings between 1 ma and 5 ma, the collector voltage swings between approximately 2 and 10 v, and the collector current varies from approximately 4.8 to 0.9 ma.

The output voltage across R_L is

$$V_{RL} = \frac{V_{CB\,\text{max}} - V_{CB\,\text{min}}}{2\sqrt{2}} \qquad (4\text{-}31)$$

$$= \frac{10 - 2}{2.828}$$

$$\cong 2.83 \text{ v rms}$$

The operating current gain is

$$A_{ib} = \frac{I_{C\,\text{max}} - I_{C\,\text{min}}}{I_{E\,\text{max}} - I_{E\,\text{min}}} \qquad (4\text{-}32)$$

where A_{ib} = the current gain in the grounded-base amplifier. Substituting the given values

$$A_{ib} = \frac{4.8 - 0.9}{5 - 1}$$

$$= \frac{3.9}{4}$$

$$= 0.975$$

If we assume an input resistance, R_i, of 200 ohms, the input voltage V_e is 0.2828 v rms ($R_i I_e$), and the circuit voltage gain, A_{eb}, is

$$A_{eb} = \frac{V_{RL}}{V_e} \qquad (4\text{-}33)$$

$$= \frac{2.83}{0.283}$$

$$= 10$$

The power gain, G, is

$$G = A_{ib} \; A_{eb} \qquad (4\text{-}34)$$

$$= 0.975 \times 10$$

$$= 9.75$$

A common-emitter amplifier is shown in Fig. 4-17(a), and the load line for R_L is shown in 4-17(b). The Q point is at $I_B = 60\ \mu a$. The sinusoidal input current causes the base current to swing between 100 and 20 μa; V_{CE} swings between approximately 2.4 and 9.8 v, and I_C swings between approximately 4.7 and 1.2 ma.

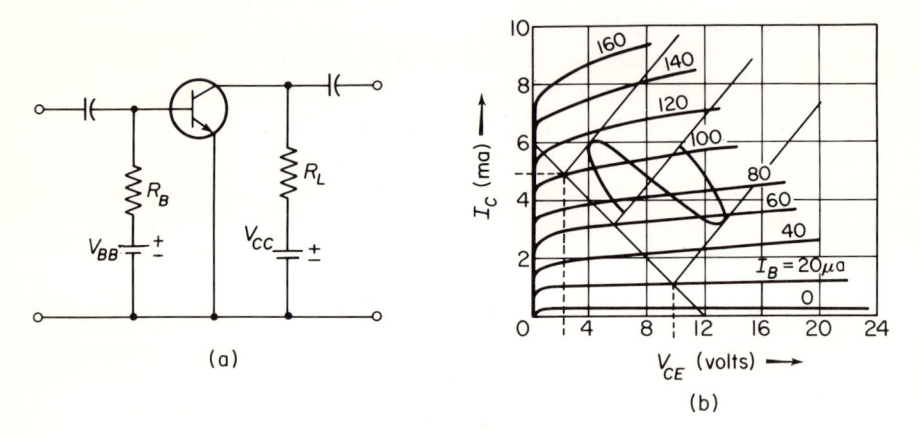

Figure 4-17 Small signal graphical analysis of a common-emitter amplifier.

The output voltage across R_L is

$$V_{RL} = \frac{V_{CE\ \max} - V_{CE\ \min}}{2\sqrt{2}} \qquad (4\text{-}35)$$

$$= \frac{9.8 - 2.4}{2.828}$$

$$= 2.616 \text{ v, rms}$$

The operating current gain, A_{ie}, (the subscript e indicates common-emitter) is

$$A_{ie} = \frac{I_{C\ max} - I_{C\ min}}{I_{B\ max} - I_{B\ min}} \quad (4\text{--}36)$$

$$= \frac{4.7 - 1.2}{100 - 20}$$

$$= \frac{3.5\ \text{ma}}{80\ \mu\text{a}}$$

$$= 43.75$$

If we assume $R_i = 1\ \text{K}$, then the input voltage $V_b = 0.0566$ v rms. (I_B in Fig. 4-17(b) has a peak-to-peak value of 80 μa. The rms value is $80 \times 0.707 = 56.56\ \mu$a. Then, since $E = IR$, we have $56.56 \times 10^{-6} \times 1 \times 10^3 = 0.0566$ v rms.)

The circuit voltage gain, A_{ee}, is

$$A_{ee} = \frac{V_{RL}}{V_b}$$

$$= \frac{2.616}{0.0566}$$

$$= 46.2$$

The power gain, G, is

$$G = A_{ie}\, A_{ee}$$

$$= 43.75 \times 46.2$$

$$= 2021$$

REFERENCES

1 Fitchen, Franklin C., *Transistor Circuit Analysis and Design*. Princeton, N. J.: D. Van Nostrand Co., Inc., 1960.

2 *General Electric Transistor Manual*, 6th ed. Liverpool, N. Y.: General Electric Company, Semiconductor Products Division, 1962.

3 *Basic Theory and Application of Transistors*, Department of the Army Technical Manual TM 11-690, March 1959.

4 Lo, A. W., et al., *Transistor Electronics*. Englewood Cliffs, N. J.: Prentice-Hall, Inc., 1955.

5 "Transistor Bias Compensation with Sensistor Silicon Resistors," *Application Notes*, Texas Instruments, Inc., Semiconductor Products Division, April 1960.

6 Shea, R. F., "Transistor Operation: Stabilization of Operating Points," *Proc. I R E*, v. 40, Nov. 1952.

7 Shea, R. F., *Principles of Transistor Circuits*. New York: John Wiley & Sons, Inc., 1953.

EXERCISES

4-1 Express α in terms of β and vice versa.

4-2 What are the most important variations to be considered in the replacement of a transistor?

4-3 What is the stability factor of a common-emitter amplifier using fixed bias if $R_E = 1,500$ ohms, $R_B = 12,000$ ohms, and $\alpha = 0.985$?

4-4 If the amplifier of Exercise 4-3 is unstabilized, and if $S = 60$, so that the ΔI_{CO} of $10\,\mu$ becomes a ΔI_C of $600\,\mu$, what approximate improvement, expressed as a percentage, is realized by the use of self-bias?

4-5 The input resistance of a common-emitter amplifier is 1000 ohms and the value of $R_L = 2000$ ohms. The manufacturer's literature indicates that the α of the transistor used is 0.96. What is the approximate power gain?

4-6 If $I_{CO} = 4\,\mu\text{a}$ and $\alpha = 0.985$, what is the value of I_{CEO}?

4-7 What does the current stability factor express? Why is it useful?

4-8 How can the undesired degenerative effect of self-biasing be eliminated?

4-9 What d-c equations describe the self-bias circuit?

4-10 What is the phase relationship of input voltage to output voltage in
    (a) the common-base configuration?
    (b) the common-emitter configuration?

5

TRANSISTOR EQUIVALENT CIRCUITS AND PARAMETERS

The concept of an equivalent circuit to replace the transistor is similar to its use for vacuum tubes. The actual transistor characteristics between its terminals are replaced by certain combinations of linear circuit components. These linear components serve to represent the transistor only in its linear region of operation or for operating ranges so small that linear approximations of the characteristic curves are adequate.

The transistor does not lend itself to as simple an equivalent circuit as the vacuum tube, and several sets of equivalent parameters are in use; two sets are described in this chapter. The first set, called r (internal resistance) *parameters*, are described primarily because they appear in much of the available literature, particularly that appearing during the early and mid-1950's. The second set, called h (hybrid) *parameters*, are described because they are most often used by the manufacturer in preparing transistor manuals, technical data sheets, etc. In general, any of the equivalent circuits and parameters may be evaluated one from the other, and this is done for the r and h parameters.

The r parameters have the advantage of maintaining the same value for a given transistor and operating point, regardless of the circuit configuration used. They have the disadvantage, however, of being difficult to measure directly from the transistor, and may have to be calculated.

The h parameters can be measured directly from the transistor, and manufacturers usually give some or all of the h-parameter values for each transistor type. Also, once gain and performance equations are derived using h parameters, they remain of the same form. The h-parameters have the disadvantage of having different values for each configuration. Thus, it is necessary to substitute the parameters for a particular configuration in the general equation to analyze gain and performance for that configuration.

5.1 The r parameters

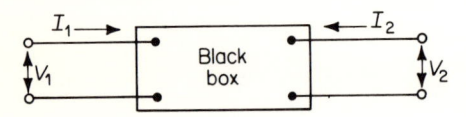

Figure 5-1 Amplifier considered as a black box.

Any amplifier, whether a single stage or a complete circuit, can be considered as a *black box* that has two input terminals and two output terminals, as shown in Fig. 5-1. Such a representation is called a *four-terminal (two-terminal pair) network*. The quantities I_1, I_2, V_1, and V_2 are measured at the input and output terminals, with I_1 and I_2 both considered positive (flowing inward) so that specification of input or output does not alter the analysis. Knowing the values of these terminal quantities, we can specify the performance of the circuit.

In the analysis, two quantities are taken as independent variables. The choice of which two quantities are considered dependent variables and which two are considered independent variables is arbitrary, but it is this choice which determines the form of the equivalent circuit.

Let us assume that the black box of Fig. 5-1 contains a transistor amplifier in the common-base configuration, and that we select the two currents as the independent variables and the two voltages as the dependent variables. Since, in this configuration, $I_1 = i_e$ and $I_2 = i_c$, the following general equations hold true.

$$v_{eb} = f(i_e, i_c) \tag{5-1}$$

and

$$v_{cb} = f(i_e, i_c) \tag{5-2}$$

Sec. 5.1   The r parameters

Equation 5-1 states that v_{eb} is a function (f) of the independent variables i_e and i_c. Equation 5-2 states that v_{cb} is also a function of the same independent variables. (Unless you have already learned a precise definition of the word "function" you may not have a sound notion of the meaning of the above statements. Specifically, "function" means that for each value of the dependent variable v_{eb} or v_{cb} there is a corresponding value of the independent variables.)

To develop an *input circuit* equation from the general equation, Eq. 5-1, we first determine the effect of each independent variable, i_e and i_c, on voltage v_{cb}, and then add the separate results. When the effect of i_e is considered, i_c is held at zero; and when the effect of i_c is considered, i_e is held at zero. (In practice, the a-c current to the transistor input or output terminals is made zero, while maintaining a d-c bias voltage, by placing a high-resistance coil in series with the terminals. The a-c voltage across the input and output terminals is made zero, while maintaining a d-c bias voltage, by placing a low-reactance capacitor across the terminals.)

For small variations of I_E and I_C from quiescent values, we can write

$$dv_{eb} = \left.\frac{dv_{eb}}{di_e}\right| di_e + \left.\frac{dv_{eb}}{di_c}\right| di_c \tag{5-3}$$

To develop an output circuit equation from Eq. 5-2, we proceed in the same manner and obtain

$$dv_{cb} = \left|\frac{dv_{cb}}{di_e}\right| di_e + \left|\frac{dv_{cb}}{di_c}\right| di_c \tag{5-4}$$

If the changes are small, the operating range on the static characteristic curves can be assumed linear, and the coefficients of the terms (partial derivatives) in Eqs. 5-3 and 5-4 become constants. If the changes are sinusoidal, the partial derivatives can be expressed as impedances and we can rewrite the equations as

$$v_{eb} = z_{ib}i_e + z_{rb}i_c \tag{5-5}$$

and

$$v_{cb} = z_{fb}i_e + z_{ob}i_c \tag{5-6}$$

where voltages and currents are in rms values. The subscripts $i, r, f,$ and o represent *input, reverse, forward,* and *output*, respectively, and the subscript b used with all of the impedances indicates that we are using the common-base configuration. The names for the subscripts $i, r, f,$ and o are apparent when we consider the following: If the output circuit is open; i.e., if $i_c = 0$, then

and
$$z_{ib} = \frac{v_{eb}}{i_e} = \text{open circuit input impedance}$$

$$z_{fb} = \frac{v_{cb}}{i_e} = \text{forward transfer impedance}$$

If the input circuit is open; i.e., if $i_e = 0$, then

$$z_{rb} = \frac{v_{eb}}{i_c} = reverse\ transfer\ impedance$$

and

$$z_{ob} = \frac{v_{cb}}{i_c} = open\ circuit\ output\ impedance$$

These four parameters are called the *open-circuit impedance* (z) *parameters*. By use of these defining equations, these parameters may be measured at low frequencies. For low frequencies, we can consider the impedances as resistances and rewrite Eqs. 5-5 and 5-6 as

$$v_{eb} = r_{ib}i_e + r_{rb}i_c \qquad (5\text{-}7)$$

and

$$v_{cb} = r_{fb}i_e + r_{ob}i_c \qquad (5\text{-}8)$$

Equation 5-7 suggests an input equivalent circuit in which v_{eb} is the sum of two voltages, one the result of i_e and the other of i_c.

Equation 5-8 suggests an output equivalent circuit in which v_{cb} is the sum of two voltages, also resulting from i_e and i_c. The combined equivalent circuit, shown in Fig. 5-2, is sometimes called a *two-generator equivalent circuit* for the common-base configuration.

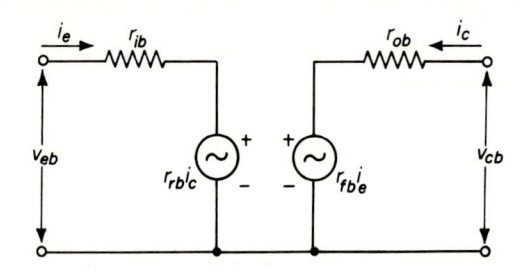

Figure 5-2 *Two generator equivalent circuit for the common-base configuration.*

Numerical subscripts, completely identical to the letter subscripts, may be used interchangeably with the transistor parameters described in this chapter. Thus, $r_i = r_{11}$; $r_r = r_{12}$; $r_f = r_{21}$; and $r_0 = r_{22}$. A letter subscript is used with the numbers to identify the configuration; b for common base; e for common emitter; and c for common collector. Letter subscripts are most often used to specify the characteristics of transistors, and numerical subscripts are most often used for general circuit analysis.

Another equivalent circuit, called the *voltage source T-equivalent circuit*, Fig. 5-3, can be derived from Eqs. 5-7 and 5-8. By rearrangement and using numerical subscripts we can write

Sec. 5.1   The r parameters

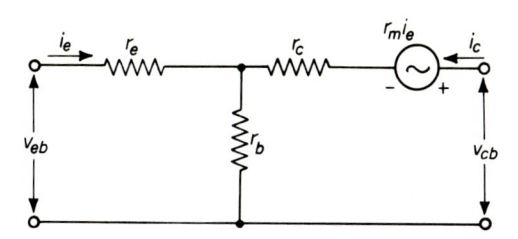

Figure 5-3 Voltage source T-equivalent circuit of a common-base amplifier.

and
$$v_{eb} = (r_{11b} - r_{12b})i_e + (i_e + i_c)r_{12b} \qquad (5\text{-}9)$$

$$v_{cb} = (r_{21b} - r_{12b})i_e + (r_{22b} - r_{12b})i_c + (i_e + i_c)r_{12b} \qquad (5\text{-}10)$$

where

$$r_{ib} - r_{rb} = r_{11b} - r_{12b} = \text{emitter resistance} = r_e$$

$$r_{rb} = r_{12b} = \text{base resistance} = r_b$$

$$r_{fb} - r_{rb} = r_{21b} - r_{12b} = \text{mutual resistance} = r_m$$

$$r_{ob} - r_{rb} = r_{22b} - r_{12b} = \text{collector resistance} = r_c$$

Equations 5-9 and 5-10 then become

$$v_{eb} = r_e i_e + (i_e + i_c)r_b \qquad (5\text{-}11)$$

and
$$v_{cb} = r_m i_e + r_c i_c + (i_e + i_c)r_b \qquad (5\text{-}12)$$

In Eq. 5-12, if $v_{cb} = 0$, we can write

$$i_e(r_b + r_m) = -i_c(r_c + r_b) \qquad (5\text{-}13)$$

We know that the external short-circuit current gain, $-\alpha$, is equal to $-i_c/i_e$ for $v_{cb} = 0$. Thus, if we divide both members of Eq. 5-13 by i_e and by $(r_c + r_b)$, it becomes

$$\frac{|i_c|}{|i_e|} = \alpha = \frac{r_b + r_m}{r_c + r_b} \qquad (5\text{-}14)$$

In a typical junction transistor, the parameters may be as follows

$$r_e = 26 \text{ ohms} \qquad r_b = 1{,}000 \text{ ohms}$$
$$r_c = 1\text{M ohm} \qquad \alpha = 0.98$$
$$r_m = 1\text{M ohm}$$

Thus in Eq. 5-14, we can disregard r_b in both the numerator and the denominator. This simplifies the equation to

$$\alpha \cong \frac{r_m}{r_c} \qquad (5\text{-}15)$$

The ratio r_m/r_c is called the *internal short-circuit forward-current gain*, and is indicated by the letter symbol *a*. Using the typical values given above, $a \cong 1$, which means that $a \cong \alpha$. Thus, we can write from Eq. 5-15

$$r_m = \alpha r_c \cong a r_c \qquad (5\text{-}16)$$

In equivalent circuits, such as that developed next, α is sometimes shown in place of *a*. This substitution is satisfactory *provided* it is kept in mind that the near-equality between the two parameters may not exist at the higher frequencies, because then r_c is effectively shunted by the collector base capacitance and its value is greatly reduced. Under these conditions, the ratio r_m/r_c no longer approaches unity.

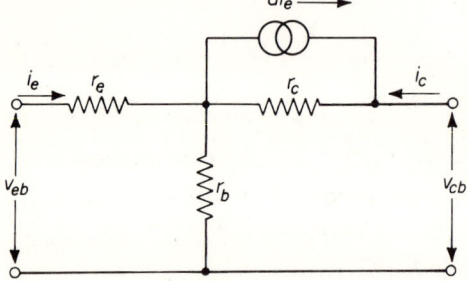

Figure 5-4 Current source T-equivalent circuit—common-base configuration.

In Fig. 5-3, the voltage generator $r_m i_e$ and series resistance r_c can be replaced by a current generator shunted by resistance r_c. The current of this generator must be equal to $r_m i_e / r_c$. Since $r_m / r_c = a$, this current is $a i_e$. The new equivalent circuit is shown in Fig. 5-4. The circuit of either Fig. 5-3 or Fig. 5-4 can be used for analysis since the two circuits are equivalent to each other.

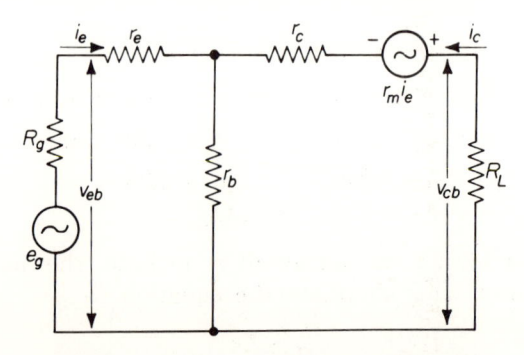

Figure 5-5 An a-c equivalent circuit for the common-base configuration with source and load terminations.

Sec. 5.1 The r parameters

An a-c equivalent circuit for the common-base configuration with source and load terminations is shown in Fig. 5-5. Using this circuit, we can derive expressions for the operating voltage gain, A_{eb}; current gain, A_{ib}; power gain, G; input resistance, R_i; and output resistance, R_o.

Kirchhoff's voltage law applied to the input and output loops, respectively, yields the following equations.

$$e_g = i_e(R_g + r_e + r_b) + i_c r_b \tag{5-17}$$

$$0 = i_e(r_b + r_m) + i_c(R_L + r_c + r_b) \tag{5-18}$$

Solving the second of these equations for i_c

$$i_c = \frac{-i_e(r_m + r_b)}{r_b + r_c + R_L} = -i_e\left(\frac{r_m + r_b}{r_b + r_c + R_L}\right) \tag{5-19}$$

Then, by substitution

$$e_g = i_e(R_g + r_e + r_b) + r_b\left[-i_e\left(\frac{r_m + r_b}{r_b + r_c + R_L}\right)\right] \tag{5-20}$$

Since $v_{cb} = i_c R_L$, and since $A_{eb} = v_{cb}/e_g$, we can write

$$A_{eb} = \frac{v_{cb}}{e_g} = \frac{-i_e\left(\dfrac{r_m + r_b}{r_b + r_c + R_L}\right) R_L}{i_e(R_g + r_e + r_b) + r_b\left[-i_e\left(\dfrac{r_m + r_b}{r_b + r_c + R_L}\right)\right]} \tag{5-21}$$

Equation 5-21 can be simplified by using the near-equality

$$a \cong \frac{r_m + r_b}{r_b + r_c + R_L} \tag{5-22}$$

Making the substitution in both the numerator and the denominator, we can rewrite Eq. 5-21 as

$$A_{eb} \cong \frac{-i_e(a) R_L}{i_e(R_g + r_e + r_b) - i_e(a) r_b} \tag{5-23}$$

By cancelling i_e in both the numerator and the denominator, the equation becomes

$$A_{eb} \cong \frac{-a R_L}{R_g + r_e + r_b - a r_b} \tag{5-24}$$

and finally

$$A_{eb} \cong \frac{-a R_L}{R_g + r_e + r_b(1 - a)} \tag{5-25}$$

Since a is less than 1, A_{eb} is positive and the common-base configuration provides no signal phase inversion.

The operating current gain, $A_{ib} = i_c/i_e$, is found very easily by recalling that Eq. 5-19 defines i_c as

$$i_c = \frac{-i_e(r_m + r_b)}{r_b + r_c + R_L}$$

Dividing both sides by $-i_e$

$$A_{ib} = \frac{i_c}{i_e} = -\frac{r_m + r_b}{r_b + r_c + R_L} \quad (5\text{-}26)$$

Since it is usual that $r_c \gg R_L$, and $r_m \gg r_b$

$$A_{ib} \cong -a = -\alpha \quad (5\text{-}27)$$

The power gain, G, is the product of $A_{eb}A_{ib}$; thus

$$G = \frac{-a(aR_L)}{R_g + r_e + r_b(1-a)} \quad (5\text{-}28)$$

or

$$G = -\frac{a^2 R_L}{R_g + r_e + r_b(1-a)} \quad (5\text{-}29)$$

The negative value of G simply indicates that power is delivered to and not taken from the load.

Again using Eqs. 5-17 and 5-18, it can be shown that the input resistance, R_{ib}, is

$$R_{ib} = \frac{v_{eb}}{i_e} = r_e + r_b r_c \left[\frac{(1-a) + R_L}{R_L + r_c + r_b} \right] \quad (5\text{-}30)$$

and with the usual assumptions that $r_c \gg R_L$, and $r_m \gg r_b$

$$R_{ib} \cong r_e + r_b(1-a) \quad (5\text{-}31)$$

With the values of a commonly found in transistors, $r_b(1-a)$ is quite small, and we can further simplify Eq. 5-31 to

$$R_{ib} \cong r_e \quad (5\text{-}32)$$

The low-frequency output resistance, R_{ob}, is equal to v_{cb}/i_c. It is calculated by assuming that $e_g = 0$ and rewriting Eqs. 5-17 and 5-18 as

$$0 = i_e(R_g + r_e + r_b) + i_c r_b \quad (5\text{-}33)$$

and

$$v_{cb} = i_e(r_m + r_b) + i_c(r_c + r_b) \quad (5\text{-}34)$$

Notice that in Eq. 5-33

$$\frac{i_e}{i_c} = -\frac{r_b}{r_e + r_b + R_g} \quad (5\text{-}35)$$

Assuming that $r_m \gg (R_g + r_e)$, if we use the relationship of Eq. 5-35 in Eq. 5-34, we find that

$$R_{ob} = \frac{v_{cb}}{i_c} = \frac{r_m r_b}{r_e + r_b + R_g} \quad (5\text{-}36)$$

An a-c equivalent circuit for the common-emitter amplifier with source and load terminals is shown in Fig. 5-6. Although this circuit may be used to calculate voltage, current, and power gains, as well as the input and output resistances, note that the generator is pro-

Sec. 5.1   The r parameters   87

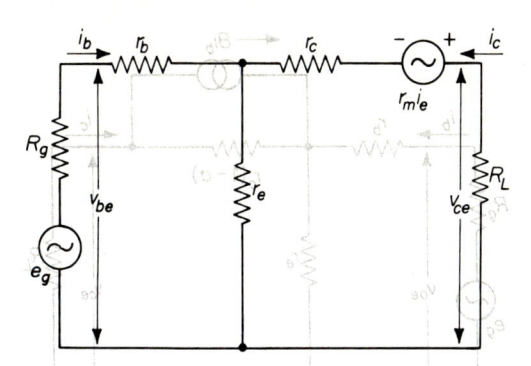

Figure 5-6 An a-c equivalent circuit for the common-emitter configuration with source and load terminations.

portional to the emitter current, whereas the input current is i_b. Thus, it is more convenient to use another equivalent circuit containing a generator that is a function of the base current.

If use is made of the fact that the sum of the three transistor currents is zero, we can use the equivalent circuit shown in Fig. 5-7. Notice that the generator $r_m i_b$ has reversed polarity compared

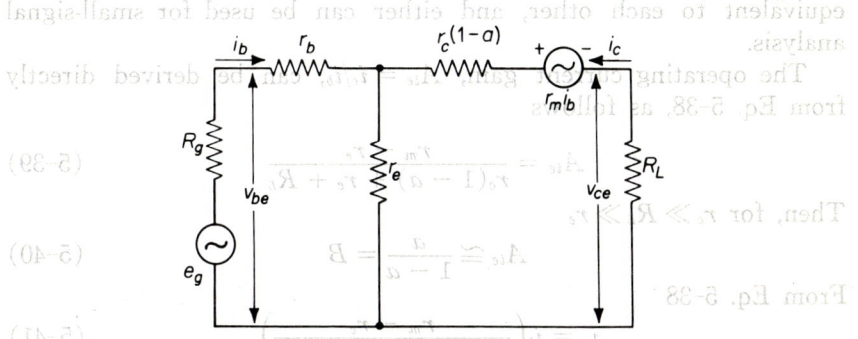

Figure 5-7 Modified equivalent circuit containing a generator $r_m i_b$ that is a function of base current.

to $r_m i_e$ of Fig. 5-6. Again making use of Kirchhoff's law we can write the equations

$$e_g = i_b(R_g + r_b + r_e) + i_c r_e \tag{5-37}$$

and

$$0 = i_b(r_m - r_e) + i_c[r_c(1 - a) + r_e + R_L] \tag{5-38}$$

Recall that r_m was previously defined as ar_c. Thus, generator $r_m i_b$ and the series resistor $r_c(1 - a)$ can be replaced with a current source, $i_b a/(1 - a) = i_b B$ (B bears the same relation to β as a does

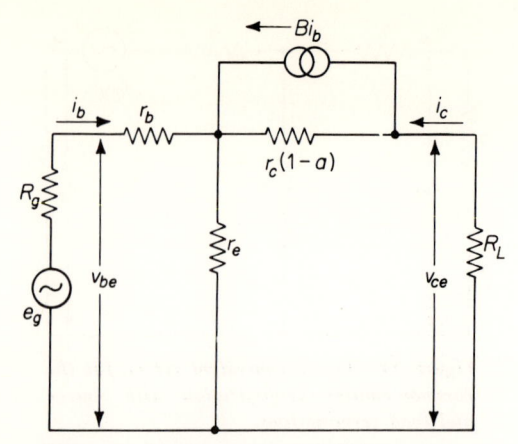

Figure 5-8 Replacement of generator $r_m i_b$ of Fig. 5-7 with a current source B_{ib} and shunting resistor $r_c(1 - a)$.

to α), and a shunting resistance $r_c(1 - a)$. If these changes are made, we can draw the equivalent circuit for the common-emitter amplifier as shown in Fig. 5-8. Figures 5-7 and 5-8 are, of course, equivalent to each other, and either can be used for small-signal analysis.

The operating current gain, $A_{ie} = i_c/i_b$, can be derived directly from Eq. 5-38, as follows

$$A_{ie} = \frac{r_m - r_e}{r_c(1 - a) + r_e + R_L} \tag{5-39}$$

Then, for $r_c \gg R_L \gg r_e$

$$A_{ie} \cong \frac{a}{1 - a} = B \tag{5-40}$$

From Eq. 5-38

$$i_c = i_b \left(\frac{r_m - r_e}{r_c(1 - a) + r_e + R_L} \right) \tag{5-41}$$

Since $A_{ee} = v_{ce}/e_g$, and since $v_{ce} = -i_c R_L$, by substituting the value for i_c from Eq. 5-41 into Eq. 5-37, we can write

$$A_{ee} = \frac{v_{ce}}{e_g} = \frac{-i_b \left(\frac{r_m - r_e}{r_c(1 - a) + r_e + R_L} \right) R_L}{i_b(R_g + r_b + r_e) + i_b \left(\frac{r_m - r_e}{r_c(1 - a) + r_e + R_L} \right) r_e} \tag{5-42}$$

By cancelling i_b in both the numerator and the denominator, and following normal algebraic manipulation, we can write Eq. 5-42 as

$$A_{ee} = \frac{-(r_m - r_e) R_L}{r_e(R_L + r_c) + (R_g + r_b)[r_c(1 - a) + r_e + R_L]} \tag{5-43}$$

and, after the usual approximations

$$A_{ee} \cong -\frac{aR_L}{r_e + (R_g + r_b)(1 - a)} \quad (5\text{-}44)$$

The power gain, G, is the product of A_{ie} and A_{ee}. Thus

$$G = \left(\frac{a}{1-a}\right)\left(-\frac{aR_L}{r_e + (R_g + r_b)(1 - a)}\right) \quad (5\text{-}45)$$

$$= -\frac{a^2 R_L}{r_e(1-a) + (R_g + r_b)(1-a)^2} \quad (5\text{-}46)$$

where the minus sign again indicates power is delivered to the load.

On examining the equivalent circuit of Fig. 5-7, it is apparent that

$$v_{be} = i_b(r_b + r_e) + i_c r_e \quad (5\text{-}47)$$

and, since $R_{ie} = v_{be}/i_b$, we can write

$$R_{ie} = r_b + \frac{r_e(r_c + R_L)}{r_e + R_L + r_c(1 - a)} \quad (5\text{-}48)$$

If we assume that $e_g = 0$, the equations for the equivalent circuit are

$$0 = i_b(R_g + r_e + r_b) + i_c r_e \quad (5\text{-}49)$$

and

$$v_{ce} = i_b(r_e - r_m) + i_c[r_e + r_c(1 - a)] \quad (5\text{-}50)$$

From these equations, we derive the output resistance, which is equal to v_{ce}/i_c.

$$R_{oe} = \frac{r_e(R_g + r_b + r_m)}{R_g + r_b + r_e} + r_c(1 - a) \quad (5\text{-}51)$$

5.2 The h (hybrid) parameters

The h parameters are most commonly used because they are easily measured with good accuracy.

In developing the h parameters, reference is again made to the black box concept. Assuming that the box contains a transistor amplifier in the grounded-base configuration, the independent variables chosen are the emitter current, i_e, and the collector-to-base voltage, v_{cb}. The general equations then assume the form

$$v_{eb} = f(i_e, v_{cb}) \quad (5\text{-}52)$$

and

$$i_c = f(i_e, v_{cb}) \quad (5\text{-}53)$$

Proceeding in the same general manner as described for the r parameters, we can develop the input and output circuit equations

$$v_{eb} = \left|\frac{dv_{eb}}{di_e}\right| di_e + \left|\frac{dv_{eb}}{dv_{cb}}\right| dv_{cb} \tag{5-54}$$

$$i_c = \left|\frac{di_c}{di_e}\right| di_e + \left|\frac{di_c}{dv_{cb}}\right| dv_{cb} \tag{5-55}$$

The coefficients of the terms in Eqs. 5-54 and 5-55 are obtained from static characteristic curves published by transistor manufacturers. The numerator of the coefficient indicates the corresponding d-c quantity to be plotted vertically (the dependent variable). The denominator indicates the corresponding d-c quantity to be plotted horizontally (the independent variable). The second independent variable, which is not permitted to vary, indicates the corresponding d-c quantity to be used on the family of curves.

The first coefficient of Eq. 5-54 is called the *input resistance*, and is measured in *ohms* because it represents the ratio of a voltage to a current. The h parameter symbol for this quantity is h_{ib}, where the subscripts i and b represent input and common base, respectively. Using numerical subscripts, h_{ib} is written as h_{11b}.

The second coefficient of Eq. 5-54 is *dimensionless* and is called the *reverse (feedback) voltage ratio*. The h parameter symbol is h_{rb}, or h_{12b}.

The first coefficient of Eq. 5-55 is *dimensionless* and is called the *forward current transfer ratio*. The h parameter symbol is h_{fb}, or h_{21b}.

The second coefficient of Eq. 5-55 is called the *output conductance* because it represents the reciprocal of resistance, and is measured in *mhos*. The h parameter symbol is h_{ob}, or h_{22b}.

We now have four h parameters for the common-base configuration: h_{ib}, h_{rb}, h_{fb}, and h_{ob}. Using the numerical forms of these parameters we can now rewrite Eqs. 5-54 and 5-54 as

$$v_{eb} = h_{11b} i_e + h_{12b} v_{cb} \tag{5-56}$$
and
$$i_c = h_{21b} i_e + h_{22b} v_{cb} \tag{5-57}$$

Equation 5-56 suggests an equivalent circuit of a resistance in series with a voltage generator, and Eq. 5-57 suggests an equivalent circuit of a current generator in parallel with a conductance. If we combine the two, for the common-base amplifier we have the equivalent circuit of Fig. 5-9. Using this equivalent circuit we can derive equations for determining A_{ib}, A_{eb}, G, R_i, and R_o, in much the same manner as for the r parameters.

Referring to Fig. 5-9, $v_{cb} = -i_c R_L$, by Ohm's law. If we substitute this quantity for v_{cb} in Eq. 5-57, we have

Sec. 5.2   The h (hybrid) parameters

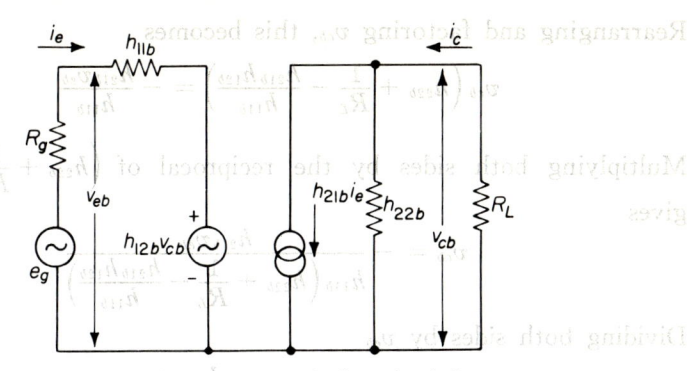

Figure 5-9 h parameter equivalent circuit for the common-base amplifier.

$$i_c = h_{21b}i_e - h_{22b}i_c R_L$$

or
$$0 = h_{21b}i_e - (1 + h_{22b}R_L)i_c \qquad (5\text{-}58)$$

Solving Eq. 5-58 for the ratio i_c/i_e gives A_i. Subtracting $h_{21b}i_e$ from both sides

$$-h_{21b}i_e = (1 + h_{22b}R_L)i_c \qquad (5\text{-}59)$$

Dividing both sides by $(1 + h_{22b}R_L)$

$$i_c = \frac{-h_{21b}i_e}{1 + h_{22b}R_L} \qquad (5\text{-}60)$$

Dividing both sides by i_e

$$A_{ib} = \frac{i_c}{i_e} = \frac{-h_{21b}}{1 + h_{22b}R_L} \qquad (5\text{-}61)$$

Again by Ohm's law, $i_c = -v_{cb}/R_L$; if we substitute this value in Eq. 5-57, we have

$$-\frac{v_{cb}}{R_L} = h_{21b}i_e + h_{22b}v_{cb}$$

or
$$0 = h_{21b}i_e + h_{22b}v_{cb} + \frac{v_{cb}}{R_L} \qquad (5\text{-}62)$$

If we eliminate i_e from Eq. 5-62, by solving Eq. 5-56 for i_e and then substituting, we can readily find A_{eb}. First, in Eq. 5-56

$$i_e = \frac{v_{eb} - h_{12b}v_{cb}}{h_{11b}} \qquad (5\text{-}63)$$

Substituting in Eq. 5-62

$$0 = h_{21b}\left(\frac{v_{eb} - h_{12b}v_{cb}}{h_{11b}}\right) + h_{22b}v_{cb}\frac{v_{cb}}{R_L}$$

or
$$0 = \frac{h_{21b}v_{eb}}{h_{11b}} - \frac{h_{21b}h_{12b}v_{cb}}{h_{11b}} + h_{22b}v_{cb} + \frac{v_{cb}}{R_L} \qquad (5\text{-}64)$$

Rearranging and factoring v_{cb}, this becomes

$$v_{cb}\left(h_{22b} + \frac{1}{R_L} - \frac{h_{21b}h_{12b}}{h_{11b}}\right) = -\frac{h_{21b}v_{eb}}{h_{11b}} \qquad (5\text{-}65)$$

Multiplying both sides by the reciprocal of $\left(h_{22b} + \frac{1}{R_L} - \frac{h_{21b}h_{12b}}{h_{11b}}\right)$ gives

$$v_{cb} = -\frac{h_{21b}v_{eb}}{h_{11b}\left(h_{22b} + \frac{1}{R_L} - \frac{h_{21b}h_{12b}}{h_{11b}}\right)} \qquad (5\text{-}66)$$

Dividing both sides by v_{eb}

$$A_{eb} = \frac{v_{cb}}{v_{eb}} = \frac{h_{21b}}{h_{11b}\left(h_{22b} + \frac{1}{R_L} - \frac{h_{21b}h_{12b}}{h_{11b}}\right)} \qquad (5\text{-}67)$$

If we multiply both the numerator and the denominator of the right side of the equation by R_L, and expand the denominator, the equation for A_{eb} is more conveniently stated as

$$A_{eb} = -\frac{h_{21b}R_L}{R_L\Delta + h_{11b}} \qquad (5\text{-}68)$$

where $\Delta = (h_{11b}h_{22b} - h_{12b}h_{21b})$.

Since the gain is the product of A_{eb} and A_{ib}

$$G = \left(\frac{h_{21b}}{1 + h_{22b}R_L}\right)\left(\frac{h_{21b}R_L}{R_L\Delta + h_{11b}}\right) \qquad (5\text{-}69)$$

or

$$G = -\frac{(h_{21b})^2 R_L}{(1 + h_{22b}R_L)(R_L\Delta + h_{11b})} \qquad (5\text{-}70)$$

Input resistance R_{ib} is the ratio v_{eb}/i_e. If we take input circuit Eq. 5–56 and again use $v_{cb} = -i_c R_L$, it is written

$$v_{eb} = h_{11b}i_e - h_{12b}i_c R_L \qquad (5\text{-}71)$$

Dividing by i_e

$$\frac{v_{eb}}{i_e} = h_{11b} - \frac{h_{12b}i_c R_L}{i_e} \qquad (5\text{-}72)$$

which can be rewritten as

$$\frac{v_{eb}}{i_e} = h_{11b} - h_{12b} R_L \left(\frac{i_c}{i_e}\right) \qquad (5\text{-}73)$$

Since the ratio i_c/i_e expresses current gain, Eq. 5–73 becomes

$$\frac{v_{eb}}{i_e} = h_{11b} - h_{12b} R_L \left(\frac{h_{21b}}{1 + h_{22b}R_L}\right) \qquad (5\text{-}74)$$

Multiplying both sides by $(1 + h_{22b}R_L)$

$$\frac{v_{eb}}{i_c}(1 + h_{22b}R_L) = h_{11b} + h_{11b}h_{22b}R_L - h_{12b}h_{21b}R_L \qquad (5\text{-}75)$$

Sec. 5.2   The h (hybrid) parameters

Factoring R_L in the right member, and dividing both sides by $(1 + h_{22b}R_L)$

$$R_{ib} = \frac{v_{eb}}{i_e} = \frac{R_L\Delta + h_{11b}}{1 + h_{22b}R_L} \quad (5\text{-}76)$$

Equation 5-76 indicates the very important fact that R_i is a function of R_L.

To determine R_{ob}, the equivalent circuit of Fig. 5-9 is changed to that of Fig. 5-10. Signal source e_g is moved from the input circuit

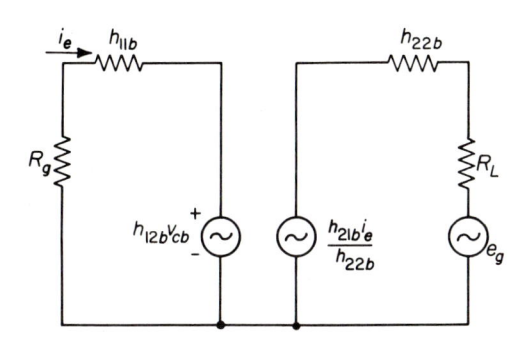

Figure 5-10 h parameter equivalent circuit for determining output resistance R_{ob}.

to the output circuit, where it is considered to have zero resistance. The source resistance (which must be considered) is, therefore, left in the circuit. Also, the current generator and shunt conductance of Fig. 5-9 are replaced by an equivalent voltage generator and series conductance. The output voltage of the voltage generator is equal to $h_{21b}i_e/h_{22b}$.

Because e_g is no longer present in the input circuit, $v_{eb} = 0$. Under these conditions we can write

$$0 = h_{12b}v_{cb} + (h_{11b} + R_g)i_e$$

or

$$i_e = -\frac{h_{12b}v_{cb}}{R_g + h_{11b}} \quad (5\text{-}77)$$

If we substitute the value of i_e in the output circuit equation, we obtain

$$i_c = h_{21b}\left(\frac{h_{12b}v_{cb}}{R_g + h_{11b}}\right) + h_{22b}v_{cb} \quad (5\text{-}78)$$

Removing the parentheses and clearing the fraction

$$i_c(R_g + h_{11b}) = v_{cb}(-h_{21b}h_{12b} + h_{22b}R_g + h_{22b}h_{11b}) \quad (5\text{-}79)$$

The ratio v_{cb}/i_c is the output resistance. Thus

$$R_{ob} = \frac{v_{cb}}{i_c} = \frac{R_g + h_{11b}}{\Delta + h_{22b}R_g} \quad (5\text{-}80)$$

Equation 5-80 shows that R_o depends on R_g. Recall that R_i depends on R_L. From these relationships, it follows that exact output matching is obtained *only* when the impedance of the signal source remains constant, and exact input matching is obtained only for one value of R_L.

Other useful information can be derived using h parameters. A formula that compares power delivered to the load to that generated at the transducer is called the *transducer gain*, G_T, and is expressed as

$$G_T = \frac{\Delta R_g R_L (h_{21})^2}{(h_{22} R_g R_L + R_g + \Delta R_L + h_{11})^2} \quad (5\text{-}81)$$

Equation 5-81 is sometimes useful when comparing several amplifiers in a circuit with fixed generator resistance.

The ratio of the available power of the transistor output to the available generator power is called the *available power gain*, G_A, and is defined as

$$G_A = \frac{R_g (h_{21})^2}{(R_g + h_{11})(\Delta + h_{22} R_g)} \quad (5\text{-}82)$$

We may also want to know the values of R_L and R_g that match the load and generator to the transistor. They are given by

$$R_g = \sqrt{\frac{h_{11} \Delta}{h_{22}}} \quad (5\text{-}83)$$

and

$$R_L = \sqrt{\frac{h_{11}}{h_{22} \Delta}} \quad (5\text{-}84)$$

The values of R_L and R_g as determined above can then be used to indicate the *maximum available gain*, MAG, under matched circuit conditions.

$$MAG = \frac{(h_{21})^2}{(\sqrt{h_{11} h_{22}} + \sqrt{\Delta})^2} \quad (5\text{-}85)$$

To achieve this optimum of operation, it is sometimes necessary to use *impedance-matching devices*. At audio frequencies the required transformers can be quite expensive, and another stage of amplification may prove to be a better solution.

In the common-emitter circuit, v_{ce} and i_b are the independent variables, and the general equations are

$$v_{be} = f(v_{ce}, i_b)$$
$$i_c = f(v_{ce}, i_b)$$

and, by following methods previously detailed, the following h parameter equations can be obtained

$$v_{be} = h_{ie} i_b + h_{re} v_{ce} = h_{11e} i_b + h_{12e} v_{ce} \quad (5\text{-}86)$$

and

$$i_c = h_{fe} i_b + h_{oe} v_{ce} = h_{21e} i_b + h_{22e} v_{ce} \quad (5\text{-}87)$$

Sec. 5.2   The h (hybrid) parameters   95

where the second subscript e indicates that we are dealing with the common-emitter configuration.

Equations 5-86 and 5-87 suggest the equivalent circuit of Fig. 5-11. Notice that this is the same form as that for the common-base

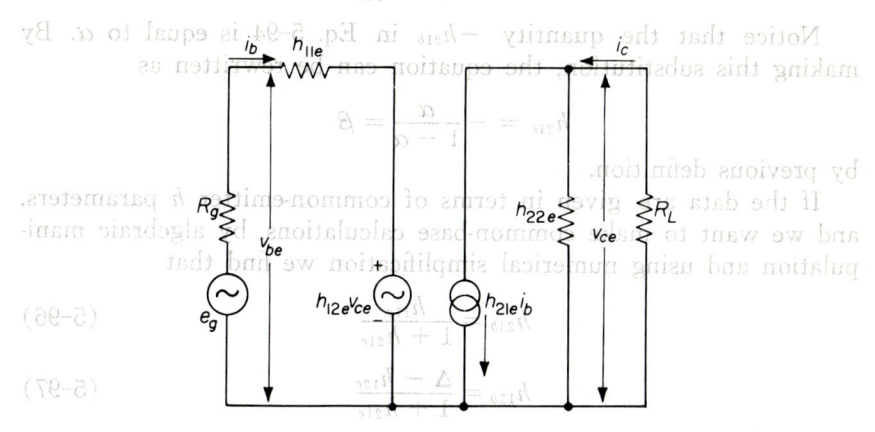

Figure 5-11 h parameter equivalent circuit for the common-emitter configuration.

amplifier but with different parameter values. Thus the gain and impedance equations derived for the common-base amplifier apply equally well to the common-emitter circuit, if the appropriate common-emitter h parameters are used.

Because manufacturer's data are usually given in terms of h_b parameters, we must relate the two.

$$h_{11e} = \frac{v_{be}}{i_b} = \frac{h_{11b}}{\Delta + h_{21b} - h_{12b} + 1} \quad (5\text{-}88)$$

$$h_{12e} = \frac{v_{be}}{v_{ce}} = \frac{\Delta - h_{12b}}{\Delta + 1 + h_{21b} - h_{12b}} \quad (5\text{-}89)$$

$$h_{21e} = \frac{i_c}{i_b} = \frac{-\Delta - h_{21b}}{\Delta + h_{21b} - h_{12b} + 1} \quad (5\text{-}90)$$

$$h_{22e} = \frac{i_c}{v_{ce}} = \frac{h_{22b}}{\Delta + 1 + h_{21b} - h_{12b}} \quad (5\text{-}91)$$

Using the numerical simplifications $(\Delta - h_{12b}) \ll (1 + h_{21b})$ and $\Delta \ll h_{21b}$

$$h_{11e} = \frac{h_{11b}}{1 + h_{21b}} \quad (5\text{-}92)$$

$$h_{12e} = \frac{\Delta + h_{12b}}{1 + h_{21b}} \quad (5\text{-}93)$$

$$h_{21e} = -\frac{h_{21b}}{1 + h_{21b}} \tag{5-94}$$

$$h_{22e} = \frac{h_{22b}}{1 + h_{21b}} \tag{5-95}$$

Notice that the quantity $-h_{21b}$ in Eq. 5-94 is equal to α. By making this substitution, the equation can be rewritten as

$$h_{21e} = -\frac{\alpha}{1 - \alpha} = \beta$$

by previous definition.

If the data are given in terms of common-emitter h parameters, and we want to make common-base calculations, by algebraic manipulation and using numerical simplification we find that

$$h_{21b} = \frac{h_{11e}}{1 + h_{21e}} \tag{5-96}$$

$$h_{12b} = \frac{\Delta - h_{12e}}{1 + h_{21e}} \tag{5-97}$$

$$h_{21b} = \frac{h_{21e}}{1 + h_{21e}} \tag{5-98}$$

$$h_{22b} = \frac{h_{22e}}{1 + h_{21e}} \tag{5-99}$$

where $\Delta = (h_{11e}h_{22e} - h_{12e}h_{21e})$.

5.3 Relating the h and r parameters

Since either the h or the r parameters can be used for small-signal analysis, it is sometimes convenient to express them in terms of each other.

$$r_b = \frac{h_{11b}}{h_{22b}} \tag{5-100}$$

$$r_e = h_{11b} - \frac{h_{12b}}{h_{22b}}(1 + h_{21b}) \tag{5-101}$$

$$r_c = \frac{1}{h_{22b}} \tag{5-102}$$

$$r_m = -\frac{h_{21b}}{h_{22b}} \tag{5-103}$$

The h_b parameters are related to the r parameters as follows:

$$h_{11b} = r_e + (1 - \alpha)r_b \tag{5-104}$$

$$h_{12b} = \frac{r_b}{r_c} \tag{5-105}$$

$$h_{21b} = -\alpha \tag{5-106}$$

$$h_{22b} = \frac{1}{r_c} \tag{5-107}$$

The h_e parameters are related to the r parameters as follows

$$h_{11e} = r_b + \frac{r_e}{1-\alpha} \tag{5-108}$$

$$h_{12e} = \frac{r_e}{(1-\alpha)r_c} \tag{5-109}$$

$$h_{21e} = \frac{\alpha}{1-\alpha} = \beta \tag{5-110}$$

$$h_{22e} = \frac{1}{(1-\alpha)r_c} \tag{5-111}$$

5.4 Transistor frequency response

As the frequency increases, consideration must be given to the effects on the internal capacitances of the transistor, and how such changes may modify the previously developed equivalent circuits. The emitter capacitance is shunted by the low-value emitter resistance, r_e, and the effect on this parameter is considered negligible. Only the output capacitance, C_o, must be included in our new equivalent circuits. Because the equivalent circuit using r parameters lends itself to quick calculation more readily than the h-parameter equivalent circuit, the former is used in the explanation that follows.

Both r_c and C_o decrease with increasing frequency. The current gain, alpha, is also frequency-dependent as a result of charge diffusion, and at higher frequencies may also be complex. Specifically, the value of alpha varies with frequency as

$$\frac{\alpha}{\alpha_o} = \frac{1}{1 + j^{f/f_\alpha}} \tag{5-112}$$

where α_o = the low-frequency value of α

f_α = the *alpha cutoff frequency*, at which the value of α is -3 db α_o.

A high-frequency equivalent circuit for the common-base amplifier is shown in Fig. 5-12. Notice that the generator is now termed $z_m i_e$ instead of $r_m i_e$ to take into account the complex nature of Eq. 5-112. Variations of the high-frequency equivalent circuit for the common-emitter equivalent circuit are shown in (a) and (b) of Fig. 5-13.

The exact analysis of the high-frequency performance of a

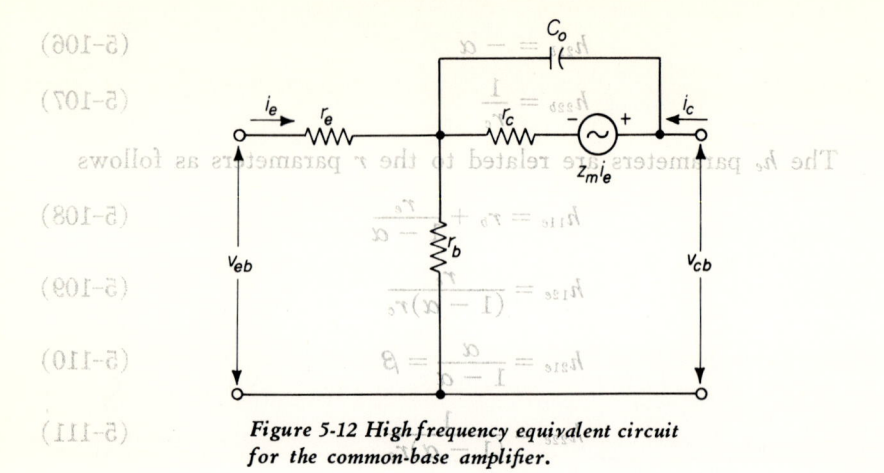

Figure 5-12 High-frequency equivalent circuit for the common-base amplifier.

transistor requires a knowledge of physics beyond the scope of this book. However, the approximate relations given below can be used to obtain a practical idea of high-frequency performance.

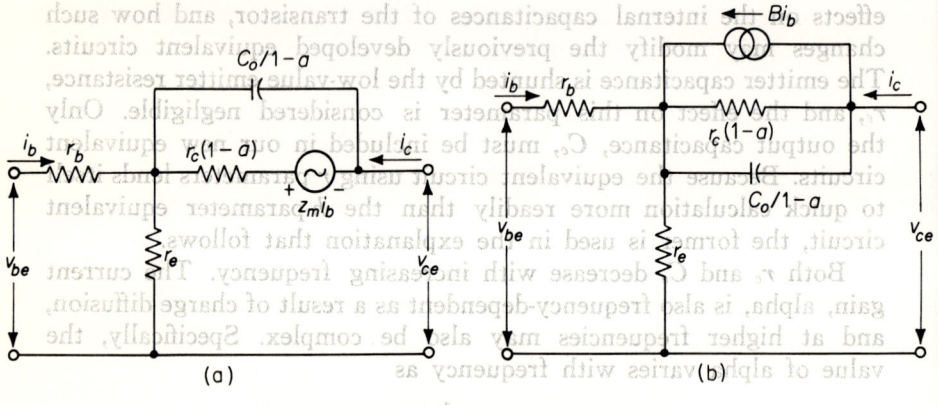

Figure 5-13 High-frequency equivalent circuits for the common-emitter configuration (a) voltage source and (b) current source.

In the common-base amplifier, the mid-frequency current gain, A_{ib}, is approximately equal to $-\alpha$. The ratio of high-frequency current gain, A_{ibo} to A_{ib}, is obtained from Eq. 5-112

$$\frac{A_{ibo}}{A_{ib}} = \frac{1}{1 + jf/f_\alpha} \qquad (5\text{-}113)$$

In the common-emitter amplifier, the mid-frequency current gain, A_{ie}, is approximately equal to $\alpha(1-\alpha)$. The ratio of high-frequency current gain, A_{ieo} to A_{ie}, is also obtained from Eq. 5-112.

Sec. 5.5   Transistor R-C amplifier   99

$$\frac{A_{ieo}}{A_{ie}} = \frac{1}{1 + j^{f/f\alpha}/(1 - \alpha_o)} \qquad (5\text{-}114)$$

5.5 Transistor R-C amplifier

Although transformer coupling enables cascading of any combination of transistor amplifier configurations, this freedom in design is generally not feasible with R-C coupling. We do not, for example, generally use two common-base stages, because the input impedance of the second stage is too low to permit the first stage to develop reasonable power gain. The same condition holds true in a common-emitter, common-base arrangement.

Despite considerable power loss because of impedance mismatch, we often use R-C coupling between common-emitter stages. The inherent power gain of this configuration permits such a cascaded arrangement to develop practically any required power amplification by using an appropriate number of stages.

A typical R-C coupled transistor amplifier is shown in Fig. 5-14(a), and the equivalent circuit for the analysis of the low- and mid-frequency ranges is shown in Fig. 5-14(b).

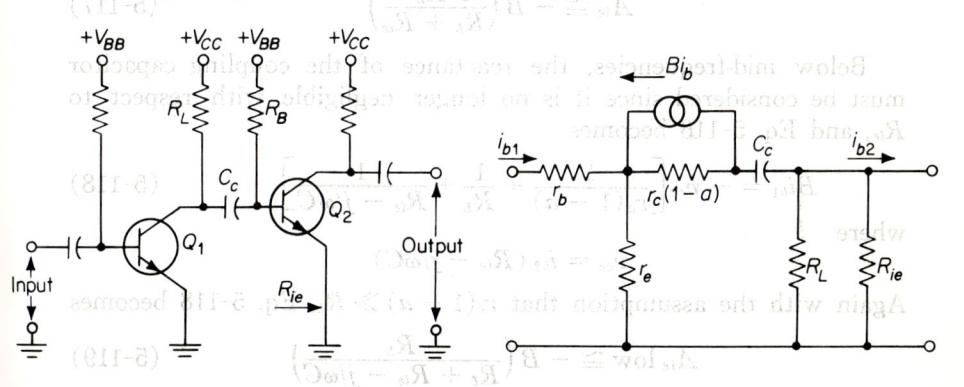

Figure 5-14 (a) Typical R-C amplifier and (b) its equivalent circuit for the low- and mid-frequency ranges.

In Fig. 5-14(a), the effective load for transistor Q_1 is low because of the low input resistance of transistor Q_2, and we can write the approximation

$$R_{ie} = r_b + \frac{r_e}{1 - a} \qquad (5\text{-}115)$$

In Fig. 5-14(a), resistor R_B is typically of the order of 100 K ohms

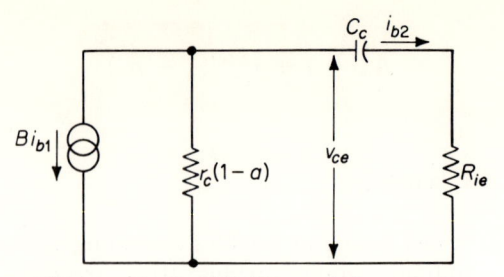

Figure 5-15 Simplified form of Fig. 5-14 (b).

and R_{ie} is typically of the order of 2 K ohms. This fact, and the usual approximations, permit the redrawing of the equivalent circuit in the form shown in Fig. 5-15.

At mid-frequencies, the reactance of coupling capacitor C_c is negligible with respect to R_{ie} and

$$Bi_{b1} = -v_{ce}\left[\frac{1}{r_c(1-a)} + \frac{1}{R_L} + \frac{1}{R_{ie}}\right] \qquad (5\text{-}116)$$

where $v_{ce} = i_{b2}R_{ie}$. Then, with the assumption that $r_c(1-a) \gg R_L$, Eq. 5-116 becomes

$$A_{ie} \cong -B\left(\frac{R_L}{R_L + R_{ie}}\right) \qquad (5\text{-}117)$$

Below mid-frequencies, the reactance of the coupling capacitor must be considered since it is no longer negligible with respect to R_{ie}, and Eq. 5-116 becomes

$$Bi_{b1} = -v_{ce}\left[\frac{1}{r_c(1-a)} + \frac{1}{R_L} + \frac{1}{R_{ie} - j/\omega C}\right] \qquad (5\text{-}118)$$

where

$$v_{ce} = i_{b2}(R_{ie} - j/\omega C)$$

Again with the assumption that $r_c(1-a) \gg R_L$, Eq. 5-118 becomes

$$A_{ie}\text{ low} \cong -B\left(\frac{R_L}{R_L + R_{ie} - j/\omega C}\right) \qquad (5\text{-}119)$$

At the low-frequency half-power point, f_1

$$f_1 = \frac{1}{2\pi C}\left(\frac{1}{R_L + R_{ie}}\right) \qquad (5\text{-}120)$$

the gain is -3 db (70.7 per cent) of its mid-range value, and the phase angle is $+45°$. Then the low-frequency current gain relative to the mid-frequency current gain is

$$\frac{A_{ie}\text{ low}}{A_{ie}} = \frac{1}{1 - j^{(f_1/f)}} \qquad (5\text{-}121)$$

Sec. 5.5   Transistor R-C amplifier   101

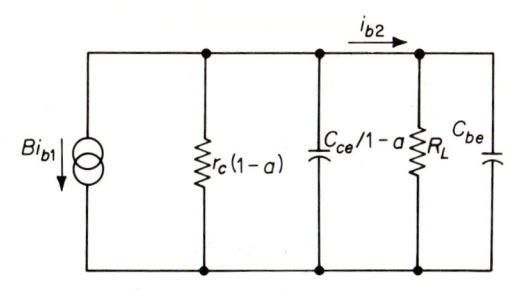

Figure 5-16 High-frequency equivalent circuit.

The high-frequency equivalent circuit for the R-C amplifier is shown in Fig. 5-16.

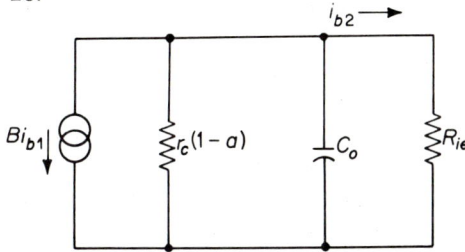

Figure 5-17 Simplified high-frequency equivalent circuit.

We can combine $C_{ce}/(1-a)$ and C_{be}, since they are effectively in parallel, and identify the combination as C_o. This permits a simplification of the equivalent circuit to that shown in Fig. 5-17. Using this figure, we can write for frequencies well below f_a

$$Bi_{b1} = -v_{ce}\left[\frac{1}{r_c(1-a)} + \frac{1}{R_L} + \frac{1}{R_{ie}} + j\omega C_o\right] \qquad (5\text{-}122)$$

where $v_{ce} = i_{b2}R_{ie}$.

With the previous simplification of $r_c(1-a) \gg R_L$, Eq. 5-122 becomes

$$A_{ie}\text{ high} = -B\left(\frac{R_L}{R_{ie} + R_L - j\omega C_o R_{ie}R_L}\right) \qquad (5\text{-}123)$$

At the high-frequency half-power point f_2

$$f_2 = \frac{1}{2\pi C_o}\left(\frac{1}{R_L} + \frac{1}{R_{ie}}\right) \qquad (5\text{-}124)$$

the gain is again -3 db of its mid-range value, and the phase angle is $-45°$. Then, the ratio of A_{ie} high to A_{ie} is

$$\frac{A_{ie}\text{ high}}{A_{ie}} = \frac{1}{1 + j^{(f/f_2)}} \qquad (5\text{-}125)$$

In developing the gain equations, it is apparent that the values of the coupling capacitor and R_{ie} primarily determine f_1, and the values of R_{ie} and C_o control f_2. Thus, to widen the frequency response band, we would use a large-value coupling capacitor ($10\mu f$ or better) and select a transistor with a small C_o.

REFERENCES

1. Peterson, L. C., "Equivalent Circuits of Linear Active Four-Terminal Networks," *Bell System Tech. Jour.*, v. 27, 1948, p. 593.
2. Lo, A. W., et al., *Transistor Electronics*. Englewood Cliffs, N. J.: Prentice-Hall, Inc., 1955.
3. Shea, R. F., *Transistor Circuits*. New York: John Wiley and Sons, Inc., 1953.
4. Millman, Jacob, *Vacuum Tube and Semiconductor Electronics*. New York: McGraw-Hill Book Company, Inc., 1958.
5. *Basic Theory and Application of Transistors*, Department of the Army Technical Manual TM11-600, March 1959.
6. Knight, G., R. A. Johnson, and R. B. Holt, "Measurement of the Small-Signal Parameters of Transistors," *Proc. I. R. E.*, v. 41, August 1953, p. 983.
7. Early, J. M., "Design Theory of Junction Transistors," *Bell System Tech. Jour.*, v. 32, 1953, p. 1271.
8. Giacoletto, L. C., "Study of P N P Alloy Junction Transistors from D-C Through Medium Frequencies," *RCA Review*, v. 15, 1954, p. 506.

EXERCISES

5-1 Derive the r parameters.

5-2 Why is the internal short-circuit forward-current gain a used in preference to α in the T-equivalent circuits?

5-3 Given a transistor in the common-base configuration with the parameters shown below, $R_L = 1.5$ K ohms and $R_g = 400$ ohms

$r_e = 25$ ohms $r_b = 1$ K ohms
$r_c = 1$ M ohms $\alpha = 0.995$

Calculate the low-frequency values of A_{eb}, A_{ib}, G, R_{ib}, and R_{ob}.

5-4 Draw the voltage source T-equivalent circuits for the common-base and common-emitter configurations.

5-5 Using the same values as given in Exercise 5-3, but with the amplifier connected in the common-emitter configuration, calculate A_{ee}, A_{ie}, G, R_{ie}, and R_{oe}.

5-6 What are the advantages and disadvantages of the r parameters? The h parameters?

Exercises

5-7 The given hybrid parameter of a transistor are

$h_{ie} = 5000$ ohms $\qquad h_{fe} = 49$
$h_{re} = 8.5 \times 10^{-4} \qquad h_{oe} = 10 \times 10^{-6}$ mhos

Also, $R_g = 250$ ohms and $R_L = 20,000$ ohms. Find the operating voltage, the current and power gains, R_i, and R_o when used as a common-base amplifier.

5-8 The common-base h parameters given in a tech data sheet are

$h_{ib} = 45$ ohms $\qquad h_{fb} = -0.99$
$h_{rb} = 3.5 \times 10^{-4} \qquad h_{ob} = 3 \times 10^{-7}$ mhos

Also, $R_g = 2000$ ohms and $R_L = 2000$ ohms. Determine the voltage, the current and power gains, R_i, and R_o when used as a common-emitter amplifier.

5-9 With the common-emitter amplifier from Exercise 5-8, what are the values of R_g and R_L for perfect matching, and what is the *MAG* of the circuit?

5-10 Under conditions of perfect matching, calculate A_{ie} and A_{ee} for the amplifier of Exercise 5-8.

6
WIDEBAND AMPLIFIERS

In the vacuum tube and transistor circuits discussed in Chaps. 3 and 5, gain falls off at both the high and low ends of the response curve. When nonsinusoidal waveforms are to be *faithfully* reproduced, the bandwidth of such amplifiers is inadequate to pass all the frequency components making up the waveform. For example, in a square wave having a fundamental frequency of 15 kilocycles, it may be necessary to reproduce all harmonics up to the 200th if the output is to be a faithful enough reproduction of the input to perform a required function. To reproduce this well, the amplifier must have a bandwidth of at least three megacycles (15,000 × 200 = 3 × 10^6 cps).

To reduce frequency distortion and to provide the needed bandwidth, certain compensating reactive elements are added to the basic R-C amplifier circuits previously studied. Such an amplifier is said to be *frequency compensated*, and it may be called a *wideband, video,* or *pulse* amplifier.

Sec. 6.1   Defects in the reproduced waveform   105

6.1 Defects in the reproduced waveform and the meaning of transient response

If an ideal square wave is applied to the input terminals of an amplifier, the characteristics of the amplifier and its associated circuit prevent the output voltage from being an exact replica of the input. Instead of occurring instantaneously, the rise of the leading edge and fall of the trailing edge of the voltage pulse occur over finite periods of time, as shown in Fig. 6-1. The flat portion

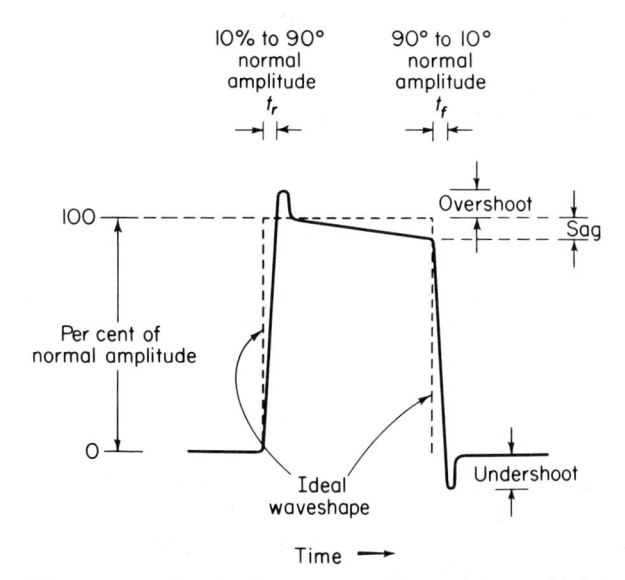

Figure 6-1 Defects in the output waveform owing to high- and low-frequency deficiencies of the amplifier.

of the waveform may sag or decrease in amplitude with time as shown. Finally, equal or varying amounts of undershoot and overshoot sometimes occur.

Because the leading and trailing edges are the most rapidly changing portions of a waveform to be amplified, they represent the highest frequency components making up the input waveform. If the amplifier bandwidth is inadequate, it is not possible for these frequency components to appear in the output, and the rise time, t_r, of the output waveform increases. Thus, the speed of rise is a good indication of the high-frequency performance of an amplifier.

Similarly, because the flat-top portion of the waveform in Fig. 6-1 is the most slowly changing portion, it represents the low-frequency components of the waveform. A decrease or sag in the amplitude of the flat-top portion is, therefore, an indication of low-frequency deficiencies in the amplifier.

Overshoot and undershoot are primarily the result of high-frequency deficiencies. The degree to which they may be present depends on the kind and degree of *frequency compensation* used.

Mention has already been made of the significance of rise time in evaluating the high-frequency capabilities of an amplifier. That portion of the output wave which includes the rise time and overshoot, if any exists, is called the *transient response*. This name is derived from the definition of a transient—that period of time required for a system to transfer from one steady state to another. The transient response indicates how an amplifier responds to a sudden change in amplitude of an applied signal.

6.2 Vacuum tube gain-bandwidth product

If the upper-3 db point of bandwith $f_1 - f_2$, is taken as the frequency f_2 at which the total capacitive reactance, X_{ct}, is equal to the coupling resistance, R_{eq}, then

$$R_{eq} = X_{ct} = \frac{1}{2\pi f_2(C_g + C_p)} + \frac{1}{2\pi f_2 C_T} \qquad (6-1)$$

It has been shown in the case of pentode amplifiers, however, that when r_p is much greater than R_L (ten or more times as great), the gain of a resistance-coupled amplifier is approximately equal to the product of $g_m R_L$.

Assuming that $R_L \cong R_{eq}$

$$A = \frac{g_m}{2\pi f_2 C_T} \qquad (6-2)$$

When Eq. 6-2 is rearranged as

$$A f_2 = \frac{g_m}{2\pi C_T} \qquad (6-3)$$

the product $A f_2$ is called the gain-bandwidth product; it is dependent only upon the tube constants and the circuit capacitance, provided that frequency f_2 is very much larger than f_1.

This formula indicates that for an increase in gain, there is a corresponding decrease in bandwidth, and for a decrease in gain there is a corresponding increase in bandwidth. Thus, in order to achieve wide bandwidth, low values of resistance are used for R_L in pulse amplifiers.

6.3 Transistor gain-bandwidth product

The high-frequency response of a transistor amplifier is generally determined by the frequency characteristic of α. For a common-emitter stage

$$f_2 = (1 - \alpha)f_\alpha \tag{6-4}$$

Since the maximum current gain is

$$A_{ie} = \frac{\alpha}{1 - \alpha} \tag{6-5}$$

the gain-bandwidth product is

$$A_{ie}f_2 = \alpha f_\alpha \tag{6-6}$$

The same product is also applicable to the common-base configuration.

Notice that in Eq. 6-4, the higher alpha becomes, the lower f_2 becomes, and the smaller the bandwidth. Although bandwidth can obviously be increased by using transistors with a very high f_α, this method is generally quite expensive. A more practical solution to the problem is to use *degenerative feedback* (discussed in Chap. 7), which reduces gain but increases bandwidth.

6.4 Vacuum tube figure-of-merit

To enable the engineer or technician to select the tube most suitable for wideband application, use is made of a *figure-of-merit*, F, which is equal to the product of gain and bandwidth. Mathematically

$$\begin{aligned} F &= \frac{g_m}{2\pi C_g} \\ &= \frac{g_m}{2\pi(C_{gk} + C_{pk})} \\ &= \frac{g_m}{2\pi(c_i + c_o)} \end{aligned} \tag{6-7}$$

where g_m = transconductance in μmhos
c_i = input capacitance of the tube in $\mu\mu$f
c_o = output capacitance of the tube in $\mu\mu$f

To provide a high figure-of-merit, a special line of pentodes having high transconductance and low values of input and output capacitance, c_i and c_o, has been developed for wideband application. Among others, this group includes such vacuum tubes as the 6CB6,

6AK5, and 6AG5. It is interesting to note that no figure-of-merit can be established for a triode since the value of c_i increases as the tube amplifies. This phenomenon is called the *Miller effect*.

In some cases, both wide bandwidth (low value of R_L) and large output voltage (high value of R_L) are needed. Because these are mutually contradictory requirements, some additional factors must be considered. If a very low value of load impedance is used to improve bandwidth, large swings in plate current are needed to produce a large output voltage. But this means that the quiescent plate current, I_{bo}, must be high if the output is to consist of equal positive and negative voltages. The value of I_{bo} under these conditions must be at least equal to the peak value of output signal divided by the load resistance. In the amplifiers studied thus far, where the load is inserted between the amplifier anode and the positive return of the supply voltage, a relatively low value of I_{bo} is satisfactory when only negative output voltage swings are required. If the positive swing also is required, the need for a high value of I_{bo} can be eliminated by the use of a circuit called a *cathode follower*, in which the load is inserted between the cathode and the negative return of the supply voltage. Cathode followers are fully described in Chap. 8.

With the assumption that the input signal can be supplied by an appropriate generator, transconductance (g_m) and input capacitance (c_i) are of secondary importance, and the tube to be used for a wide bandwidth and large output voltage is selected on the basis of maximum plate-current rating and minimum output capacitance (c_o).

What this boils down to in practice is that *all* factors must be considered. In general, if a tube is to be used for wideband application and relatively *small* values of output voltage are required, it is wise to choose a tube having the highest values of g_m and lowest values of c_i and c_o. If wideband application and relatively large output voltages are required, use a tube having the highest possible values of g_m, the necessary maximum plate-current rating, and the smallest values of c_i and c_o.

6.5 Transistor figure-of-merit

We may define a figure-of-merit for transistors as

$$F_{tr} \cong \frac{\alpha r_c/r_b r_c (1-\alpha)}{C_c/(1-\alpha)} \qquad (6\text{-}8)$$

where C_c = collector capacitance, and r_m, r_b, and r_c are equivalent circuit r parameters.

On performing the indicated operations, Eq. 6-8 becomes

$$F_{tr} \cong \frac{\alpha}{r_b C_c} \qquad (6\text{-}9)$$

Equation 6-9 indicates that for wideband application a transistor with a low $r_b C_c$ product should be selected. (Recall that $r_b = h_{12b}/h_{22b}$ when the manufacturer's information is given in terms of h parameters.)

Notice that Eq. 6-9 is an approximation which results from the use of r_b. As you know, the base region is very thin. Thus, current entering the base region passes through a long narrow path, of much smaller cross-sectional area than the emitter or collector, to reach the external base terminal. Between the internal base terminal, termed b', and the external base terminal, there exists a d-c ohmic resistance called the *base spreading resistance*, $r_{b'b}$. Using this quantity

$$F_{tr} = \frac{\alpha}{r_{b'b} C_c} \qquad (6\text{-}10)$$

During manufacture, $r_{b'b}$ can be reduced by making the base region extremely thin. Special transistors of this type are described later.

6.6 Need for compensation

Although the uncompensated R-C amplifiers previously discussed provide fair bandwidth and rise time without overshoot, their characteristics are unsatisfactory for pulse systems. At high frequencies, shunting capacitances cause a fall-off in response; and at low frequencies, the presence of the coupling capacitor also causes a fall-off in response.

To extend the range of the R-C amplifier (both vacuum tube and transistor types), various compensating networks are used.

6.7 Methods of high frequency compensation[1,2,3,4]

The simplest way to improve the high-frequency response of a vacuum tube or transistorized resistance-capacitance amplifier is to connect a small inductance L_1 in series with R_L, as shown in (a) and (b) of Fig. 6-2. This arrangement is called *shunt compensation* or *shunt peaking*.

At high frequencies, the load impedance increases as a result of

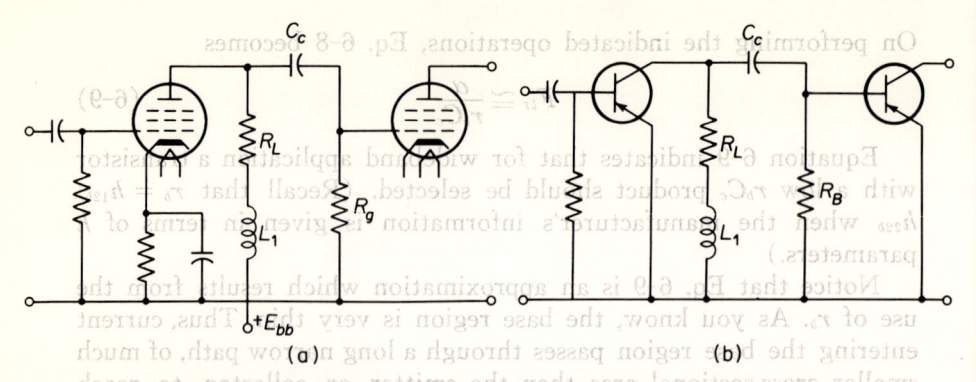

Figure 6-2 Shunt peaking: (a) vacuum-tube amplifier and (b) transistor amplifier.

the inductive reactance of the shunting coil, and the normal fall-off in amplification does not occur. This simple circuit addition provides an appreciable reduction in rise time of the output waveform, provided the bandwidth demands are not too great and only a few stages of amplification are required.

The response of the amplifier depends on the parameter $Q = X_{L1}/R_L$, where $X_{L1} = 2\pi f_2 L_1$, and f_2 is the frequency at which the amplification of the uncompensated amplifier is down 0.707 (3 db) of the mid-frequency value.

A series of curves showing the amplification characteristics of a shunt-compensated amplifier for different values of Q is shown in Fig. 6-3.

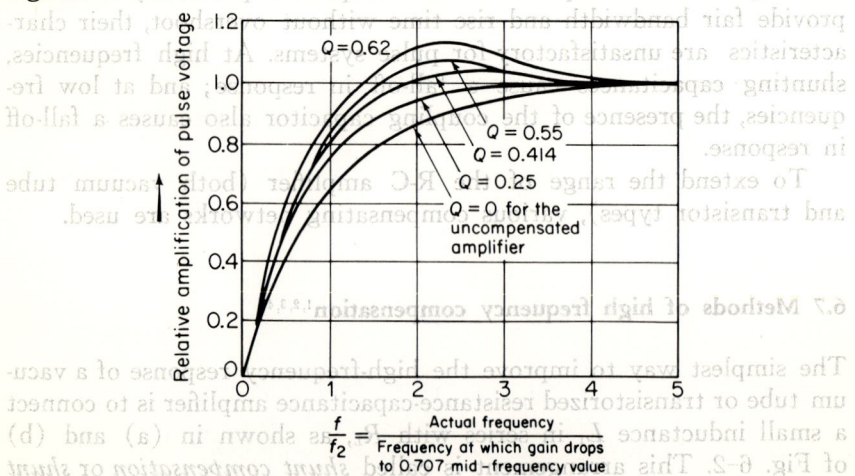

Figure 6-3 Transient response for different values of Q in a shunt-compensated amplifier.

Sec. 6.7   Methods of high frequency compensation   111

In the case of the uncompensated ($Q = 0$) amplifier, when $f/f_2 = 1$, the relative voltage amplification drops to 0.707 of the mid-frequency value.

For a Q of 0.55, the response is even greater than at its mid-frequency value, so that overshoot exists.

For a Q of 0.414, maximum flatness is obtained (no overshoot) with the rate of rise being approximately 1.7 times that of the uncompensated amplifier.

The high-frequency response of an R-C coupled amplifier may also be improved by *series compensation*, sometimes called *series peaking*. A small inductance coil L_2 is connected in series with coupling capacitor C_c, as shown in (a) and (b) of Fig. 6-4. At high frequencies, L_1 forms a resonant circuit with the input capacitance, c_i, of the next stage. This, in turn, causes increased voltage across c_i, as well as increased gain.

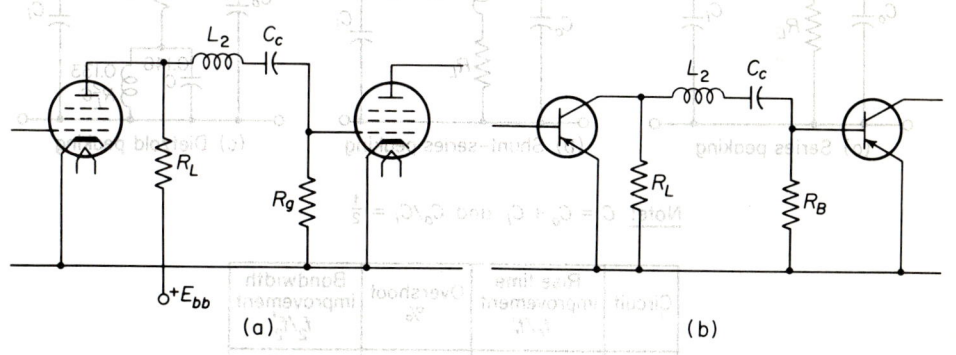

Figure 6-4 *Series-compensated amplifiers: (a) vacuum-tube type and (b) transistor type.*

The advantages of series and shunt compensation may be combined by using the high-frequency peaking effect of a shunt coil, L_1,

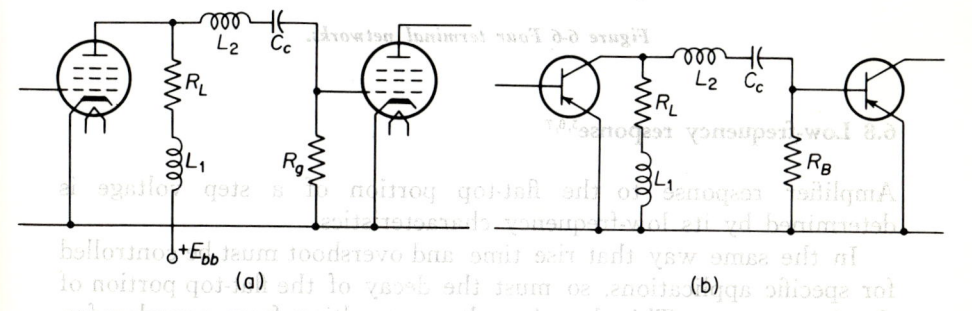

Figure 6-5 *Shunt-series compensated amplifiers: (a) vacuum-tube type and (b) transistor type.*

and the resonant-peaking effect of a series-peaking coil, L_2, in shunt-series compensation. [See (a) and (b) of Fig. 6-5.]

The high-frequency response of an R-C amplifier may be further improved by *four-terminal networks*[4,5] in which the output capacitance of one stage and the input capacitance of the succeeding stage are separated.

Three typical four-terminal network arrangements are shown in (a), (b), and (c) of Fig. 6-6. The rise time and percentage overshoot, for each arrangement, are compared in part (d) of the figure.

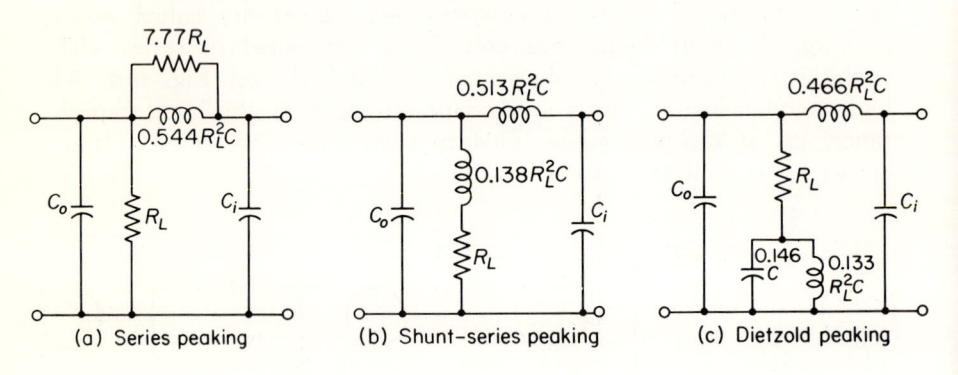

(a) Series peaking

(b) Shunt-series peaking

(c) Dietzold peaking

Note: $C = C_o + C_i$ and $C_o/C_i = \frac{1}{2}$

Circuit	Rise time improvement t_r/t_r'	Overshoot %	Bandwidth improvement f_2/f_2'
(a)	1.9	3.0	2.07
(b)	2.21	3.0	2.28
(c)	2.47	0.3	2.48

(d) Comparison of above arrangements
f_2 = upper −3db frequency

Figure 6-6 *Four terminal networks.*

6.8 Low-frequency response[5,6,7]

Amplifier response to the flat-top portion of a step voltage is determined by its low-frequency characteristics.

In the same way that rise time and overshoot must be controlled for specific applications, so must the decay of the flat-top portion of the input wave. This drop in voltage, resulting from poor low-frequency response, is called *sag* (droop).

Sec. 6.8   *Low-frequency response*   113

In vacuum tube circuits, sag is the result of the action of the R_gC_c network and the impedance of the bypass capacitors in the screen and cathode circuits.

In a common-emitter transistor amplifier, sag is the result of the action of the R_BC_c coupling network.

To compensate for the effect of R_gC_c, or R_BC_c, it is necessary to add another network, R_fC_f, in series with R_L, as shown in (a) and (b) of Fig. 6-7. At low frequencies, the reactance of C_c increases

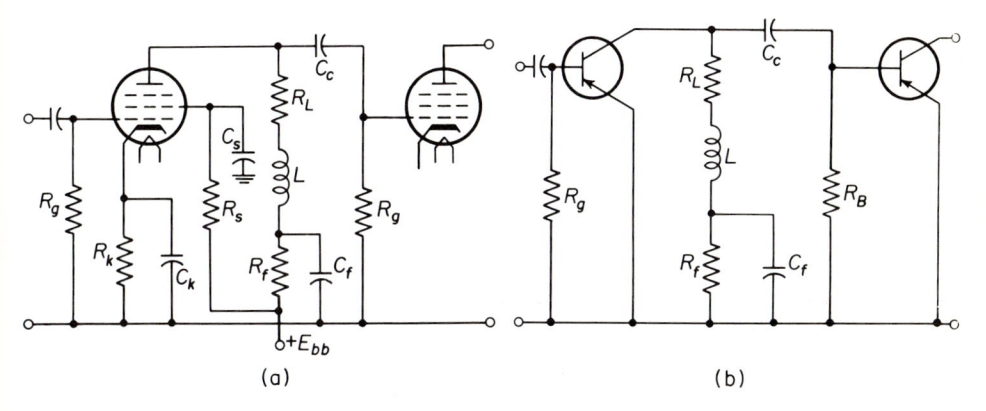

Figure 6-7 Low frequency compensation: (a) vacuum-tube amplifier and (b) transistor amplifier.

and causes a loss of input to V_2 or Q_2, as the case may be. The reactance of C_f increases at the same time. As a result, the R_fC_f network acts as an impedance added to R_L, and the output voltage is increased. It has been shown experimentally that best results are usually obtained when the time constants R_fC_f and R_gC_c or R_BC_c are equal. The value of R_f is chosen to be about twenty times the reactance of C_f at the *lowest* frequency to be bypassed. The value of R_f should be as large as possible while still permitting the required quiescent plate voltage to be applied to V_1 or collector voltage to Q_1.

It can also be shown mathematically that capacitor C_c causes a leading phase shift. The R_fC_f network tends to cancel this phase lead by introducing a phase shift in the lagging or opposite direction.

In addition to the R_gC_c network, in vacuum tube circuits, two other R-C combinations cause a reduction in gain and a leading phase shift at low frequencies; these are the cathode-bias network, R_kC_k, and the screen network, r_sC_s. In the case of the last mentioned network, r_s refers to the *dynamic screen resistance* which is analogous to dynamic plate resistance. It does not represent the screen voltage-dropping resistor, designated R_s, in Fig. 6-7(a).

At very low frequencies, the bypassing efficiency of C_k in Fig. 6-7(a) is reduced by its increased reactance. Signal variations are then developed across R_k and are applied to the grid-cathode circuit along with the d-c bias that establishes the quiescent operating point. The polarity of this signal opposes the input signal and results in a loss of stage gain. This action is referred to as degeneration; it is discussed at length in the next chapter.

This undesired effect can be compensated for by maintaining the following relation between the various resistors and capacitors

$$C_k R_k = C_f R_f$$

$$R_f = R_k(g_m R_L)$$

$$C_f = \frac{C_k}{g_m R_L}$$

You know from previous discussion that the $g_m R_L$ product is the mid-frequency stage gain. Its presence in the above equations suggests a low-frequency compensation method without an exact knowledge of g_m or the stage gain at high frequencies.

Referring to Fig. 6-7, select both C_k and C_f to have negligible reactance at a frequency of 10 kilocycles. Select R_k to provide the desired grid bias. Connect an oscilloscope or vacuum tube voltmeter from plate to ground of V_1. Apply a voltage to the grid of V_1 having a frequency of 10 kilocycles and sufficient magnitude to produce a readable deflection on the indicating instrument. With R_L selected for the best high-frequency performance, note the instrument reading obtained. This is a measure of $g_m R_L$, the high-frequency gain. Now remove C_f and C_k, and with a constant-input voltage maintained at the grid, select a value of R_f that gives the same indication previously obtained. Next place across R_k a value of capacitor that provides satisfactory performance in this circuit. Finally, select a value of C_f that makes the time constant $R_k C_k$ equal to $R_f C_f$.

Because cathode-bias operation permits the use of a larger grid leak than fixed bias, the low-frequency characteristics of the grid-coupling circuits are improved. It is, therefore, generally not necessary to compensate, in the plate circuit of one stage, for deficiencies in grid-coupling and cathode-bias circuits used in the same stage.

It should be noted that the amount of sag permissible in a given amplifier depends on the application for which the amplifier is used. In a television receiver, for example, the sag in the video amplifier

Sec. 6.9   Direct-coupled vacuum tube amplifiers

must not exceed five per cent in one-sixtieth of a second, the time of one picture field.

In pulse applications, the amount of sag occurring in a time interval corresponding to the length of the pulse must often be limited to one per cent, or even less. Obviously, great care must be taken in the design of amplifiers which are required to meet specifications of this order.

6.9 Direct-coupled vacuum tube amplifiers

Low-frequency amplification down to zero frequency (d-c), without attenuation or phase shift, can be achieved by using a carefully designed direct-coupled vacuum tube amplifier.

A direct-coupled vacuum tube amplifier is one in which the plate of one stage is directly connected to the grid of the next. An elementary form of direct-coupled amplifier is shown in Fig. 6-8.

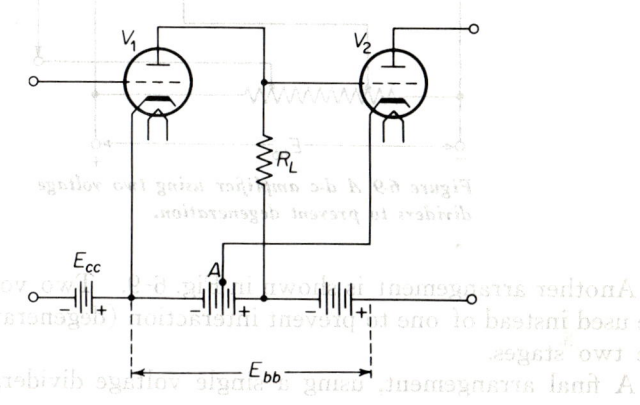

Figure 6-8 Simple vacuum tube d-c amplifier.

At first glance, it appears that the grid of V_2, in being connected directly to the plate of V_1, is positive with respect to its cathode. This would normally result in heavy grid current and probable destruction of the tube. Notice, however, that the cathode of V_2 is connected to point A on battery E_{bb}. Assume that the d-c potential at this point is +110 volts, that the bottom end of R_L is connected to a potential of +200 volts, and that the quiescent plate current through V_1 produces a 100-volt drop across R_L.

Under these conditions, the plate voltage of V_1 is +100 volts and, therefore, the grid voltage of V_2 is also +100 volts. Because the cathode is at a potential of +110 volts, however, the grid is

10 volts negative with respect to the cathode. By proper selection of point A for the cathode connection of V_2, any required value of $-E_{cc}$ can be achieved.

The plate of V_2 must be positive in respect to its cathode before conduction and the process of amplification can occur. If the quiescent plate voltage of V_2 is to be $+100$ volts with respect to its cathode, and if 100 volts is again dropped across the plate load resistor, then the plate must be returned to a source voltage of $+310$ volts. From this it is seen that a d-c amplifier requires an increasingly higher value of source voltage.

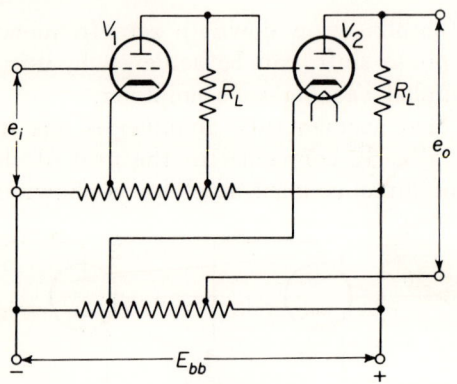

Figure 6-9 A d-c amplifier using two voltage dividers to prevent degeneration.

Another arrangement is shown in Fig. 6-9. Two voltage dividers are used instead of one to prevent interaction (degeneration) between the two stages.

A final arrangement, using a single voltage divider, is shown in

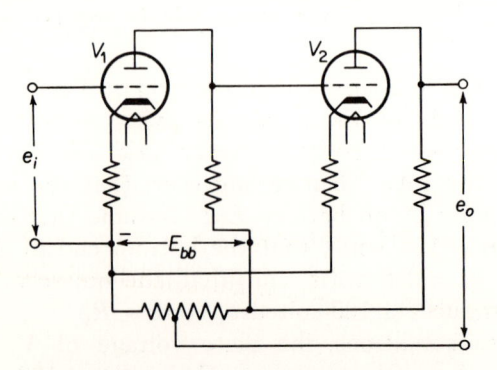

Figure 6-10 A d-c amplifier using degeneration introduced by voltage developed across unbypassed cathode resistors.

Sec. 6.10 Direct-coupled transistor amplifiers 117

Fig. 6-10. The voltage across the cathode resistor of V_2 must equal the plate voltage of V_1 minus the grid voltage of V_2. The output of each stage is actually somewhat less than its input because of the degeneration introduced by the unbypassed cathode resistors. The fidelity of this circuit is, however, excellent.

Because of practical difficulties in maintaining highly regulated d-c supply voltages as well as the unavoidable changes in tube characteristics with aging, d-c amplifiers are generally limited to two stages.

6.10 Direct-coupled transistor amplifiers

It is possible to make use of the reversed polarity and current directions of the NPN and PNP types of transistors to construct the simple d-c transistor amplifier shown in Fig. 6-11. The positive-

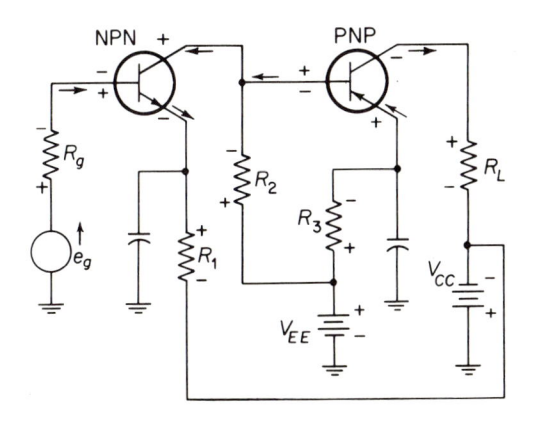

Figure 6-11 Simple two stage d-c amplifier.

current directions of the NPN type are out of the emitter, into the base, and into the collector. The reverse directions are true of the PNP type. The current directions and voltage polarities are shown in Fig. 6-11. Thus, a portion of the collector current of the first stage is the base current of the second stage. Additional collector current for the first stage is applied through resistor R_2. The divided power supply is used to supply both transistors by means of the dropping resistors R_g, R_1, R_2, R_3, and R_L.

The equivalent of a d-c amplifier using a single d-c supply is quite easy to achieve by the arrangement of Fig. 6-12, called *complementary symmetry*.

A transistor of one type is used to drive a second transistor of

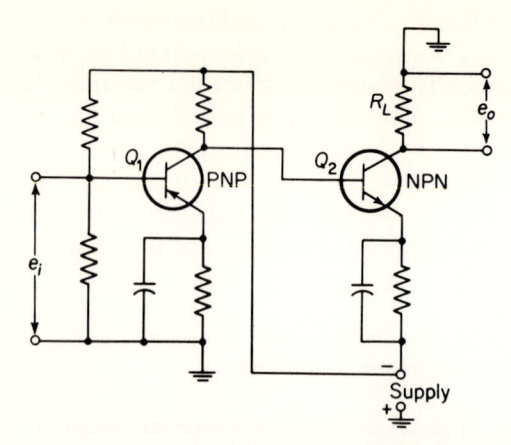

Figure 6-12 Equivalent of a Loftin-White d-c amplifier by using two transistors in a complementary circuit arrangement.

the other type. Both stages are connected in the common-emitter configuration. Because the transistors are of opposite type, the same current biases the collector-base junction of Q_1 and the base-emitter junction of Q_2.

REFERENCES

1 Kallman, H. E., R. E. Spencer, and C. P. Singer, " Transient Response," *Proc. IRE*, v. 33, 1945, p. 169, and correction, v. 33, 1945, p. 482.
2 Bedford, A. V., and G. L. Fredenall, " Transient Response of Multistage Video Frequency Amplifiers," *Proc. IRE*, v. 27, 1939, p. 277.
3 Wheeler, H. A., " Wideband Amplifiers for Television," *Proc. IRE*, v. 27 1939, p. 429.
4 Seeley S. W., and C. N. Kimball, " Analysis and Design of Video Amplifiers," *RCA Review*, v. 2, 1937, p. 171, and v. 3. 1939, p. 290.
5 RCA Industry Service Lab Report LB-930.
6 Bereskin, A. B., " Cathode-Compensated Video Amplification," *Electronics*, v. 22, 1949, p. 98; and v. 23, 1949, p. 104.
7 Miller, J.M. Jr., " Cathode Neutralization of Video Amplifiers,"*Proc.IRE*, v. 37, 1949, p. 1070.

EXERCISES

6-1 What is meant by the term " transient response " of an amplifier?
6-2 What does the vacuum tube gain-bandwidth product indicate, and how does this affect the choice of load resistors in pulse amplifiers?

Exercises

6-3 How does alpha affect the transistor gain-bandwidth product?

6-4 If wideband application and relatively large output voltages are required, what factors are most important in the selection of a tube?

6-5 Explain how shunt peaking extends the frequency range of an R-C amplifier.

6-6 Explain how series peaking extends the frequency range of an R-C amplifier.

6-7 Draw diagrams of three four-terminal networks used for frequency compensation, and compare the rise time and percentage overshoot for each.

6-8 What causes sag in R-C amplifiers (both vacuum tube and transistor types), and how is it compensated? In your explanation include any necessary diagrams.

6-9 In the vacuum tube d-c amplifier of Fig. 6-8, what modification can be made to prevent degeneration? Draw the new circuit.

6-10 Draw the circuit of a direct-coupled transistor amplifier and explain its operation.

7

FEEDBACK AMPLIFIERS

A feedback amplifier is one in which the amplifier input signal is derived (in part) from the amplifier output and (in part) from an external source. The transfer of energy from output to input is called *feedback*. When the phase of the feedback is such that it increases the effective value of the input, it is known as *positive feedback, regenerative feedback*, or, briefly, *regeneration*. When the phase of the feedback is such that it decreases the effective value of the input, it is known as *negative feedback, inverse feedback, degenerative feedback*, or, briefly, *degeneration*.

Negative feedback is used to reduce frequency distortion, harmonic distortion, and phase distortion, and results in an extended range of flat frequency response. The use of negative feedback also provides greater operating stability by making the amplifier appreciably more independent of variations in amplifier characteristics and of variations in operating voltages.

Positive feedback is used in amplifier circuits primarily to accentuate the gain markedly for one particular frequency, or for a band of frequencies. Under certain conditions this positive feedback may be excessive and cause the amplifier to

Sec. 7.1   Understanding feedback

oscillate. The amplifier then produces an output at one frequency even though no external signal is applied.

Primarily, negative feedback is considered here. Positive feedback is taken up in more detail later in the discussion of oscillators.

7.1 Understanding feedback

The operation of a feedback amplifier can be understood by referring to Fig. 7-1. The over-all gain of the amplifier without feedback is termed A. A fraction, β, of the output voltage, e_o, is returned to the input through the feedback network. The actual input to the amplifier, e_g, is, therefore, the sum of input voltage e_i plus the feedback voltage βe_o; this sum is written as

$$e_g = e_i + \beta e_o \quad (7-1)$$

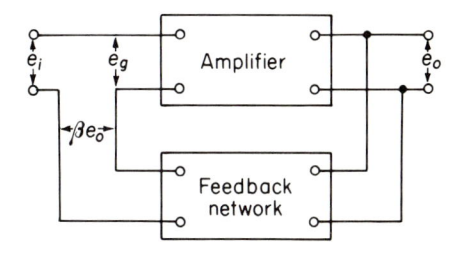

Figure 7-1 Simplified block diagram of a feedback amplifier.

Since the amplification of the amplifier without feedback is A, then

$$e_o = A e_g \quad (7-2)$$

If we substitute $(e_i + \beta e_o)$ from Eq. 7-1 for e_g, Eq. 7-2 becomes

$$e_o = A(e_i + \beta e_o) \quad (7-3)$$

Performing the multiplication

$$e_o = A e_i + A \beta e_o \quad (7-4)$$

Subtracting $A \beta e_o$ from both sides

$$e_o - A \beta e_o = A e_i \quad (7-5)$$

Factoring e_o

$$e_o(1 - A\beta) = A e_i \quad (7-6)$$

Dividing both sides by $(1 - A\beta)$ and by e_i

$$\frac{e_o}{e_i} = \frac{A}{1 - A\beta} \quad (7-7)$$

Since e_o/e_i is the amplification with feedback, A', Eq. 7-7 can be written as

$$A' = \frac{A}{1 - A\beta} \quad (7-8)$$

where the product $A\beta$ is called the *feedback factor* and the sign convention is such that when feedback opposes the input voltage β is *negative*. As a result, the denominator $(1 - A\beta)$ is normally positive for amplifiers using negative feedback. Then, Eq. 7-8 becomes

$$A' = \frac{A}{1 + A\beta} \qquad (7\text{-}9)$$

If there are large amounts of feedback, and the feedback factor $A\beta$ is large compared to 1, Eq. 7-9 becomes

$$A' = \frac{A}{A\beta} = \frac{1}{\beta} \qquad (7\text{-}10)$$

The significance of the above equations is best understood by using a numerical example. Assume an amplifier wherein $A = 50$, $\beta = -1/5$, $e_g = 1$ millivolt (mv), and $e_o = 50$ mv. The feedback factor $A\beta$ is $50(-1/5) = -10$ mv, and e_i from Eq. 7-1 must equal $(e_g - \beta e_o)$ $= [1 - (-10)] = (1 + 10) = 11$ mv. By Eq. 7-1, the voltage gain with feedback is $e_o/e_i = 50/11 = 4.545$. This represents a considerable reduction compared to the gain without feedback.

7.2 Improved stability

The reduction in gain with negative feedback may seem objectionable, but other effects result from the use of negative feedback that tend to offset this disadvantage.

Suppose the gain without feedback in the above amplifier drops from 50 to 40 as a result of tube aging, voltage variations, or a change of some component value. The gain with feedback then becomes

$$A' = \frac{40}{1 - 40(-1/5)} = \frac{40}{1 + 8} = 4.444$$

This shows that a 20 per cent drop (from 50 to 40) in voltage gain *without* feedback causes only a 1.01 per cent drop $(4.545 - 4.444)$ in the amplifier *with* feedback. Negative feedback, therefore, increases amplifier stability by greatly reducing variations in amplifier performance. The stabilizing effect improves with greater amounts of negative feedback.

7.3 Using decibels(db)

Feedback applied to an amplifier is occasionally expressed in terms

Sec. 7.4   Reduced nonlinear distortion

of a decibel (db) ratio. If the output power of an amplifier changes from P_1 to P_2 watts, then the db change is

$$\text{db} = 10 \log_{10}\left(\frac{P_2}{P_1}\right) \tag{7-11}$$

If the change is expressed in terms of voltages, then, since $P_2 = E_2^2/R$ and $P_1 = E_1^2/R$, the db change is

$$\text{db} = 20 \log_{10}\left(\frac{E_2}{E_1}\right) \tag{7-12}$$

For example, suppose an amplifier having a gain of 30 is used with 15 per cent feedback. The over-all gain with feedback is

$$A' = \frac{A}{1+\beta} = \frac{30}{1+(30 \times 0.15)} = \frac{30}{5.5} = 5.45$$

and the feedback level in db for this example is

$$\frac{A}{A'} = 20 \log_{10}\left(\frac{30}{5.5}\right) = 14.75 \text{ db}$$

Thus, an amplifier with 14.75 db feedback is one in which the gain is reduced 14.75 db by feedback.

As proof, the gain without feedback in db is

$$\text{db} = 20 \log 30 = 29.5$$

and the gain with feedback in db is

$$\text{db} = 20 \log 5.45 = 14.75$$

Then the reduction in gain in db is

$$29.5 - 14.75 = 14.75 \text{ db}$$

which agrees with the answer obtained.

7.4 Reduced nonlinear distortion

Nonlinear distortion in an amplifier arises mainly in the output stage, where the signal voltage amplitudes are large and the operating point of the tube moves into the nonlinear portions of its characteristic curve. Distortion from this source is greatly reduced by the use of negative feedback. As an example, let the distortion generated in the amplifier without feedback be termed D, and the distortion with feedback, D'. The distortion returned to the amplifier input through the feedback network is then βD, and it must be amplified A times. The total distortion in the output is, then

$$D' = D + A\beta D' \tag{7-13}$$

Subtracting $A\beta D'$ from both sides of the equation

$$D' - A\beta D' = D \qquad (7\text{-}14)$$

Factoring D'

$$D'(1 - A\beta) = D \qquad (7\text{-}15)$$

Multiplying both sides of the equation by $1/(1 - A\beta)$

$$D' = \frac{D}{1 - A\beta} \qquad (7\text{-}16)$$

Compare Eq. 7-16 with Eq. 7-8. It is apparent from Eq. 7-16 that the amplitude distortion appearing in the output, D, is reduced by by the factor $(1 - A\beta)$, which is the same factor by which stability is improved.

It should be noted that Eq. 7-16 applies only to distortion generated in the output stage. Negative feedback is less effective in reducing nonlinear distortion generated in earlier stages and has no effect on distortion that arises from nonlinearity in the input circuit of the first amplifier, because this distortion is amplified to the same extent as the feedback.

The reduction in nonlinear distortion can perhaps be best understood by the series of drawings in Fig. 7-2. Although the amplifier

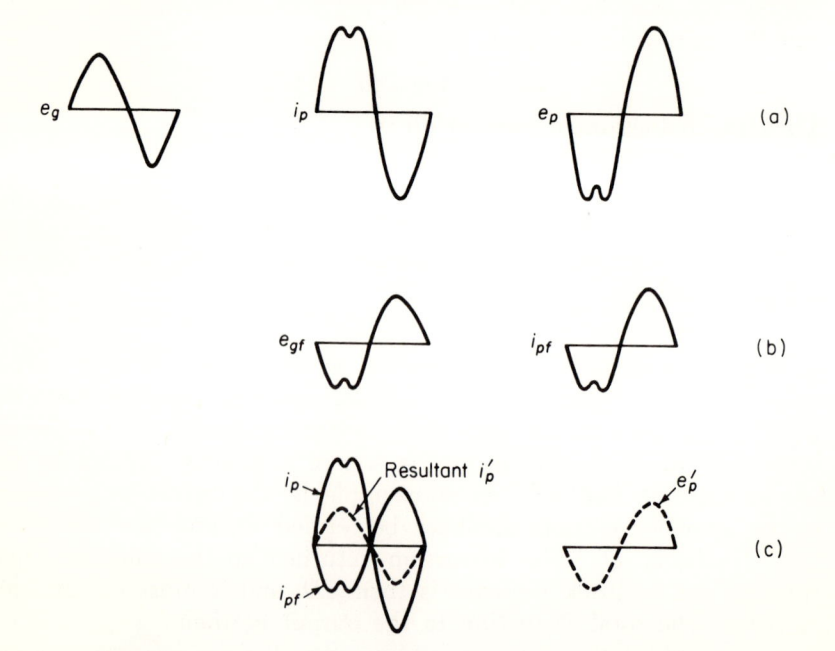

Figure 7-2 Reduction of nonlinear distortion by using feedback.

Sec. 7.5   Maintaining stability over a wide frequency range   125

in this case is a vacuum tube, a similar action in a transistor amplifier is readily apparent. The additional subscript f indicates feedback.

The undistorted input voltage (e_g) and the distorted plate current (i_p) and plate voltage (e_p) waveforms without feedback are shown in Fig. 7-2(a). The feedback waveforms are shown in Fig. 7-2(b).

The sum of i_{pf} and i_p produces the resultant undistorted i'_p and e'_p waveforms shown by the dashed line in Fig. 7-2(c).

Although the feedback results in the cancellation of harmonic voltages as well as in reduction of the resultant plate current i'_p and reduction of amplification, it also reduces distortion. The *greater* the feedback, the *greater* is the reduction of distortion.

Hum and other extraneous noises introduced into an amplifier from an external source are treated by negative feedback just as though they were distortion generated within the amplifier. Thus, a reduction by the factor $(1 - A\beta)$ occurs. From the practical standpoint, this means that the power supply used with a feedback amplifier need not be filtered as carefully as that used with an amplifier that does not employ negative feedback.

Negative feedback does not improve the amplifier signal-to-noise ratio, because internally generated noises and any signal components are reduced by the same factor. The signal-to-noise ratio is, however, altered by other effects of feedback, but the improvement, in general, is not great.

7.5 Maintaining stability over a wide frequency range

From the equation $A' = A/(1 - A\beta)$ for amplification with negative feedback at mid frequencies, it is apparent that both A and βA are vector quantities whose magnitude and phase angle at very high and very low frequencies are different than at mid frequencies. If βA is *real and negative*, the feedback is negative and amplification is decreased. If βA is *real, positive, and less than 1*, the feedback is positive and amplification is increased. If $\beta A = 1$, amplification is theoretically infinite and the system is unstable. If βA is *real, positive, and greater than 1*, the system will probably oscillate initially and then become stable. Such systems are said to be *conditionally* stable.

In analyzing any particular amplifier for stability, the amplification and phase angle over a wide range of frequencies is first calculated and then plotted in diagrams similar to those shown in (a) and (b) of Fig. 7-3. At the mid frequency termed f, the phase angle is zero; at some lower and higher frequencies, termed f_1 and f_2, respectively, the angles of lead and lag are 90°; at some still lower and higher

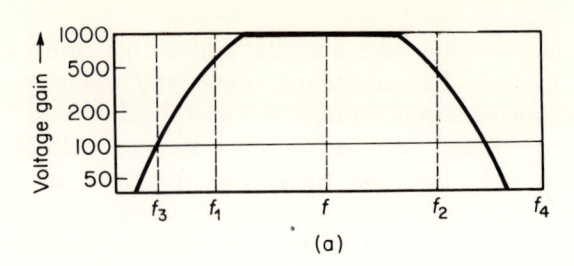

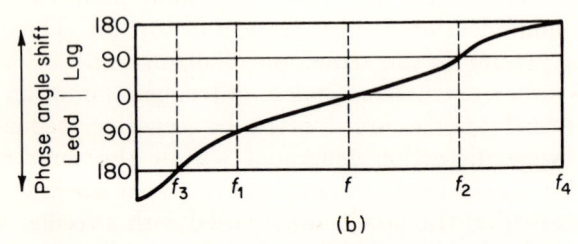

Figure 7-3 (a) Typical amplification and (b) phase angle diagrams over a wide frequency range.

frequencies, termed f_3 and f_4, respectively, the angles of lead and lag are 180°.

For a single stage R-C amplifier with fixed bias, we can plot the *locus* (path followed by a moving point) of the values of βA from zero frequency to infinity with polar coordinates. The result, known as a *Nyquist Diagram*, is shown in Fig. 7-4.

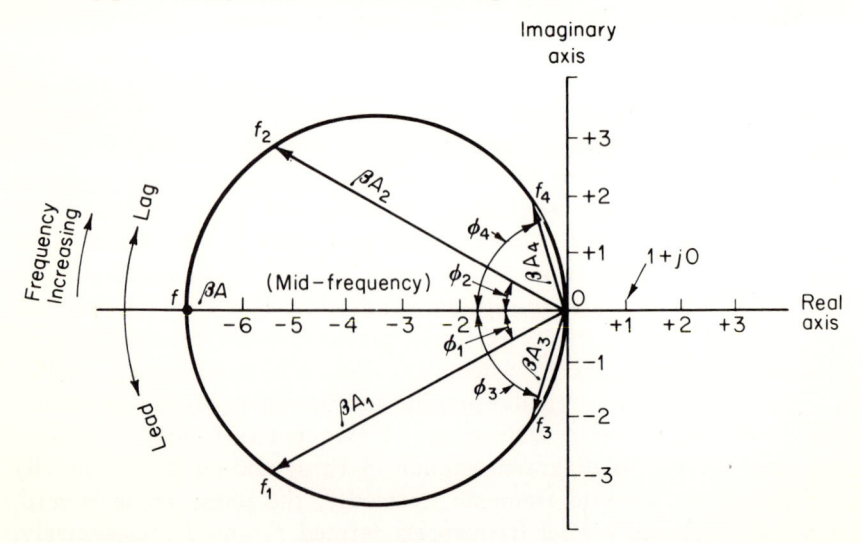

Figure 7-4 Nyquist diagram.

Sec. 7.5 Maintaining stability over a wide frequency range

If A is the gain at mid frequency f, this is normally the maximum value of βA, and the phase angle displacement is zero. It is, of course, *negative* for negative feedback, as shown in Fig. 7-4; i.e., displaced 180° on the real axis as shown.

At frequency f_1, the value of amplification, termed A_1, is taken from the curve of Fig. 7-3(a). The value of β is considered a fixed negative fraction which is less than 1. The product of these two quantities gives the value of βA_1 and the phase angle, ϕ_1, is determined from the chart of Fig. 7-3(b). This is plotted as shown in Fig. 7-4, with βA_1 extending from the origin to f_1. The angle ϕ_1 indicates that the phase of the feedback voltage *leads* the mid-frequency feedback voltage.

Following a similar procedure, we can construct βA_2 in Fig. 7-4, with a lagging phase angle displacement ϕ_2.

As the amplifier frequency bandwidth increases from f_2 to f_3, the length of βA_3, decreases, with a larger phase angle displacement, ϕ_3. At the extreme limit, βA equals zero, and the phase angle is 90° *lagging*.

Similarly, at frequency f_4, the length of βA_4 decreases, with a larger phase angle, ϕ_4; and at the extreme, βA again equals zero, and the phase angle is 90° *leading*.

For the above case, the shape of the locus is a circle lying entirely in the negative region, and the amplifier is always stable.

The βA locus of any amplifier can be plotted by the method outlined above. Typically, the plot for a two-stage R-C amplifier resembles that shown in Fig. 7-5. Here βA_1 still has an appreciable length when ϕ_1 is greater than 90°; but βA is zero when ϕ is 180° either leading or lagging.

The distance from any point on the locus to the point $(1 + j0)$ on the positive real axis, corresponding to unity, is equal to $(1 - \beta A)$. It follows from this fact that $|1 - \beta A| > 1$, and the feedback is negative for all frequencies corresponding to portions of the βA locus that lie outside the circle of unit radius centered at point $(1 + j0)$. Also, the feedback is positive for all frequencies corresponding to points lying inside the unit circle of Fig. 7-5.

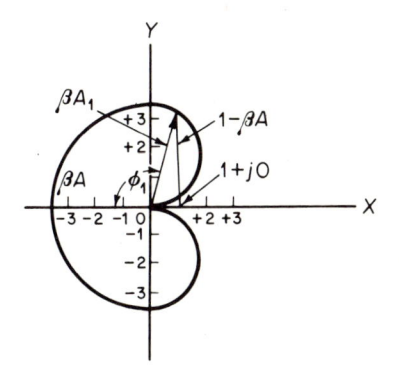

Figure 7-5 *Typical Nyquist diagram of a two stage R-C amplifier.*

7.6 Voltage and current feedback

The circuit of Fig. 7-6(a) is known as a *voltage feedback* circuit, because the feedback voltage is proportional to the voltage across the output. A circuit using *current feedback* is shown in Fig. 7-6 (b).

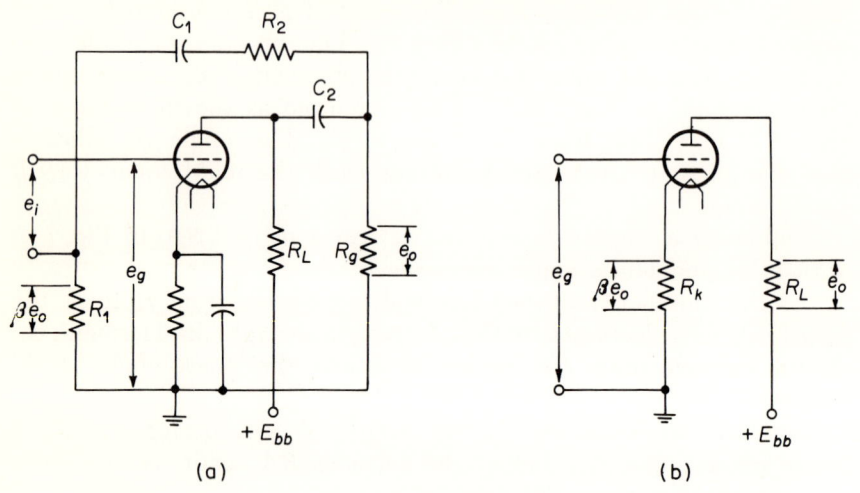

Figure 7-6 Illustration of (a) voltage feedback and (b) current feedback.

Here, the feedback voltage is proportional to the current in the output circuit. The feedback factor in this case is $(R_L/R_L + R_K)$. Generally, this type of feedback reduces amplification to about 1/3 its value without feedback; i.e., when R_k is adequately bypassed. Combinations of voltage and current feedback are sometimes used.

7.7 Two-stage feedback

A two-stage R-C coupled amplifier using negative voltage feedback is shown in Fig. 7-7. If the reactance of C_1 is negligible compared to $(R_1 + R_2)$, the feedback factor is the ratio of $R_1/(R_1 + R_2)$. Because a 180° phase shift is introduced in each tube, the feedback voltage must be returned to the cathode of V_1 instead of to its grid if the feedback is to be negative.

One difficulty with the arrangement shown in Fig. 7-7 is that the circuit tends to oscillate at very low and very high frequencies. To illustrate this, use is made of the Nyquist diagram shown in Fig. 7-8. The locus of $A\beta$ as a function of frequency is no longer circular, but is heart-shaped (cardioid), because the phase of $A\beta$

Sec. 7.7   Two-stage feedback

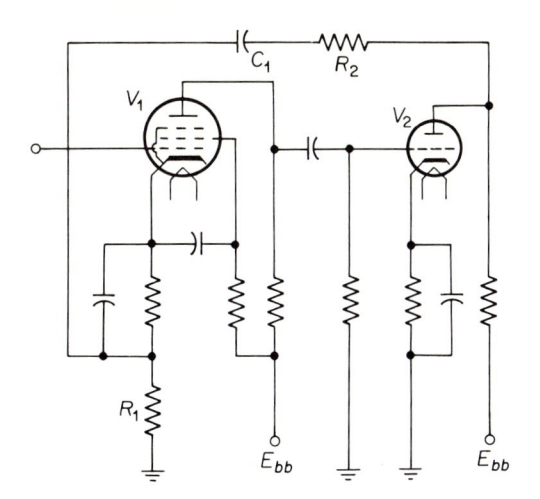

Figure 7-7 Two stage amplifier using negative voltage feedback.

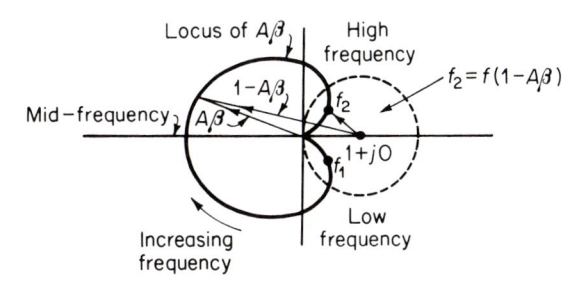

Figure 7-8 Nyquist diagram for circuit of Fig. 7-7.

may be $\pm 180°$ the mid-frequency value. This occurs because each coupling network may introduce a phase shift of $\pm 90°$. At very high and very low frequencies (f_2 and f_1, respectively), the $A\beta$ contour lies within the circle of unit radius about point $(1 + j0)$, shown by the dashed line, and the value of $(1 - A\beta)$ may, therefore, be less than one (unity). The feedback then becomes positive, and the voltage amplification with feedback, A, becomes greater than the voltage amplification, A, without feedback. Thus, the feedback may change from negative to positive over the complete frequency range. The circuit of Fig. 7-7 is still stable, because the value of $(1 - A\beta)$ cannot be zero even though it may be less than one at the very low and very high frequencies. It should be noted, however, that the possibility of instability increases as more stages are added to the feedback loop, because each stage will produce phase shifts which vary from $180°$.

7.8 Three-stage feedback

A three-stage amplifier using negative voltage feedback is shown in Fig. 7-9(a), with the Nyquist diagrams for small and large values of β in Fig. 7-9(b). Because three R-C coupling networks are used, the phase shift may become $\pm 270°$ relative to that at mid-frequency.

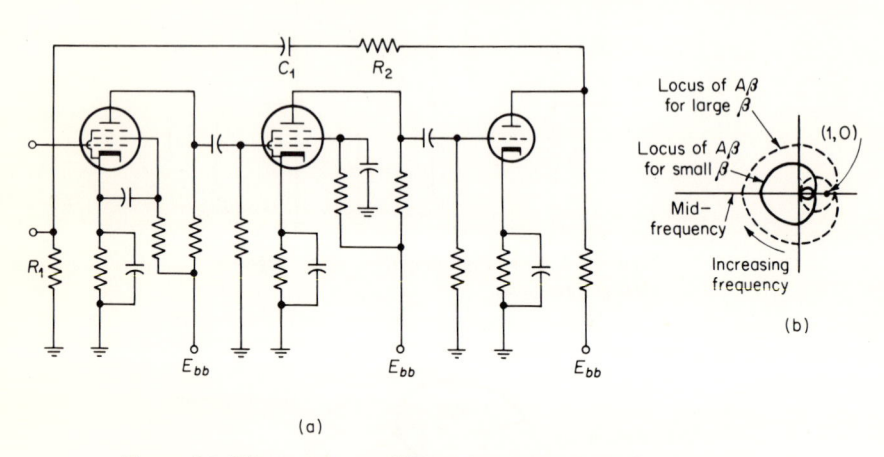

Figure 7-9 Three stage amplifier using (a) negative feedback and (b) Nyquist diagrams for small and large β.

It is seen that for small values of β (the solid line), the point $(1 + j0)$ lies outside the locus of $A\beta$, so that the quantity $(1 + A\beta)$ can never become zero. Thus, the circuit is stable, and oscillation cannot occur even though the feedback is positive at very high and very low frequencies. For large values of $A\beta$ (the dashed line), the point $(1 + j0)$ is enclosed by the locus of $A\beta$ and the circuit is still stable. If, however, the value of $(1 - A\beta)$ is zero at some particular frequency, the locus of $A\beta$ passes through the point $(1 + j0)$ and the circuit oscillates. This may occur at some value of β between the small and large values of β shown in Fig. 7-9(b).

When feedback is used with a three-stage amplifier, a form of coupling which does not exceed $\pm 90°$ per stage must be used. The frequency response of the first two stages is made sufficient only for the purpose at hand. The third stage is designed to have a flat frequency response with negligible phase shift at frequencies much lower and much higher than the first two stages. The third stage, therefore, introduces a negligible phase shift (except the 180° between grid and plate) until the frequency is so high or so low that the first two stages cause the amplification to drop greatly. As a result,

Sec. 7.9 Effect of voltage feedback on output impedance

feedback factor $A\beta$ can be made to drop to less than unity due to the reduction in amplification, A, before the additional phase shift introduced by the third stage is sufficient to make the total phase shift of all three stages reach $\pm 180°$.

7.9 Effect of voltage feedback on output impedance

The familiar constant-voltage equivalent circuit of an amplifier without feedback is shown in Fig. 7-10(a).

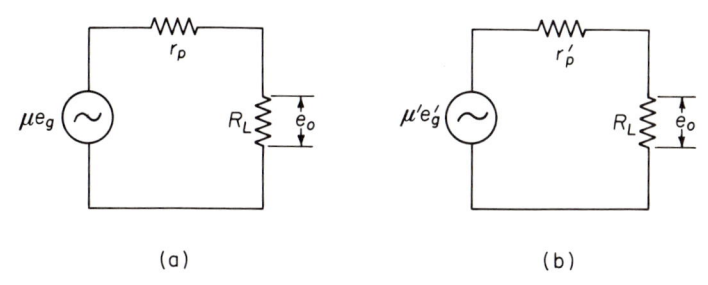

Figure 7-10 Constant voltage equivalent circuits of amplifier (a) without feedback and (b) with feedback.

In the circuit of Fig. 7-10(a)

$$e_o = \mu e_g \left(\frac{R_L}{r_p + R_L} \right) \tag{7-17}$$

and

$$A = \frac{e_o}{e_g} = \frac{\mu R_L}{r_p + R_L} \tag{7-18}$$

The equivalent circuit *with* feedback is shown in Fig. 7-10 (b). In this circuit

$$e_o = \mu' e'_g \left(\frac{R_L}{r'_p + R_L} \right) \tag{7-19}$$

Now, recall that

$$A' = \frac{A}{1 - \beta A} \tag{7-20}$$

which can be rewritten as

$$A' = \frac{1}{(1/A) - \beta} \tag{7-21}$$

Going back to Eq. 7-19

$$e_o = A' e'_g = \frac{e'_g}{(1/A) - \beta} \tag{7-22}$$

from Eq. 7-21.

By substituting the value of A from Eq. 7-18, Eq. 7-22 becomes

$$e_o = \frac{\mu R_L e'_g}{r_p + R_L(1 - \mu\beta)} \qquad (7\text{-}23)$$

which can be rearranged as

$$e_o = \left(\frac{\mu e'_g}{1 - \mu\beta}\right)\left(\frac{R_L}{r_p/(1 - \mu\beta) + R_L}\right) \qquad (7\text{-}24)$$

By definition

$$\mu' = \frac{\mu}{1 - \mu\beta} \qquad (7\text{-}25)$$

and

$$r'_p = \frac{r_p}{1 - \mu\beta} \qquad (7\text{-}26)$$

and if we substitute μ' and r'_p in Eq. 7-24, it becomes Eq. 7-19.

Equations 7-25 and 7-26 give the effective amplification factor and effective plate resistance of an amplifier with feedback. Each is equal to the corresponding value, μ and r_p, respectively, without feedback divided by $(1 - \mu\beta)$. The gain, distortion, and internally produced hum or noise are divided by $(1 - \beta A)$.

Thus, μ and r_p are both *effectively* reduced by negative voltage feedback. The actual tube characteristics are not changed, however, since the feedback is external to the tube. It can also be shown that the output voltage is independent of the value of R_L, and this indicates good regulation.

7.10 Effect of voltage feedback on input impedance

The effect of voltage feedback on amplifier input impedance depends on the method of application. Two cases are illustrated here. In the circuit of Fig. 7-11(a), the feedback voltage is applied in series with

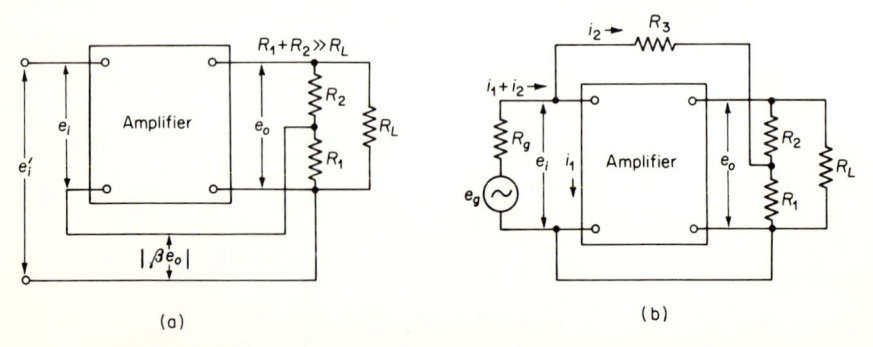

Figure 7-11 Illustration of the effect of feedback on the input resistance. Feedback voltage (a) in series with the input and (b) in shunt with the input.

Sec. 7.11   Effect of negative current feedback   133

the input voltage, and if the input resistance without feedback is R_i, then

$$R_i = \frac{e_i}{i_1} \tag{7-27}$$

With feedback, the input current is the same; but the input voltage becomes

$$e_i' = \frac{e_i}{1 - \beta A} \tag{7-28}$$

With this substitution, the input resistance with feedback, R_i', becomes

$$R_i' = \frac{e_i'}{i_1} = \left(\frac{e_i}{i_1}\right)(1 - \beta A) \tag{7-29}$$

and $$R_i' = R_i(1 - \beta A) \tag{7-30}$$

In this case, then, R_i is increased in the same proportion as the gain is decreased. This holds true whenever the feedback voltage is applied in series with the input voltage even for current feedback, as shown later.

In the circuit of Fig. 7-11(b), the feedback voltage is applied in shunt with the input voltage. The input resistance without feedback is

$$R_i = \frac{e_i}{i_1} \tag{7-31}$$

With feedback, i_1 is the same; but there is an additional current, i_2, through resistor R_3, in such a direction as to increase current from the source, which becomes $(i_1 + i_2)$. Voltage e_i is unchanged. Thus

$$R_i' = R_i \frac{i_1}{i_1 + i_2} \tag{7-32}$$

and the input resistance is *decreased*.

7.11 Effects of negative current feedback on output and input impedances

Fig. 7-12 is a block diagram of an amplifier using negative current feedback. The feedback voltage is developed across resistor R in series with R_L, and $R \ll R_L$.

With R included in the circuit, but without feedback

$$A = \frac{e_o}{e_i} = \frac{\mu R_L}{r_p + R + R_L} \tag{7-33}$$

With current feedback

$$A' = \frac{e_o}{e_i'} = \frac{e_o}{e_i + Ri_o} = \frac{e_o}{e_i + e_o(R/R_L)} \tag{7-34}$$

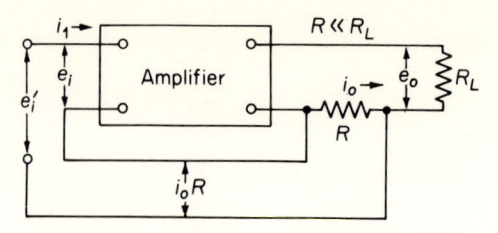

Figure 7-12 Amplifier using negative current feedback.

which can be written as

$$A' = \frac{1}{\left(\dfrac{e_i}{e_o}\right) + \left(\dfrac{R}{R_L}\right)} \tag{7-35}$$

Therefore

$$A' = \frac{\mu R_L}{(\mu + 1)R + r_p + R_L} \tag{7-36}$$

$$= \frac{A}{1 + AR/R_L} \tag{7-37}$$

$$= \frac{A}{1 + \dfrac{\mu R}{R + r_p + R_L}} \tag{7-38}$$

Equation 7-36 indicates that the amplification factor with current feedback is the same (μ) as without feedback, and Eq. 7-35 indicates the decrease in gain.

The effective plate resistance r'_p with feedback is

$$r'_p = r_p + R(\mu + 1) \tag{7-39}$$

and

$$\frac{r'_p}{r_p} = \frac{1 + (\mu + 1)R}{r_p} \tag{7-40}$$

Thus, the plate resistance increases with negative current feedback, and the increase is proportionately greater than the decrease in gain.

If μ is very large, it can be shown that A' is practically independent of amplifier characteristics, and also that the output current is practically constant, irrespective of R_L.

In Fig. 7-12, without feedback

$$R_i = \frac{e_i}{i_1} \tag{7-41}$$

and with feedback

$$R'_i = \frac{e'_i}{i_1} = \frac{e_i + e_o(R/R_L)}{i_1} \tag{7-42}$$

and

$$R'_i = \frac{e_i + Ae_i(R/R_L)}{i_1} \tag{7-43}$$

Sec. 7.12 *Illustrating improved frequency response* 135

Therefore
$$R'_i = R_i \left(\frac{1 + AR}{R_L} \right) \qquad (7\text{-}44)$$

which shows that R_i is increased with negative current feedback in the same proportion that gain is decreased.

7.12 Illustrating improved frequency response

Negative feedback increases amplifier bandwidth because it tends to maintain the output voltage constant. This is illustrated by referring to Fig. 7-13. The maximum gain of the amplifier without

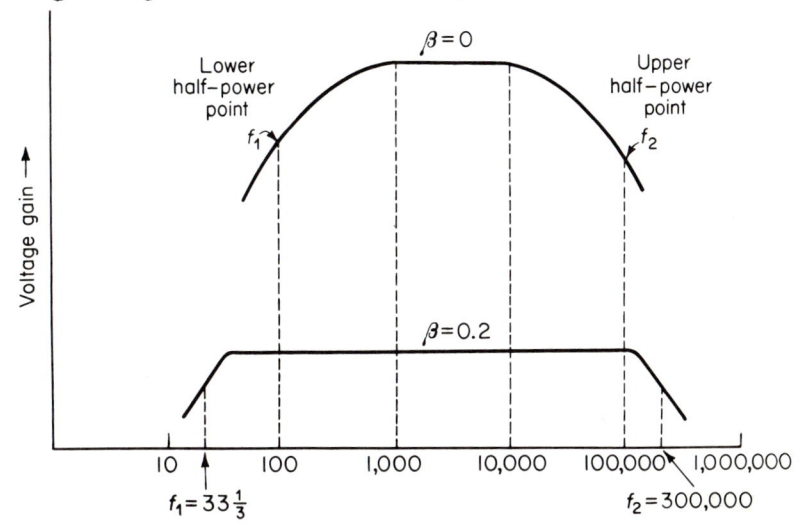

Figure 7-13 Illustration of the improvement of frequency response with negative feedback.

feedback is arbitrarily given the value 100, and the response is assumed flat between 1,000 and 10,000 cps. The half-power points, f_1 and f_2, occur at 100 and 100,000 cps, respectively. In a single-stage amplifier, frequency f_1 is reduced by the factor $1/(1 - A\beta)$, and frequency f_2 is increased by the factor $(1 - A\beta)$. If β is 0.02, then $1/(1 - A\beta)$ is

$$\frac{1}{1 - (0.02)(-100)} = \frac{1}{1+2} = \frac{1}{3}$$

Thus, frequency f_1 is reduced to $100/3 = 33\frac{1}{3}$ cps. Since $(1 - A\beta) = 3$, frequency f_2 is increased by this factor, 3, to $100,00 \times 3 = 300,\text{-}$

000 cps. Thus, although the effective gain or maximum amplitude per stage is decreased, the bandwidth has been increased.

7.13 Feedback in transistor amplifiers

The well-developed basic concepts of feedback in vacuum-tube amplifiers are applicable to transistor amplifiers with feedback. In general, negative feedback decreases gain, distortion, and noise, and improves frequency response. In most cases, the reduction in distortion is about equal to gain reduction. The amplifier input and output impedances can be either increased or decreased, depending upon the manner in which feedback is applied.

A single-stage transistor amplifier using voltage feedback is shown in Fig. 7-14. The resistor R_{fb} provides both feedback and operating bias. A decrease in voltage gain occurs, but it is the result of the parallel connection of R_{fb} and R_L, not of the feedback. Also, since R_L is typically of the order of several thousand ohms, while R_{fb} is usually about 100 K, the decrease in voltage gain is not as great as might be expected. However, current gain is greatly reduced by the degenerative current feedback through R_{fb}. Both R_i and R_o are reduced by voltage feedback.

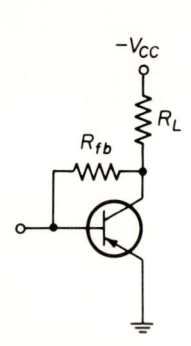

Figure 7-14 Single-stage transistor amplifier with voltage feedback.

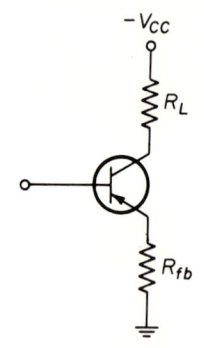

Figure 7-15 Single-stage transistor amplifier using current feedback.

A single-stage transistor amplifier using current feedback is shown in Fig. 7-15. Since $R_L \gg R_{fb}$, current gain is not greatly affected. The voltage gain is, however, greatly reduced. Both R_i and R_o are increased by current feedback.

When a feedback loop is used over several stages, the usual

Exercises 137

arrangement is voltage feedback. A typical arrangement is shown in Fig. 7-16.

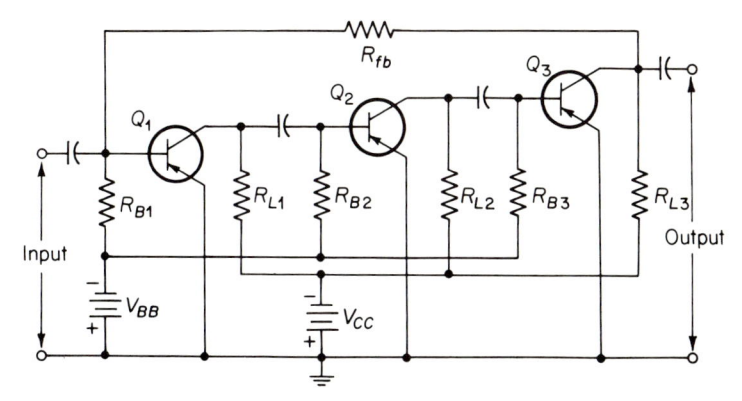

Figure 7-16 Voltage feedback loop over several stages.

REFERENCES

1. Langford-Smith, F., *Radiotron Designer's Handbook*, 4th. ed. Harrison, N. J.: RCA, 1953.
2. Black, H. S., "Feedback Amplifiers," *Bell Labs. Rec.*, June 1934, p. 290.
3. Crowhurst, N. H., *Audio Handbook No. 2-Feedback*. England: Norman Price Ltd., 1952.
4. Tellegen, B. D. H., "Inverse Feedback," *Philips Tech. Rev.*, Oct. 1937, p. 289.
5. Erhorn, P. C., "Notes on Inverse Feedback," *QST*, June 1943, p. 13.
6. Nyquist, H., "Regeneration Theory," *Bell System Tech. J.*, Jan. 1932, p. 126.
7. West, J. C., "The Nyquist Criterion of Stability," *Elec. Eng.*, May 1950, p. 169.
8. Lynch, W. A., "The Stability Problem in Feedback Amplifiers," *Proc. IRE*, Sept. 1951, p. 1000.
9. Pratt, J. H., "The Equivalent Characteristics of Vacuum Tubes Operating in Feedback Circuits," *RCA Review*, July 1941, p. 102.

EXERCISES

7-1 In a given amplifier, $A = 80$, $\beta = -1/4$, $e_g = 2$ mv, and $e_o = 160$ mv. What is the voltage gain with feedback?

7-2 What is the feedback level of Prob. 7-1 in terms of db?

7-3 Prove your solution to Prob. 7-2.

7-4 Suppose the gain of the amplifier in Prob. 7-1 drops from 80 to 65 as a result of tube aging. What is the drop in voltage gain, expressed as a percentage?

7-5 Explain what effect negative feedback has on distortion that arises from nonlinearity in the input circuit of the first amplifier.

7-6 (a) If βA is real and negative, what effect does it have on amplification and stability? (b) If βA is real, positive, and less than 1? (c) If βA is real, positive, and greater than 1? (d) If $\beta A = 1$?

7-7 What is feedback voltage proportional to in the circuit of Fig. 7-6(a)? In Fig. 7-6(b)?

7-8 Draw the schematic diagram of a three-stage R-C coupled transistor amplifier using negative-voltage feedback.

7-9 Define mathematically the effective amplification factor and the effective plate resistance of an amplifier with feedback.

7-10 If the feedback voltage is applied in series with the input voltage, what effect does it have on input resistance: (a) when using negative voltage feedback; (b) when using negative current feedback?

7-11 If feedback voltage is applied in shunt with the input signal, what effect does it have on the input impedance: (a) when negative voltage feedback is used; and (b) when negative current feedback is used?

7-12 How does the use of negative current feedback affect the amplification factor μ?

7-13 How does the use of negative current feedback affect plate resistance?

7-14 The maximum gain of a single-stage amplifier without feedback is 150. The response is flat from 200 to 8000 cps, and the upper and lower half-power points occur at 80,000 cps and 20 cps, respectively. If feedback is applied, and if $\beta = 0.3$, what are the new upper and lower half-power points?

ized cathode-follower circuit is shown in Fig. 8-1.

8

CATHODE-FOLLOWER CIRCUITS, EMITTER FOLLOWERS, AND PHASE INVERTERS

Low effective input capacitance and low effective load impedance are required to achieve uniform response over a wide frequency range. It has been shown that response is improved by the use of negative feedback. These qualities are inherent in the *cathode-follower circuit* (grounded-plate amplifier).

A cathode-follower circuit is a single-stage degenerative amplifier in which the output is taken across an unbypassed cathode resistor. It is essentially an impedance-matching device for matching a high-impedance circuit to a low-impedance circuit without frequency discrimination. Since it is a negative feedback amplifier, it has all the advantages associated with that circuit configuration, including excellent frequency response, low distortion, high stability, and low effective internal plate impedance. A disadvantage is that the voltage gain is always less than unity; but this is compensated for by an appreciable power gain.

Typically, a cathode-follower might be used between a pulse generating stage and a transmission line whose effective shunt capacitance might be great enough to cause objectionable effects. More power, of course, is delivered when the source is matched to the load. For example, a grounded-cathode amplifier having a high output impedance could supply less power to a low-impedance coaxial line than would a cathode follower having an output impedance equal to the load impedance.

As the name of the cathode-follower implies, the output voltage follows the input voltage; that is, it has not only the same waveform but the same instantaneous phase (polarity).

The *emitter-follower circuit* (common-collector amplifier) resembles the cathode-follower. It is useful as an impedance transformation device between a moderate input impedance level and a very low output impedance level.

The discussion in this chapter of the common-collector amplifier follows the same general order of development as used in Chaps. 4 and 5 for the common-base and common-emitter amplifiers.

The two tubes or transistors in a push-pull system require input voltages of equal magnitude but opposite phase. Thus, some method is needed to obtain an exciting voltage that is symmetrical with respect to ground from a single-ended amplifier.

Although transformers can be used for this purpose they are expensive and possess a limited frequency range. A better solution is to use push-pull exciting arrangements based on the R-C coupled amplifier. Such systems are usually called *phase inverters*; they are discussed in the concluding topics of this chapter.

8.1 The cathode-follower circuit[1,2,3]

The circuit arrangement of a cathode-follower is shown in Fig. 8-1. Notice these features of the circuit.
  1. Cathode resistor R_k is common to the output and input.
  2. Output voltage e_o is in series with input signal e_s.
  3. Since the amount of feedback voltage developed across R_k depends on the plate current, this is current feedback.
  4. Under quiescent conditions, plate current through R_k establishes normal bias.
  5. Under dynamic conditions, since R_k is unbypassed, degeneration occurs both on the positive half-cycle when plate current through R_k increases the bias and on the negative half-

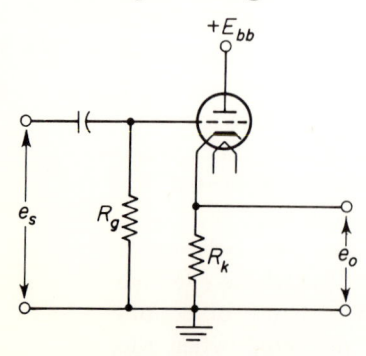

Figure 8-1 Cathode follower.

Sec. 8.2   Voltage gain   141

cycle when plate current through R_k decreases the bias. During the positive half-cycle, the increased bias *subtracts from* e_s and reduces the amplitude of the grid-to-cathode voltage. During the negative half-cycle, the bias *adds to* e_s, and the accompanying decrease in bias again reduces the amplitude of the grid-to-cathode voltage. Thus, in both half-cycles, the peak value of the a-c component of plate current is decreased and output voltage e_o is less than it would be without the *negative current feedback*.

8.2 Voltage gain

The constant-voltage equivalent circuit of a cathode-follower, Fig. 8-2, can be used to determine voltage gain. Since the maximum possible gain is $1/\beta$, and since β, in this case, is 1, the maximum possible gain of a cathode-follower is 1.

In practice, the maximum possible gain cannot be achieved. The output voltage is always less than the input voltage, since the equivalent generator voltage $\mu e_s/(\mu + 1)$ is divided across the series voltage divider consisting of $r_p/(\mu + 1)$ and R_k.

The voltage gain of the amplifier is equal to the ratio e_o/e_s. As an example of gain calculation, suppose $\mu = 43$, $r_p = 6300$ ohms and $R_k = 750$ ohms. Referring to Fig. 8-2, the effective value of r_p is

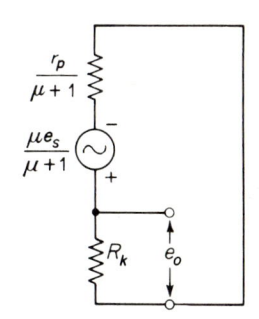

Figure 8-2 Constant-voltage equivalent circuit of a cathode follower.

$$\frac{r_p}{\mu + 1} = \frac{6300}{43 + 1} = 143 \text{ ohms}$$

If the input signal is one volt, the equivalent generator voltage is

$$\frac{\mu e_s}{\mu + 1} = \frac{43 \times 1}{43 + 1} = 0.977$$

This voltage then divides between the effective value of r_p and R_k; the output voltage is approximately

$$\frac{750 \times 0.977}{893} = 0.821$$

and the gain is

$$\frac{0.821}{1} = 0.821$$

Gain calculation is considerably simplified by using the constant-

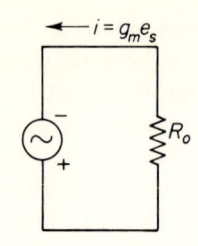

Figure 8-3 Constant-current equivalent circuit of a cathode follower.

current equivalent circuit of Fig. 8-3, which takes into account the transconductance of the tube. The output voltage using this circuit becomes

$$e_o = g_m e_s R_o \qquad (8\text{-}1)$$

where $R_o = 1/g_m$ and R_k in parallel. Thus

$$A = \frac{e_o}{e_s}$$

$$= \frac{g_m e_s R_o}{e_s}$$

$$= g_m R_o \qquad (8\text{-}2)$$

8.3 Power gain

The power delivered to the 750-ohm resistor is

$$\frac{e_s^2}{R_k} = \frac{(0.821)^2}{750} = 0.899 \text{ mw}$$

If $R_g = 1$ megohm, the input power is

$$\frac{e_s^2}{R_g} = \frac{1^2}{1 \times 10^6} = 0.001 \text{ mw}$$

The power output is seen to be 899 times the input power, which represents a very appreciable power gain.

8.4 Input resistance and capacitance

If the bottom of the grid resistor is returned directly to ground as

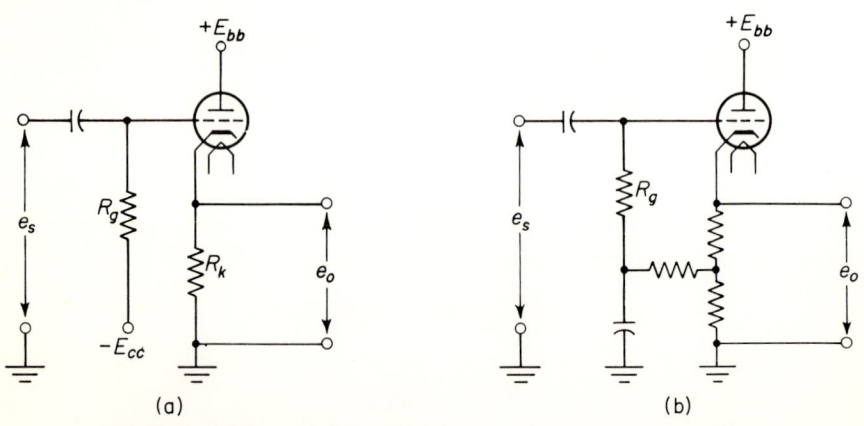

Figure 8-4 Resistor R_g (*a*) returned to $-E_{cc}$ and (*b*) bypassed to ground. In both cases $R_i = R_g$.

Sec. 8.5   Output impedance

in Fig. 8-4(a), or if it is returned to a negative bias voltage or bypassed to ground as shown in Fig. 8-4(b), the input resistance is equal to the value of R_g. If the circuit is rearranged as shown in Fig. 8-5, the input resistance, R_i, becomes

$$R_i = \frac{R_g}{1 - A} \qquad (8\text{-}3)$$

For example, if $R_g = 1$ megohm and $A = 0.821$

$$R_i = \frac{1 \times 10^6}{1 - 0.821}$$

$$\cong 5.59 \text{ megohm}$$

From the above, it is apparent that a large value of R_i is obtained by using a large value of grid resistance, R_g, and a gain, A, as close to unity as possible.

The effect of interelectrode capacitances in the amplifier is greatly reduced in a cathode-follower. The input capacitance is $C_{gp} + C_{gk}(1 - A)$; but since A has a value of 1 at most, C_{gk} has hardly any effect. Thus, the input capacitance is reduced approximately to C_{gp}. From this it is apparent that a small input capacitance is obtained by making A as large as possible and by using a tube having small values of interelectrode capacitances.

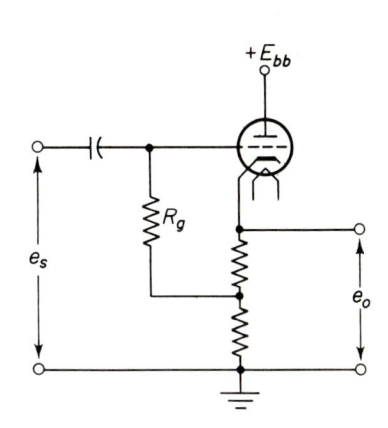

Figure 8-5 Resistor R_g returned to a tap on the cathode resistor. With this arrangement, $R_i = R_g/1 - A$.

8.5 Output impedance

Reference is again made to the constant-current equivalent circuit of Fig. 8-3, to determine the output impedance of a cathode-follower. This knowledge is essential if the cathode-follower is to match correctly the impedance of, perhaps, a transmission line. As an example, suppose the transconductance of a tube connected as a cathode-follower is 2000 μmhos, and the value of R_k is 500 ohms. The effective internal resistance in parallel with R_k is $1/(2000 \times 10^{-6})$ = 500 ohms, and the output impedance is

$$\frac{500 \times 500}{500 + 500} = 250 \text{ ohms}$$

8.6 Frequency response

The frequency response of the cathode-follower is generally excellent, which makes the circuit suitable for wideband applications.

The response falls off at the very-low-frequency end because of the reactance of the output filter capacitor in the plate voltage supply source. This capacitor must, therefore, be large if prohibitive low-frequency attenuation is to be prevented.

The high-frequency response is affected somewhat by the plate-to-cathode shunting capacitance, but far more important is the capacitance of the output circuit connected across R_k. If this output load assumes the form of a cable, for example, every precaution should be taken to minimize cable capacitance.

By using a tube with a high g_m (a value of high μ and low value of r_p), a low output impedance, Z_o, is obtained. A low Z_o results in a high upper-frequency response because any shunting effect due to tube capacitance or output circuit capacitance is negligible.

Since current feedback is used, the input impedance is increased; this in turn provides less of a shunting effect on the previous stage.

The negative feedback results in circuit stability which compensates for variations in line voltage, aging of tubes, etc.

8.7 The common-collector amplifier-general[4,5]

A common-collector amplifier circuit is shown in Fig. 8-6. The collector is now the common element between input and output. When a positive-going signal is applied to the input, the increase in base-to-emitter voltage v_{be} causes an increase in emitter current i_e. The increase in i_e causes emitter-to-collector voltage v_{ec} to increase. Thus, a positive-going input signal produces a positive-going output

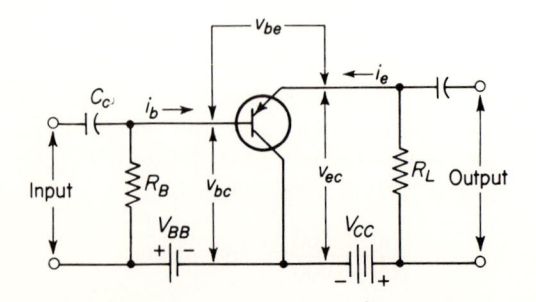

Figure 8-6 Common-collector amplifier.

signal, which indicates that the input and output signal voltages are in phase.

The common-collector (emitter-follower) circuit has a relatively high input impedance and a low output impedance. Also, the voltage gain is less than unity, as in the cathode-follower. The current gain is approximately β/α, which is of the same order as β in the common-emitter amplifier, so power gain is appreciable. As an example, suppose $\beta = 50$, $R_i = 100,000$ ohms, and $R_2 = 2000$ ohms. Then

$$G = \frac{I_c^2 R_L}{I_B^2 R_i} = \frac{\beta^2 R_L}{R_i} = 50$$

Since beta and alpha are short-circuit current amplification factors, and since R_L is not zero in the above example, it represents an approximation. More exact expressions are derived shortly.

8.8 Common-collector r parameters

Recall that the common-emitter equivalent circuit was derived from the common-base equivalent circuit. Thus, it is logical that the equivalent circuit for the common-collector amplifier can be derived from either. Because the generator $r_m i_b$ in the common-emitter equivalent circuit is already a function of base current, however, it is easiest to rearrange its elements into the form of the common-collector circuit. This is done in (a) and (b) of Fig. 8-7, with source and load terminations included. Either circuit can be used to derive operating gain and impedance expressions.

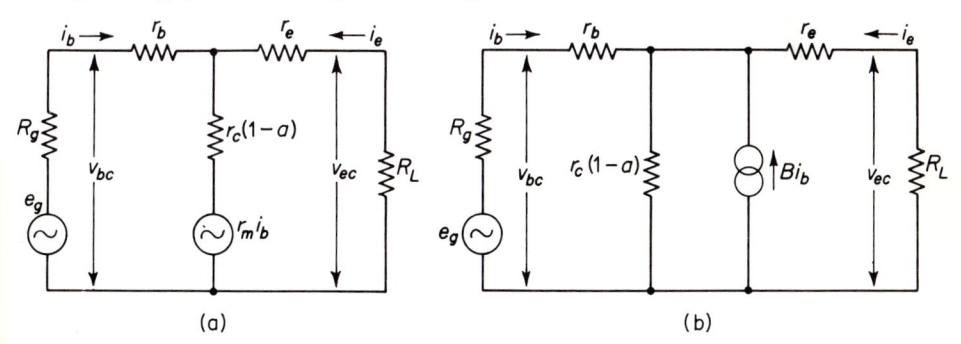

Figure 8-7 Common-collector r parameter equivalent circuit (a) constant voltage and (b) constant current.

Kirchhoff's voltage law applied to the input and output loops, respectively, of Fig. 8-7(a) yields the following equations

$$e_g = i_b(R_g + r_b + r_c) + r_c(1-a)i_e \qquad (8\text{-}4)$$

$$0 = i_b r_c + i_e[r_e + r_c(1-a) + R_L] \tag{8-5}$$

The operating current gain, $A_{ic} = i_e/i_b$, which can be derived directly from Eq. 8-5, is

$$A_{ic} = \frac{-r_c}{r_e + r_c(1-a) + R_L} \tag{8-6}$$

When r_e and R_L are small with respect to $r_c(1-a)$, this becomes

$$A_{ic} = \frac{-1}{1-a} \tag{8-7}$$

By Ohm's law, $v_{ec} = -i_e R_L$, and the operating gain $A_{ec} = -i_e R_L/e_g$. From the lengthy explanations given in Chap. 5, the reader should have little difficulty in performing the necessary algebraic manipulation to determine that

$$A_{ec} = \frac{r_c R_L}{(R_g + r_b)[r_e + r_c(1-a) + R_L] + r_c(r_e + R_L)} \tag{8-8}$$

With the usual numerical simplification, Eq. 8-8 can be rewritten in simplified form as

$$A_{ec} \cong \frac{1}{1 + \frac{(1-a)(R_g + r_b)}{R_L}} \tag{8-9}$$

which is normally close to but less than unity.

The power gain, G, is simply the product of $A_{ic} A_{ec}$, so that

$$G = \frac{1}{1 - a + \frac{(1-a)^2(R_g + r_b)}{R_L}} \tag{8-10}$$

or, in simplified form

$$G \cong \frac{1}{(1-a)} \tag{8-11}$$

From Fig. 8-7(a) we can write

$$v_{bc} = i_b(r_b + r_c) + r_c(1-a)i_e \tag{8-12}$$

and, since $R_{ic} = v_{bc}/i_b$

$$R_{ic} = r_b + \frac{r_c(r_e + R_L)}{r_e + r_c(1-a) + R_L} \tag{8-13}$$

or, in simplified form

$$R_{ic} \cong \frac{R_L}{1-a} \tag{8-14}$$

If we assume $e_g = 0$, the equivalent circuit equations are

$$0 = i_b(R_g + r_b + r_c) + r_c(1-a)i_e \tag{8-15}$$

$$v_{ec} = i_b r_c + i_e[r_e + r_c(1-a)] \tag{8-16}$$

Sec. 8.9   Common-collector h parameters   147

From these equations we derive the output resistance, R_o, which is equal to v_{ec}/i_e

$$R_{oc} = r_e + \frac{r_c(1-a)(R_g + r_b)}{R_g + r_b + r_c} \qquad (8\text{-}17)$$

and, for r_c large

$$R_{oc} = r_e + (1-a)(R_g + r_b) \qquad (8\text{-}18)$$

8.9 Common-collector h parameters

The equivalent circuit for the grounded-collector amplifier, using the h parameters, is shown in Fig. 8-8. This is the same form, of course,

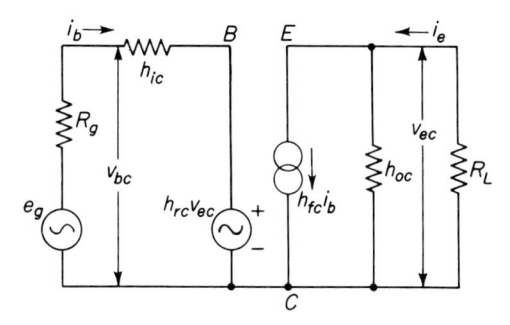

Figure 8-8 h parameter common-collector equivalent circuit.

as the other transistor configuration but with different parameter values. Thus, the gain and impedance equations derived for the common-base amplifier apply equally well to the common-collector circuit, if the appropriate common-collector h parameters are used.

Because manufacturer's data are usually given in terms of common-base h parameters, we must relate the two

$$h_{11c} = \frac{v_{bc}}{i_b} = \frac{h_{11b}}{\Delta - h_{12b} + 1 + h_{21b}} \qquad (8\text{-}19)$$

$$h_{12c} = \frac{h_{21b} + 1}{\Delta - h_{12b} + h_{21b} + 1} \qquad (8\text{-}20)$$

$$h_{21c} = -\frac{1 - h_{12b}}{\Delta - h_{12b} + 1 + h_{21b}} \qquad (8\text{-}21)$$

$$h_{22c} = \frac{h_{22b}}{\Delta - h_{12b} + h_{21b} + 1} \qquad (8\text{-}22)$$

Typical numerical values show that

$$(1 + h_{21b}) \gg (\Delta - h_{12b})$$

With this numerical simplification

$$h_{11c} = \frac{h_{11b}}{1 + h_{21b}} \tag{8-23}$$

$$h_{12c} = \frac{h_{21b} + 1}{h_{21b} + 1} = 1 \tag{8-24}$$

$$h_{21c} = -\frac{1}{1 + h_{21b}} \tag{8-25}$$

$$h_{22c} = \frac{h_{22b}}{1 + h_{21b}} \tag{8-26}$$

If the data are given in terms of common-emitter h parameters, and if we want to make common-collector calculations, we can use the simplified expressions

$$h_{11c} = h_{11e} \tag{8-27}$$

$$h_{12c} = 1 \tag{8-28}$$

$$h_{21c} = -(1 + h_{21e}) \tag{8-29}$$

$$h_{22c} = h_{22e} \tag{8-30}$$

8.10 Relating the r and h parameters

Since either the r or the h parameters can be used for small-signal analysis, it is sometimes convenient to express them in terms of each other.

$$r_b = \frac{h_{12b}}{h_{22b}} \tag{8-31}$$

$$r_e = h_{11b} - \frac{h_{12b}}{h_{22b}}(1 + h_{21b}) \tag{8-32}$$

$$r_c = \frac{1}{h_{22b}} \tag{8-33}$$

$$r_m = -\frac{h_{21b}}{h_{22b}} \tag{8-34}$$

and

$$h_{11c} = r_b + \frac{r_e r_c}{r_e + r_c(1 - a)} \tag{8-35}$$

$$h_{12c} = \frac{r_c(1 - a)}{r_e + r_c(1 - a)} \tag{8-36}$$

$$h_{21c} = -\frac{r_c}{r_e + r_c(1 - a)} \tag{8-37}$$

$$h_{22c} = \frac{1}{r_e + r_c(1 - a)} \tag{8-38}$$

8.11 Phase inverters-general

A phase inverter, more appropriately termed a *polarity inverter*, is a circuit that simply changes a positive-going signal to a negative-going signal, or *vice versa*, without introducing distortion or changing the original waveform. Perhaps the simplest type of phase inverter is the transformer, with which the instantaneous polarity of the load may be reversed with respect to the source by reversing either the connections of the secondary leads to the load or the primary leads to the load.

In pulse applications, many harmonics must generally be passed unattenuated and undistorted, so that little use is made of the transformer inverter whose ratio of transformation varies with frequency. For this reason, it is not discussed here.

A conventional electron-tube amplifier (untuned and R-C coupled) also produces an output of polarity opposite to the input, and if no gain is desired, various methods (some of which are discussed below) may be used to produce unity gain. Occasionally, both the positive-going and negative-going pulses of a waveform are desired. Either single or two-tube amplifiers may be made to convert one input waveform into two output waveforms of opposite polarity. Such amplifiers are called *phase splitters* or *paraphase amplifiers*.

Transistor phase inverters are also used; they are discussed following the vacuum tube type.

8.12 Single-tube phase inverters

A single-tube phase inverter is shown in Fig. 8-9; it is readily recognized as a simple R-C amplifier. A positive-going input signal causes an increase in plate current and a decrease in plate voltage. The output, therefore, appears as a negative-going signal. The advantages of this type of circuit are the gain realized from the tube, and the isolation it affords between the input and output circuits.

If a gain of unity or less is satisfactory, all that is necessary to accomplish this is to remove the cathode-bypass capacitor which introduces cathode degeneration. This method is preferred over a voltage divider because it reduces distortion and improves frequency response. The degeneration is the result of a rise in cathode potential with the rise in grid potential. By adjusting the value of R_k, the gain of the tube can be reduced to a desired value.

If unity gain is necessary, the arrangement shown in Fig. 8-10

150　　　　　　　　　　　　　　*Cathode-follower circuits*　　Chap. 8

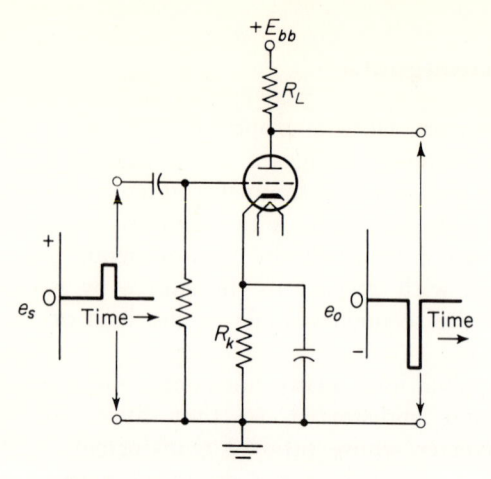

Figure 8-9 Single-tube phase inverter.

can also be used. A series voltage divider, R_{g1} and R_{g2}, reduces the voltage of the input signal applied to the grid. To prevent the loss of harmonics in the signal to be inverted, capacitor C_1 is connected across resistor R_{g1}. The reactance of this capacitor should bear the same numerical relation to the reactance of input capacitance, C_i, as R_{g1} does to R_{g2}. For example, if the value of R_{g1} is 220,000 ohms and that of R_{g2} is 22,000 ohms, their ratio is 10:1. If now, $C_i = 15\,\mu\mu\text{f}$, C_1 should be 15/10 1.5 $\mu\mu\text{f}$.

A single-tube paraphase amplifier (phase splitter) is shown in Fig. 8-11. The input signal, e_i, produces two output signals, e_{o1} and e_{o2}. Notice that $R_L = R_k$, and $R_{g1} = R_{g2}$. Because the same current

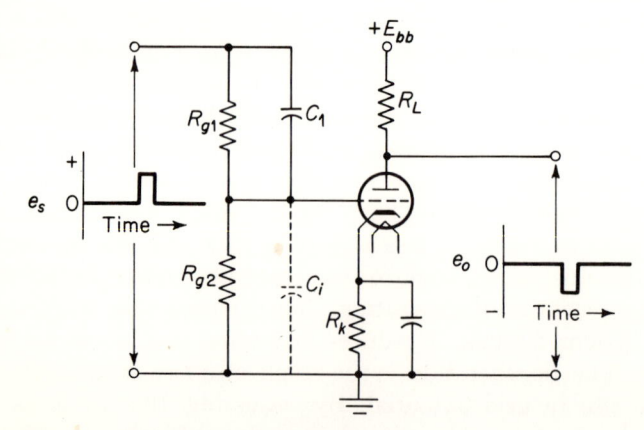

Figure 8-10 By proper arrangement of R_{g1}, R_{g2}, and C_1, output voltage $e_0 = e_s$, but it is of opposite polarity.

Sec. 8.12   Single-tube phase inverters

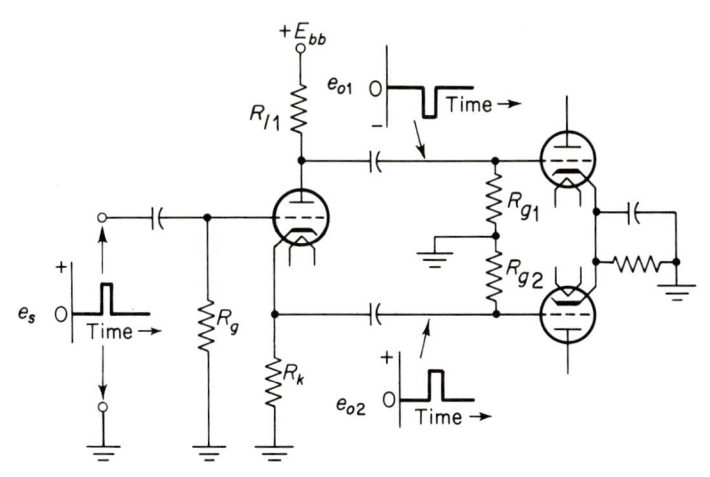

Figure 8-11 Simple phase-splitter circuit.

flows through R_k and R_L from $B+$ to ground, the same voltage should be developed across each resistor. Since one output is taken directly from the plate and the other directly from the cathode, they are opposite in phase.

Because R_L and R_k are shunted by unequal values of interelectrode capacitance, however, they are loaded unequally and produce unequal output voltages at high frequencies. This produces some deterioration in the high-frequency response. High-frequency unbalance is also caused by the fact that voltage e_{o1} is derived from the plate of the tube, using a negative-current feedback circuit, and e_{o2} is derived from the cathode of the same tube, using negative-voltage feedback. The difference in feedback is important at high frequencies where it is necessary to consider the effects of reactances which shunt the two output circuits of the tube. Because voltage feedback tends to maintain the output voltage constant, a variable reactance shunting the cathode circuit has negligible effect on the constancy of e_{o2}. The equivalent generator for e_{o1}, however, tends to maintain a constant-load current, and therefore its value varies with variations in load impedance. As a result, e_{o1} varies with frequency since the constant-load current passes through a parallel combination of R_{g1} and a capacitance whose reactance varies with frequency.

The voltage gain of this paraphase amplifier is always less than unity because output voltage e_{o2}, appearing across R_k of the cathode-follower circuit, is always less than unity.

A modified form of paraphase amplifier is shown in Fig. 8-12. Grid bias is obtained across R_{k1}. The a-c output voltage e_{o2} is taken

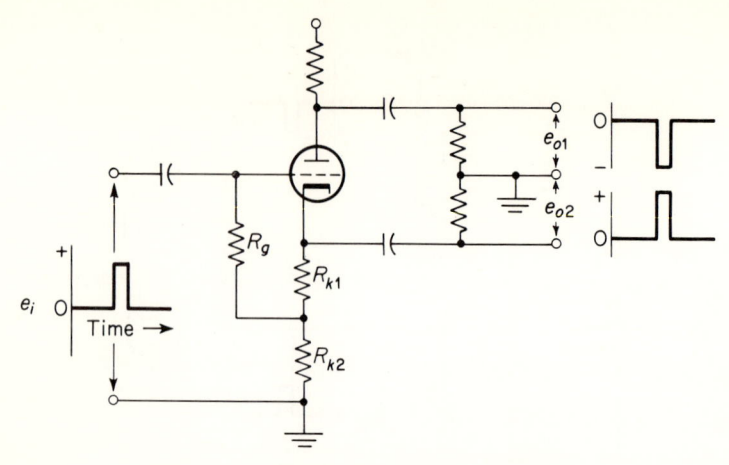

Figure 8-12 Modified paraphrase amplifier.

across the entire load consisting of $R_{k1} + R_{k2}$. This circuit is useful when the total d-c drop between cathode and ground exceeds the d-c bias required for proper operation of the tube.

8.13 Two-tube phase inverters

A paraphrase amplifier using two tubes and providing greater gain than the single-tube type is shown in Fig. 8–13. Tube V_1 functions as an ordinary R-C amplifier and produces an amplified and inverted version of the input signal across the series network composed of R_{g1} and R_{g2}. That portion of the output signal developed across R_{g2} is applied to the grid of V_2. This amplifier then produces an amplified and inverted output signal across R_{g3}. If the voltage developed across R_{g2} and applied to V_2 is equal in amplitude to the input voltage to V_1, and if the amplifiers are otherwise identical, the two output voltages, e_{o1} and e_{o2} are identical except for polarity.

A great disadvantage in this circuit is that it can be balanced perfectly over only a narrow band of frequencies. This is because phase shift is introduced at the low-and high-frequency ends of the band. For example, phase shift is caused at high frequencies by the shunt reactances of stray capacitances across the output circuit of V_1. At low frequencies, phase shift is caused by the series reactance of coupling capacitor C_2. These effects are multiplied in the output of V_2, because its output circuit is also shunted by stray reactances and this shunting results in phase shift at high frequencies. The series reactance of C_3 produces low-frequency phase shift. Thus, the

Sec. 8.13   Two-tube phase inverters

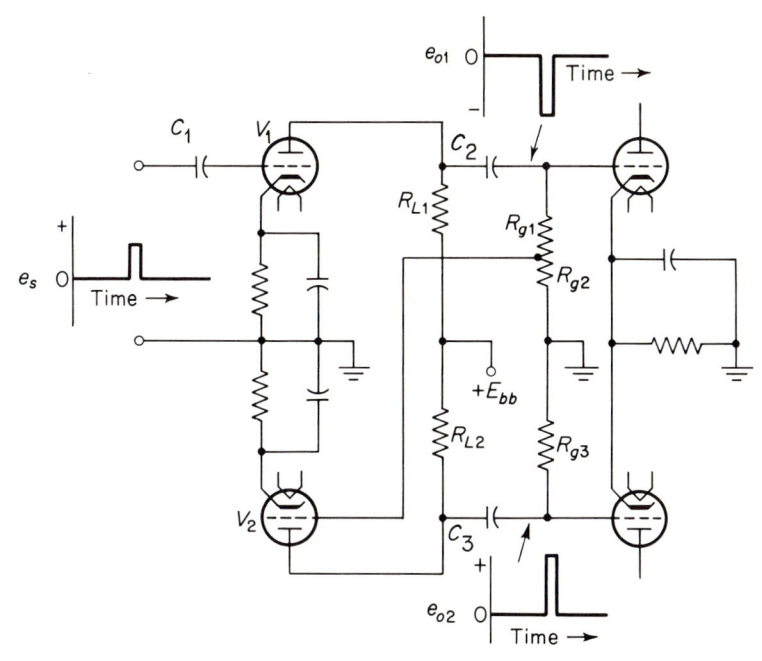

Figure 8-13 Form of paraphrase amplifier using two tubes.

phase shift between e_{o1} and e_{o2} is considerable at the high and low frequencies. It is also to be noted that e_{o2} has greater amplitude distortion than e_{o1} since it passes through both tubes.

An improved form of two-tube paraphase amplifier is shown in Fig. 8-14. The input signal to V_2 is derived from the differential in voltage between the outputs of the two tubes. Degenerative current feedback is applied to V_1 by omitting the cathode-bypass capacitor. This appreciably reduces distortion.

Let us first assume that V_1 and V_2 and their associated output circuits are identical, and that $R_{g1} = R_{g2} = R_{g3}$. Tube V_1 develops an output voltage, e_{o1}, across R_{g1} and R_{g3} in series when a signal is applied to its grid. One-half of e_{o1}, developed across R_{g3}, is equal to e_i and is used as the input to the grid of V_2. This stage then develops an output voltage, e_{o2}, across R_{g2} and R_{g3} in series that is equal to e_{o1} but opposite in phase. Under these conditions, the net signal voltage across R_{g3} would be zero. Without an input signal, V_2 cannot function; therefore, that portion of e_{o1} developed across R_{g3} is made slightly larger than that portion of e_{o2} across the same resistor. This is easily accomplished by making the value of R_{g1} slightly less than the value of R_{g2}, or by making R_{L1} slightly larger than R_{L2}.

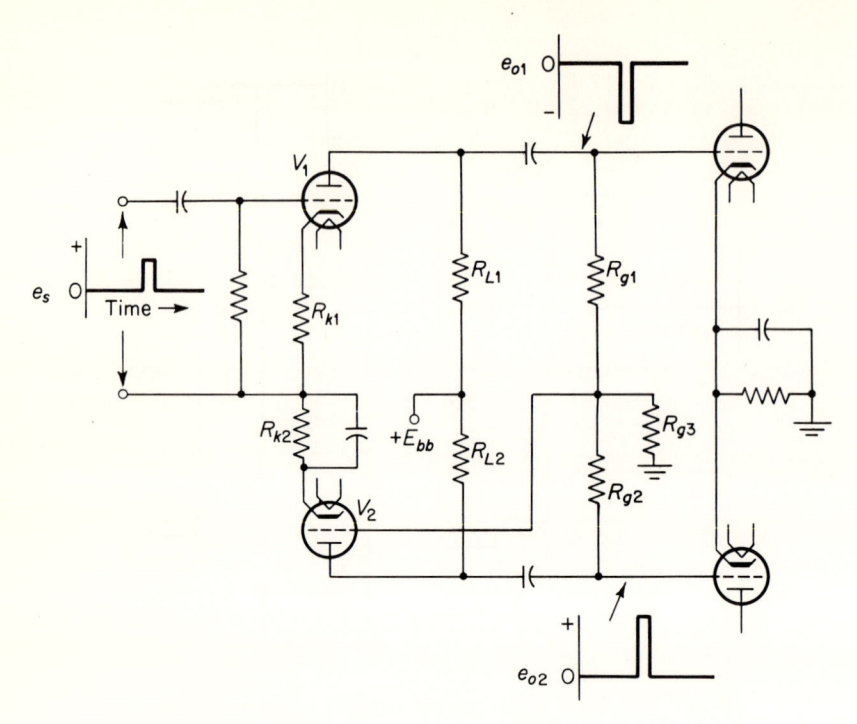

Figure 8-14 Differential voltage developed across R_{g3} used as input to V_2.

A form of cathode-coupled paraphase amplifier is shown in Fig. 8-15. Tube V_1 is a degenerative amplifier, using an unbypassed cathode resistor, R_k, which carries the total a-c and d-c plate currents

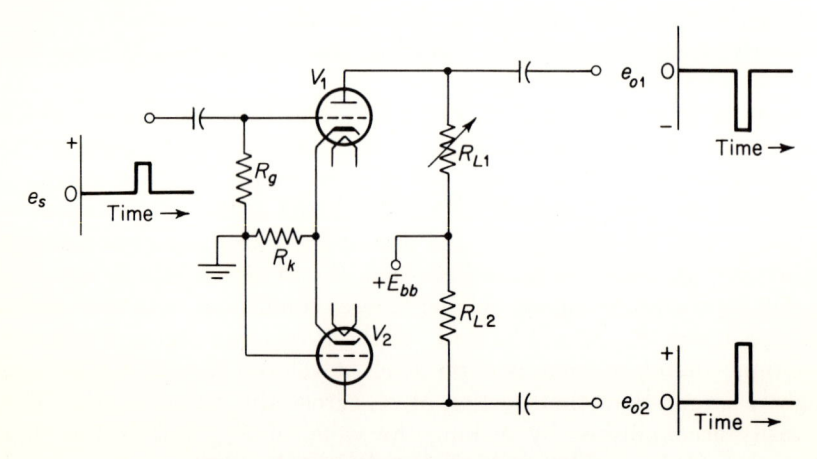

Figure 8-15 Cathode-coupled phase inverter.

of both tubes. Since R_k is common to both tubes, a change in plate current of either is reflected immediately into the grid circuit of both. The value of R_k is such that it reduces the grid-to-cathode voltage of V_1 to approximately one-half e_i. About one-half e_i, therefore, appears across R_k. As a result, the a-c voltage developed between the cathode and ground is in phase with voltage e_{g1}, because an increase in grid voltage increases plate current making the cathode more positive in respect to ground. Of course, the cathode voltage also decreases as the grid voltage decreases.

The voltage developed across R_k is applied between the grid and cathode of V_2. As a result, the voltages applied to the grids of V_1 and V_2 are opposite in polarity and almost equal in amplitude. When these two signals are amplified, the outputs are of opposite polarity. Experiment has shown that the a-c plate current in V_1 is greater than the a-c plate current in V_2, and that e_{o1} is usually larger than e_{o2}. To make the two outputs exactly equal, R_{L1} is made adjustable.

8.14 Transistor split-load inverters

Figure 8-16 shows a split load inverter Q_1, driving push-pull stages Q_2 and Q_3. Resistors R_E and R_L are equal in value, and resistor R_B establishes the required bias.

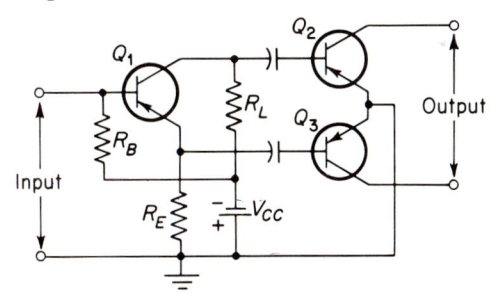

Figure 8-16 Split-load transistor phase inverter.

A positive-going input signal increases the output current, and causes the collector voltage to swing negative and the emitter voltage to swing positive, both with respect to ground. The output signals taken across R_L and R_E are, therefore, 180° out of phase with each other.

In this circuit, two conditions of unbalance exist. First, since i_e is always greater than i_c, Q_3 receives slightly more signal than Q_2.

156   *Cathode-follower circuits*   Chap. 8

Second, the collector output impedance of Q_1 is greater than its emitter output impedance.

To equalize the drive impedance, resistor R_1 is added as shown in Fig. 8-17, and the values of R_E and R_L are changed because of the loss of power in R_1, with resistor R_E being of higher value than R_L.

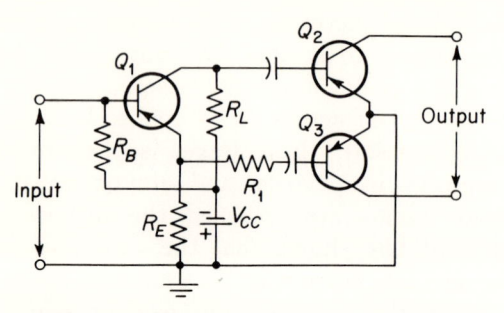

Figure 8-17 Split-load inverter with equalized output impedance.

8.15 Two-stage transistor phase inverters

Figure 8-18 shows a two-stage phase inverter consisting of Q_1 in the common-emitter configuration and Q_2 in the common-base configuration. Resistor R_1 provides bias for Q_1, and resistor R_2 provides bias for Q_2. Capacitor C_1 places the base of Q_2 at a-c ground potential. Resistors R_{L1} and R_{L2} are the load resistors for Q_1 and Q_2, respectively, and resistor R_E develops a small signal used to drive Q_2.

Two signals, out of phase with each other by 180°, are developed

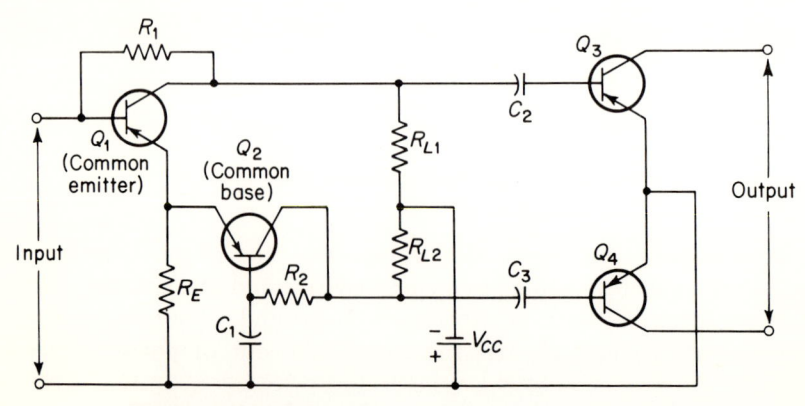

Figure 8-18 Phase inverter using common-emitter and common-base amplifiers.

Sec. 8.15 Two-stage transistor phase inverters

across R_{L1} and R_{L2}; these signals are used to drive push-pull amplifiers Q_3 and Q_4. Capacitors C_2 and C_3 perform the usual functions of blocking d-c while passing the signal voltages.

A negative-going input signal produces a positive-going output signal across R_{L1} because of the inherent phase reversal of the common-emitter configuration. The same signal current also passes through R_E and develops a signal voltage that follows the input. Thus, the signal voltage polarities across R_{L1} and R_E are 180° out of phase.

The signal voltage developed across R_E acts as the input to Q_2; and, since the common-base configuration does not produce signal inversion, the output developed across R_{L2} is in phase with the input. Thus, the signals across R_{L1} and R_{L2} are 180° out of phase. The signals applied to Q_3 and Q_4 are balanced by selecting the correct values for resistors R_{L1}, R_{L2}, and R_E.

Although there is more negative feedback at the input of Q_1, caused by the presence of R_E, the input resistance to Q_1 remains low because R_E is shunted by the low resistance of the Q_2 input circuit.

Figure 8-19 shows another form of phase inverter; this type uses two common-emitter stages Q_1 and Q_2 to drive the push-pull amplifiers Q_3 and Q_4.

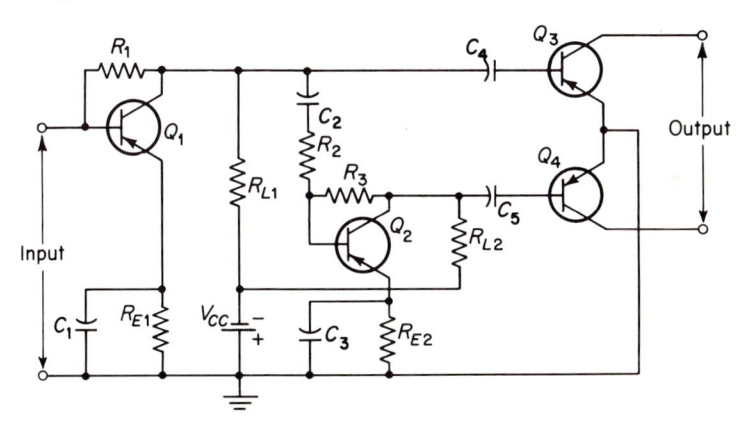

Figure 8-19 Two stage common-emitter phase inverter.

A positive-going input signal to Q_1 produces a negative-going output signal across R_{L1}. A portion of this signal is applied to the base of Q_3 through capacitor C_4, and a portion is coupled through capacitor C_2 and attenuating resistor R_2 to the base of Q_2. The negative signal at the base of Q_2 causes a positive-going signal across

R_{L2}. This signal is coupled to Q_4 through capacitor C_5. Thus, by proper selection of R_2, signals of equal and opposite polarity are applied to Q_3 and Q_4.

This inverter has the advantage of providing equal source impedances for Q_3 and Q_4, since both drivers are identical common-emitter configurations.

REFERENCES

1. Williams, E., "The Cathode-Follower Stage," *Wireless World*, July 1941, p. 176.
2. Greenwood, H. M., "Cathode-Follower Circuits," *QST*, Nov. 1945.
3. Cathode Ray, "The Cathode-Follower—What It Is and What It Does," *Wireless World*, Nov. 1945, p. 322.
4. Ryder, John D., *Electronic Fundamentals and Applications*, Second Edition. Englewood Cliffs, N. J.: Prentice-Hall, Inc. 1959.
5. Fitchen, Franklin C., *Transistor Circuit Analysis and Design*. Princeton, N. J.: D. Van Nostrand Co., Inc., 1960.
6. Wheeler, M. S., "An Analysis of Three Self-Balancing Phase Inverters," *Proc. IRE*, Jan. 1946, p. 67.

EXERCISES

8-1 In a cathode-follower, circuit $\mu = 59$, $r_p = 720$ ohms, $R_g = 1$ megohm and $R_k = 500$ ohms. The value of e_i is 1 volt. What is; (a) the voltage gain; (b) the power gain; (c) the input impedance: and (d) the output impedance?

8-2 What steps should be taken to insure the maximum frequency response of a cathode-follower?

8-3 Draw the constant-current and constant-voltage equivalent circuits of the cathode-follower.

8-4 Derive an equation for the operating current gain of a common-collector amplifier using r parameters. (Show the equivalent circuit used.)

8-5 Given the following common-emitter h parameters:

$h_{ie} = 3,000$ ohms $\qquad h_{fe} = 49$
$h_{re} = 3 \times 10^{-4} \qquad h_{oe} = 12 \times 10^{-6}$ mhos

Also, $R_g = 200$ ohms and $R_L = 1500$ ohms. Determine A_{ic}, A_{ec}, G, R_i, and R_o for the transistor used as a common-collector amplifier.

8-6 Given the following common-base h parameters:

$h_{ib} = 45$ ohms $\qquad h_{fb} = 0.98$
$h_{rb} = 3.25 \times 10^{-4} \qquad h_{ob} = 2.5 \times 10^{-7}$ mhos

Exercises

Also, $R_g = 10{,}000$ ohms and $R_L = 200$ ohms. Determine the value of all the r parameters, and calculate A_{ic}, A_{ec}, and G using r parameters.

8-7 Are the two output voltages of a split-load vacuum-tube phase inverter equal at all frequencies? Explain your answer.

8-8 What are the disadvantages of the paraphase amplifier shown in Fig. 8-13?

8-9 Explain the operation of the phase inverter shown in Fig. 8-18.

8-10 What is the purpose of resistor R in Fig. 8-17? Explain your answer fully.

9

THE VACUUM TUBE AND THE TRANSISTOR AS SWITCHES

The remainder of this book deals with circuits wherein rapid transitions between full on *and* full off *occur. Such circuits are collectively referred to as* electronic switching *circuits. Included in this group, to mention a few, are* multivibrators, blocking oscillators, *and* sweep generators.

The major advantages of electronic switches over their mechanical counterparts are: (1) much faster switching speeds; (2) absence of mechanical wear; and (3) less susceptibility to damage from vibration and mechanical shock.

A study of vacuum-tube and semiconductor switching characteristics is, therefore, imperative. Only by understanding the limitations of such devices is it possible to undertake intelligently the minimization of undesirable effects.

9.1 Ideal switches

In the simple circuit of Fig. 9-1(a), the switch and its contacts are assumed to possess zero resistance. As shown in Fig. 9-1(b), when the switch is open, from time t_0 to time t_1, the switch acts as a perfect open circuit and the entire applied

Sec. 9.2   Practical forms of electronic switching devices   161

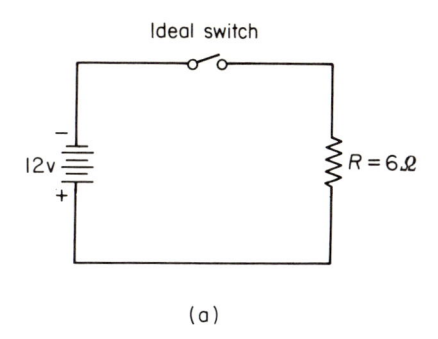

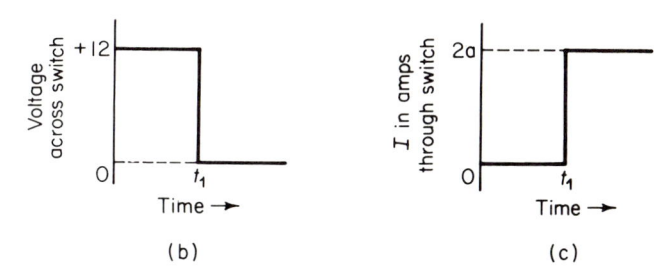

Figure 9-1 Simple circuit containing (a) an ideal switch, (b) voltage drop across switch, and (c) current through switch.

voltage appears across its terminals. At time t_1, the switch is closed. It then provides a perfect short-circuit and the voltage drop across its terminals becomes zero. Current through the switch during the "opened" and "closed" periods is illustrated as Fig. 9-1(c).

The ideal diode rectifier has a voltage-current characteristic identical to that of the ideal switch of Fig. 9-1(a). When the ideal diode is biased in the forward direction, it conducts maximum current and the voltage drop across its terminals is zero. When the ideal diode is reverse-biased, conduction cannot occur and the voltage drop across its terminals is equal to the applied voltage. In a voltage-current plot of the ideal diode, the point at which the transition from *on* to *off* occurs, is called the *breakpoint*.

9.2 Practical forms of electronic switching devices

Unfortunately, the actual electronic devices used as switches do not exhibit the same voltage-current characteristics as the ideal devices.

Inspection of typical voltage-current characteristics of diodes of both the high-vacuum thermionic and semiconductor junction types, Fig. 9-2, immediately indicates two defects of the actual devices: (1) in neither device is the transition from *off* to *on* instantaneous; and (2) neither device represents a perfect short-circuit in the forward direction. A scale expansion of Fig. 9-2 in the vicinity of the origin would also indicate that neither device represents a perfect open-circuit in the reverse direction.

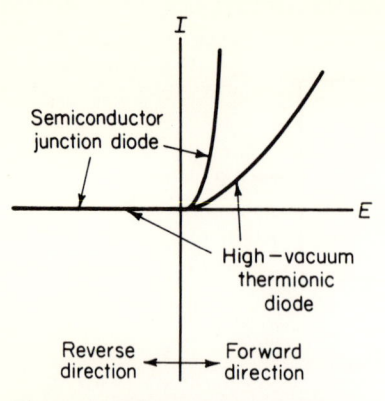

Figure 9-2 Voltage-current characteristics of a semiconductor junction diode and a high-vacuum thermionic diode.

9.3 The high-vacuum thermionic diode

The noted defects of an actual high-vacuum thermionic diode compared to a theoretical ideal diode are minimized by making the forward current several milliamperes or by using a reverse voltage of several volts, and by making the external circuit resistance very large compared to the diode resistance.

The vacuum diode has an interelectrode capacitance typically of the order of several micromicrofarads. Thus, in the reverse direction, the diode and the circuit resistance together form a series R-C circuit, and a finite time is required before any change in magnitude of the stored energy can occur. This explains why the diode cannot switch from *off* to *on* instantaneously. In the forward direction, the effect of interelectrode capacitance is negligible since it is short-circuited by the low resistance of the diode. For switching applications, diodes having very low interelectrode capacitance are, therefore, desirable. *Electron transit time* must also be considered when it is appreciable compared to the switching time, but this is not the usual case.

9.4 The semiconductor junction diode and breakdown

Semiconductor junction diodes possess the following advantages compared to thermionic diodes: (1) the semiconductor diodes have lower forward resistance; (2) they are smaller in size and weight; (3) they

Sec. 9.4 *The semiconductor junction diode and breakdown*

have no need for heater power and, therefore, no power frequency pickup; and (4) in some types, semiconductors have lower shunt capacitance.

Semiconductor junction diodes also have certain disadvantages compared to the thermionic type. These disadvantages include: (1) they generally have lower values of reverse breakdown voltage (to be discussed shortly); (2) their voltage-current characteristics depend on operating temperature; and (3) except for *breakdown* diodes, their response to abrupt changes in voltage is usually slower.

A typical semiconductor junction diode current-voltage characteristic is shown in Fig. 9-3. Reverse current in the reverse-voltage region is

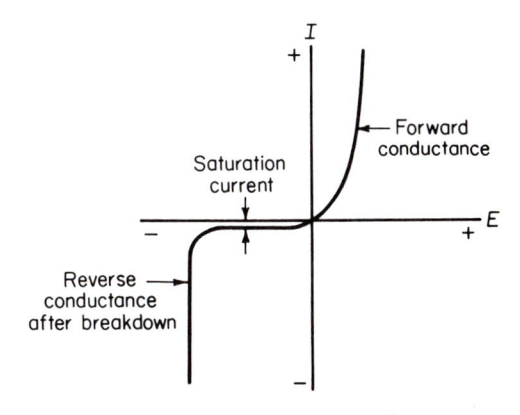

Figure 9-3 A typical semiconductor junction diode current-voltage characteristic.

caused by the presence of holes in the n-type and electrons in the p-type semiconductor. The presence of these carriers is primarily the result of *thermal excitation* of some of the intrinsic semiconductor atoms. Thus, saturation current increases with temperature and can become large enough to make the diode useless. The saturation current of silicon diodes at room temperature is very small (of the order of 10^{-10} amp); that of germanium diodes is considerably greater. The maximum operating temperatures without excessive reverse current is about 300°C for silicon diodes and 80°C for germanium diodes.

The reverse current of an ideal PN junction would remain constant at the saturation value even at large values of negative voltage. In the actual semiconductor diode, however, the reverse characteristic has appreciable slope throughout the negative voltage range, and a value of voltage is reached beyond which the slope of the characteristic becomes very large. The finite slope of the characteristic throughout the reverse-voltage range (caused by shunt-leakage conductance) is small in good diodes. The rise in reverse-current magnitude above a critical value of reverse voltage is caused by an *avalanche multiplication* of carriers as the result of a high electric field within the diode junction. This phenomenon, called *breakdown*, is somewhat similar to the breakdown of a gas diode.

If reverse current is limited to a value that does not cause the allowable dissipation of the diode to be exceeded, breakdown is non-destructive. Silicon diodes exhibit the most abrupt breakdown with respect to voltage, and the slope of the characteristic after breakdown is very steep.

Some diodes, called *breakdown (Zener) diodes* are designed to make use of the abrupt breakdown or the steepness of the characteristic in the reverse-current range beyond breakdown. The breakdown voltage is usually specified at an arbitrary value of current above the saturation value. The schematic symbol for a breakdown diode is shown in Fig. 9-4.

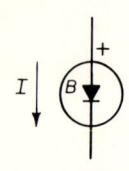

Figure 9-4 Schematic symbol for a breakdown diode indicating polarity convention. The arrow points in the direction of forward current and the letter B indicates breakdown.

By proper selection of manufactured units, breakdown can be set to occur at some low voltage value and can be used for switching action. Similarly, it can be set at some high voltage level where it does not occur during normal circuit operation.

9.5 Transient response of junction diodes[1]

The *forward transient* occurs when a diode is switched from a zero-bias condition to a forward-conducting condition. It is the result of a change of the semiconductor resistance caused by *minority-carrier* injection across the rectifying junction.

Typical voltage waveforms for very large, low, and intermediate forward currents are illustrated in Fig. 9-5. In the case of very large current values, Fig. 9-5(a), the diode behaves like an R-L circuit, with the voltage across the diode rising sharply to form a high-level spike, then decaying to the normal forward drop for the diode.

For low values of forward current, Fig. 9-5(b), the diode acts like an R-C circuit in which the voltage builds up to a steady value.

Between these two extremes, Fig. 9-5(c), the voltage first rises sharply to a spike, but then instead of decaying to the normal drop for the diode, the voltage may oscillate.

Experiment has shown that the magnitude and duration of the

Sec. 9.5 *Transient response of junction diodes* 165

forward transient can be appreciably decreased by placing a slight forward bias on the diode before applying a current pulse.

The *reverse transient* occurs when a diode is switched from a forward-conducting condition to a reverse-biased condition. Referring back momentarily to Fig. 9-3, it is seen that, in the forward direction, a nearly constant voltage is developed across the diode, since the impedance is relatively low and current through the external circuit is controlled by circuit resistance. The very low value of saturation current in Fig. 9-3 also indicates that the diode exhibits high reverse impedance.

If an input is applied that suddenly reverse-biases the diode, as in Fig. 9-6, the high impedance reappears only after a considerable delay. This delay, represented in Fig. 9-6 by the abnormally large reverse current following the switching action, is the result of *minority carrier storage* and *junction capacitance*. The flat portion of the reverse-current response curve in Fig. 9-6 is caused by the minority carrier storage, and the decaying portion by the junction capacitance.

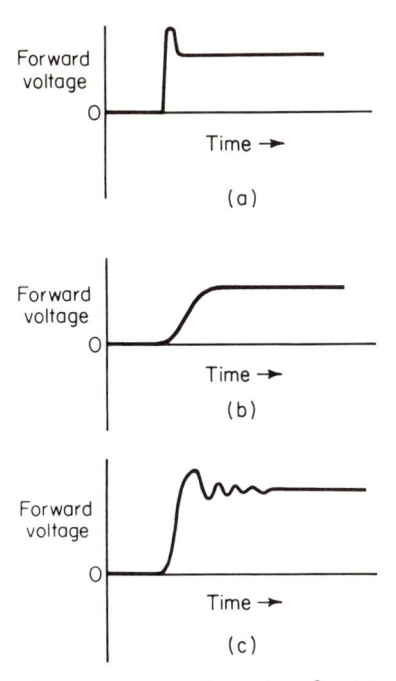

Figure 9-5 Forward transient for (*a*) high forward current, (*b*) low forward current, and (*c*) intermediate forward current.

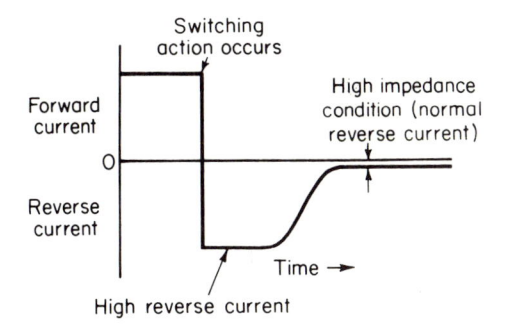

Figure 9-6 An illustration showing that the high impedance condition does not appear instantaneously when a semiconductor diode is reverse biased.

During the period when the junction is forward-biased, an excess of minority carrier builds up on either side of the junction region. When the switching action occurs, the minority carriers are swept from the junction, causing a high reverse current. The magnitude and duration of this transient reverse current is determined primarily by: (1) the magnitude of the initial forward current; (2) the magnitude of the applied reverse voltage; and (3) the series resistance of the external circuit.

When the transient reverse current resulting from minority carrier storage ceases (at the end of the flat portion of the reverse response curve in Fig. 9-6), junction capacitance prevents any immediate change in current to the normal reverse-current value.

A voltage across the junction of two semiconductor materials is always associated with the presence of a space charge on either side of the junction. As voltage increases in the forward direction, the *depletion region* is narrowed, corresponding to the replacement of the charge layers in the transition region. Electrons are built up on the n side, and holes on the p side. Small changes in voltage add a proportionate amount of charge, corresponding to the action of a capacitor with a cross-sectional area and a plate separation equal to that of the junction. A reduction in applied voltage widens the depletion layer. Physically, this corresponds to pulling apart the plates of a capacitor. The junction capacitance can be shown to be a nonlinear function of the junction voltage; it increases with forward bias and decreases with increasing reverse bias.

The junction capacitance, together with the series resistance of the circuit, determines the effective time constant of the reverse voltage across the diode and, therefore, the decay of the reverse current in Fig. 9-6 to the saturation current.

The exact effect of temperature upon recovery time is not well defined, but noticeable increases do occur with increases in temperature. Recovery time is directly proportional to the lifetime of the carriers in the device, and this lifetime increases with temperature. This increase in carrier lifetime is probably one of the largest contributors to increases in recovery time with temperature.

Other factors which may contribute to the temperature dependence of the switching speed of a diode are junction capacitance, which increases with temperature, and back resistance, which decreases with temperature.

For best switching speeds, diodes are operated with the following precautions: (1) the forward current pulse should have as short a duration as possible; (2) the magnitude of the forward current should be as small as circuit application requires; (3) the reverse current pulse should be the maximum attainable in a specific application; (4) the impedance of the external circuit should be held to a minimum; and (5) the environmental temperature in which the device is to operate should be as low as possible.

It should also be noted that, by using breakdown diodes and arranging the circuit to switch at the reverse breakdown point, the action can be made comparable to that of a high-vacuum diode.

9.6 The high-vacuum thermionic triode[2]

The triode as an electronic switch has certain advantages and disadvantages as compared to the diode. Its major advantage is that it provides amplification, and thereby permits regenerative switching in which a slight initiating stimulus provides a switching action of much greater amplitude with power provided by the d-c supply. This principle is used in many of the circuits you will study in later chapters, and a detailed discussion is delayed until that point is reached. Another advantage of the triode switch is that it can be arranged to draw very little current or power from a source of control voltage. The disadvantages include a high tube-voltage drop, and distortion of the output because of the nonlinearity of the tube characteristics.

When a triode is conducting, it can be driven beyond plate-current cutoff simply by applying to the control grid a negative voltage of sufficient amplitude. The triode can then again be made to conduct by permitting the grid voltage to become less negative than the plate-current cutoff value. The inclusion of a control grid thus permits the tube to act as a separately actuated switch.

As long as the grid is negative with respect to the cathode, no grid current is drawn. Under this condition, if the signal applied to the grid is of such amplitude as to cause alternate cutoff and conduction, negligible power is drawn from the actuating source.

If the actuating signal does drive the grid positive with respect to the cathode, grid current results. The grid-to-cathode circuit then acts as a diode. Although the current-voltage characteristic of this *diode* is affected somewhat by the triode plate voltage, the effect is small if the plate voltage is much greater than the grid voltage. Under this condition, grid current is a constant fraction of total space current, which in turn varies directly with grid voltage.

A full discussion of how grid conduction can serve a useful function is left to Chap. 10. It should be noted at this point, however, that grid conduction must be kept to a value that does not cause the permissible dissipation of the grid to be exceeded.

In switching operations, triodes, pentodes, and transistors operate in three distinct regions. When the plate or collector current is cut off, the device is operating in the *cutoff region*. When the device

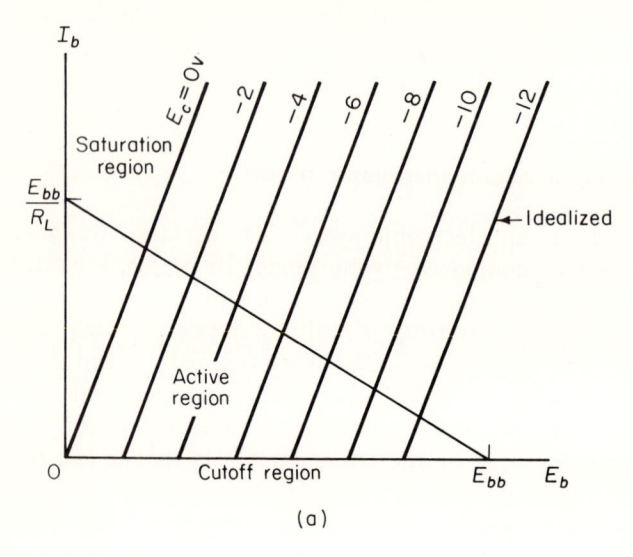

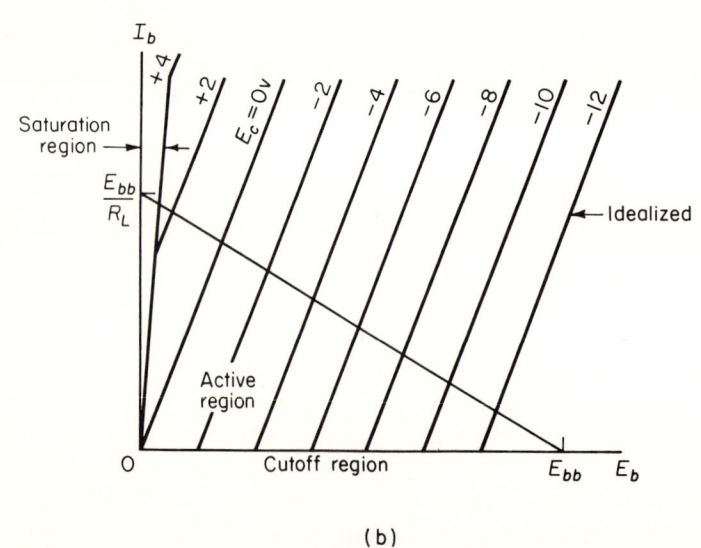

Figure 9-7 Three regions of a triode (a) under grid current conditions with a high resistance in grid and (b) when saturation region boundary results from internal plate current saturation only.

operates over that portion of its characteristic which permits amplification, it is operating in the *active region*. When, for a given voltage and load resistor, the plate or collector current is at its maximum value at the boundary of the active region, the device is operating in the *saturation region*.

Figure 9-7(a) illustrates the three regions of a triode under grid-current conditions with a high series resistance in the grid circuit. The combined effects of the external circuit and tube are such that the saturation region boundary is along the line $E_c = 0$.

Figure 9-7(b) illustrates the three regions of the triode which exist when the saturation region boundary is the result of internal plate current saturation only. Notice that the positive grid-voltage lines all merge into a single line representing the saturation region boundary. Thus, in this region, the tube acts like a diode that is independent of the exact value of E_c.

It is interesting to investigate this phenomenon further. The positive grid permits a heavy flow of cathode-emitted electrons. Many of these electrons pass through the grid but are not attracted to either the plate or the grid. Instead, they form a virtual cathode from which the plate draws electrons as though it were an actual cathode. The presence of the space charge causes the potential of the region to drop to a low value. As long as there is an excess of charge, the exact value of grid voltage is of no consequence. If a load line is drawn on the triode characteristics, the point where it intersects the saturation region boundary is often called the *bottomed region*, because the plate voltage is at its minimum value at this point.

Factors acting to reduce the switching speed of a triode include the interelectrode capacitances of the tube, the capacitance of the tube socket, and the stray capacitance to ground of the circuit wiring.

The effects of capacitance are illustrated in Fig. 9-8. When e_i swings negative at time t_1 and drives the grid beyond cutoff, the immediate effect upon e_0 is to drive the plate also slightly negative and to produce a small negative-going spike in e_0. This results from feed-through from input to output by way of the grid-to-plate interelectrode capacitance C_{gp}, which, together with C_{pk} and C_w, forms a capacitive voltage divider. The small negative-going spike in e_0 is followed by an exponential rise to the level of supply voltage E_{bb}. The R-C time constant of the exponential rise is equal to $R_L(C_{gp} + C_{pk} + C_w)$; the time constant is an indication of the relative speed of the circuit.

A reduction in interelectrode capacitance by careful selection of the tube used in a given application, good wiring techniques, and

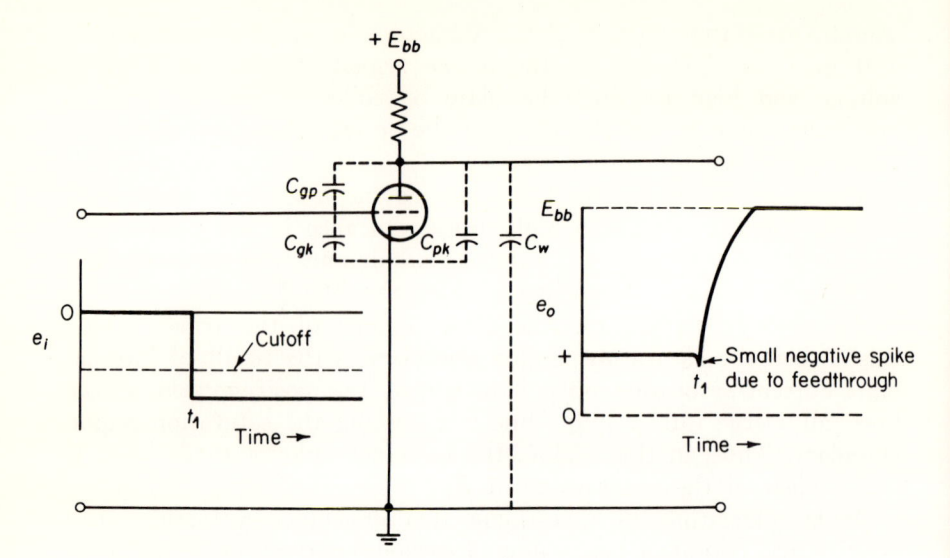

Figure 9-8 The effects of capacitance on the switching speed of a triode.

consideration of the values of R_L and E_{bb}, all lead to circuit improvement. An increase in switching speed also results when the tube is driven from the *off* state to the *on* state rather than from *on* to *off*, since the tube current then contributes to faster charging or discharging of the capacitance.

9.7 The junction transistor[3,4,5]

Ideal collector characteristics for a common-emitter amplifier are shown in Fig. 9-9. The cutoff, active, and saturation regions are indicated, with the cutoff region intentionally exaggerated to distinguish it from the V_{cc} axis of the graph.

In the cutoff region, the base-to-emitter voltage, V_{BE}, is negative by a few hundredths of a volt, and both the emitter and collector are reverse biased. When $I_E = 0$, there is still a small collector current (several microamperes), which is I_{CO}.

In the active region, collector current is controlled by the base current, and the relationship is expressed by the familiar parameter β. This parameter is a measure of the vertical separation of a family of collector-current curves. The slope of these curves is the inverse of the dynamic collector resistance r_c.

The saturation region boundary occurs where the base-current curves merge. Both junctions are forward biased in the saturation

Sec. 9.7   The junction transistor

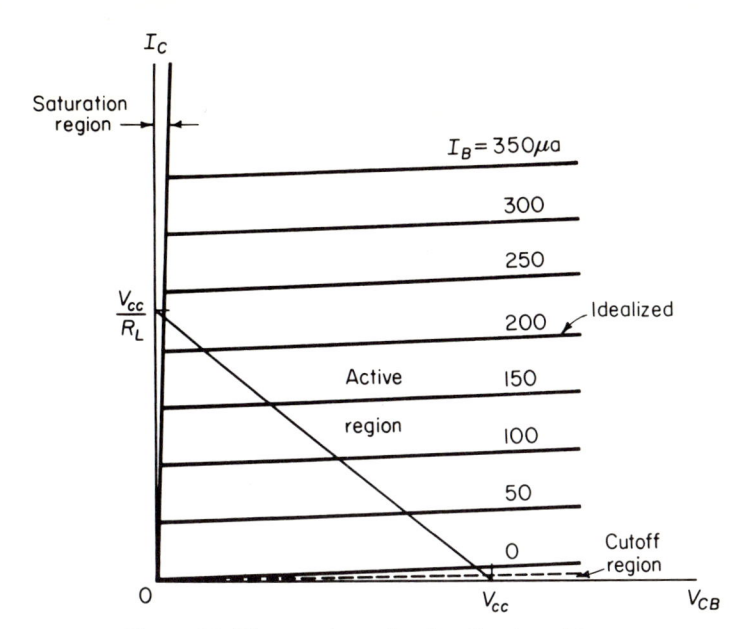

Figure 9-9 Three regions of a junction transistor.

region, and excess electrons are, therefore, emitted into the base. The collector collects these electrons in proportion to its own voltage rather than to the exact condition at the base terminal.

The base-emitter junction of the common-emitter amplifier acts like a forward-biased diode influenced, to some extent, by collector voltage. In switching applications, if V_{CB} is greater than one-tenth volt, the effect of collector voltage is negligible. As in a junction diode with reverse voltage applied, there is a small and virtually constant current at the base input terminal for negative voltages. Thus, the dynamic resistance is essentially infinite.

As in other electronic switching devices, the transistor requires a finite time to switch from *off* to *on*, or *vice versa*. As in semiconductor diodes, switching delay is due primarily to minority carrier storage in the transistor base region.

To illustrate transistor transient response, assume that the rectangular waveform e_i, of Fig. 9-10(b) is applied to the input of a common-emitter amplifier (a), and that at time t_1, voltage e_i rises to a positive value. The resulting base and collector currents are shown in Figs. 9-10(c) and (d), respectively. The collector current rise time, t_r, varies inversely with the amount of base current drive. If the base current is sufficient to drive the transistor into saturation, the final value of i_c is $V_{cc}R_L$.

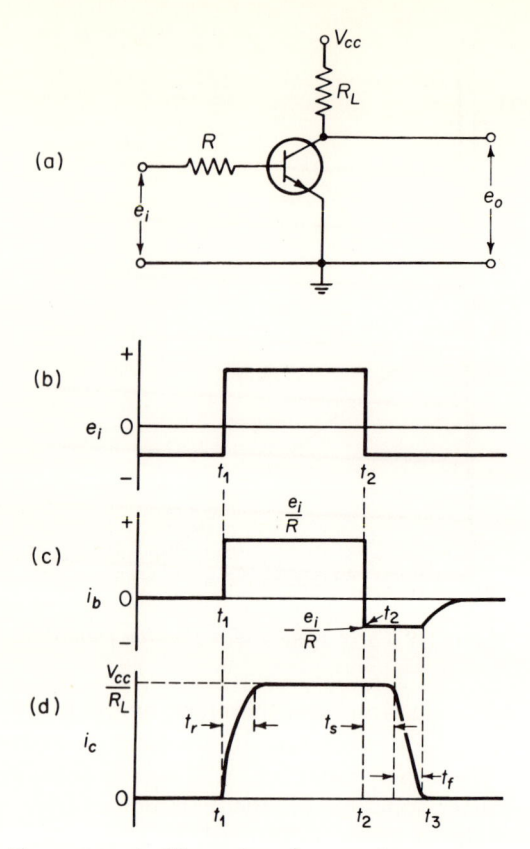

Figure 9-10 An illustration of a transistor response.

At time t_2, the polarity of e_i reverses, but there is an appreciable delay before any change occurs either in i_b or i_c. This is the *storage delay time*, t_s, required to clear the base of excess charge accumulated during operation in the saturation region. During this time, the low resistance of the base input results in a large base current and continued high collector current. Storage time is reduced by keeping the transistor from becoming heavily saturated and by making the reverse current pulse the maximum attainable in a specific application.

At time t_3 in Fig. 9-10(d), the charge concentration in the transistor has adjusted itself to a new value, and i_c returns to its original *off* level in fall time t_f, which varies inversely with the value of the turnoff current $-e_i/R$ in Fig 9-10(c).

If the load resistance is low, the collector capacitance of a transistor has little effect on the transient response, but in general tends to increase t_r and t_f.

Transistor storage time can be eliminated by preventing saturation during the *on* period. In practice, this is accomplished by using a breakdown diode having negligible storage. Numerous examples of such circuits are described later.

Better switching time is also achieved by using transistors having thinner base regions. The reduced base volume reduces charge build-up and storage. Numerous types of transistors have been developed for this purpose and are described later. They are usually referred to collectively as *high-frequency transistors*. Thus, the "frequency cutoff" of a transistor serves as a measure of its base thickness and its suitability to switching applications.

9.8 Pentodes

An idealized set of curves for a pentode are shown in Fig. 9-11, and the three operating regions are again identified. The physical construction of this type of tube results in a high dynamic plate resistance r_p, so that the curves are nearly horizontal.

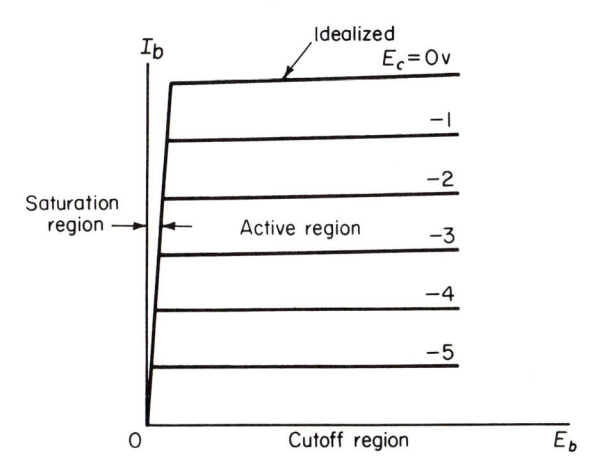

Figure 9-11 Three operating regions of a pentode.

The cutoff region is determined by the combination of control-grid and screen voltages which cut off space current.

The saturation region boundary is the line at which the curves merge. At such low plate voltages, the total space current does not flow to the plate. This results in the formation of a virtual cathode from which the plate draws in proportion to its voltage, thus acting like a simple diode. Since the screen current rises during plate current saturation, care must be taken not to exceed the allowable dissipation of the screen. The control grid draws no current since it is still negative during operation in the saturation region.

In the pentode, switching speed is reduced by the grid-to-cathode and plate-to-cathode interelectrode capacitances. The effects of other interelectrode capacitances are negligible because of the inherent shielding between control grid and plate of such devices. In the output waveform, this eliminates the spike produced by feed-through in the triode.

REFERENCES

1. "Diode Recovery Time," *Application Notes*, Staff of Texas Instruments Incorporated, Components Division, Dallas, Texas, Feb. 1961.
2. Pettit, Joseph M., *Electronic Switching, Timing, and Pulse Circuits*. New York: McGraw-Hill Book Co., Inc., 1959.
3. "A Discussion of Storage Time," *Application Notes*, Staff of Texas Instruments Incorporated, Components Division, Dallas, Texas, Oct. 1960.
4. Moll. J. L., "Large-Signal Behavior of Junction Transistors," *Proc. IRE*, Dec. 1954, p. 1773.
5. Shive, John M., *The Properties, Physics, and Design of Semiconductor Devices*. Princeton, N. J.: D. Van Nostrand Co., Inc., 1959.

EXERCISES

9-1 How do the characteristics of actual electronic switches differ from the ideal devices?

9-2 What is the effect of interelectrode capacitance on the switching response of a high-vacuum thermionic diode?

9-3 Name the advantages and disadvantages of the semiconductor diode compared to a thermionic diode when used as an electronic switch.

9-4 What process is responsible for the rise in reverse current of a semiconductor diode above a critical value of reverse voltage?

9-5 When a semiconductor junction diode is switched from *on* to *off*, does the reverse current immediately drop to the value of I_{CO}? If not, give a detailed explanation.

9-6 Explain how temperature affects diode recovery time.

9-7 For best switching speeds, how should a semiconductor diode be operated?

9-8 Is the saturation boundary of a triode the same both with and without a high series resistance in the grid circuit? Explain.

9-9 In switching circuits, what is meant by the term "bottomed region"?

Exercises 175

9-10 A triode switch is driven from *on* to *off* by a rectangular waveform input voltage. Draw the output waveform, and explain why it has the form shown.

9-11 Draw and explain the waveform resulting from the application of a rectangular pulse to the input terminals of a junction transistor.

9-12 How can transistor storage time be eliminated in switching circuits?

10

NONLINEAR WAVESHAPING

In linear waveshaping networks, a sine wave is passed without distortion. This is not the case when nonlinear components such as thermionic diodes, crystal diodes, or triodes are used. Such devices are nonlinear because their volt-ampere characteristics are nonlinear (not a straight line). Nonlinear characteristics are used in the development of additional wave forms that are useful in pulse applications.

In this chapter, a type of nonlinear waveshaping circuit called a *clipper* is studied. (Such a circuit is also referred to as a *limiter, voltage selector* or *amplitude selector*.) A clipper is defined as a circuit which limits the amplitude of a voltage by removal of a part of the waveform. Either the positive or negative amplitude, or both, may be limited. In the clipping circuits to be described, a sinusoidal waveform is used, but the operation is similar for any *a-c* input waveform. It is also assumed that the amplitude of the input waveform is sufficient to prevent operation on the knee of the characteristic curve.

10.1 Series-diode clipper

The circuit of Fig. 10-1 is used to remove the negative half of the input waveform. Either a thermionic or semiconductor diode, indicated by dashed lines, can be used. The input voltage is applied

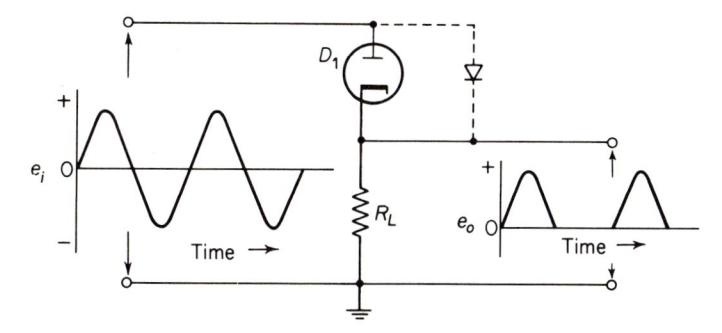

Figure 10-1 A series diode clipper that removes negative half-cycle of e_i. A semiconductor may be used in place of the thermionic diode. This is indicated by dashed lines to the right of D_1.

to the diode in series with resistor R. During the positive half-cycle of voltage e_i, the diode plate is positive with respect to its cathode, and conduction occurs. Current passes through the series network, consisting of the resistor and diode, and develops a voltage across R_L. The input voltage divides between R_L and D_1 in the ratio of their resistance. During conduction, the resistance of diode D_1 is very low compared to the resistance of R_L. The output voltage, therefore, has essentially the same shape and amplitude as the positive half-cycle of e_i.

When the negative half-cycle of e_i is applied to the input terminals, it makes the plate of D_1 negative with respect to its cathode. Under these conditions, conduction cannot occur and the diode acts as an infinite resistance. Because there is no current through R_L, no output voltage is developed, and the negative half-cycle of input voltage is removed. During conduction, therefore, the applied voltage appears across R_L, whereas during non-conduction it switches and appears across the diode.

It should be noted that the clipping action described above assumes a perfect diode; i.e., one in which no reverse current occurs when the plate is negative with respect to its cathode. As noted in Chap. 9, a small reverse current does occur through the inverse diode resistance and a small voltage is, therefore, produced across R_L. The amplitude of this voltage, however, is small enough to be insignificant, and e_o is considered to be zero.

If the diode is reversed, as shown in Fig. 10-2, the positive half-cycle of e_i is removed instead of the negative half-cycle. Conduction occurs during the negative half-cycle because the cathode of D_1 is negative with respect to the anode, and an output voltage is produced across R_L. During the positive half-cycle, however, the cathode is positive with respect to the anode, and the diode acts as an infinite impedance. As before, no output is produced under these conditions.

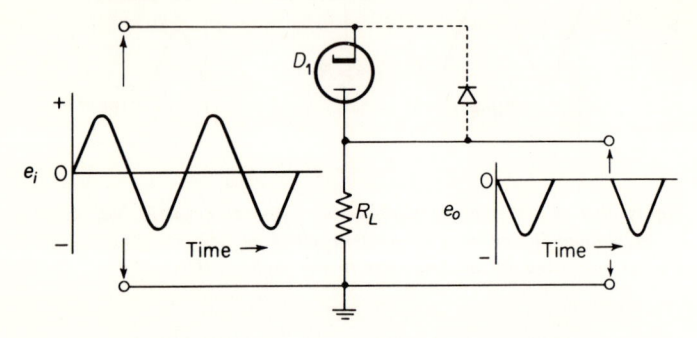

Figure 10-2 A series diode clipper that removes positive half-cycle of e_i.

10.2 Parallel-diode clipper

Clipping can also be accomplished by connecting the output circuit in parallel with the diode as shown in Fig. 10-3. Resistor R is again selected so that its value is large compared to the resistance of the diode during periods of conduction. In addition, the impedance of the load connected to the output terminals must be large in comparison with the impedance of the diode and resistor R.

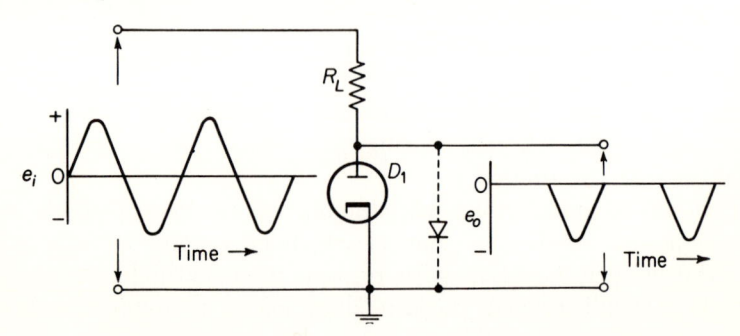

Figure 10-3 A parallel diode clipper that removes positive half-cycle of e_i.

Sec. 10.3   Biased-diode clipper

The positive half-cycle of e_i makes the diode plate positive with respect to its cathode. The resulting conduction causes a voltage of essentially the same amplitude as e_i to appear across R. The small forward resistance of the diode produces a voltage of negligible amplitude across the output terminals.

When the negative half-cycle of e_i occurs, the diode cannot conduct. Since the impedance of the diode is then large compared to the resistance of R, practically the entire input voltage appears across the diode.

By reversing the diode, as shown in Fig. 10-4, the output waveform is reversed with only the positive half-cycles appearing.

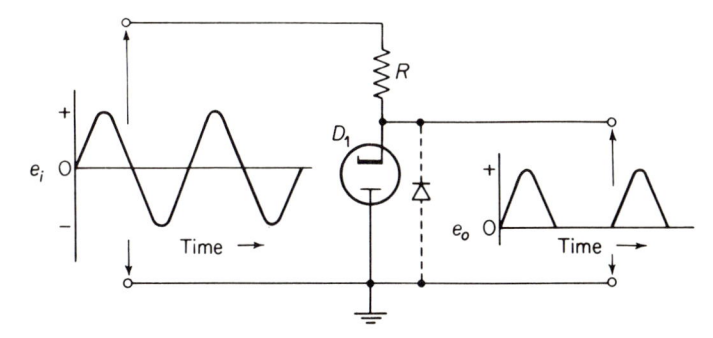

Figure 10-4 *A parallel diode clipper that removes positive half-cycle of e_i.*

10.3 Biased-diode clipper

In the clipping circuits described thus far, the input voltage is limited with respect to the zero level. It is often necessary to clip the input waveform at a certain positive or negative amplitude. One way in which this can be accomplished is by using the circuit shown in Fig. 10-5. The bias battery inserted in series with the cathode makes it 3 v positive with respect to ground.

When voltage e_i is applied to the input terminals, the diode does not conduct immediately as before. Instead, conduction on the positive half-cycle is delayed until the positive amplitude of e_i exceeds 3 v. As long as e_i is less than this value, the plate of the diode is negative with respect to its cathode. The resistance of D_1 is very high as compared to R during periods of nonconduction, and the output voltage waveform is essentially the same as e_i.

When the positive amplitude of e_i exceeds $+3$, the diode conducts and becomes a very low resistance. Any increase of e_i in excess of

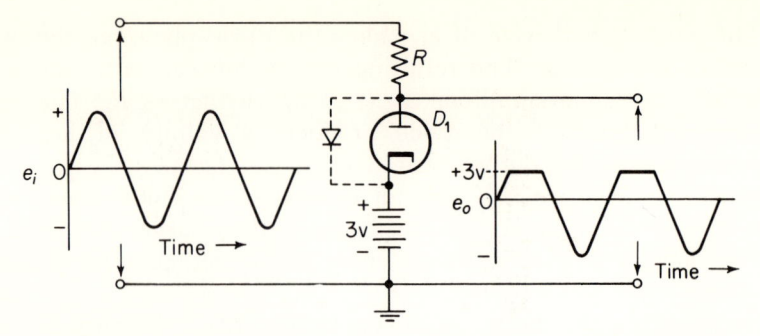

Figure 10-5 A parallel clipper with a biased diode to limit amplitude of positive portion of e_0 to the value of the battery.

$+3$ v is developed across R. Thus, the output voltage during the positive half-cycle of e_i is limited in amplitude to 3 v, the voltage of the bias battery.

On the negative half-cycle of e_i, the diode does not conduct, no voltage is developed across R, and the output waveform follows the input. The final output voltage, as shown in Fig. 10-5, consists of a flat-topped positive half-cycle of 3 v maximum amplitude and a negative half-cycle of the same shape and amplitude as the input. If the voltage of the bias battery is increased, the width of the flat-topped portion of the waveform is decreased, but the amplitude of the positive half is increased to the amplitude of the battery.

To limit the amplitude of the output voltage during the negative half-cycle, the diode and battery connections may be reversed as shown in Fig. 10-6. On the positive half-cycle of e_i, the diode does not conduct, and the output waveform is the same as the input. On the negative half-cycle, the output waveform is the same as the input until e_i exceeds -3 v. When the input goes more negative than -3 v, the diode conducts and clipping occurs.

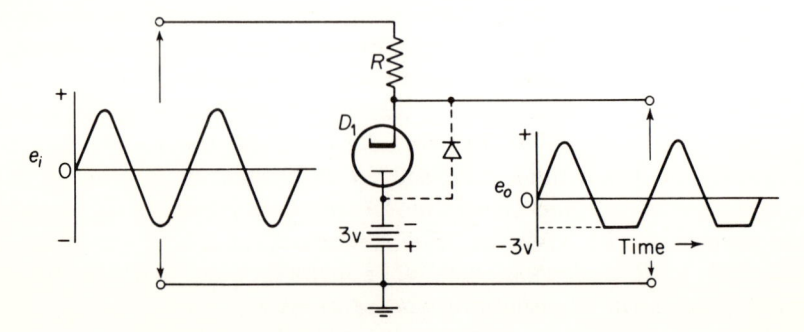

Figure 10-6 Negative clipping using a biased diode.

Sec. 10.3 Biased-diode clipper   181

Both extremes of e_i can be clipped simultaneously by using two diodes arranged as shown in Fig. 10-7. Diode D_1 and resistor R limit the positive amplitude of e_o to the value of bias battery E_1. Diode D_2 and resistor R limit the negative amplitude of e_o to the value of bias battery E_2. This circuit is often used to produce a substantially square-wave output waveform from a sine-wave input.

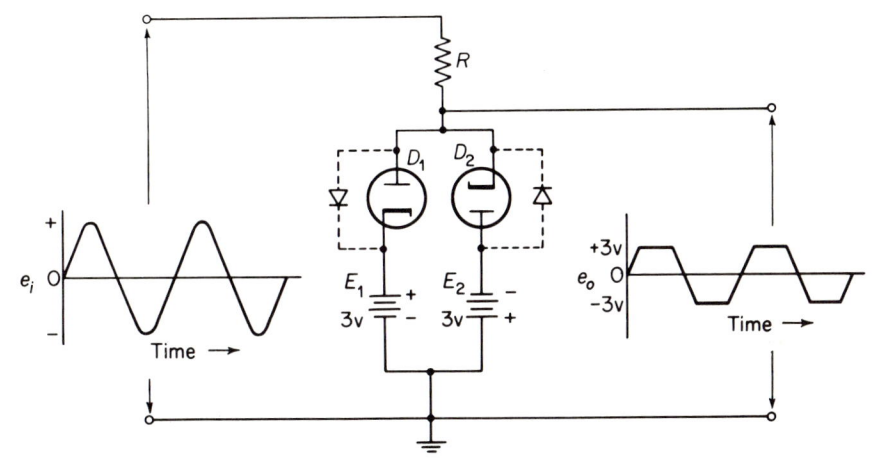

Figure 10-7 Clipping both extremes of e_i by using two diodes.

Another variation of the parallel-diode clipper is shown in Fig. 10-8. This circuit has the advantage of passing either the positive or the negative peak portions of the input waveform and rejecting the portions of lower amplitude. It is very similar to the circuit of Fig. 10-5 except that the bias battery is reversed. The cathode is

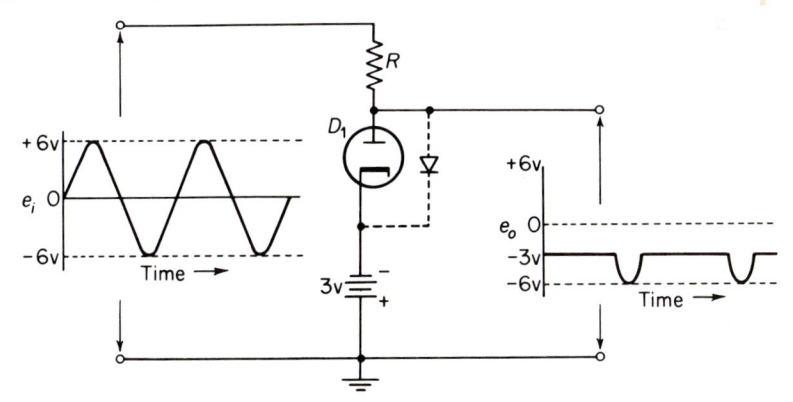

Figure 10-8 A modification of the parallel diode clipper that permits only a portion of the negative alternation of e_i to appear in the output.

182   Nonlinear waveshaping   Chap. 10

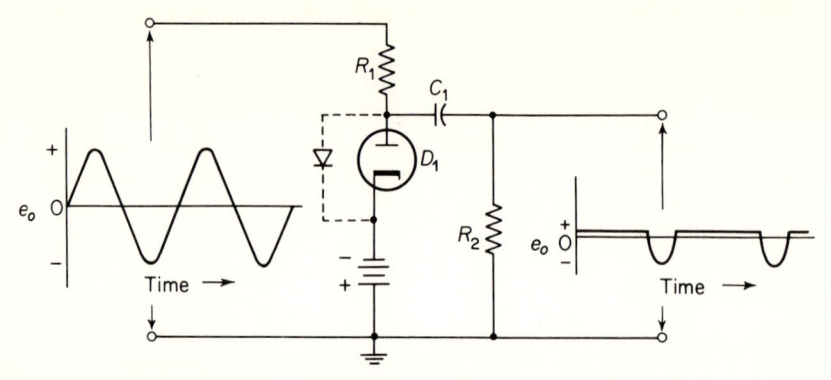

Figure 10-9 The arrangement for obtaining an output that is essentially a series of negative pulses.

now negative with respect to the anode during quiescent operation. The diode, therefore, conducts and develops a voltage across resistor R that is almost equal to the bias voltage since the impedance of the conducting diode is very small compared to the resistance of R. Thus, an output voltage of -3 v appears between the anode plate terminal and ground. When the instantaneous positive-going input voltage is applied, conduction increases, causing the voltage across R to increase correspondingly. The output voltage remains at -3 v during the entire positive half-cycle of e_i and during that portion of the negative half-cycle when e_i is less negative than -3 v. When e_i exceeds -3 v, the diode no longer conducts and the voltage drop across the resistor disappears. The remainder of the negative peak of e_i in excess of -3 v is developed across the diode and appears at the output terminals.

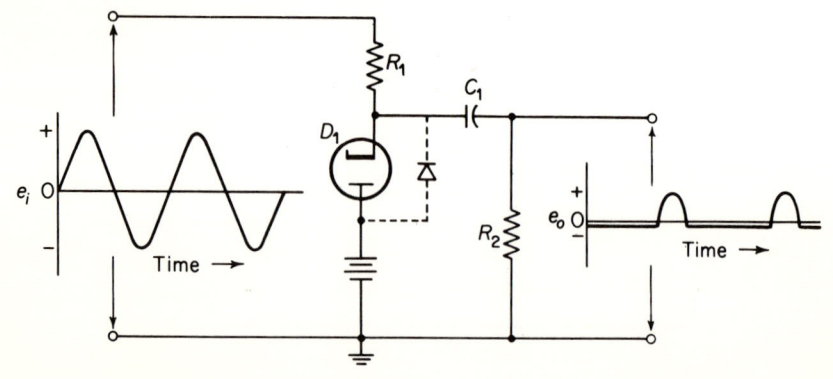

Figure 10-10 The output of this arrangement is essentially a series of positive pulses.

Sec. 10.4   Grid limiting

An output voltage having a larger *a-c* component, causing the output to alternate positively and negatively, can be obtained from this circuit by using an additional resistor and capacitor as shown in Fig. 10-9. The positive portion of the output waveform is of very low amplitude so that the output is essentially a series of negative pulses. By reversing both the diode and the bias battery as shown in Fig. 10-10, a series of positive pulses is obtained in the output.

10.4 Grid limiting

The grid-cathode circuit of a triode, tetrode, or pentode may be used as a limiter circuit in exactly the same way as the plate-cathode circuit of the diode limiter illustrated in Figs. 10-4 and 10-5. This is accomplished by connecting, in series with the grid, a resistor R that is very large compared to the grid-to-cathode resistance, R_{gk}, of the tube during periods of grid current, i.e., when the grid is positive with respect to the cathode. (See Fig. 10-11.)

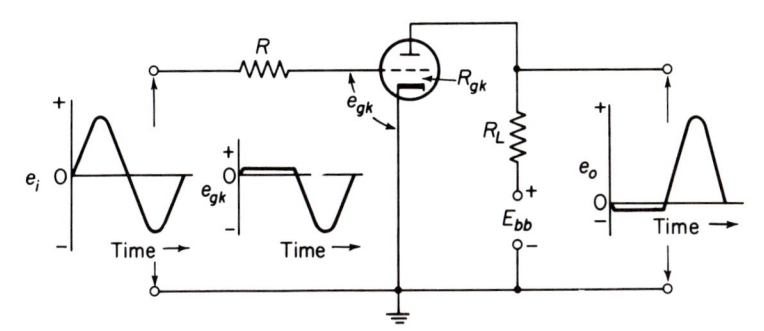

Figure 10-11 A grid limiter.

To illustrate the circuit action, let us assume that $R = 1$ megohm and that $R_{gk} = 1000$ ohms during periods of grid conduction and is infinite during periods of non-conduction.

The grid-limiter circuit is held normally at zero. During the positive alternation of e_i, the grid draws instantaneous current i. This current develops voltage drops iR and iR_{gk} across resistors R and R_{gk}, the polarity of which opposes e_i. Since $e = (iR + iR_{gk})$, and since during grid conduction R is very large compared to R_{gk}, voltage e_{gk}, appearing across resistor R_{gk}, assumes a low constant value. The iR drop may be considered as an automatic bias developed during the positive alternation of e_i.

During the negative alternation of e_i, no grid current is drawn, and $e_{gk} \cong e_i$ since $R_{gk} \gg R$. The output voltage e_o appearing across load resistor R_L is, of course, an amplified and inverted reproduction of voltage e_{gk}.

In Fig. 10-12(a), the cathode is grounded and the grid is biased negative with respect to the cathode by battery E_{cc}. In this arrangement, there is no grid current until the positive alternation e_i exceeds E_{cc} and effectively removes the bias. As long as the grid is negative with respect to the cathode, $e_{gk} \cong e_i$. When the amplitude of the positive alternation of e_i exceeds E_{cc}, the grid draws current, and limiting action again occurs as the result of the iR drop across resistor R. During the negative alternation of e_i, there is no grid current drawn and $e_{gk} \cong e_i$.

In Fig. 10-12(b), the same limiting action is achieved, but the tube employs *self-bias* provided by the plate current through R_k, instead of fixed battery bias. Capacitor C_k is assumed to be large enough to prevent degeneration.

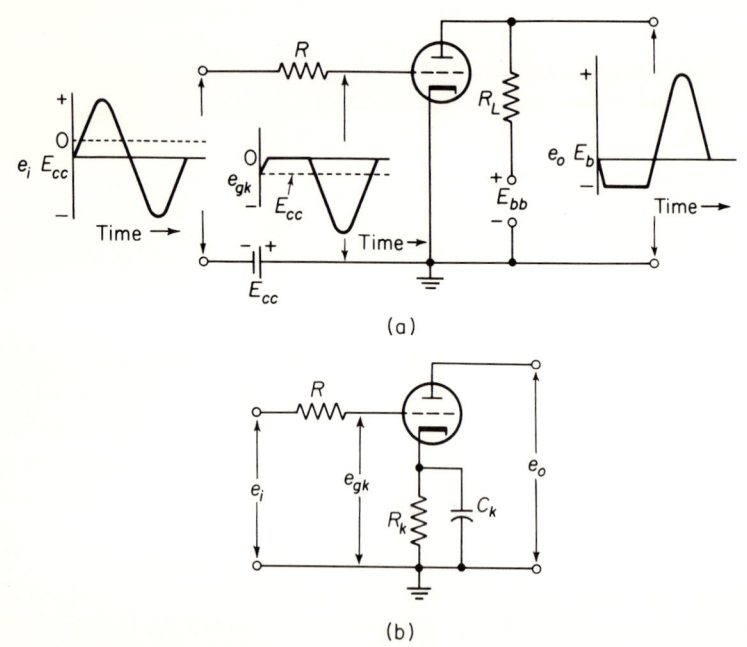

Figure 10-12 Biased grid limiters. Waveforms for (b) are identical to those shown for (a).

10.5 Saturation limiting

In the saturation or grid limiter of Fig. 10-11, the grid never rises to an appreciable positive voltage, and maximum plate current is determined by supply voltage E_{bb} and the resistance of the plate circuit at zero bias. Thus, the minimum value of plate voltage is determined by the grid-circuit limiting action.

If a low impedance high-power source provides the input signal, the grid-limiting resistor can be omitted and limiting can be achieved in the plate circuit. This condition is caused by plate-current saturation and is, therefore, called *saturation* limiting. Plate-current saturation is not to be confused with emission saturation. Tubes using oxide-coated cathodes have no definite saturation value of emission current.

By using a large value of plate load resistance and a low value of plate supply voltage, saturation limiting may be produced by a relatively low amplitude of positive grid voltage.

Fig. 10-13(a) shows the circuit arrangement of a saturation limiter, and Fig. 10-13(b) illustrates the effect of saturation limiting on the plate voltage.

The grid is normally at zero bias. The negative alternation of input signal e_i is not of sufficient amplitude to drive the tube to cutoff, and an amplified and inverted version of the alternation appears in the output, as shown in Fig. 10-13(b). When the positive alternation is applied, however, the tube is driven to saturation. Under any circumstances, the maximum plate current, i_p, cannot exceed the value of E_{bb}/R_L no matter how high the amplitude of the positive grid signal. In practice it is less than E_{bb}/R_L since the plate-to-cathode resistance at saturation does not decrease to absolute zero and this resistance is in series with R_L. The maximum plate current defines the lowest value to which voltage can fall, as shown in Fig. 10-13(b). During the remaining portion of the input cycle, the grid controls the plate current, and this in turn determines the shape of the plate-voltage waveform.

The results of saturation limiting are similar to those of grid limiting in that the negative portion of the plate-voltage waveform is affected. Saturation limiting has the advantage of producing an output waveform of greater amplitude, but it has the disadvantage of requiring considerably more grid driving power.

186 Nonlinear waveshaping Chap. 10

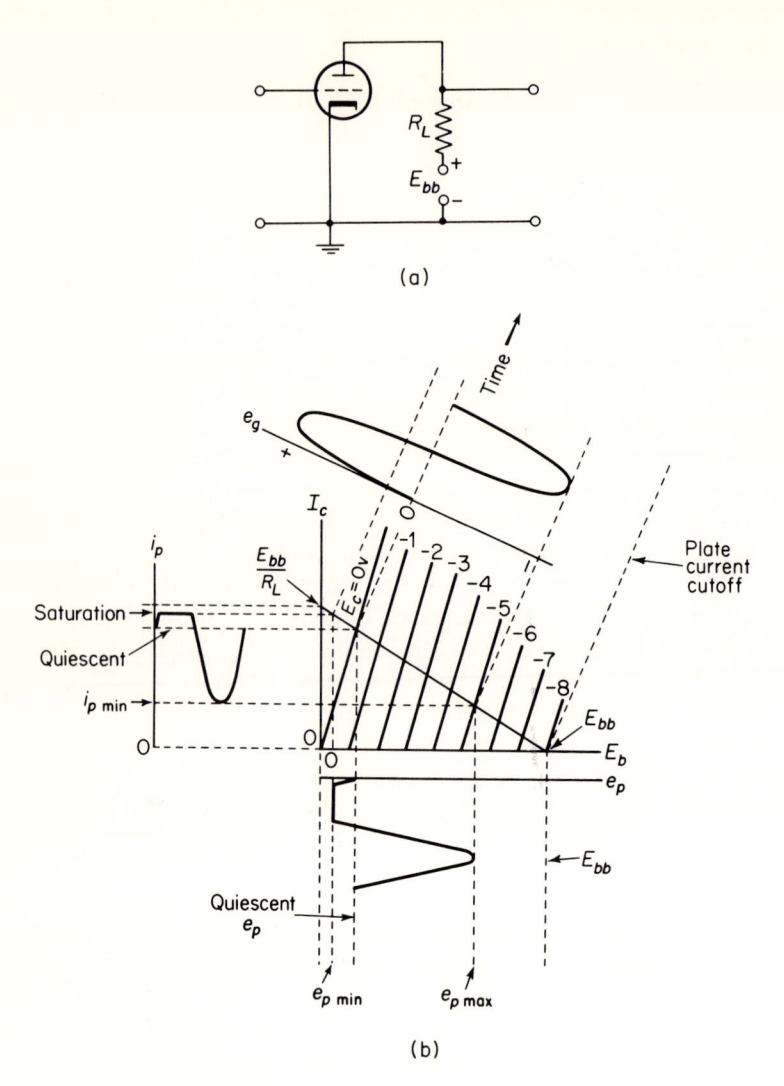

Figure 10-13 (a) *Saturating limiter and* (b) *waveforms illustrating action.*

10.6 Cutoff limiting

When the grid of a vacuum tube is driven to cutoff, the plate current drops to zero and, because there is no voltage drop across R_L, the plate is maintained at the full value of E_{bb}. Thus, by driving

Sec. 10.6   Cutoff limiting   187

the grid beyond cutoff, a type of limiting is achieved in which the positive extreme of the plate voltage waveform is flattened.

In Fig. 10-14(a), an amplifier is biased by the voltage drop across resistor R_k. The circuit is adjusted so that saturation does not occur during the positive alternation of e_i, and an amplified and inverted version of this alternation appears across the plate load resistor. On the negative alternation, the grid is driven beyond cutoff. When cutoff occurs, the plate voltage rises to the level of E_{bb} and remains there until conduction again occurs.

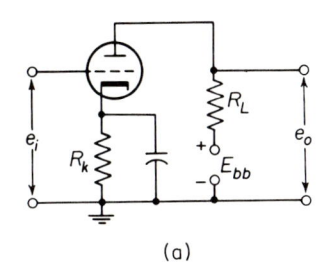

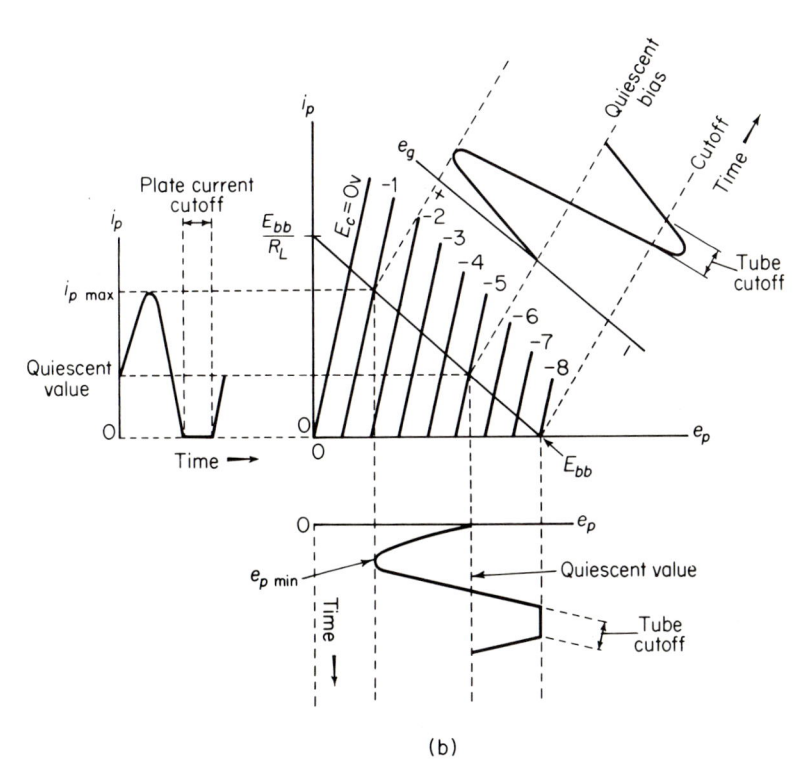

Figure 10-14 (a) Cutoff limiter and (b) waveforms.

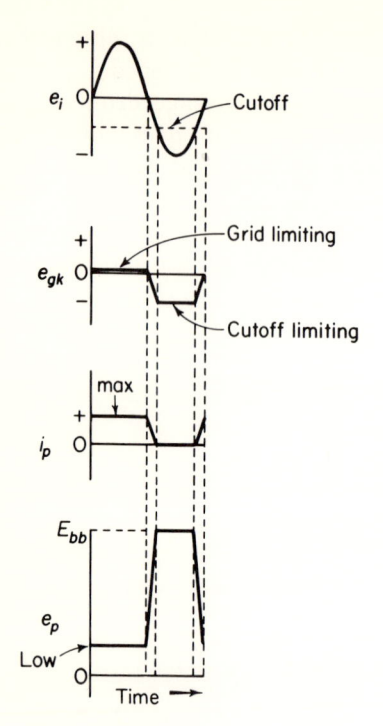

Figure 10-15 Producing a square wave by combination grid and cutoff limiting.

A combination of grid and cutoff limiting may also be used in an amplifier circuit to clip both extremes and thus produce a square wave from a sine wave, as shown in Fig. 10-15. The amplitude of the input voltage is sufficiently high to hold the grid beyond cutoff for most of the negative alternation. The high-value grid resistor limits the grid voltage essentially to zero during the positive alternation. Overdriven amplifiers and cathode-coupled clippers, described below, illustrate the production of square waves by grid and cutoff limiting.

10.7 Overdriven amplifiers

An overdriven amplifier, Fig. 10-16(a), uses both saturation and cutoff limiting to produce a rectangular wave from a sine wave, as illustrated in Fig. 10-16(b). The driving circuit must have a relatively low impedance and must be capable of delivering power during periods of grid conduction.

10.8 Cathode-coupled clipper

In the cathode-coupled clipper, Fig. 10-17, V_1 is a cathode-follower with its output circuit directly connected to the cathode of V_2, a grounded-grid amplifier. Both tubes are biased by the voltage drop across resistor R_k which carries the *d-c* plate current of both tubes. The bias developed across R_k tends to be constant because the plate current of one tube increases as that of the other tube decreases.

When the positive peak of the input-signal voltage, e_i, exceeds the *d-c* voltage drop across R_k, clipping occurs in V_2. When the negative peak of e_i exceeds the voltage drop across R_k, V_1 is driven to plate-current cutoff. Thus, clipping occurs on both the positive and negative peaks of input signal voltage. The clipping is sym-

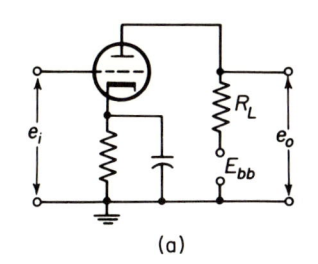

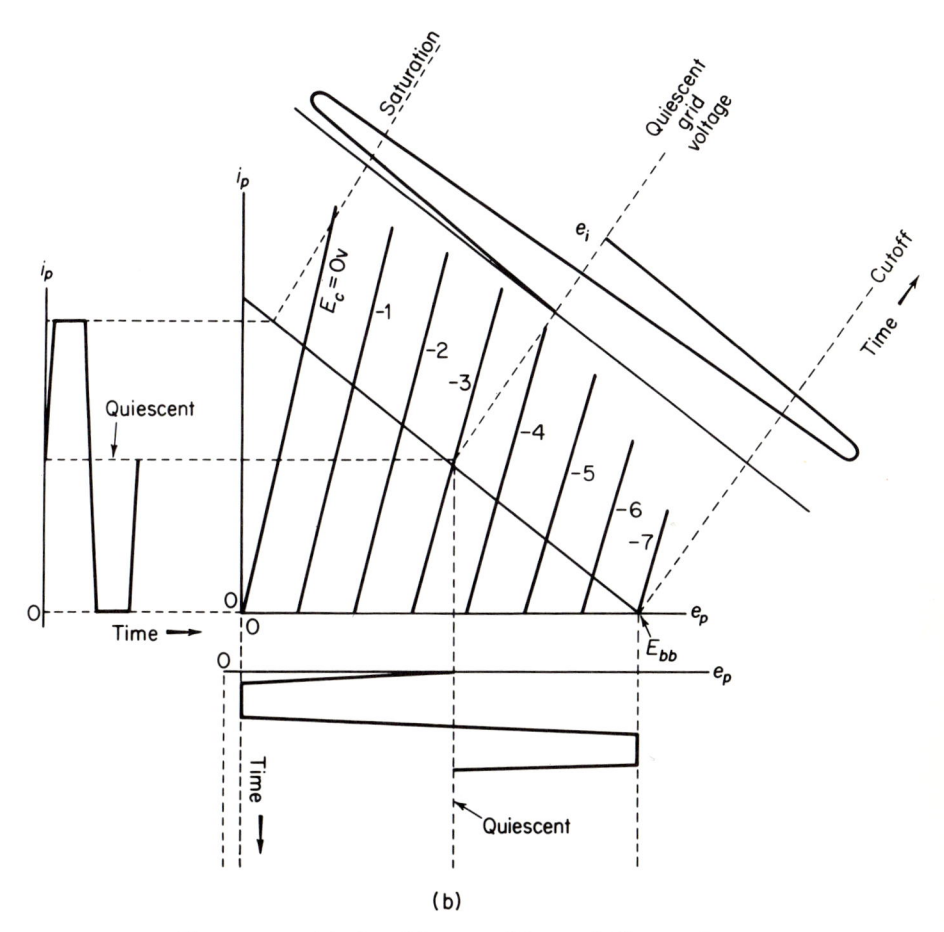

Figure 10-16 (a) Overdriven amplifier and (b) waveforms.

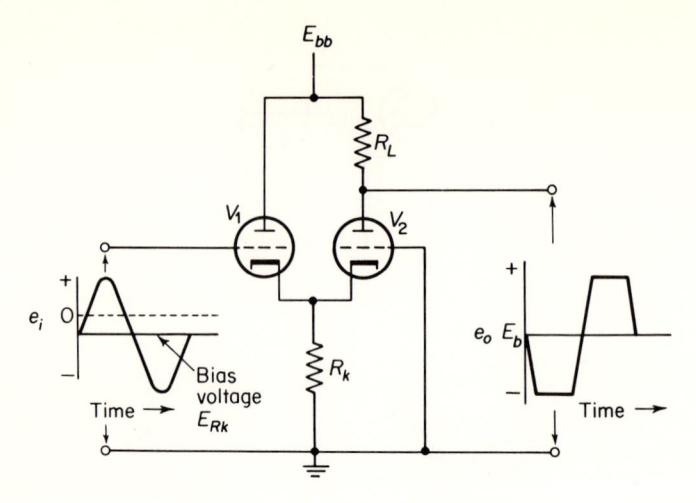

Figure 10-17 Cathode-coupled clipper.

metrical, provided that the *d-c* voltage drop across R_L is small enough so that the operating conditions of the two tubes are substantially the same. For signal voltages below the clipping level, the circuit operates as a normal amplifier with low distortion.

10.9 Transistor limiters

From the information on transistor switches presented in Chap. 9, it should be readily apparent that, by driving such devices between

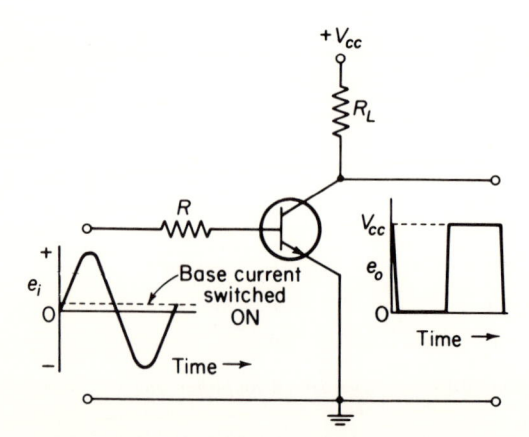

Figure 10-18 Transistor clipper that produces a square wave output from a sine wave input.

Sec. 10.10   Limiter applications   191

saturation and cutoff, the same clipping actions can be achieved as were described for vacuum tubes.

A typical transistor clipper circuit which produces a square-wave output from a sine-wave input is shown in Fig. 10-18. The accompanying waveforms should help to clarify the operation of the circuit. When the base current is switched on by e_i, base-emitter rectification analogous to grid rectification in tube circuits occurs.

Clipping by collector saturation has the disadvantage that minority carrier storage slows transistor response when the magnitude of the base voltage falls below the value at which saturation occurs and reduces the sharpness of the trailing edge of e_o.

10.10 Limiter applications

A typical *FM* limiter circuit is shown in Fig. 10-19. A low value of plate voltage (about 50 v) is used so that the limiter tube reaches saturation with a relatively low applied grid voltage. Even when large input signals are applied, the output remains essentially constant.

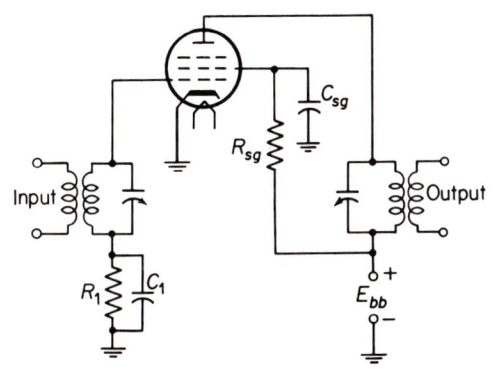

Figure 10-19 FM limiter.

An increase in input signal voltage causes increased bias to be developed across R_1. Bias voltage is maintained by capacitor C_1 until the input level changes markedly. Thus, amplitude modulation is removed from the signal regardless of its source. No interference with the frequency modulation is caused because limiter action has no effect on the instantaneous frequency of the wave.

Limiters are also used in radar pulse-forming networks called *squarers* and *peakers*. In these applications, an input sine wave is first changed to a square wave, then to a peaked wave, and finally

192   Nonlinear waveshaping   Chap. 10

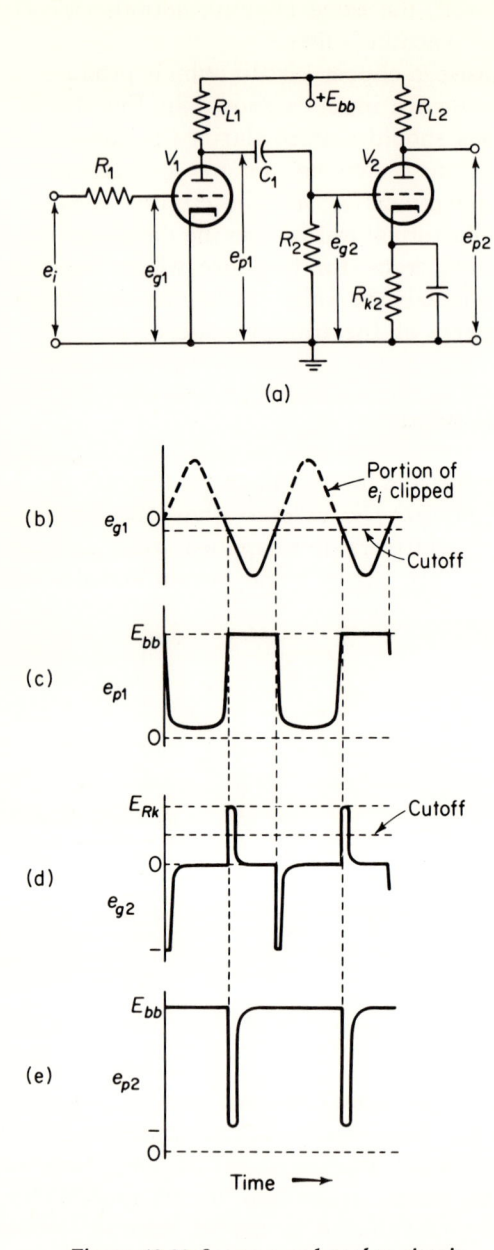

Figure 10-20 Squarer and peaker circuit.

Sec. 10.10   Limiter applications   193

to a pulse. The circuit arrangement is shown in Fig. 10-20(a) with the waveforms in (b), (c), (d), and (e) to help clarify the operation. Under quiescent conditions, tube V_1 has zero bias, and V_2 is biased beyond cutoff by the voltage drop across resistor R_{k2}.

During the positive alternation of the input sine wave, grid limiter V_1 draws grid current, and essentially the entire applied voltage is dropped across resistor R_1, since $R_1 \gg R_{gk1}$. The grid voltage waveform e_{g1} of tube V_1 is shown in Fig. 10-20(b), and the plate voltage waveform, e_{p1}, in (c). The plate voltage during this period is quite low since the plate current is not limited by the grid.

On the negative alternation of the input voltage, the grid draws no current and there is no voltage drop across R_1. Thus, the entire input voltage appears at the grid. Assume that the amplitude of the input signal is sufficient to cause plate current cutoff on the negative alternation. When this point is reached, the plate current of V_1 drops to zero and no voltage drop occurs across R_{L1}. The plate voltage, therefore, rises to the value of the supply voltage, E_{bb}, and remains at that level for the duration of the negative alternation.

The process described above is repeated for each cycle of input voltage. Thus, a sine-wave input appears as a square-wave output at the plate of V_1.

This square wave is then applied to an *R-C* differentiating network consisting of capacitor C_1 and resistor R_2.

Under quiescent conditions, capacitor C_1 is charged to the voltage level appearing at the plate of V_1. When the negative-going square wave is produced as described above, capacitor C_1 must recharge to the new lower value of e_{p1}. The charge on the capacitor cannot change instantaneously, however, and the full negative-going voltage e_{p1} initially appears across R_2, as shown in Fig. 10-20(d). Capacitor C_1 then discharges exponentially through R_2 and V_1 at the rate determined by the *R-C* time constant. Since this is a differentiating circuit, the time constant is very short; and, in a very brief time equal to five time constants, the process is complete.

The current through R_2 then drops to zero. Since there is no voltage developed across R_2 under these conditions, e_{g2} drops to zero with respect to ground and remains at that level until such time as the voltage across C_1 again changes. Notice, however, that C_{g2} is not zero *with respect to the cathode*, but is negative by the amount of the voltage drop developed across R_{k2}.

Since the negative excursion of e_{g2} simply drives the grid of V_2 further into the cutoff region, it has no effect on the plate current, which remains at zero. Thus the output voltage appearing across R_{L2} remains at the value of E_{bb}.

At the end of the first alternation of input voltage, the plate voltage of V_1 swings in the positive direction and capacitor C_1 must recharge to a new higher level. Again since the charge on the capacitor cannot change instantaneously, the full positive-going voltage appears across R_2. Nothing happens, however, until this voltage reaches the cutoff level of V_2, at which time that tube conducts. As the voltage across R_2 continues to increase, e_{g2} rises to zero voltage *with respect to the cathode.* As shown in Fig. 10-20(d), this level corresponds to E_{RK}.

Referring to the plate voltage waveform of V_2, shown in Fig. 10-20(e), it is seen that, during the first half-cycle of input voltage, this tube was cut off and the plate remained at the level of the supply voltage. When the plate current begins, however, during the next alternation, e_{p2} drops to the value of the voltage at which e_{g2} is zero with respect to the cathode; that is, $e_{g2} = E_{RK}$.

Returning to waveform e_{g2}, when it rises above the value where the grid is at zero voltage with respect to the cathode, the grid becomes positive and starts to draw current. Thus, e_{g2} levels off at this point. Only a small part of the remaining voltage rise appears across the grid to cathode. This is owing to the fact that the grid current passes through a voltage divider consisting of the high reactance of C_1 and the comparatively low resistance of R_{L1}. The time constant of this circuit is very fast compared to the time constant of the discharge path, and therefore the capacitor charges very quickly. The voltage at the grid, e_{g2}, then drops back to zero with respect to ground and remains at that level until e_{p1} again swings negative.

Tube V_2 amplifies and inverts that portion of e_{g2} that is positive with respect to ground. Thus, the output consists of a series of negative pulses, one for each complete cycle of the sine wave input.

EXERCISES

10-1 Draw schematic diagrams of the following circuits and explain their operation with the aid of waveforms: (a) series diode positive clipper; (b) parallel diode negative clipper; (c) grid limiter; (e) overdriven amplifier; and (e) cathode-coupled limiter.

10-2 Draw diagrams indicating three methods of obtaining bias for a grid limiter.

10-3 Draw a circuit that will clip an input sine wave of 18 v amplitude peak-to-peak and provide an output that consists of the negative peaks of the input waveform between -6 and -9 v.

Exercises

10-4 What would be the effect of using a low-value resistor in the grid circuit of a grid limiter?

10-5 What is the disadvantage of clipping by collector saturation in the transistor clipping circuit?

10-6 What condition must be met to produce symmetrical clipping in a cathode-coupled clipping circuit?

10-7 What would be the effect of an increase in the size of capacitor C_1 in the circuit of Fig. 10-20?

10-8 What conditions must be met to produce saturation limiting by a relatively low amplitude of positive grid voltage in a vacuum-tube limiter circuit?

10-9 What important operating conditions exist between diode and triode limiters?

10-10 A sawtooth wave of maximum amplitude 30 v is applied to a series diode positive clipper designed to clip at a level of -10 v. Draw the Output Waveform.

11

CLAMPING
CIRCUITS

A clamping circuit may be defined as one which holds either extreme of an a-c voltage to a definite level without distorting the waveform. It is also referred to as a *d-c restorer* or a *baseline stabilizer.*

11.1 General

There are two major classifications of clamping circuits: (a) *diode* and *grid clamping* circuits, which clamp either the positive or negative amplitude extreme and permit the output to extend in only one direction from the reference level; and (b) *keyed (synchronized) clamping* circuits, which maintain the output at some fixed level until a synchronizing pulse is applied and the output is permitted to follow the input waveform.

11.2 Action of an R-C coupling network

Before you can understand the operation of clamping circuits, it is first necessary to review certain aspects of the operation

Sec. 11.2   Action of an R-C coupling network   197

of a resistance-capacitance coupling network. Refer to Fig. 11-1. Quiescent operation places the plate voltage of the tube V_1 at a level of $+20$ v as shown in Fig. 11-1(c). When a negative-going pulse is applied to the input terminals, at $t = 20\,\mu\text{sec}$ as shown in Fig. 11-1(b), it causes a reduction of plate current and an abrupt rise in plate voltage to $+100$ v, where it remains for the duration of the pulse. At

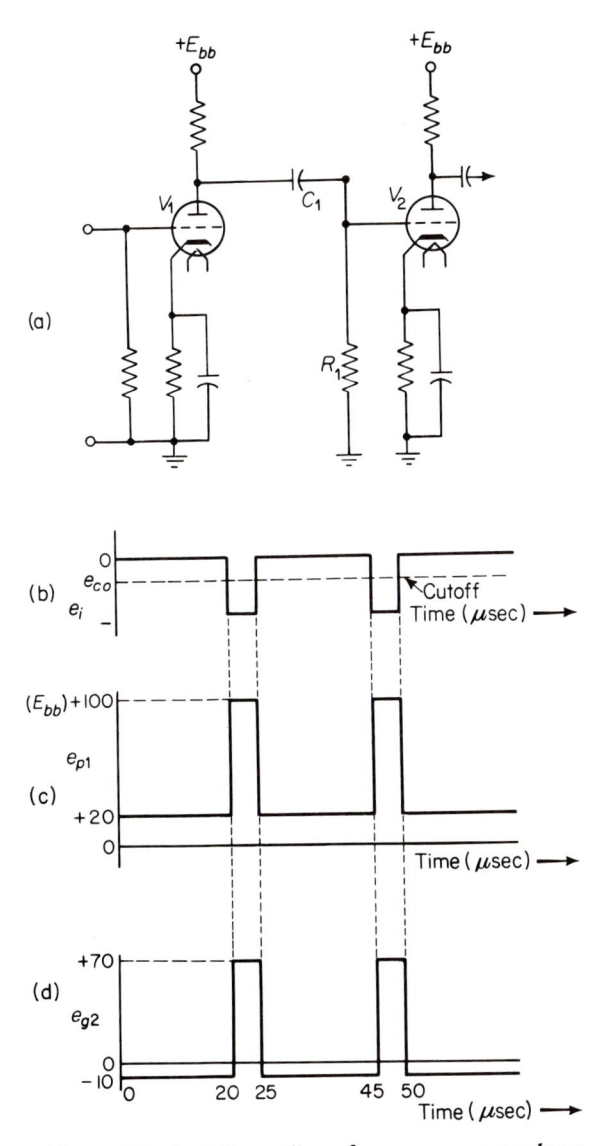

Figure 11-1 An R-C coupling of a square wave voltage.

$t = 25\,\mu\text{sec}$, e_i returns to zero. The plate voltage of V_1 then drops to its quiescent value and remains at that level until the next pulse is applied to the input terminals.

The coupling network, consisting of capacitor C_1 and resistor R_1, is selected to have a long time constant in comparison to the 25 μsec period of the input cycle. Because C_1 and R_1 form a series voltage divider, with only that portion of the voltage developed across R_1 serving as useful input to V_2, the value of C_1 is selected so that its reactance is small (in comparison with the resistance of R_1) at the fundamental and harmonic frequencies included in the rectangular waveform. The R-C time constant is long, and the output across the resistor, R_1, should follow the input waveform. An ideal coupling network, therefore, exists between the two triodes.

The charge on capacitor C_1 during the interval $t = 0$ to $t = 20$ μsec is equal to the plate voltage of V_1, 20 v, plus a residual charge owing to the more positive portion of the preceding cycle. Let us suppose that the residual charge is 10 v and the total charge is 30 v; and also that C_1 is discharging through R_1, causing the grid voltage e_{g2} of V_2 to be -10 v as shown in Fig. 11-1(d). When e_{p1} rises to $+100$ v, the voltage across the C_1-R_1 network increases by 80 v. The voltage across the capacitor cannot change instantaneously; therefore, the full 80 v appears across R_1. This means that capacitor C_1 stops discharging and begins charging through R_1, producing the 80 v across this resistor. At $t = 20\,\mu\text{sec}$, grid voltage e_{g2}, therefore, changes from -10 to $+70$ v. Because the R_1C_1 time constant is long in comparison to the period of the applied input waveform, the voltage across these two elements does not change greatly during the period from $t = 20$ to $t = 25\,\mu\text{sec}$. At $t = 25\,\mu\text{sec}$, the plate voltage of V_1 suddenly returns to its original value of $+20$ v. Now, C_1 stops charging and begins to discharge, because the plate voltage is less than the voltage across its terminals. In discharging, C_1 again brings the upper end of R_1 to 10 v negative with respect to ground.

Notice that the nature of input wave e_{p1} to the R-C network is a pulse having an a-c component as well as a d-c level. That is

$$e_{p1} = E_{dc} + E_{\max} \sin \omega t + \frac{E_{\max}}{3} \sin 3\omega t, \text{etc.} \qquad (11\text{-}1)$$

where $\quad E_{dc} =$ the d-c component

$E_{\max} \sin \omega t =$ the fundamental

$\dfrac{E_{\max}}{3} \sin 3\omega t =$ the third harmonic

From the description of circuit operation given, it is seen that the d-c level of the applied positive-going, unidirectional pulse has been reduced.

The voltage applied to the grid of V_2 is an alternating voltage, with an average value of zero, from which the d-c level has been removed. It is equal to the voltage drop across R_1 at every instant. When the plate voltage of V_1 is high, its output voltage is large and positive for a short time; and when the plate voltage of V_1 is low, its output voltage is small and negative for a longer time. It should also be noted that the capacitor must receive the same amount of charge during the positive half of the cycle as it loses during the negative half. Thus, the positive and negative portions of the a-c waveform applied to the grid of V_2 are equal. This condition is fulfilled in the explanation given. It is impossible, however, to maintain perfectly square waves of voltage in practical circuits. The plate voltage of V_1 cannot rise and fall instantaneously because of the interelectrode capacitances of the tube. This effective capacitor must be charged or discharged each time the plate voltage changes. Further, the rate of change in the plate voltage of V_1 depends on the value of the resistance in the circuit as well as the capacitance. Finally, the time constant of the R-C coupling network can never be made infinitely long. Consequently, the flat portion of the ideal grid-voltage waveforms are actually sections of exponential curves which tend to approach the zero axis.

In the R-C coupling network just considered, the output voltage applied to the grid of V_2 is an a-c voltage which alternates about ground potential. The coupling capacitor C_1 has prevented the d-c component from being transferred to the grid circuit of V_2. This condition is not always desirable. It may be necessary for the output voltage of the R-C coupling network to swing entirely in the positive or negative direction in respect to ground, and the d-c component must, therefore, be restored to the grid voltage. This d-c restoration is accomplished by using a *clamper*. The latter name is derived from the fact that the voltage is clamped to one side of the waveform in respect to ground.

11.3 Diode clampers

The simplest type of clamper uses a diode in conjunction with an R-C coupling network as shown in Fig. 11-2. The diode acts as a simple switch. When the input voltage applies negative potential to the diode cathode, it makes the diode plate positive with respect to the cathode. The diode has a low forward resistance, and any charge

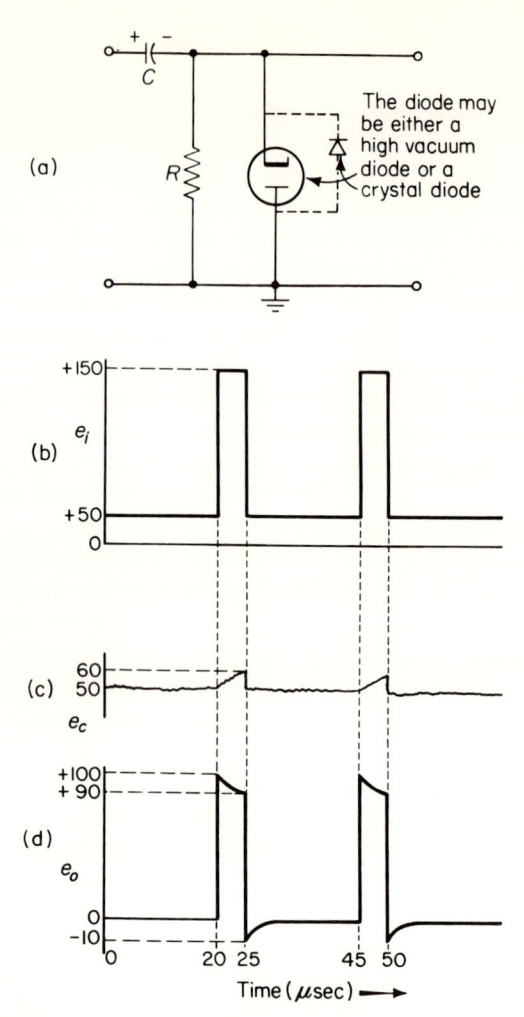

Figure 11-2 (a) *A positive clamping circuit showing* (b) *waveforms at input,* (c) *capacitor, and* (d) *output.*

in the coupling capacitor is rapidly discharged to ground. From this it can be seen that, theoretically, a negative output cannot be obtained. For positive input signals, the diode acts as a high resistance and does not conduct. A positive voltage then appears across the output terminals.

Suppose a constant d-c voltage of +50 v has been applied to the circuit shown in Fig. 11-2(a) so that the capacitor has charged to the full 50 v through resistor R. When the first rectangular pulse,

Sec. 11.3   Diode clampers   201

Fig. 11-2(b), is applied to the input terminals at $t = 20\,\mu\text{sec}$, the applied voltage is increased to 150 v. Because the charge on the capacitor cannot change instantaneously, the difference between the applied pulse and the charge on the capacitor, 100 v, appears across R.

A very long R-C time constant is used in this circuit, so that the current through R is small and the change in charge on the capacitor is small over the duration of the input cycle from 20 to 625 μsec, as shown by the capacitor voltage waveform e_c, in Fig. 11-2(c). For example, if the time constant is ten times the applied pulse duration, the capacitor charges from 50 v to approximately 60 v. The output during the same period drops from 100 v to 90 v as shown by the output voltage waveform, in Fig. 11-2(d).

At $t = 25\,\mu\text{sec}$, the input pulse voltage decays to 50 v. The capacitor is charged to 60 v at this time, with the output side negative in polarity. The voltage appearing across R is, therefore, $-60 + 50 = -10$ v. When the voltage across the capacitor drops to 50 v, the voltage across R returns to zero as shown in the output waveform. This procedure is repeated for each cycle. The relative ability of this circuit to duplicate the input waveform therefore depends on a long time constant and a diode having a high ratio of inverse to forward resistance.

Now, let us study the action of the same circuit when a series of negative pulses is applied to the input as shown in Fig. 11-3. Suppose that the constant d-c voltage is initially zero, and that no charge exists across the capacitor. At $t = 20\,\mu\text{sec}$, a pulse is applied and voltage e_i drops to -100 v. Since the capacitor cannot change its charge instantaneously, the voltage appearing across resistor R also drops to -100 v. This initial transient is not repeated in successive alternations of input. The diode now conducts, and the capacitor charges rapidly to 100 v through the low forward resistance. The output voltage across R drops to zero. At $t = 25\,\mu\text{sec}$, e_i returns to 0 v. The capacitor is charged to 100 v at this time, with the output side positive in polarity. The diode cannot conduct under these conditions, and C discharges slowly through R. With a long time constant, the capacitor voltage drops very little in the period between $t = 20$ and $t = 45\,\mu\text{sec}$. For the sake of illustration, it is assumed to drop to 90 v.

At $t = 45\,\mu\text{sec}$, the input voltage again abruptly changes to -100 v. With $+90$ v charge remaining on the capacitor, and an applied input voltage of -100 v, a negative 10 v now appears across resistor R. The diode conducts, and the capacitor again charges to 100 v through this low-resistance path. After the first cycle, the output waveform

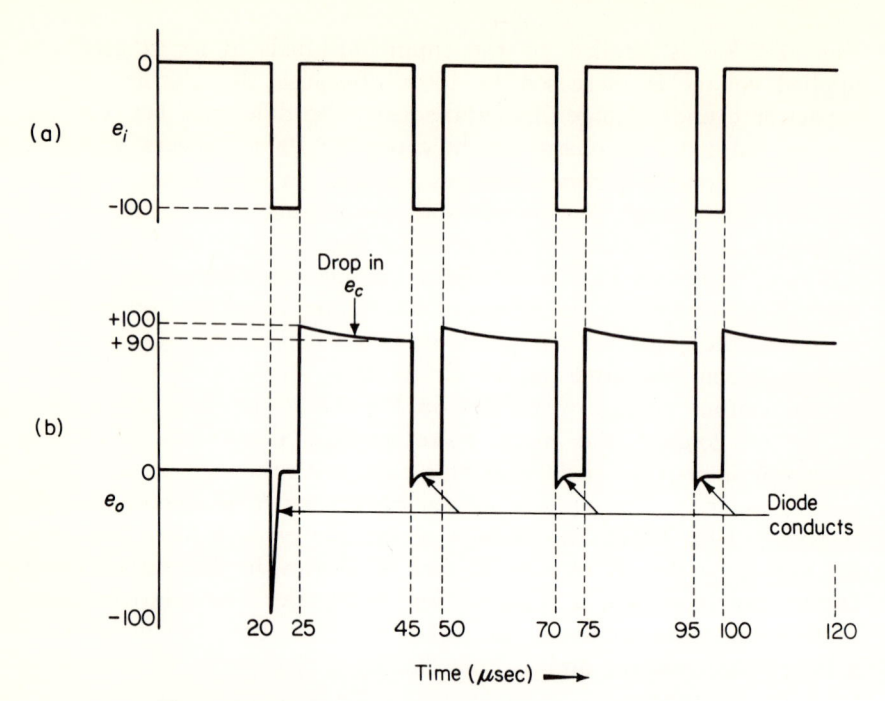

Figure 11-3 Response of a positive clamper to a negative going input pulse train.

is quite similar to the input waveform with one essential difference: the clamper shifts the zero level from the top to the bottom of the waveform.

With regard to the d-c levels in the two waveforms of Fig. 11-3, it should be noted that the d-c level of the input waveform e_i is -20 v, since $E_{dc} = 5/25(-100) = -20$ v; and the d-c level of the output waveform is $+80$ v, since $E_{dc} = 20/25 (100) = 80$ v. This is illustrated in Fig. 11-4, with e_o drawn as a square wave. It can be seen that the d-c level has changed from -20 v to $+80$ v; in other words, it has shifted 100 volts as expected.

Fig. 11-5(a) illustrates a diode clamping circuit capable of causing the output waveform to vary between some *negative* value and the zero reference voltage. The only difference between this circuit and the one shown in Fig. 11-2(a) is that the diode is inverted with its cathode grounded. The diode, therefore, conducts whenever the plate rises above ground.

It is again assumed that a constant d-c voltage of $+50$ v has been applied to the circuit of Fig. 11-5(a) so that the capacitor has charged to the full 50 v through resistor R.

Sec. 11.3   Diode clampers

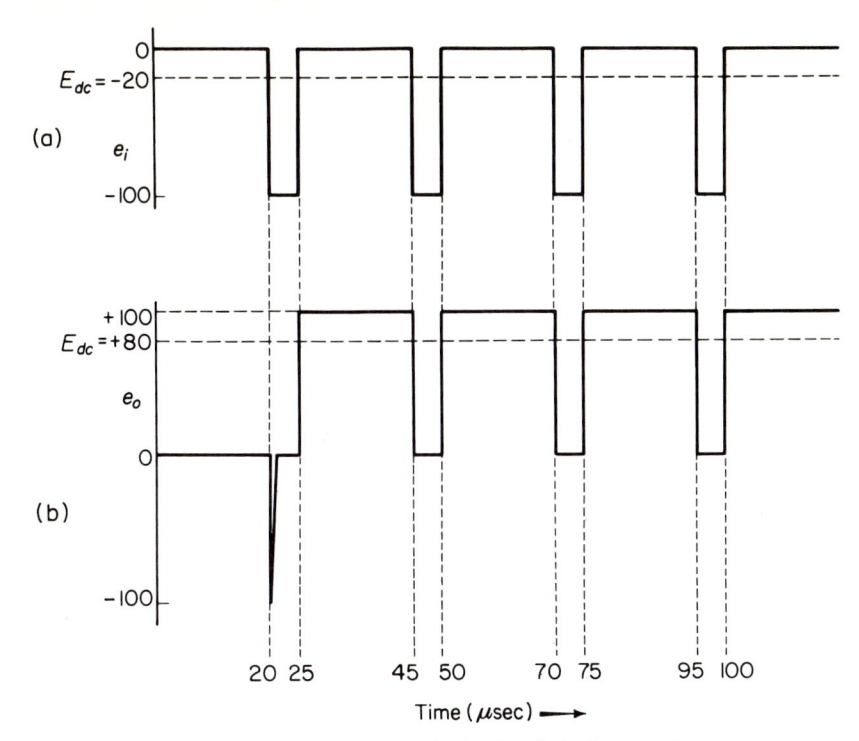

Figure 11-4 Indication of the d-c levels of the input and output voltage waveforms of the positive clamper.

When the first pulse of the input wave train, Fig. 11-5(b), is applied to the input terminals at $t = 20\ \mu\text{sec}$, the applied voltage is increased to 150 v. Since the charge on capacitor C cannot change instantaneously, the full increase in applied voltage appears across R and makes the plate of the diode 100 v positive with respect to its cathode. This causes the diode to conduct, and capacitor C charges very rapidly to 150 v through its low forward resistance, bringing the output voltage back to zero quickly. Notice in Fig. 11-5 that the initial transient in waveform e_o is not repeated in successive alternations of input.

At $t = 25\ \mu\text{sec}$, the input signal suddenly drops 100 v. Since capacitor C is charged to 150 v, and since this charge cannot change instantaneously, the full 100 v drop appears across resistor R. Thus, when e_i drops from $+150$ to $+50$ v, e_o drops from 0 v to -100 v, as shown in Fig. 11-5(c).

From $t = 25$ to $t = 45\ \mu\text{sec}$, the capacitor discharges a small amount through the high resistance path of R, producing a decrease in e_o. For the sake of illustration, e_o is assumed to drop to -90 v.

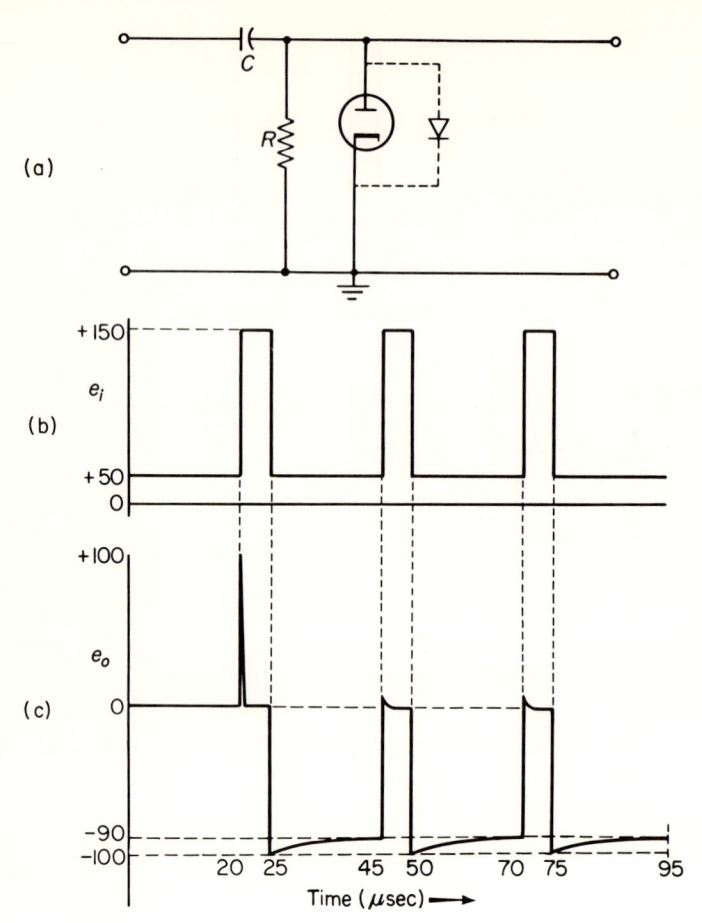

Figure 11-5 (*a*) *Negative clamper and its* (*b*) *input and* (*c*) *output waveforms.*

At $t = 45$ μsec, the input voltage again abruptly changes to $+150$ v, an increase of 100 v. This full change appears, of course, across R and makes the plate of the diode 10 v positive with respect to its cathode, causing conduction. (Remember that, when this pulse is applied, the plate of the diode is 90 v negative with respect to its cathode, and the sum of $-90 + 100 = +10$ v.) The capacitor then discharges very rapidly through the low forward resistance of the diode, bringing the output voltage back to zero quickly.

Diode clamps are often used in the sweep circuits of cathode-ray tubes (CRT). The necessity for such circuits is illustrated in Fig. 11-6. Parts (a) and (c) of this illustration shows *a-c* sweep voltages of the same base-to-peak amplitude and frequency, but with different

Sec. 11.3   Diode clampers   205

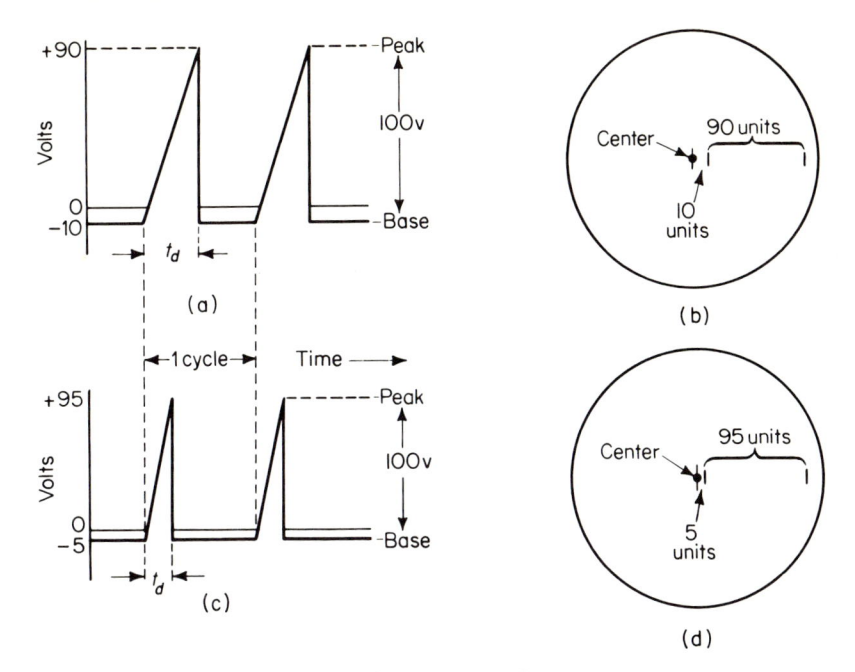

Figure 11-6 An illustration of the necessity for clamps in CRT sweep circuits. The waveforms of (a) and (c) are the same base-to-peak amplitude and frequency, but with different duration times.

duration times. Also, the waveform in (a) swings from -10 v to $+90$ v with respect to the zero reference line, while the waveform in (c) swings from -5 v to $+95$ v with respect to the zero reference line.

Suppose both of these waveforms are to be applied in turn to the right horizontal deflection plate of a CRT.

When the waveform in (a) is applied, it causes the trace to be positioned as shown in (b). When the waveform in (c) is applied, the trace is observed to be shifted to the right as shown in (d). The portion of sweep voltage during the sweep duration is narrower in this case, causing the voltage to rise 95 v in the positive direction and fall to -5 v in the negative direction. At the start and at the end of the sweep, the deflection plate is 5 v more positive in (d) of Fig. 11-6 than in (b). This causes the trace to shift sideways when the change is made from one sweep voltage to the other.

This condition can be avoided by applying the sweep voltages to the deflection plate through a clamper, as shown in Fig. 11-7. The clamping circuit ensures that each sweep voltage operates a full 100 v on the positive side of the ground reference level and does not swing in the negative direction. The trace is therefore made to start at

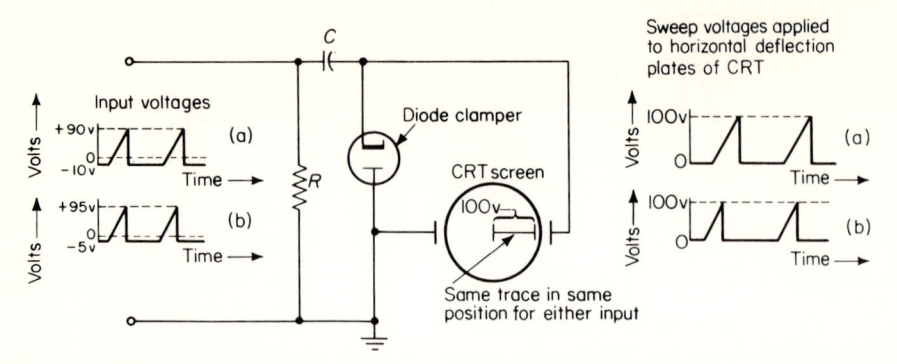

Figure 11-7 A clamping circuit used to prevent horizontal trace shift in an unbalanced CRT circuit.

the same point on the screen regardless of the input waveform. In practice, the *horizontal centering controls* set the position of this point so that the trace is properly centered. This is especially important in radar sets in which the trace must remain aligned with a scale fixed to the outside of the tube face.

Although the above explanation is for an unbalanced CRT circuit, the necessity for clamping is equally important in balanced circuits. The method of connecting diode clamping circuits in a balanced CRT sweep circuit is shown in Fig. 11-8. The a-c sweep voltages are applied to the horizontal deflection plates by way of the clamping

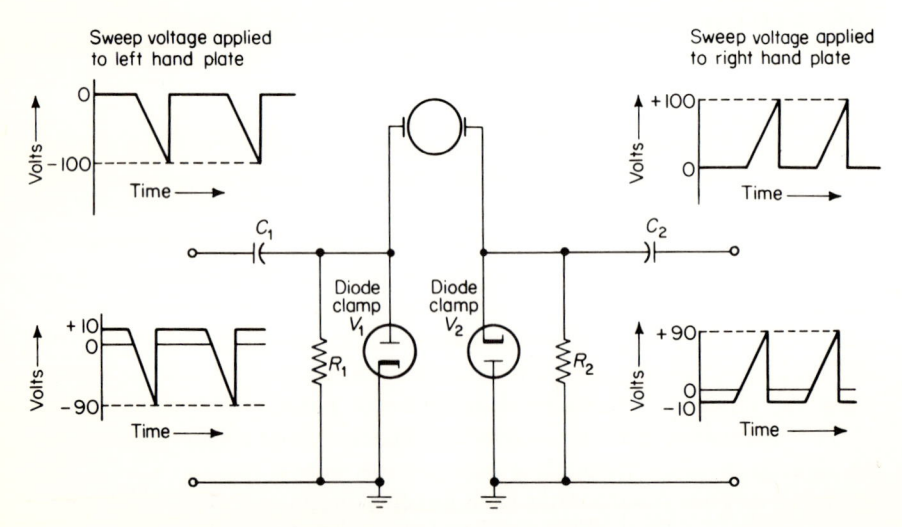

Figure 11-8 A clamping circuit used to prevent horizontal trace shift in a balanced CRT circuit.

Sec. 11.4   Biased diode clamps

circuits. Diodes D_1 and D_2 are connected in parallel with resistors R_1 and R_2 to ensure that the trace starts at the same point for different sweep voltages. The horizontal centering control (not shown) is adjusted to set the position of the starting point in the trace.

11.4 Biased diode clamps[1]

The diode clamping circuits studied thus far have used ground as the reference level. By using a bias voltage, it is possible to restore (place) the signal at any desired level.

One form of *biased diode clamp* is illustrated in Fig. 11-9. The junction point of resistor R and the cathode of diode D_1 is returned to a given potential other than zero, with respect to ground. This potential, labeled E_{bias} in Fig. 11-9, may be either positive or negative.

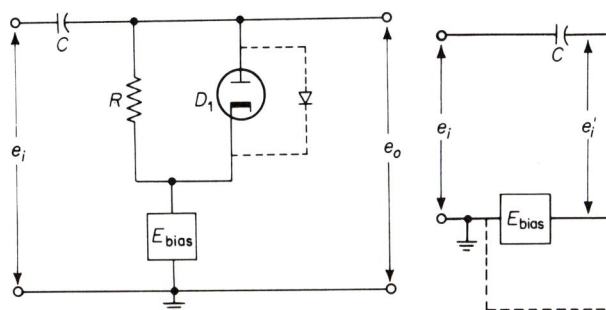

Figure 11-9 One form of biased diode clamp.

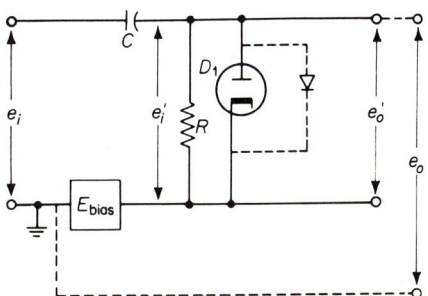

Figure 11-10 Action of the biased diode clamp of Fig. 11-9.

For the sake of illustration, the circuit of Fig. 11-9 can be redrawn as shown in Fig. 11-10. The bias voltage, E_{bias}, is temporarily considered part of input voltage e'_i, and output voltage e'_o is considered to be taken across resistor R. The study of unbiased clamps has shown that the positive peaks of e'_o are clamped at zero regardless of the average value of input voltage. Thus, both e_i and e'_i in Fig. 11-10 will produce the same e'_o. Since the positive peaks of e'_o, taken across resistor R, are clamped at zero, the positive peaks of e_o, taken across the points indicated in Fig. 11-10, must have the positive peaks clamped at E_{bias}.

The diode can be reversed, as shown in Fig. 11-11, to provide clamping of the negative peaks. Voltage E_{bias} can again be either

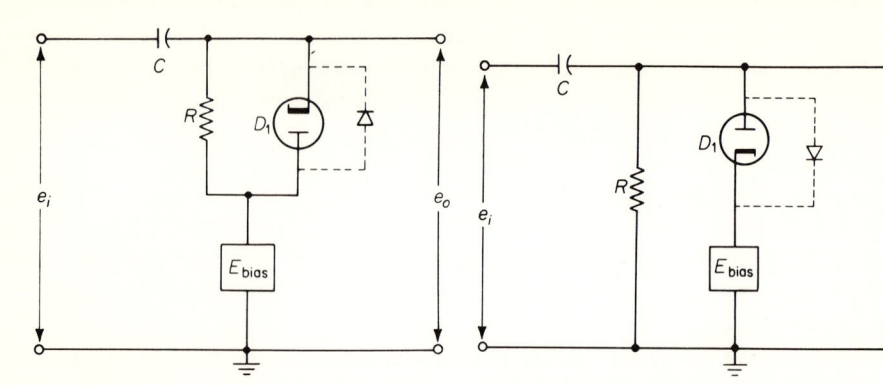

Figure 11-11 The biased diode clamp with diode D_1 inverted.

Figure 11-12 Biased diode clamp with resistor R returned to ground and the cathode of D_1 to potential E_{bias}.

positive or negative. The negative peaks of the output signal will be fixed at potential E_{bias} above ground.

Another form of biased diode clamper is shown in Fig. 11-12. In this circuit, resistor R is returned to ground, but the cathode of the diode is returned to a fixed potential, E_{bias}, which may hold the cathode either positive or negative with respect to ground.

If the cathode is at some positive potential with respect to ground, there is a possibility that it may never conduct, in which case the device is a coupling circuit but not a clamper. If the cathode is at some negative potential with respect to ground, the diode will always conduct for some part of the cycle, and the positive peaks will be clamped at the potential level of E_{bias}.

Similarly, if the diode is reversed and the anode is held at a negative voltage with respect to ground, there is a possibility it will never conduct and, again, the device will act as a coupling circuit only. If, however, the anode is held at some positive voltage with respect to ground, the circuit will clamp the negative peaks at reference level E_{bias}.

In studying the switching characteristics of diodes, it was shown that both thermionic and semiconductor diodes have nonlinear characteristics in the forward direction, and that the conducting resistance of either diode decreases as the applied voltage across the diode increases.

With this in mind, it is apparent that if signals of approximately the same shape but of different magnitude are applied to a diode clamper, operation may occur on different portions of the diode

Sec. 11.5   Notes on using semiconductor diodes in clamping circuits   209

characteristic, resulting in different values of resistance. On some waveforms, particularly narrow pulses, the output voltage may then become quite different.

This undesirable effect can be minimized by using the circuit arrangement of Fig. 11-13, where a positive bias voltage is inserted in series with resistor R. The total d-c voltage developed across R is the sum of E_{bias} and the signal-developed voltage.

If E_{bias} is made large compared to the signal-developed potential, the average current through the resistance will not change appreciably with changes in magnitude of the input voltage. Because there is no average current through the capacitor, the average diode current will also remain relatively constant. It is assumed, of course, that the forward resistance of the diode is very small compared to resistor R.

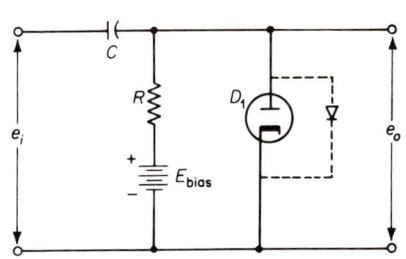

Figure 11-13 Circuit used to minimize changes in the conducting resistance of the diode when operation may occur on different portions of the diode characteristic.

11.5 Notes on using semiconductor diodes in clamping circuits

Semiconductor diodes (both germanium and silicon) are often used in clamping applications. The possibility of this use is indicated by dashed lines in the circuits already studied.

In certain applications, germanium diodes may not be satisfactory since they have a relatively low back resistance. Since this back resistance is in parallel with input resistor R (Figs. 11-9 through 11-13) during the portion of input waveform when a thermionic high-vacuum diode would normally not conduct, it may limit the effective value of R in the circuit.

When an increasing reverse voltage is applied across a semiconductor diode, the reverse current remains essentially constant, which indicates an increasing back resistance with an increasing inverse voltage. At some value of inverse voltage, however, the diode reverse current will increase appreciably with any further increase in inverse voltage, which indicates a decrease in diode back resistance. In junction diodes, the increase in reverse current resulting from the decrease in back resistance is nearly a step function. This region is, of course, the Zener region previously described.

In germanium diodes, the back resistance varies as an exponential function of temperature, and it may range typically from about 100,000 ohms to a few megohms. Silicon diodes exhibit much higher back resistances and, as a consequence, the problem of low back resistance does not occur.

The delay in response resulting from minority carrier storage in the base region imposes a limitation on the use of junction diode clampers. When the device is driven from "on" to "off," the initial high-value reverse current indicates a low back resistance that is in parallel with resistor R and that effectively limits the value of R until the reverse transient is complete.

Silicon junction diodes have an additional limitation. The effective voltage at which forward conduction begins varies more with temperature in silicon junction diodes than in germanium point-contact diodes. This sensitivity to temperature is of particular importance when the diode is used in low-level circuits.

11.6 Grid clamps

Just as the grid-cathode circuit of a triode (or a tetrode, a pentode, or a transistor) was used for clipping and limiting, clamping can also be accomplished in the grid-cathode circuit of most tubes. A triode-grid clamper is shown in Fig. 11-14(a). As we have previously seen, the grid-cathode circuit of the triode acts like a diode. When the applied voltage is positive, the grid-cathode resistance is low, and the circuit has a short time constant. When the applied voltage is negative, the grid-cathode resistance is very high, and the time constant is long. Thus, the action is the same as that of the negative clamping circuit previously discussed. It should be noted, however, that the grid circuit shown cannot be used for positive clamping.

From $t = 0$ to $t = 20\,\mu\text{sec}$ on the time base of the waveforms shown in Figs. 11-14(b), (c), and (d), the input voltage is zero and there is no charge on the capacitor. At $t = 20\,\mu\text{sec}$, the input voltage rises abruptly to $+10$ v. The voltage across the capacitor cannot change instantaneously, and the full 10 v appears across resistor R, making the grid 10 v positive with respect to the cathode. This causes capacitor C to charge rapidly to a potential of 10 v through the low-resistance cathode-to-grid path of the tube, and the grid voltage drops rapidly to zero. The output voltage has the same waveform as the grid voltage except that it is inverted and amplified. (A gain of 10 is assumed in Fig. 11-14(d).)

Sec. 11.6   Grid clamps   211

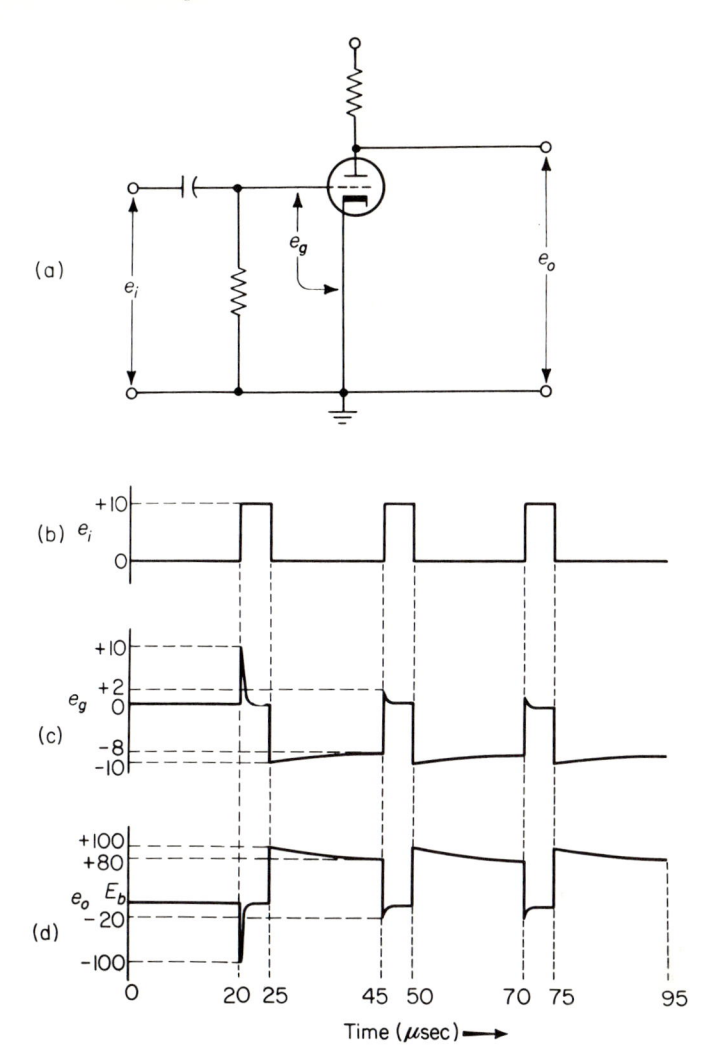

Figure 11-14 Grid clamper.

At $t = 25\,\mu\text{sec}$, the input voltage decays abruptly to zero, causing a negative 10 v to appear across the grid resistor and driving the grid beyond cutoff. Because the tube no longer conducts, capacitor C discharges through the high-value resistance R and slowly loses part of its charge. Assume that the new time constant is such in comparison with the pulse period, that capacitor C loses 2 v of its charge during the period between $t = 25$ and $t = 45\,\mu\text{sec}$. This means that the grid voltage moves in the positive direction from -10 to -8 v during this period.

At $t = 45\,\mu$sec, the second input pulse rises abruptly to $+10$ v. Grid voltage rises from -8 to $+2$ v. This action once again makes the grid positive in respect to the cathode, and grid current results, rapidly charging capacitor C to a value equal to the maximum amplitude of the input voltage waveform, 10 v, and the grid voltage drops rapidly to zero. The rapid change of capacitor C produces the small positive pip in the voltage waveform developed across R, which is the input to the triode. The cycle is then repeated.

As with the diode clamper, the reference level for a grid-circuit clamper can be set at any desired level through the use of a bias voltage. When a negative bias is required, it is usually obtained through the use of a cathode bias resistor.

11.7 Disadvantages of clamping circuits

Clamping circuits can be used only when the value of input voltage that corresponds to the reference level is either the minimum or the maximum amplitude value of the pulse. For example, for the positive-pulse series in Fig. 11-2, the reference level is either the $+50$ or $+150$ v amplitude, depending on whether positive or negative clamping is used. If it is necessary to reinsert the d-c component at a midpoint reference level, this circuit cannot be used.

The clamper is a unilateral device. That is, if the peak voltage is greater than the voltage across the capacitor, the diode conducts (a negative clamper) and the capacitor charges to the peak value. If, however, the reference voltage falls below the voltage existing across the capacitor, the capacitor cannot discharge to the reference level, and the incorrect value of d-c is added to the circuit.

Fig. 11-15 illustrates the unilateral characteristics of a clamping circuit. The leading edge of the first pulse of the input voltage series, in Fig. 11-15(b), causes $+10$ v to be developed across resistor R, as shown in Fig. 11-15(c), because the voltage across the capacitor cannot change instantaneously. This voltage causes the diode to conduct. The capacitor charges rapidly to 10 v through this low-resistance path, and the voltage across the tube falls quickly to zero. When the input pulse decays to zero, a negative 10 v is developed across R. The tube is unable to conduct under these conditions, and the capacitor discharges slowly through the high-value resistance of R. As a result of this discharge, the voltage across the resistor rises to -9 v before the application of a second pulse.

At the beginning of the second 10 v pulse, the voltage across the resistor rises from -9 to $+1$ v. The tube conducts, charging the

Sec. 11.7   *Disadvantages of clamping circuits*

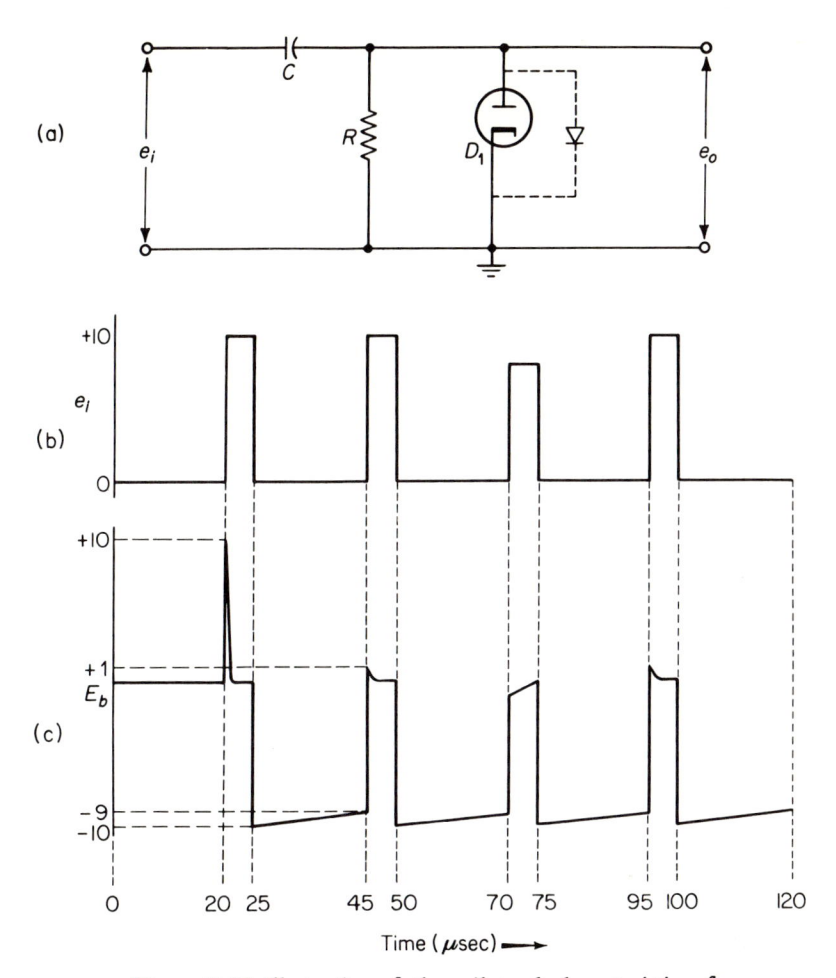

Figure 11-15 Illustration of the unilateral characteristics of clamping circuits.

capacitor to 10 v and causing the voltage across the resistor to fall rapidly to zero. The decay of the second pulse causes the voltage across the resistor to drop to -10 v. Again, the discharge of the capacitor causes the voltage across the resistor to rise to -9 v.

The third pulse has an amplitude of only 8 v, and when it is applied, the voltage across the resistor can rise to only -1 v. This is one volt below the zero reference level. Under these conditions, the tube cannot conduct and charge the capacitor to the required 10 v. Instead, the capacitor continues to discharge slowly. This means that the voltage across the resistor rises slowly toward the zero reference level. It is obvious that the third pulse has not been properly restored. If the intent is not to reproduce faithfully the original pulse but to provide an output in which all pulses are identical, then clamping may be an advantage rather than a disadvantage.

Another disadvantage is that clamping circuits readily charge to noise peaks, but recover from them very slowly. This effect is reduced by increasing the circuit-charging time constant so that the capacitor does not charge up to the full value of the noise input. If, however, the charging time constant is too large, the capacitor may not charge to the desired value when a large-level difference exists between two successive pulses. Some compromise between these two factors must be made.

11.8 Keyed (synchronized) clamping circuits[2]

In the clamping circuits discussed thus far, the effective clamping time is controlled by the signal itself. Sometimes, however, it is necessary to modify the circuit in such a manner that the effective clamping time may or may not be directly related to the signal. *Keyed (synchronized) clamping* circuits may be further classified as one-way or two-way, depending on whether conduction occurs unilaterally or bilaterally when the clamp is closed.

The modification necessary to convert a simple biased-diode clamp

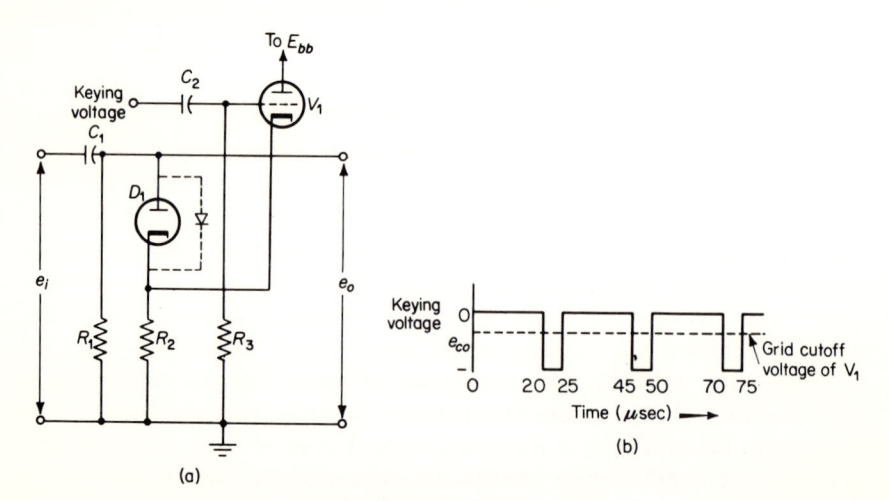

Figure 11-16 Keyed one-way clamp.

Sec. 11.8   Keyed (synchronized) clamping circuits   215

into a keyed one-way clamp is shown in Fig. 11-16. Triode V_1 acts as a switch that turns on the clamping action of diode D_1 each time that a keying pulse is applied to its control grid. When the amplitude of the keying voltage is zero, the triode plate current through resistor R_2 makes the cathode of D_1 positive with respect to its plate. The magnitude of this biasing voltage is made greater than the maximum positive amplitude of e_i, and D_1 cannot conduct.

Suppose that, at time $t = 20\,\mu\text{sec}$, the keying voltage of Fig. 11-16(b) drops to a negative value that cuts off V_1. Plate current is no longer available to apply a biasing voltage to D_1, the diode conducts during positive excursions of e_i, and clamping occurs. At time $t = 25\,\mu\text{sec}$, the clamping action is removed, and then reinstated at $t = 45\,\mu\text{sec}$, etc.

A common application of a one-way clamp is in a *positive sawtooth waveform generator* (discussed in a later chapter). The circuit arrangement is shown in Fig. 11-17.

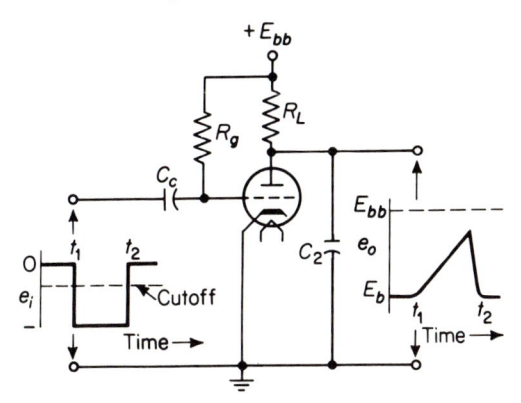

Figure 11-17 Positive sawtooth waveform generator.

Under conditions of no signal, the grid, which is returned to B+ through resistor R_g, is slightly positive, and the plate current is limited primarily by the voltage drop across R_L. If a high-value resistor is used for R_L, the quiescent plate voltage is quite small since most of the supply voltage, E_{bb}, is then dropped across the load.

When a negative-going pulse, of sufficient amplitude to drive the grid beyond cutoff, is applied to the input terminals at time t_1, capacitor C_2 in the output starts to charge to the level of E_{bb} through R_L. The time constant $R_g C_c$ is made long compared to the time interval of the negative-going input pulse, so that capacitor C_c cannot discharge appreciably during the pulse. This prevents the grid from rising above the cutoff value and producing plate current.

The voltage rise across C_2 during plate-current cutoff is exponential. If the $R_L C_2$ time constant is long compared to the input pulse duration, the peak value of e_o will be small compared to E_{bb}, and the sawtooth waveform will be quite linear.

The rate of return to the quiescent value of e_o provides a good indication of the quality of the clamp. In an ideal clamp circuit of this type, a constant initial value of e_o would be achieved independent of the repetition frequency and duration of the input pulse.

The circuit of Fig. 11-17 can be rearranged for use in a negative sawtooth generator circuit by connecting the triode as a cathode-follower, returning the grid to ground and the cathode to a negative voltage source, as shown in Fig. 11-18. Circuit operation is essentially

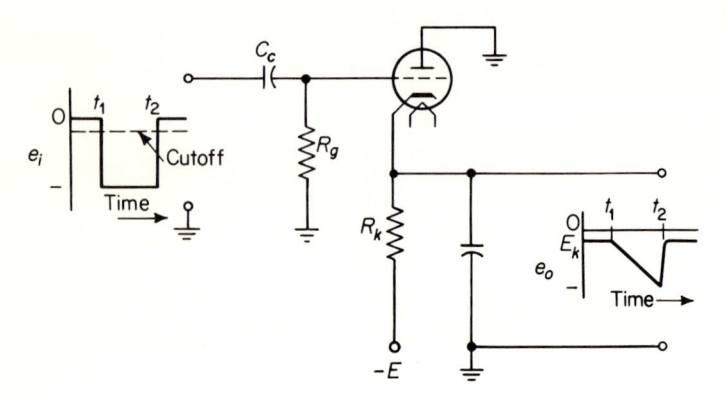

Figure 11-18 Negative sawtooth waveform generator.

identical to that described for positive clamping, but there is a disadvantage in that a much larger driving pulse is required. The amplitude of the pulse must now exceed the sum of the peak value of e_o, the tube cutoff bias, and the voltage decay at the tube grid during the pulse interval.

A two-way clamp is shown in Fig. 11-19. From $t = 0$ to $t = 20$ μsec, the series-connected triodes, V_1 and V_2, conduct. During this period, the grid potential of V_3 is maintained at a constant reference level determined by V_1 and V_2 in series, and capacitor C_1 discharges through V_1. At $t = 20$ μsec, V_1 and V_2 are driven beyond cutoff by the negative keying pulses. This action prevents C_1 from discharging, and the signal voltage is, therefore, transferred to the grid of V_3. At $t = 25$ μsec, the keying pulse returns abruptly to zero, V_1 and

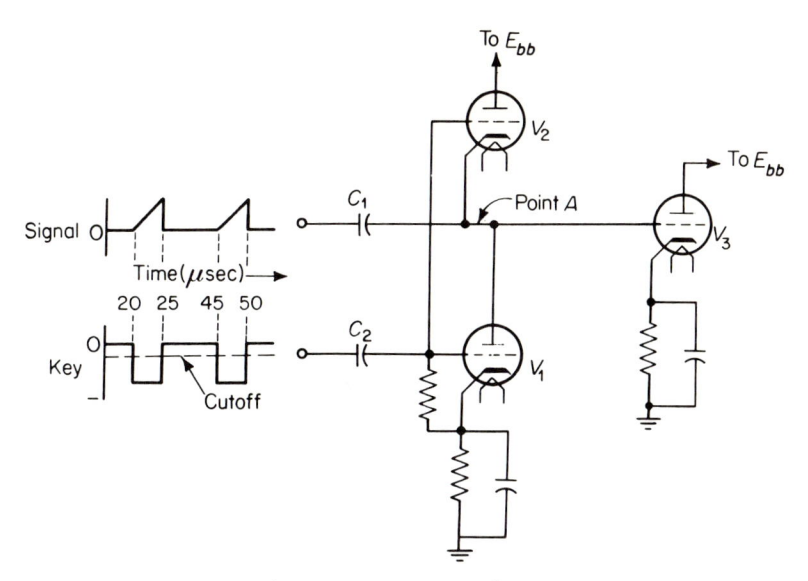

Sec. 11.8   Keyed (synchronized) clamping circuits   217

Figure 11-19 Two-way clamp.

V_2 conduct, and the grid of V_3 returns to the reference voltage existing at point A.

If a change in potential from the reference level appears on the grid of V_3 at any time when a keying pulse is not applied to V_1 and V_2, the clamping triodes act to eliminate this change. If the potential at point A decreases from the reference level, the plate current of V_1 decreases, causing a decrease in bias on V_2. The reduced bias permits V_2 to conduct more. Thus, the conduction through the two tubes in series remains essentially constant and offsets any change in potential at point A. The reverse action occurs if the potential at point A increases from the reference level.

Clamping circuits of the type shown in Fig. 11-19 can be used in the sweep circuits of a radar PPI (Plan-Position Indicator). If the sweep voltages do not always start from the same reference point, the trace on the screen of the PPI (a special form of display unit) does not begin at the same point each time the cycle is repeated, causing a jittery display. By placing a clamper in the grid circuit of the final sweep amplifier, the voltage from which the sweep starts can be regulated by adjusting the d-c voltage applied to the clamping circuit.

Although the PPI equipment uses both vertical and horizontal deflection amplifiers, only the vertical amplifier and clamper will be shown for the sake of clarity. (See Fig. 11-20.)

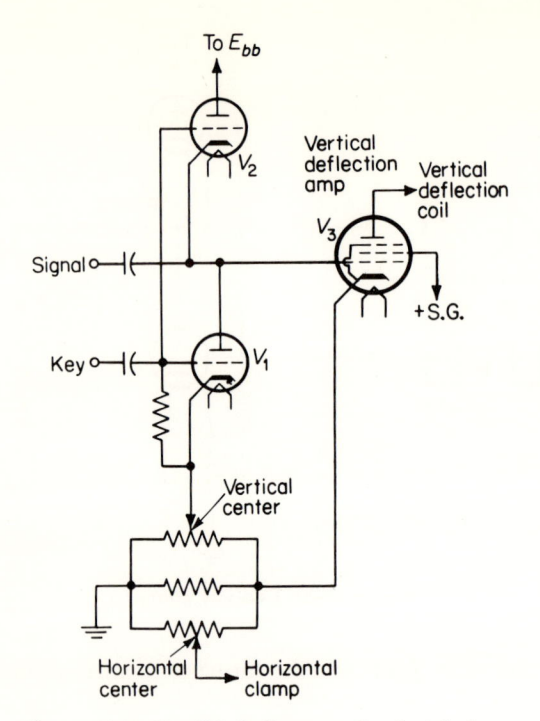

Figure 11-20 Simplified diagram of a keyed clamper used in the sweep circuits of a radar PPI.

In this arrangement, notice that the cathode of triode V_1 is returned to a variable resistor, rather than to a fixed resistor as in Fig. 11-19. The purpose of this variable resistor is to center the start of the PPI vertically. The clamping action of V_1 and V_2, which is identical to that already described, sets the reference level at the grid of the vertical deflection amplifier, V_3.

The circuit shown in Fig. 11-21 uses two keying pulses of equal amplitude but opposite polarity. When these pulses are simultaneously applied to diodes D_1 and D_2 at $t = 20\ \mu\text{sec}$, both diodes conduct, and the opposite ends of potentiometer R_2 are at the same potential. Capacitors C_1 and C_2 assume equal but opposite charges. When the keying pulses are simultaneously removed at $t = 25\ \mu\text{sec}$, the diodes no longer conduct and the capacitors discharge through R_2. Because both the circuit and the keying pulses are balanced, D_1 and D_2 reach the same potential during the pulse intervals. This is the reference potential that exists at point A during the interval between pulses.

To make the charge on capacitors C_1 and C_2 remain almost constant, the R_2C_1 and R_2C_2 time constants are made large compared

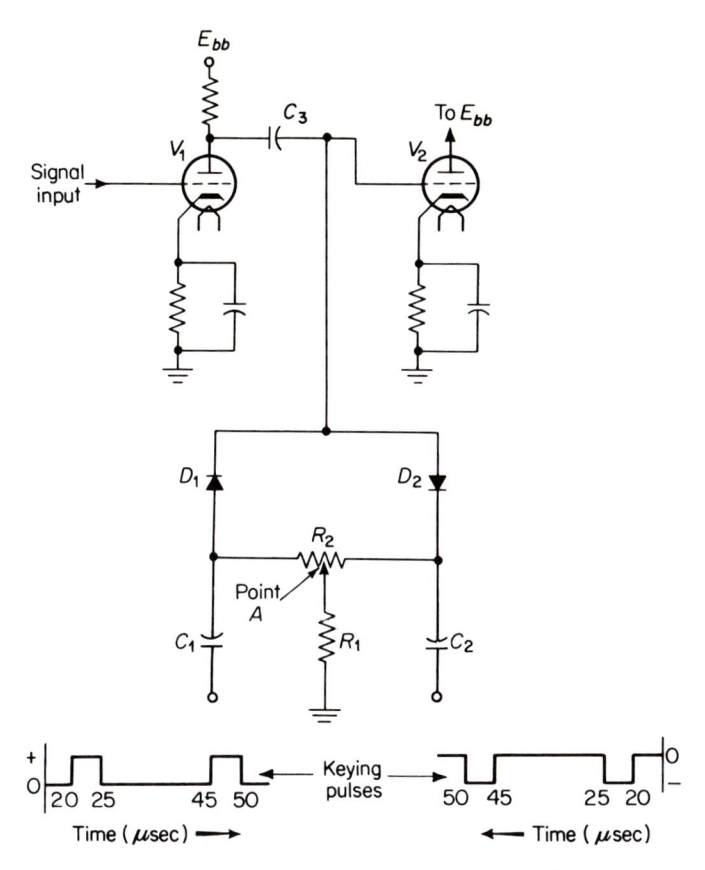

Figure 11-21 Keyed clamp using keying pulses of equal amplitude but opposite polarity.

to the duration of the pulses. Resistor R_1 is inserted in the arm of the potentiometer to minimize unbalance and help maintain the shape of the pulse.

When keying pulses are applied, coupling capacitor C_3 cannot change, and the low-frequency response of the coupling circuit between V_1 and V_2 is not seriously affected.

REFERENCES

1. Von Tersch, L. W., and A. W. Swago, *Recurrent Electrical Transients.* Englewood Cliffs, N. J.: Prentice-Hall, Inc., 1953.
2. Wendet, K. R., "Television D-C Component," *RCA Review*, March 1948, p. 85.

EXERCISES

11-1 How would you define a clamping circuit?

11-2 Draw a circuit that can clamp the input voltage to -6 v. Use waveform sketches to clarify the circuit operation.

11-3 What are the disadvantages of clamping circuits?

11-4 How may keyed clamps be classified? Explain the meaning of each classification.

11-5 Draw the diagram of a grid clamper and show waveforms explaining its operation.

11-6 Why is the time constant in the circuit of Fig. 11-2 very high?

11-7 How is negative bias usually achieved in a grid-circuit clamper?

11-8 How can the tendency of a clamping circuit to charge rapidly to noise peaks be offset?

11-9 Draw the circuit of a two-way clamp; include waveforms that illustrate its operation.

11-10 How does the circuit of Fig. 11-19 tend to prevent changes in the reference level potential?

11-11 Explain how different sawtooth sweep voltages of the same amplitude and frequency, but with different sweep times, can be applied to a CRT so that the trace starts from the same position on the screen for each sweep voltage.

11-12 Explain the conditions under which clamping may occur in R-C coupled amplifiers.

11-13 When signals of approximately the same shape but of different magnitude are applied to a diode clamper, operation may occur on different portions of the diode characteristic, resulting in different values of resistance. How can this undesirable effect by minimized?

11-14 How may the use of a germanium diode in place of a vacuum diode effect clamper operation?

11-15 If you were going to use a semiconductor diode in a low-level clamping circuit, would you select a germanium or a silicon diode? Explain your answer.

11-16 What is meant by the statement that a clamper is a unilateral device? Use waveform sketches to illustrate your answer.

11-17 Give three applications of clamping circuits and explain each.

12

ASTABLE MULTIVIBRATORS

Pulse-generation circuits are divided into two general classifications: *passive* (pulse-shaping) and *active* (self-oscillating). In pulse generators of the passive type, a sine-wave oscillator is used as the basic generator. The output of this oscillator is then passed through pulse-shaping circuits to obtain the desired waveform. For example, a sine wave may be passed through a clipper that removes both extremes to obtain a trapezoidal pulse output.

Active pulse generators are circuits which generate a pulse waveform directly, and most of the active pulse generators use the relaxation principle. This method consists of building up energy in a capacitor and then, when a certain level of voltage is reached, discharging the capacitor. The multivibrator is the most common type of relaxation oscillator.

Before undertaking a study of multivibrator circuits, some of the terms to be used require definitions.

12.1 Definitions

Multivibrator (*MV*). A type of relaxation oscillator consisting of a two-stage resistance-coupled amplifier with the output of each stage regeneratively coupled to the other. In operation, the plate or collector current of one stage is at a maximum when the plate or collector current of the other is cut off. At regular intervals, or when properly triggered, switching from one state to the other occurs.

State. The relative condition of each of the two stages of the MV; i.e., conduction in the first stage and cutoff in the second or *vice versa*.

Stable State. The condition in which the MV may remain indefinitely until the circuit is triggered (pulsed or keyed) by some external signal.

Astable MV. A multivibrator in which neither stage is at a stable state and the stages are switched from one state to the other at regular time intervals without any triggering or actuating voltage.

Bistable MV. A multivibrator in which one stage remains stable in one state, with either stage conducting and the other cut off, until a triggering pulse is applied to initiate the switching action to reverse the stability condition.

Monostable MV. A multivibrator which maintains current in one stage until it is triggered, at which time the other stage is made to conduct for a predetermined length of time and is then automatically switched back to its original state.

The operation of the astable MV only is considered in this chapter. The bistable and monostable types are studied in following chapters.

12.2 Astable MV

The circuit diagram of a vacuum-tube astable (free-running) MV is shown in Fig. 12-1. It is a simple R-C amplifier with the plate of each tube capacitively coupled to the grid of the other. Because there is a 180° phase reversal in signal between the grid and plate circuits of each tube, the feedback through the capacitor is regenerative. Any phase shift introduced by the R-C components is negligible and for practical purposes is disregarded.

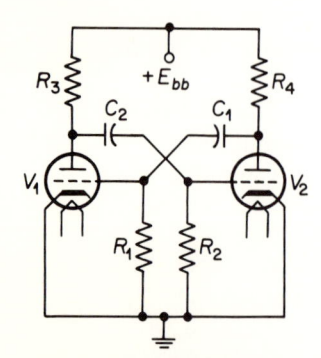

Figure 12-1 Astable MV. Output can be taken from either plate.

Sec. 12.2   Astable MV   223

When plate voltage is applied, each tube conducts, and capacitors C_1 and C_2 take on charge as the voltage increases. A perfect balance of the initial plate currents is impossible. One tube always conducts more than the other, however slightly, due to such factors as inherent tube "noise" and unequally distributed emission along the length of the cathode. This unbalance, regardless of how small it may be, starts multivibrator action. The condition of unbalance is cumulative, and eventually results in the plate current of one tube reaching its maximum while the plate current in the other is cut off, as described below.

Suppose that the initial conduction of V_1 is slightly greater than that of V_2. The voltage drop across R_3 increases, with a corresponding decrease in the plate voltage of V_1. Because this voltage is applied across the R_2C_2 network, and because the charge across C_2 cannot change instantaneously, the full negative change in e_{b1} appears across R_2. This negative-going voltage, applied between the grid and the cathode of V_2, causes a decrease in the plate current and an increase in the plate voltage of V_2. The increase in plate voltage (e_{b2}) is, in turn, applied across the R_1C_1 coupling network and is of such polarity as to make the grid of V_1 swing more in the positive direction. This further increases the plate current of V_1 and amplifies the action outlined above. Operation continues as described until V_2 is cut off ultimately and the plate current of V_1 is at its maximum value. Because V_2 is cut off, its plate-to-cathode voltage rises to the value of the supply voltage, E_{bb}, and C_1 charges through the low-resistance cathode-to-grid path of V_1.

Thus, the slight initial unbalance starts a cumulative feedback action that cuts off one tube and causes maximum plate current in the other. The action described is extremely fast because of the regenerative feedback between the tubes.

With V_2 cut off and V_1 conducting, a switching action begins. The coupling capacitor C_2 must now discharge since the plate-to-cathode voltage of V_1, which is applied to R_2 and C_2, has been abruptly reduced. The capacitor-discharge path is through R_2, and the grid voltage of V_2 rises exponentially toward zero. When the grid voltage, which is moving in a positive direction, reaches cutoff, V_2 starts to conduct. The increase in the plate current of V_2 causes a corresponding decrease in its plate-to-cathode voltage. The cumulative action from this point on is exactly the same as described for the previous half-cycle. In a very short time, V_1 is cut off and V_2 is conducting its maximum plate current. The plate-to-cathode voltage of V_1 rises to the value of the supply voltage, E_{bb}, and C_2 charges through the low-resistance cathode-to-grid path of V_2. A complete cycle has now occurred.

The output voltage, which may be taken from either plate, is a series of rectangular pulses for reasons described below. These pulses have a maximum amplitude when the tube is cut off and a minimum amplitude when the tube is at maximum conduction.

A transistor astable multivibrator also is shown in Fig. 12-2. The circuit is a simple R-C coupled common-emitter amplifier with the output of each transistor coupled to the input of the other. Because the common-emitter configuration provides signal inversion, the feedback is regenerative.

As in the vacuum-tube circuit, one transistor always conducts more than the other, and this unbalance starts multivibrator action. Although only positive direct voltages are applied to the circuit shown, switching action produces voltages which are negative with respect to ground.

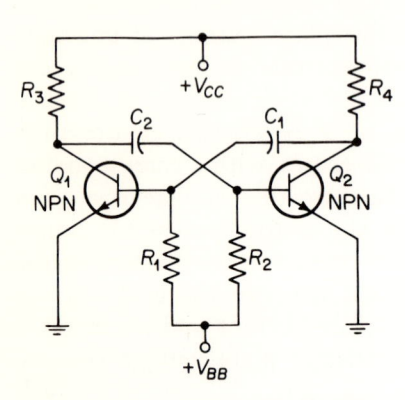

Figure 12-2 Transistor astable MV.

Assuming that Q_1 initially conducts more than Q_2, the voltage drop across R_3 increases, with a corresponding decrease in the collector voltage of Q_1. This voltage is applied across the R_2C_2 network and, because the charge across C_2 cannot change instantaneously, the full negative change appears across R_2. A negative-going voltage, applied between the base and the emitter of Q_2, decreases the forward bias on that stage, resulting in a decrease in collector current and an increase in collector voltage. This increase in collector voltage is in turn applied across the R_1C_1 coupling network, and is of such polarity as to increase the forward bias of Q_1. This further increases the collector current of Q_1 and amplifies the action outlined above. Operation continues as described until Q_2 is *full off* and Q_1 is *full on*. When Q_2 is full off, its collector-to-emitter voltage is essentially equal to the supply voltage, V_{cc}, and C_1 charges rapidly to this value through the low-resistance emitter-to-base path of Q_1.

At the instant Q_1 is turned *on*, capacitor C_2 holds a positive charge on its collector terminal. Assuming that Q_1 is bottomed by the switching action, however, its collector potential becomes $v_{c1} \cong 0$. Since the base of Q_2 is attached to the right terminal of C_2, its

Sec. 12.3   Equivalent circuit analysis   225

potential must then become $-V_{cc}$; that is, v_{b2} becomes negative with respect to ground.

With Q_2 off and Q_1 on a switching action begins. Coupling capacitor C_2 now starts to discharge exponentially through R_2. When the charge on C_2 reaches zero, the capacitor attempts to continue charging to the value of V_{BB}; but this re-establishes forward bias on Q_2 and that stage starts to conduct. The increase in collector current, i_{c2}, of transistor Q_2 causes a corresponding decrease in its collector-to-emitter voltage, v_{c2}. The cumulative action from this point on is exactly the same as described for the previous half-cycle. In a very short time, Q_1 is *full off* and Q_2 is *full on*. The collector-to-emitter voltage, v_{c1}, of Q_1 rises to the value of V_{cc}, and C_2 charges through the low-resistance emitter-to-base path of Q_2. The base voltage of Q_1 becomes $-V_{cc}$ as a result of the switching action. A complete cycle has now occurred.

The output voltage, which may be taken from either collector, is essentially a series of rectangular pulses, having a maximum amplitude when the transistor is off and a minimum amplitude when it is on, as shown below.

12.3 Equivalent circuit analysis

The equivalent circuits of each R-C network in the vacuum-tube astable multivibrator during the charge and discharge periods are shown in Fig. 12-3. The charge path is through the low-resistance cathode-to-grid circuit of the tube. The discharge path is through the high-resistance of the grid resistor, the cathode-to-grid path being closed during periods of non-conduction. For example, during the charging period of C_1, the grid-to-cathode resistance, R_{g1}, of tube V_1, is very low. This resistance is normally much lower than R_1, and the charging current is mainly

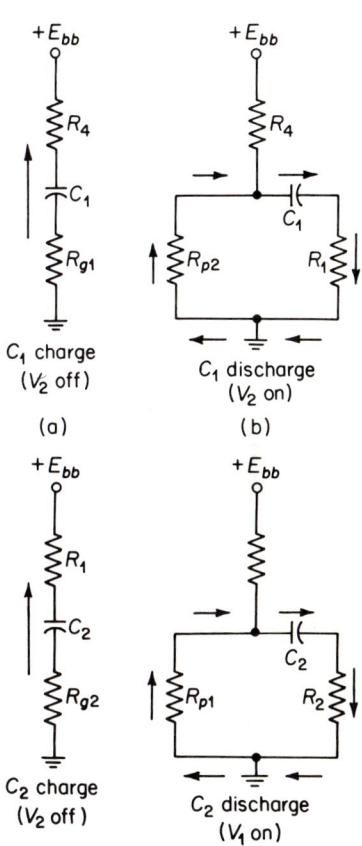

Figure 12-3 Equivalent circuits for vacuum tube astable MV.

through this resistance. For this reason, R_{g1} is shown alone in the equivalent circuit, and the relatively higher parallel resistance of R_1 is disregarded. Tube V_2 is cut off during this period, so that its plate resistance, R_{p2}, is infinite. A similar analysis is shown for the charging period of capacitor C_2.

During the discharge period of capacitor C_1, the grid-to-cathode resistance R_{g1} is high compared to R_1 and is neglected. Tube V_2 is conducting, and R_{p2} has a finite value. The equivalent circuit, therefore, now includes this resistance. A similar analysis is shown for the discharge period of capacitor C_2.

The equivalent circuits for each R-C network in the transistor multivibrator are similarly shown in Fig. 12-4.

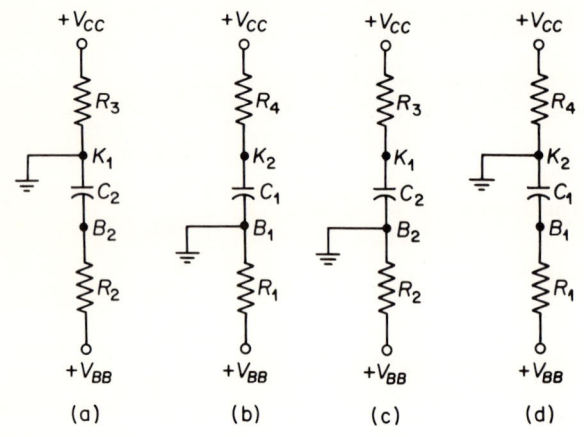

Figure 12-4 Equivalent circuits of transistor astable MV (a) Q_1 on (b) Q_2 off (c) Q_1 off (d) Q_2 on.

With Q_1 on, it is assumed to be bottomed, and $v_c \cong 0$. Under these conditions, capacitor C_2 attempts to charge to V_{BB} through resistor R_2. (The fact that Q_1 is bottomed is indicated in Figs. 12-4(a) and (b) by short-circuiting the collector, K_1, and base, B_1, of transistor Q_1 to the grounded emitter.) The exponential charging process is not completed, however, for as soon as the voltage on B_2 gets slightly positive, switching occurs.

With Q_1 off, it is assumed that $v_{c1} \cong V_{cc}$. Under these conditions, capacitor C_2 charges very rapidly to V_{CC} through the negligible base-to-emitter resistance of Q_2. (The fact that Q_1 is now *off* and Q_2 is *on* is indicated in Figs. 12-4(c) and (d) by connecting the collector, K_2, and base, B_2, of transistor Q_2 to ground, and removing the grounds from electrodes K_1 and B_1.)

12.4 Symmetrical waveforms

The waveforms appearing at the plate and grid of each tube in Fig. 12-5 are produced for symmetrical (balanced) operation; that is, the time constants R_1C_1, and R_2C_2, the tubes used, and the corresponding applied voltages are identical. The conducting and nonconducting periods for each tube are, therefore, essentially the same.

At $t = 1$ on each waveform, assume that the current through V_1 is maximum and that V_2 is cut off. This makes the plate voltage of V_1 minimum and that of V_2 maximum. Capacitor C_1 is charging through the grid-to-cathode resistance of V_1 to the value of the applied voltage. At the same time, there is no voltage drop across R_4 since V_2 is cut off, and the full supply voltage is applied at the plate of this tube. Capacitor C_1 is fully charged within a small fraction of the total conduction period of V_1. Capacitor C_2 begins to discharge slowly through R_2 and the plate resistance of V_1. (Refer to the equivalent circuits of Fig. 12-3.) The grid voltage of V_1 momentarily increases to a positive value. This lowers the grid-to-cathode resistance and provides a low-resistance path to charge C_1. The brief positive voltage at the grid of V_1 causes current to be drawn. As the capacitor becomes fully charged, the grid voltage and current drop rapidly to zero and remain at this level for the remainder of the V_1 conduction period. The grid voltage of V_2 is driven well beyond cutoff and no grid current is drawn. The grid voltage of V_2 starts to rise at $t = 1$ as C_2 discharges through R_2 and the plate resistance of V_1. Between $t = 1$ and $t = 2$, V_2 is cut off, and its plate voltage and current remain essentially constant. Similarly, because the grid voltage of V_1 is held at 0 v for this period, the plate voltage and current of V_1 remain essentially constant, except for the short pip during the charge of C_1 at $t = 1$. The only factor that changes during this period is the grid voltage of V_2, which increases exponentially toward cutoff. The equivalent circuit shows that C_2 discharges through R_2 and the plate resistance of V_1. Because R_2 is usually much larger than R_{p1}, however, the time constant of the circuit is primarily a function of the value of R_2C_2.

At $t = 2$, the grid voltage of V_2 exceeds the cutoff value and the tube conducts. Within a very short time, the plate current of V_2 reaches its maximum and the plate voltage drops to its minimum value. Grid voltage momentarily swings positive, causing grid current, and then both grid voltage and current drop to zero. When V_2 starts to draw current, the grid voltage of V_1 becomes negative,

228 Astable multivibrators Chap. 12

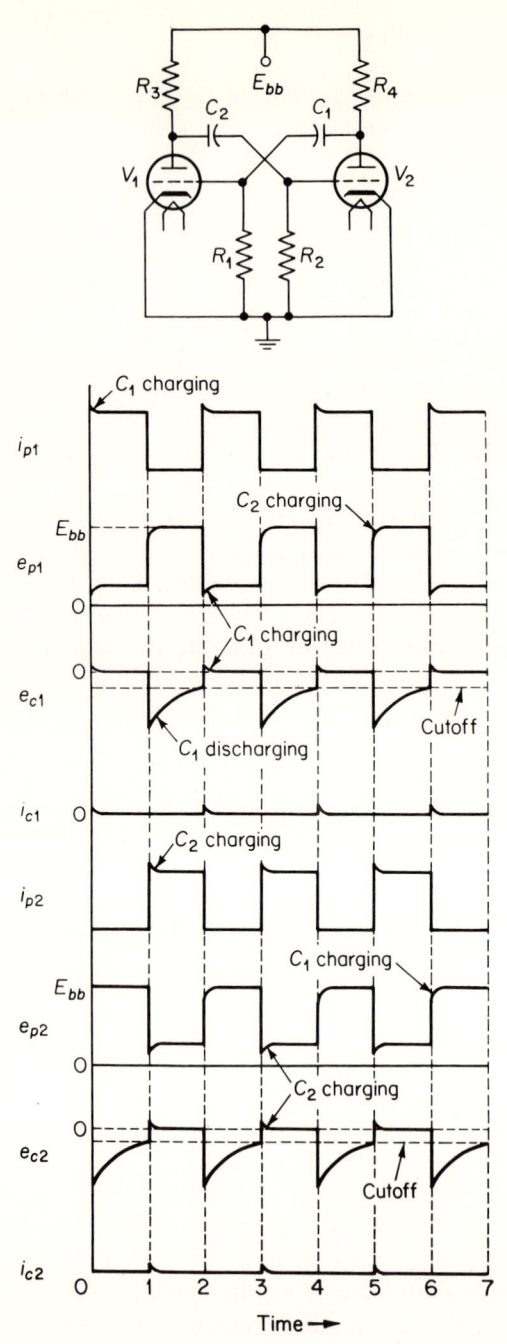

Figure 12-5 Waveforms of astable MV for symmetrical operation.

Sec. 12.4   Symmetrical waveforms   229

causing a reduction in the plate current and an increase in the plate voltage. The grid voltage of V_1 is quickly driven beyond cutoff, the plate voltage reaches a maximum, and the plate current is cut off. Also at $t = 2$, the grid voltage of V_2 becomes positive and the capacitor C_2 is charged through the low grid-to-cathode resistance to the full value of supply voltage E_{bb}. At this time there is no drop across R_3 because V_1 is cut off and the full supply voltage appears at the plate. Capacitor C_2 is fully charged within a small fraction of the total V_2 conduction period. The charging current through R_2 produces a small positive voltage at the grid which, when amplified and inverted, causes the small peaks in the plate-voltage waveform of V_2. Capacitor C_1 discharges exponentially through R_1 and the plate resistance of V_2. Between $t = 2$ and $t = 3$, V_1 is cut off, causing the plate current and voltage to remain essentially constant. Similarly, with the exception of the small pips at $t = 2$, caused by the charging of C_2, the plate voltage and current of V_2 remain essentially constant over this same period. Only the grid voltage of V_1 is changing as it moves exponentially toward cutoff. Because R_1 is usually much smaller than R_{g2}, the time constant of the discharge curve is primarily a function of R_1C_1. At $t = 3$, the grid voltage of V_1 exceeds the cutoff value and the entire cycle starts again.

The duration of the symmetrical multivibrator depends on the time required for the grid voltage of the nonconducting tube to recover and reach the cut-off value. The cut-off period depends on how far beyond cutoff the grid was previously driven, the cut-off point of the tube, and the time constant of the capacitor discharge circuit. The further the grid is driven beyond cutoff, the longer it takes for recovery to the cut-off value. The magnitude of grid voltage beyond cutoff depends on the voltage drop across the plate load resistance of the conducting tube. The higher this drop, the greater the decreases in plate-to-cathode voltage and the more negative the grid voltage swings. The more rapidly the capacitor discharges, the sooner the grid reaches cutoff. The discharge time constant is primarily a function of R_1C_1 and R_2C_2. The duration of the symmetrical multivibrator, therefore, is essentially directly proportional to the difference between the time required for the grid voltage to reach the cut-off value and the time constant of the discharge circuit of the grid capacitor.

The waveforms of each transistor in Fig. 12-6 are produced for symmetrical operation.

At $t = 1$ on each waveform, assume that Q_1 is full on and Q_2 is

230　　　　　　　　　　　　　　　　*Astable multivibrators*　　*Chap. 12*

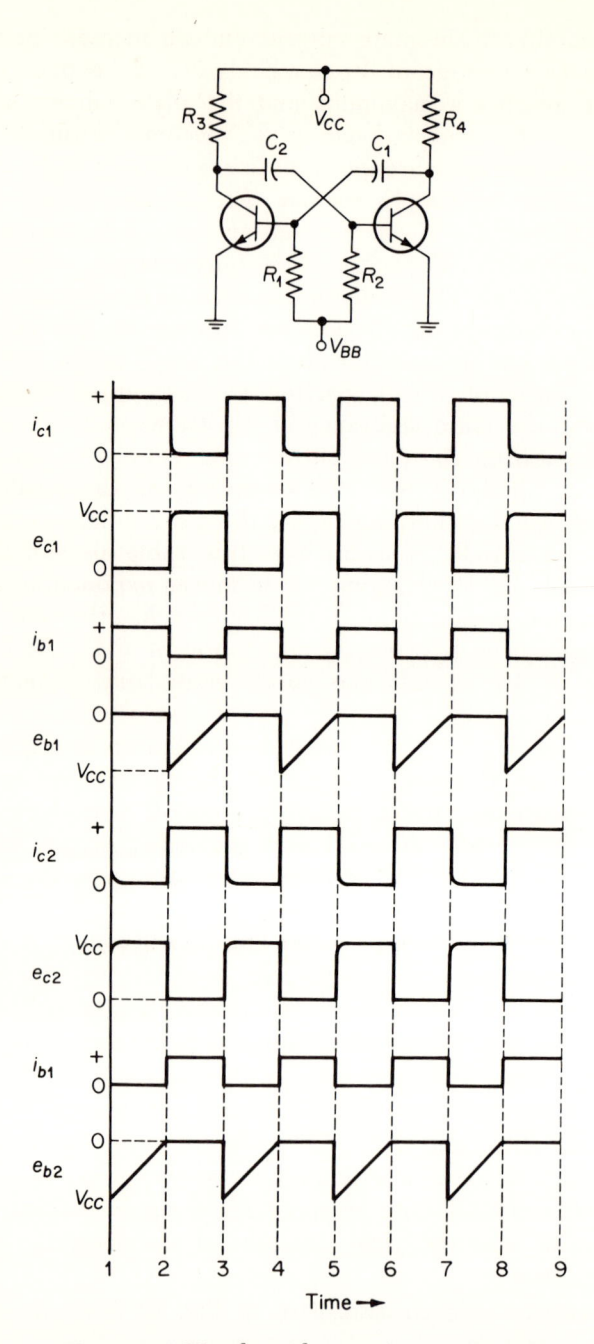

Figure 12-6 Waveforms for transistor astable MV.

Sec. 12.4   Symmetrical waveforms

full off. This makes the collector voltage of Q_1 minimum and that of Q_2 maximum. Capacitor C_1 is charging through the emitter-to-base resistance of Q_1 to the value of V_{CC}. At the same time, there is no voltage drop across resistor R_4 since Q_2 is cut off, and the full supply voltage is applied to the collector of this transistor. Capacitor C_1 is fully charged within a small fraction of the total conduction period of Q_1. Capacitor C_2 begins to charge toward V_{BB} through R_2. (Refer to the equivalent circuits of Fig. 12-4.)

Between $t = 1$ and $t = 2$, Q_2 is cut off and its collector voltage and current remain essentially constant. The only factor that changes during this period is the base voltage of Q_2, which increases exponentially toward V_{BB}.

At $t = 2$, the base voltage of Q_2 exceeds zero and the transistor conducts. Within a very short time, the collector current of Q_2 reaches its maximum and the collector voltage drops to zero. When Q_2 starts to draw current, the base voltage of Q_2 becomes negative. The base voltage of Q_1 is quickly driven beyond cutoff, the collector voltage reaches a maximum, and the collector current is cut off. Within a small fraction of the total Q_2 conduction period, capacitor C_2 is fully charged to the value of V_{CC} through the low-resistance emitter-to-base path of Q_2. Capacitor C_1 charges exponentially toward V_{BB} through R_1.

Between $t = 2$ and $t = 3$, Q_1 is cut off, causing its collector current and voltage to remain constant. Similarly, the collector current and voltage of Q_2 remain constant over the same period. Only the base voltage of Q_1 is changing as it moves exponentially toward V_{BB}. At time $t = 3$, the base voltage of Q_1 exceeds zero and the entire cycle starts again.

The duration of the transistor symmetrical multivibrator depends on the time required for the base voltage of the cut-off transistor to reach the forward bias value. The cut-off period depends on the magnitude of the applied reverse bias and the time constant of the capacitor charge (toward V_{BB}) circuit. This time constant is primarily a function of R_1C_1 and R_2C_2.

By using equivalent circuits more refined than those of Fig. 12-4, it can be shown that, just after time $t = 1$, the base circuit of Q_1 is not actually a simple open circuit and the collector circuit of Q_2 is not a true short circuit. This was pointed out in a previous chapter. Using these new equivalent circuits, a slight *positive* voltage jump then appears on waveforms e_{c1} and e_{b2} at time $t = 1$. For practical purposes, however, the equivalent circuits used are entirely satisfactory and the voltage pips are not shown on the waveforms.

The amplitude of these pips is only a few tenths of a volt, and their duration is very short compared to the interval from $t = 1$ to $t = 2$.

12.5 Asymmetrical operation

If either the R_1C_1 or R_2C_2 time constant of either a tube or transistor symmetrical multivibrator is changed, the resulting waveforms are no longer of equal duration. Suppose, for example, the R_1C_1 time constant is reduced. The discharge curve of capacitor C_2 remains unchanged, but the discharge curve of C_1 is steeper, causing the cut-off stage to conduct sooner than it did before. Stage V_1 or Q_1 is, therefore, cut off for a shorter period than V_2 or Q_2, and the output waveform at the plate of V_1 or the collector of Q_1 is of shorter duration than that appearing at the plate of V_2 or the collector of Q_2. The multivibrator is now operating in an asymmetrical (unbalanced) manner.

12.6 Astable cathode-coupled multivibrator

The circuit of an astable cathode-coupled multivibrator is shown in Fig. 12-7. The plate voltage of V_2 is coupled to the grid of V_1 through capacitor C_1. A cathode resistor, R_k, is common to both cathode circuits, and biases the grids of both tubes. Notice that the grid of of V_1 obtains bias by two paths: the voltage developed across R_k, and the voltage applied through capacitor C_1 from the plate of V_2. On the other hand, V_2 is biased only by the voltage drop across R_k.

Assume a temporary steady-state condition in which V_2 is conducting and V_1 is cut off as a result of the two biasing voltages noted above. Of particular importance is the fact that the voltage across R_k by itself is not enough to cut off V_1. When the second steady-state condition occurs, V_1 conducts, and the current through R_k is sufficient to produce cut-off voltage for V_2. The simultaneous cutoff of V_1 is prevented by the high positive voltage at the plate of V_2 being coupled to its grid through C_1.

Initially, V_1 is cut off as a result of the dual bias and the following conditions exist: The plate voltage of V_1 is at its maximum value and the grid voltage is at its most negative value beyond cutoff. The plate voltage of V_2 is at its lowest value, and the grid

Sec. 12.6   Astable cathode-coupled multivibrator   233

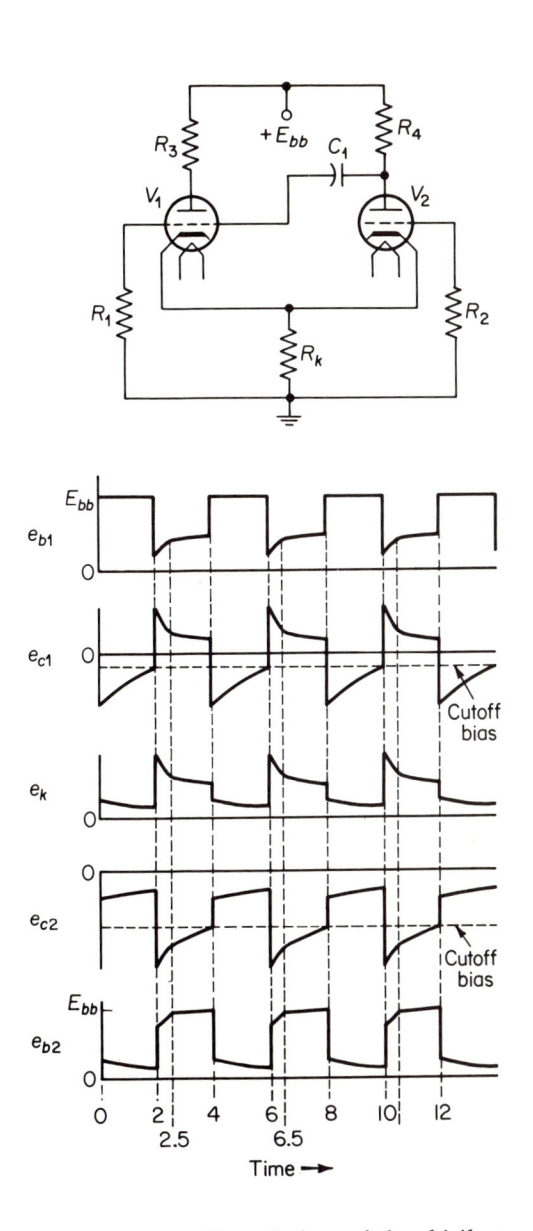

Figure 12-7 An astable cathode-coupled multivibrator.

is at its least negative bias value. Because V_1 is cut off, the voltage across R_k is approximately at its minimum value.

Before V_2 starts conducting, capacitor C_1 is charged to the value of supply voltage E_{bb}. When V_2 conducts, the voltage across the capacitor drops rapidly. The charge across C_1 cannot change instantaneously. This means that the full voltage drop exists across R_1, and is sufficient to drive V_1 beyond cutoff. Capacitor C_1 then starts to discharge to the lower value of the V_2 plate voltage. As the capacitor discharges, the voltage across resistor R_1 decreases and the grid bias of V_1 becomes progressively less negative until a point is reached where it exceeds cutoff and the tube conducts.

The voltage waveforms indicated at the various electrodes are for symmetrical operation. At $t = 0$, the grid of V_1 is at its most negative value, the plate voltage of V_1 is maximum, the cathode voltage is close to minimum, the grid voltage of V_2 is close to its maximum value, and the plate voltage of V_2 is near its minimum.

First consider the waveforms of V_2 for the time interval $t = 0$ to $t = 2$. Both the plate current of V_2 and the discharge current of C_1 pass through resistor R_k. As the capacitor discharge current decreases, the voltage across R_k drops, lowering the negative bias applied to V_2. This increases the plate current and causes a decrease in the plate voltage. The magnitude of the discharge current, however, is small in comparison to the plate current, and the changes in cathode, grid, and plate voltages are relatively small.

Now consider the waveforms of V_1 for the same period. Capacitor C_1 discharges through R_1, R_k, and the low plate resistance of V_2 at a rate determined by the circuit time constant. Because R_k is a small-value resistance, this constant is primarily a function of the $R_1 C_1$ network. The grid voltage increases as the current decreases exponentially. In a practical circuit, the grid voltage deviates somewhat from the discharge curve shown because of the change in bias voltage developed across R_k. Because the tube is operating beyond cutoff, the plate voltage remains at a constant value equal to the supply voltage, E_{bb}.

At time $t = 2$, the grid voltage of V_1 reaches cutoff and the tube conducts. This causes a greater drop across R_k which, in turn, causes a decrease in the plate current of V_2. The plate voltage of V_2 increases, and this increase is coupled back to the grid of V_1. This starts another cycle; and, within a very short period of time, the second steady-state condition is reached, with V_1 conducting and V_2 cut off.

During their respective conduction periods, V_1 passes a greater

Sec. 12.6   Astable cathode-coupled multivibrator   235

current than V_2 because the grid of V_1 is driven more positive by the additional feedback through capacitor C_1. The negative-bias voltage developed across R_k is, therefore, greatest when V_1 conducts, and becomes sufficiently high to cut off V_2.

In this circuit, the tubes are identical and are operated at the same plate voltage. Neither tube can cut itself off by means of cathode bias when that bias is the result of plate current through the cathode resistor and where the value of plate current is determined only by cathode bias. The plate voltage of V_2 is, however, coupled to the grid of V_1. When the plate voltage of V_2 is high, the grid of V_1 is at a higher potential than the grid of V_2. Under this condition, the cathode bias across R_k can cut off V_2 but not V_1. When the plate voltage of V_2 is low, the grid of V_1 is at a lower potential than the grid of V_2. It is, therefore, possible for V_2 to conduct and for V_1 to be cut off. Thus, at $t = 2$, with V_1 conducting and V_2 cut off, e_{c1} immediately rises to a higher positive value than the cathode voltage. Capacitor C_1 starts to charge through the low grid-to-cathode resistance toward the high value of the V_2 plate voltage. The plate voltage of V_1 simultaneously drops to its minimum value. The cathode voltage across R_k increases to a maximum value as the result of the high value of e_{c1}. This high cathode voltage makes the grid of V_2 highly negative with respect to the cathode. The plate voltage of V_2 increases to a high value, but is prevented from reaching its maximum value, even though the tube is cut off, by the capacitor-charging current through resistor R_4.

At $t = 2$, the large positive voltage coupled through C_1 from the plate of V_2 drives the grid of V_1 positive. A low-resistance grid-to-cathode path exists, and capacitor C_1 charges rapidly to the plate voltage of V_2 through resistors R_4 and R_k. As C_1 charges, the positive bias on the grid of V_1 decreases exponentially. At $t = 2.5$, the positive bias on V_1 is equal to the voltage drop across R_k, and the net grid-to-cathode voltage is zero. The grid no longer draws current, and the grid-to-cathode resistance path for the charging of C_1 ceases to exist.

It is shown in a later discussion of the equivalent circuit that, after time $t = 2.5$, the capacitor-charging time constant becomes greater than the charging time constant from $t = 2$ to $t = 2.5$. Capacitor C_1, therefore, charges at a slower rate. The plate voltage of V_1 follows a similar curve, but in the opposite direction, first dropping to a low value and then gradually increasing. The voltage developed across R_k reaches its maximum value at $t = 2$ and declines as the grid voltage and charging current become smaller. The plate voltage of V_2 increases gradually to the value of E_{bb} as the current through R_4 decreases. Because there is no current through R_2, the grid-to-ground voltage of V_2 remains at zero during this entire period. There is, however, a large positive voltage across R_k, and the grid-to-cathode voltage of V_2 is beyond cutoff. As the voltage across R_k declines, the effective grid-to-cathode voltage of V_2 rises.

At $t = 4$, the grid-to-cathode voltage of V_2 exceeds the cut-off value, and V_2 starts to conduct. The plate voltage of V_2 drops rapidly and another cycle begins.

12.7 Equivalent circuit analysis

Two separate charging paths and one discharging path must be considered for the astable cathode coupled multivibrator, as shown in Fig. 12-8. The discharge path of C_1 is shown in (a); this path exists when V_2 is conducting. The discharge time constant is determined primarily by R_1C_1, since R_1 is normally much larger than R_k and R_{p2}.

The equivalent circuit for the charging period $t = 2$ to $t = 2.5$ is shown in (b). The time constant in this circuit depends on C_1R_4, since R_{g1} and R_k are small values of resistance which shunt R_1. When the grid-to-cathode voltage of V_1 is zero at $t = 2.5$, however, R_{g1} becomes high in value and the equivalent circuit changes to that shown in (c). The time constant is primarily dependent on R_1C_1.

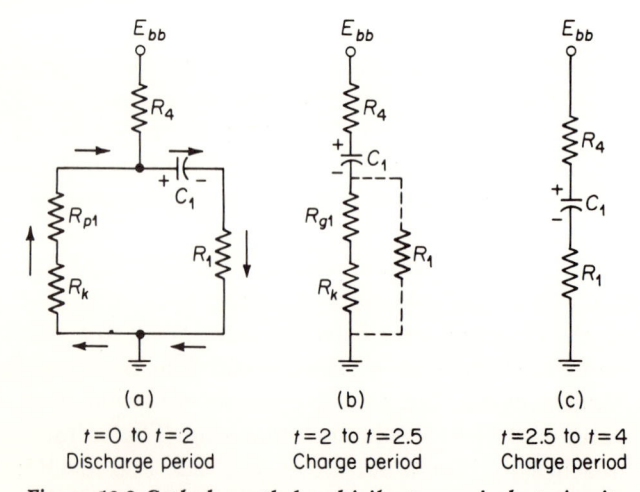

Figure 12-8 Cathode-coupled multivibrator equivalent circuits.

Sec. 12.8   *Electron-coupled astable multivibrator*   237

The waveforms shown for the cathode-coupled multivibrator are for symmetrical operation. Any desired asymmetrical waveform can be obtained by changing the constants of this circuit. This is usually done by changing the load resistors and/or using different power supplies.

12.8 Electron-coupled astable multivibrator

It is sometimes necessary to provide isolation between a load and the oscillating circuit. An electron-coupled multivibrator, Fig. 12-9,

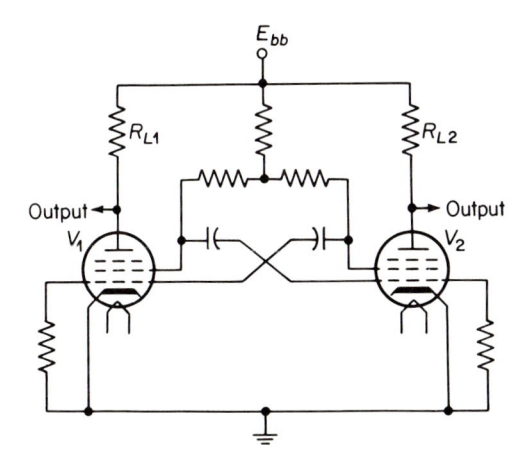

Figure 12-9 Electron-coupled multivibrator.

can be used for this purpose. Load resistors R_{L1} and R_{L2} are coupled to the oscillatory circuit through the electron streams of their associated tubes. The screen grids of the pentodes act as the plates of triodes. The circuit shown is an astable MV with coupling from the screen grid of one tube to the control grid of the other. The operation is identical to that described using triodes. Although the plates are not connected to the multivibrator circuit physically, a portion of the electron stream in the conducting tube reaches the plate because of its high positive potential. This furnishes the output signal to the load. Changes in the load do not affect the operation of the oscillatory circuit. The frequency of operation depends on the multivibrator time constants.

12.9 Frequency of multivibrator

It has been mentioned that the duration of the vacuum-tube astable multivibrator is determined primarily by the time constants R_1C_1 and R_2C_2, the grid cut-off potential, and the amplitude of negative voltage applied to the cut-off stage.

Referring back momentarily to Fig. 12-5, it can be seen that a complete cycle of operation occurs between times $t = 1$ and $t = 3$. If we label the period T, it is apparent that a cycle is the sum of two intervals: T_1, which equals $(t_2 - t_1)$, and T_2, which equals $(t_3 - t_2)$. For symmetrical operation $T_1 = T_2$. Provided that R_1 and R_2 are returned to the cathode as in Fig. 12-5, the period T is given approximately by the expressions

$$T = T_1 + T_2$$
$$= (R_1C_1 + R_2C_2) \ln \frac{\mu E_o}{E_{bb}} \qquad (12\text{-}1)$$

where μ = the amplification factor of the tube

E_o = the drop in plate voltage; i.e., $E_{bb} - e_p$ minimum

E_{bb} = the supply voltage

The frequency of the multivibrator is simply the reciprocal of the period; that is

$$f = \frac{1}{T} = \frac{1}{T_1 + T_2} \qquad (12\text{-}2)$$

The frequency of an astable multivibrator using a single set of components can be varied over a range of about ten to one by returning the grid resistors R_1 and R_2 to an adjustable positive bias voltage source. This circuit arrangement is shown in Fig. 12-10(a).

As Figs. 12-10(b) and (c) indicate, the positive bias shortens the time intervals T_1 and T_2 by causing the instantaneous grid potential to reach the cut-off value earlier in the discharge period. The period of the circuit of Fig. 12-10(a) is given by

$$T = (R_1C_1 + R_2C_2) \ln \frac{E_{cc} + E_o}{E_{cc} + (E_{bb}/\mu)} \qquad (12\text{-}3)$$

where E_{cc} = the bias voltage.

Noise voltages cause small random variations in the period T of successive cycles of the astable multivibrator. The total noise voltage superimposed on the instantaneous grid voltage of the *off* stage slightly affects the time at which the *off* stage conducts. (The total noise voltage consists of the tube noise of the *on* stage, and the resistance noise of the plate load of the *on* stage and the grid resistor

Sec. 12.9   Frequency of multivibrator   239

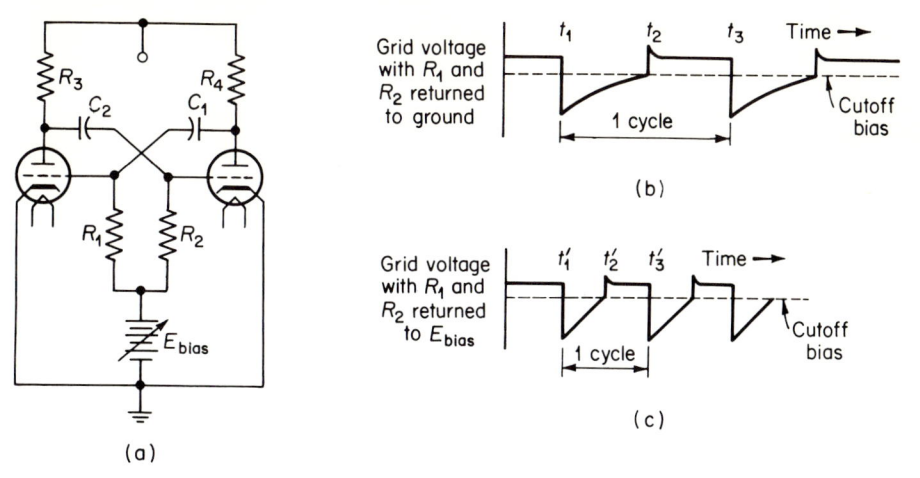

Figure 12-10 Frequency variation by returning R_1 and R_2 to an adjustable positive bias voltage source.

of the *off* stage.) This phenomenon is called *jitter*, and is minimized by the use of a positive bias voltage.

In concluding the discussion of vacuum-tube astable multivibrators, it should be pointed out that by using a suitable switching arrangement to obtain different R_1C_1 and R_2C_2 time constants, and by using a variable positive bias voltage, a great deal of versatility can be realized.

In the transistor astable multivibrator circuit of Fig. 12-6, if the time constants R_3C_2 and R_4C_1 are short compared to R_2C_2 and R_1C_1, respectively, the collector voltages rise to steady-state values rapidly.

Assuming that Q_1 is *off* and that Q_2 is bottomed, the collector voltage during the *off* interval is described by the expression

$$v_{c1} = V_{cc}(1 - \epsilon^{-t/t_1}) \tag{12-4}$$

where
$$t_1 = R_3C_2$$

When Q_1 switches *on*, its collector is at ground potential (for practical purposes) and the base voltage of Q_2 becomes $-V_{cc}$ with respect to ground, and can be described by the expression

$$v_{b2} = (V_{BB} + V_{cc})(1 - \epsilon^{-t/t_2}) - V_{cc} \tag{12-5}$$

where
$$t_2 = R_2C_2$$

If v_{b2} in Eq. 12-5 is set to zero, the *off* time interval T_1 of Q_1 may be determined by

$$0 = (V_{BB} + V_{cc})(1 - \epsilon^{-t/t_{12}}) - V_{cc} \tag{12-6}$$

and

$$T_1 = -t_2 \ln\left(\frac{V_{BB}}{V_{BB} + V_{CC}}\right) \tag{12-7}$$

Similarly, with Q_2 *off* and Q_1 *bottomed*

$$v_{c2} = V_{CC}(1 - \epsilon^{-t/t_3}) \tag{12-8}$$

where

$$t_3 = R_4 C_1$$

Also

$$v_{b1} = (V_{BB} + V_{CC})(1 - \epsilon^{-t/t_4}) \tag{12-9}$$

where

$$t_4 = R_1 C_1$$

Also

$$0 = (V_{BB} + V_{CC})(1 - \epsilon^{-t/t_4}) - V_{CC} \tag{12-10}$$

and

$$T_2 = t_4 \ln\left(\frac{V_{BB}}{V_{BB} + V_{CC}}\right)$$

The period T is then equal to

$$T = T_1 + T_2 \tag{12-11}$$

and the frequency is simply the reciprocal of the period; that is

$$f = \frac{1}{T} \tag{12-12}$$

12.10 Synchronization of the MV

Astable multivibrators do not have good frequency stability because a change in tube characteristics, or a change in the value of capacitance or resistance, affects frequency stability. To obtain a stable frequency, it is necessary to synchronize (lock-in) the multivibrator frequency with a sine wave or pulse series of stable frequency. The synchronizing signal is often referred to simply as a *sync* or *sync signal*.

Figure 12-11 shows the grid-vóltage waveform of one stage of an astable plate-coupled multivibrator during the cutoff interval. In the absence of sync, in (a) of Fig. 12-11, a certain time interval is required for the capacitor to discharge sufficiently for the control-grid voltage to reach cutoff. When a sine wave, as in (b), is applied to the control grid, however, cutoff is reached sooner and the tube conducts because the positive alternation of the sync signal adds to the grid bias. The multivibrator output may therefore be synchronized by applying a sine wave to the control grid of either stage.

In (b) and (c) of Fig. 12-11, the sync signal increases the frequency of the multivibrator. The frequency of the sync signal in

Sec. 12.10   Synchronization of the MV   241

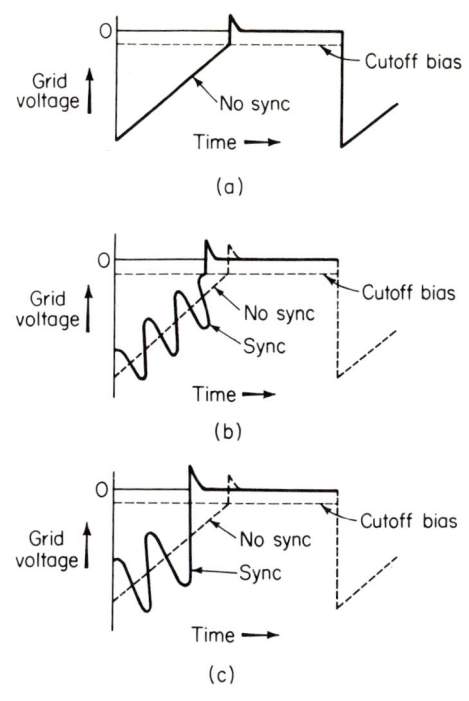

Figure 12-11 Grid voltage waveform at one stage of an astable plate coupled MV during the cutoff interval (a) with no sync (b) effect of applying sync (c) effect of increasing sync amplitude.

(b) is three times the frequency of a synchronized half-cycle of the multivibrator. The frequency of the sync signal is, therefore, six times the frequency of the multivibrator. Consequently, the frequency of the multivibrator is one-sixth the frequency of the sync signal. For example, assume that the normal free-running frequency of the multivibrator is 12,000 cps. If a sync signal having a frequency of 73,800 cps is applied, the output frequency is 12,300 cps (73,800 ÷ 6).

The ratio of multivibrator frequency to sync signal frequency can be changed by changing the amplitude of the sync signal. For example, if the sync-signal amplitude is increased, cutoff is reached earlier, as shown in Fig. 12-11(c). With an increase in sync-signal amplitude, the frequency increases because cutoff is reached after the second cycle of sync signal instead of after the third as in part (b). The frequency of the multivibrator using the sync signal of Fig. 12-11(c) is only one-fifth the frequency of the applied signal, since three sync cycles still occur during the conduction period. If the sync frequency is 73,800 cps, the output frequency of the multivibrator is $73,800 \div 5 = 14,760$ cps for the waveform shown in (c). Synchronization to improve frequency stability may be applied to any type of multivibrator. Also, the sync voltage does not have to be applied to the grid of a tube, but may just as readily be applied to the cathode, provided the polarity of the sync signal is reversed.

A more effective method than sine-wave synchronization is synchronization by a pulse series. The improvement results from the fact that there is a sudden well-defined change of voltage when a pulse is superimposed upon the grid voltage as shown in Fig. 12-12. In sine-wave synchronization, the change in voltage is gradual; as a result, the cut-off time may be reached at slightly different times in successive cycle. In pulse synchronization, however, the time at which the grid voltage rises above cutoff is very definite, and the frequency is more stable.

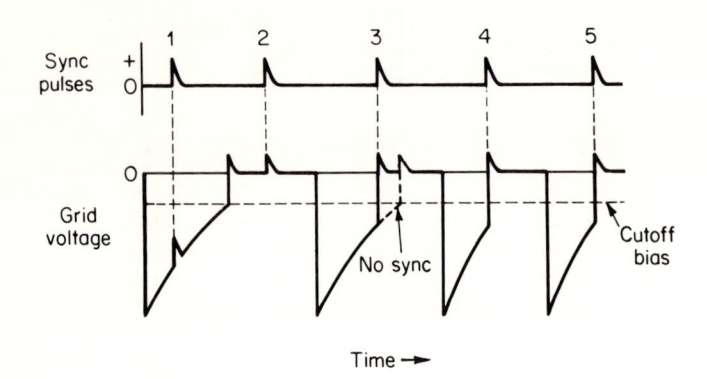

Figure 12-12 Synchronization by positive pulses.

Figure 12-12 shows the synchronization of a multivibrator output by means of positive pulses applied to the grid of either stage of a symmetrical multivibrator. Only pulses that cause the tube to conduct affect the frequency and duration of the multivibrator output. For example, the first sync pulse does not increase the grid voltage above cutoff and the nonconduction period is unchanged. The second pulse occurs during the conduction period and does not affect the frequency or duration of the multivibrator output. The third pulse, however, causes the grid voltage to rise to cutoff earlier than it would otherwise and does affect the frequency of the multivibrator. Notice that for positive synchronizing pulses, synchronization can occur only during the cut-off period of the input tube, and the sync pulse must be large enough to raise the grid to cut-off level.

Sec. 12.10 Synchronization of the MV   243

Multivibrators may also be synchronized by means of negative sync pulses. When a negative sync pulse is superimposed on the grid voltage, it produces synchronization when the stage is conducting. This is opposite to the action described above for a positive sync pulse. One method of producing synchronization is to cause the sync pulses to have sufficient amplitude to cut off the conducting tube when it is applied, but this method is not generally used. The usual method of sync does not require the pulse to have sufficient amplitude to cut off the input tube directly, but only sufficient amplitude to start the switching action.

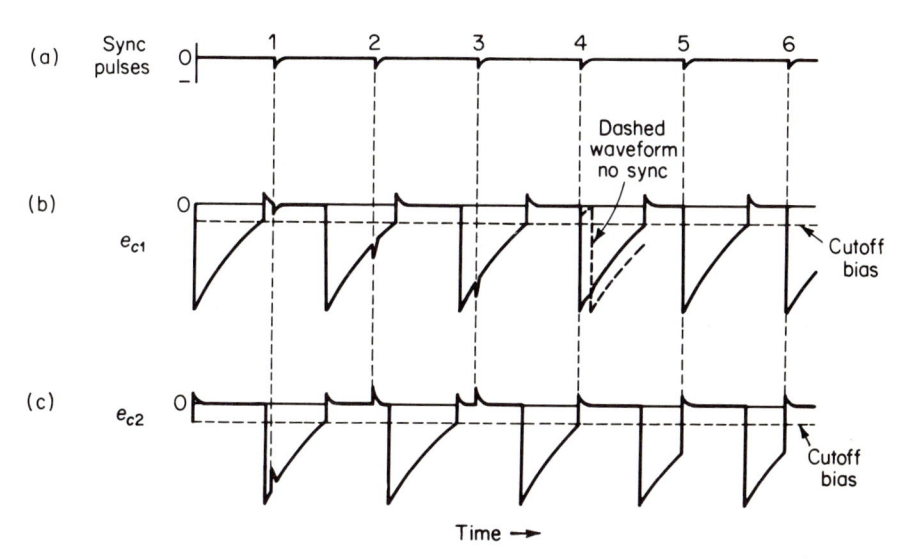

Figure 12-13 Use of negative sync pulses.

For example, refer to Fig. 12-13. The first sync pulse occurs when V_1 is conducting. The resulting decrease in e_{c1} does not produce cutoff, nor is the amplified and inverted pulse at the grid of V_2 sufficient to bring that tube out of cutoff. The second pulse does not affect multivibrator operation because it cannot produce a change in V_1 plate current. The third pulse also occurs during V_1 cutoff and has no effect on operation. The fourth pulse does not reduce e_{c1} to cutoff, but the amplified and inverted pulse at the grid of V_2 is of sufficient amplitude to cause conduction of tube V_2. Each pulse thereafter produces a synchronized switching action.

In the examples of pulse synchronization considered thus far, the multivibrator frequency equals the sync signal frequency when synchronization has been accomplished. It is also possible to synchronize

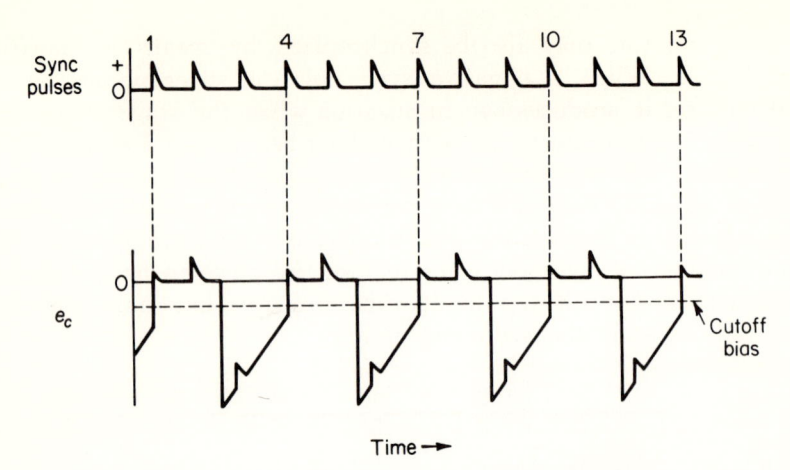

Figure 12-14 Synchronization with every third pulse.

a multivibrator by sync pulses that are a multiple of the natural multivibrator frequency. In Fig. 12-14, every third positive sync pulse produces a switching action. All other pulses occur when the tube is either cut off or conducting, and they do not affect the the multivibrator frequency. In this case, as shown in Fig. 12-14, the multivibrator frequency is equal to one-third of the sync frequency. For example, if the sync frequency is 120 kc, the multivibrator output frequency is $120 \div 3 = 40$ kc.

It is also possible to obtain synchronization when the multivibrator frequency is an integral or a whole-number multiple of the sync frequency. For example, if the multivibrator frequency is 120 kc and the synchronizing frequency is 60 kc, only every other cycle of multivibrator oscillation is controlled, but the other cycles tend to fall in line.

The principles of synchronization as described for vacuum tubes are also applicable of course to transistor circuits.

12.11 Typical applications

An unsymmetrical free-running multivibrator can be used to generate short rectangular pulses which may be used, for example, to initiate the sweep in a cathode-ray oscilloscope.

The circuit is arranged identically to that shown in Fig. 12-1, but the time constant $R_1 C_1$ is made small and the time constant $R_2 C_2$ is made large (or *vice versa*). Assuming the former arrangement, the time interval $(t_2 - t_1)$ is short compared to the time

Sec. 12.11   Typical applications   245

interval $(t_3 - t_2)$, and the output waveform, taken from the plate of V_2, is a negative pulse of short duration. To remove the pip produced by the rapid charge of C_2, the output signal can be passed through a clipper.

With the output pulse occurring during the time interval $(t_2 - t_1)$, its length is determined by the $R_1 C_1$ time constant and the potential to which R_1 is returned. For a very short pulse, R_1 can be returned to a positive bias voltage.

Another common use for the astable multivibrator is in oscilloscope electronic switching, which permits multiple displays to be shown simultaneously. A typical circuit arrangement is shown in Fig. 12-15.

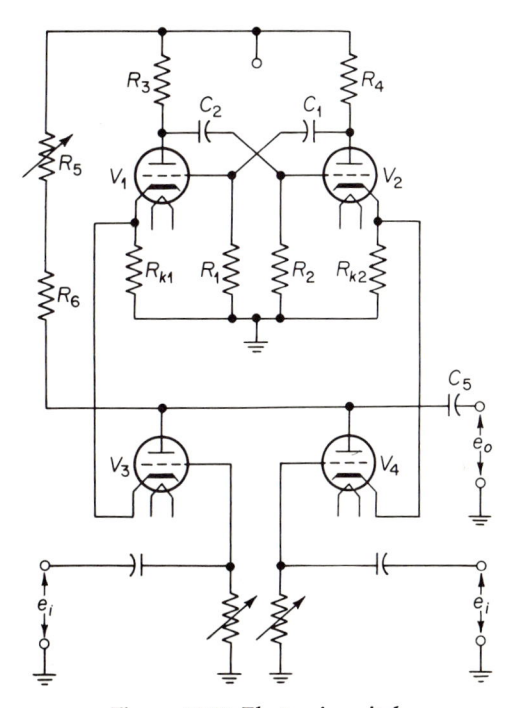

Figure 12-15 Electronic switch.

Stages V_1 and V_2 form an astable MV. The cathode of V_1 and V_3 are tied together, as are the cathodes of V_2 and V_4. When either MV stage is cut off, its cathode and that of the stage tied to it are at ground potential since there is no current through either R_{k1} or R_{k2} under cut-off conditions. When the cathode of V_3 is at ground potential (V_1 cut off), that stage conducts; similarly, V_4 conducts when its cathode is at ground potential (V_2 cut off). When the cathode of either V_3 or V_4 is positive with respect to ground (when V_1 or V_2 conducts), these stages are cut off. Since V_1 and V_2 are *on* and *off* alternately, V_3 and V_4 are also *off* and *on* in the same manner.

When V_3 is *on*, the input signal, which is applied at all times to its input terminals, is developed across R_6 and applied through C_5 to the vertical plates of the cathode ray tube of an oscilloscope. When V_4 is *on*, the input signal, applied at all times to its input terminals, is applied to the vertical plates of the CRT.

The switching action is sufficiently fast so that the signals appearing on the screen of the CRT do not decay markedly during the time interval between sweeps. Thus, both input signals are displayed simultaneously and can be subjected to any desired form of comparison.

A disadvantage of using an astable MV in the electronic switch is that the MV must be synchronized to the scope sweep, requiring frequent adjustments and often producing a serrated trace. A better switching arrangement is illustrated in the next chapter.

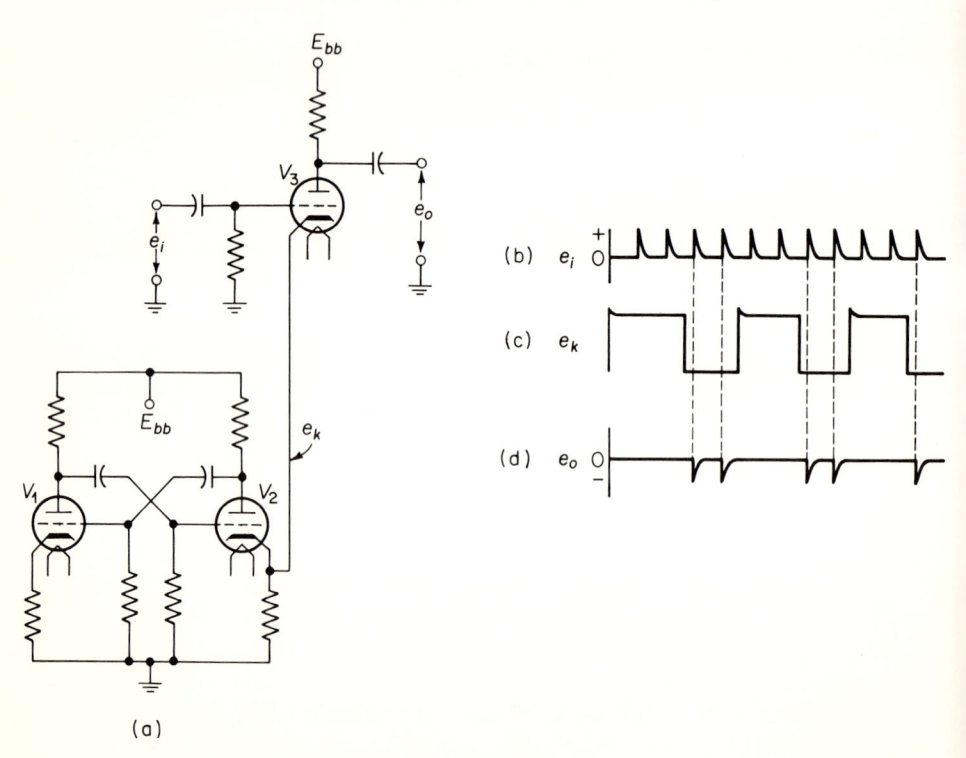

Figure 12-16 MV gating circuit.

Sec. 12.11   Typical applications   247

A multivibrator may also be used as a *gate* to pass signals to the output only during a desired time interval. The multivibrator gating circuit is shown in Fig. 12-16(a), with pertinent waveforms in (b), (c), and (d).

The pulse train, shown in Fig. 12-16(b), is applied to the input of V_3. When V_2 is cut off, V_3 conducts because its cathode is at ground potential, and pulses appear in the output, as shown in Fig. 12-16(d). When V_2 conducts, the voltage developed across R_{k2}, called the *gate voltage*, shown in Fig. 12-16(c), drives V_1 beyond cutoff. Obviously, for successful operation, the amplitude of the input pulse train must not exceed the positive bias applied to the cathode of V_3 during its *off* period.

A transistor astable MV that may be used for square-wave generation, frequency division and multiplication, and time-base generation is shown in Fig. 12-17. It is available from the manufacturer as a plug-in-module.

In this package, transistors Q_1 and Q_2 are connected in a MV circuit similar to the free-running vacuum-tube arrangement. The

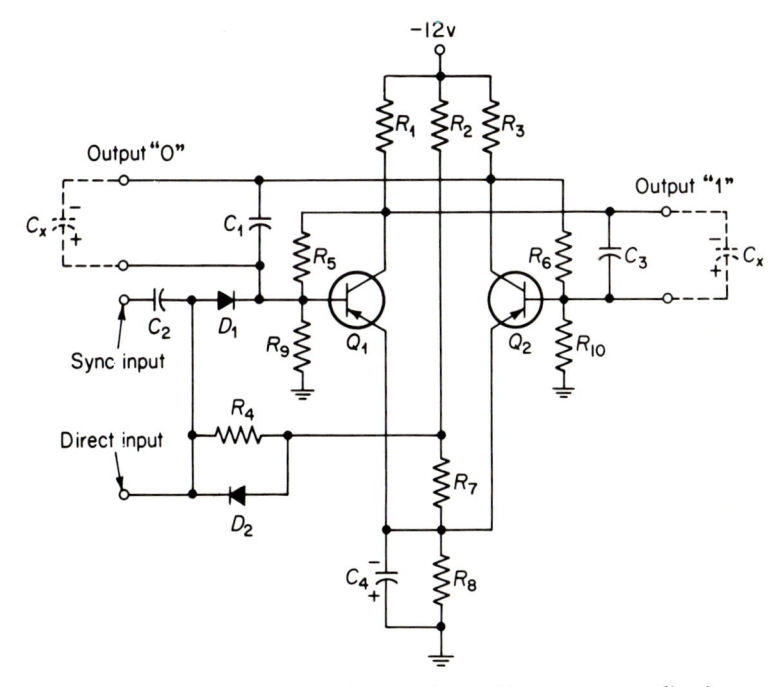

Figure 12-17 Transistor MV that can be used in numerous applications. (Courtesy of Engineered Electronics Co. and Kezer, C. F. & M. H. Aronson, One Hundred Electronic Circuits, Instrument Publishing Co., Inc., Pittsburgh, Pa., 1960.)

base of each transistor is biased negatively with respect to its emitter. When power is applied, however, one transistor, say Q_1, conducts first, and its collector potential rises to -3 v. A positive-going pulse is passed through capacitor C_3 to the base of Q_2, cutting off current in Q_2. As current decays logarithmically in capacitor C_3, the base voltage of Q_2 becomes equal to its emitter potential, base current flows, and Q_2 conducts. The change in collector voltage of Q_2 produces a positive-going pulse at the base of transistor Q_1, via capacitor C_1, cutting off current flow in Q_1. The cycles then continue in the free-running condition.

With the values of C_1 and C_2 shown, the circuit operates at its maximum free-running frequency of 380 kc. Equal-valued external capacitors, C_x, may be added to decrease the operating frequency to any desired value—down to one cycle per minute. The value of the external capacitors (in $\mu\mu$f) may be computed from the formula

$$C_x = 45 \; (T - 2.6)$$

where T is the desired period of oscillation in microseconds. The formula applies for periods as long as one second; for longer periods the capacitor values should be determined experimentally. Input triggers (pulses) can be used to synchronize the MV to a frequency somewhat higher than its free-running frequency. Synchronizing pulses are applied to the base of Q_1 through capacitor C_2 and diode CR_1. These pulses must be 6 to 9 v in peak-to-peak amplitude, positive-going, with rise time 0.1 to 1 μsec. Pulses of poorer rise time can be used by coupling through a larger external capacitor to the direct input terminal.

Triggering on noise pulses is prevented by a reverse bias of 1.5 v obtained across R_7, on diode CR_1. Positive-going input pulses of amplitude less than 1.5 v thus cannot pass through CR_1 to the transistor base, and have no effect.

The multivibrator can be synchronized to operate at a *harmonic* of the input trigger frequency (frequency multiplication) or at a *subharmonic* (frequency division). In either event, the input frequency must remain reasonably constant, and the MV free-running frequency must be set correctly because positive synchronization is obtained only over the range shown in the following table:

Harmonic	Locking Range	Harmonic	Locking Range
f	0 to $+25\%$	$f/3$	0 to $+4.5\%$
$2f$	0 to $+12\%$	$f/4$	0 to $+3.5\%$
$f/2$	0 to $+ 7\%$	$f/5$	0 to $+2.5\%$

Exercises 249

Outputs of opposite polarity are available from the two collectors. The nominal output is an 8 v shift from -11 v to -3 v, d-c. The output rise time (positive-going) is 0.4 μsec under a typical load; the fall time is approximately 25 per cent of the period.

REFERENCES

1. Chance, B., et al., *Waveforms*. New York: McGraw-Hill Book Co., Inc., 1949.
2. Neeteson, R. A, *Junction Transistors in Pulse Circuits*. New York: The Macmillan Co., 1959.
3. Kezer, Chas. F., and M. H. Aronson, *One Hundred Electronic Circuits*, v. 2. 2 Circuits Pittsburgh, Pa.: Instruments Publishing Co., Inc., 1960.

EXERCISES

12-1 Draw the circuit of a vacuum-tube astable MV and the voltage waveforms appearing at the grid and plate of each stage.

12-2 What factors determine the duration of an astable symmetrical MV?

12-3 Draw the circuit of an astable cathode-coupled MV, and show the voltage waveforms appearing at the plate and grid of each stage and across the common-cathode resistor.

12-4 Draw equivalent circuits of both the vacuum-tube and transistor types of astable MV and label each clearly.

12-5 Explain by means of equivalent circuits the operation of an astable cathode-coupled multivibrator.

12-6 If the period of an astable multivibrator is 100 μsec, what is its frequency of operation?

12-7 Using a single set of components, draw a circuit arrangement by which the frequency of an astable MV can be varied over an appreciable range.

12-8 Explain how the ratio of multivibrator frequency to sync signal frequency can be changed.

12-9 Draw a circuit illustrating how an astable MV may be used to permit the display of multiple traces on the screen of a cathode-ray oscilloscope.

12-10 In the astable multivibrator of Fig. 12-1, what factors can contribute to an initial unbalance and produce a switching action?

12-11 Explain how pulse generation circuits are generally classified.

12-12 Draw the circuit of an electron-coupled astable multivibrator and explain how it provides isolation between the load and oscillating circuit.

12-13 What prevents the simultaneous cutoff of both tubes in the cathode-coupled astable MV?

12-14 How does the use of a positive bias voltage minimize jitter?

12-15 Why is synchronization by a pulse series more effective than synchronization by a sine-wave signal?

13

BISTABLE
MULTIVIBRATORS

In the bistable multivibrator, an output pulse is obtained only if a driving (triggering) pulse is applied to the input. A full cycle of output is produced for every two triggering pulses properly applied and of correct polarity and amplitude.

The basic circuit, known as the *Eccles-Jordan trigger circuit* after its inventors, is shown in Figs. 13-1(a) and (b). Resistance coupling is used between the plates and grids of the two tubes in (a) and between the collectors and bases of the transistors in (b). The circuit has two stable states or conditions of balance; one when V_1 or Q_1 is conducting and V_2 or Q_2 is cut off, and the other when V_2 or Q_2 is conducting and V_1 or Q_1 is cut off. The circuit remains in one or the other of these stable states. There is no action to cause any of the electrode potentials to change while the circuit is in either steady-state condition.

The nonconducting stage is caused to conduct by the proper application of a triggering pulse. A rapid reversal then occurs from one steady state to the other. Because of this rapid reversal, the circuit is also referred to by various other names such as the *flip-flop, flip-flip,* and *flop-over*. The names

251

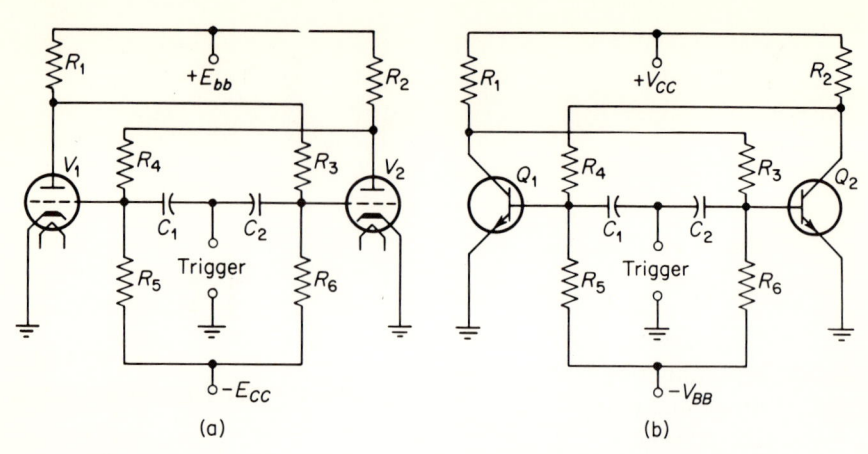

Figure 13-1 Bistable MV (a) vacuum-tube type and (b) transistor type.

regenerator, *binary counter*, *locking circuit*, and *frequency divider* have also been used in connection with this circuit.

13.1 Detailed operation

Let us first consider the vacuum-tube circuit shown in Fig. 13-1(a). Initially, it is assumed that no trigger pulses exist at the trigger input terminals and that plate voltage is applied. An unbalance in current then exists for the same reasons described for the astable circuit, and starts circuit operation.

If V_1 conducts the greater current, the voltage drop across R_1 increases and the plate voltage of V_1 decreases. This decrease, applied to the grid of V_2 through the R_3R_6 voltage-divider network, reduces the plate current and increases the plate voltage of V_2. This rise in plate voltage is, in turn, applied through voltage divider R_4R_5 to the grid of V_1, where it causes a further increase in plate current and decrease in plate voltage. The action is cumulative. When the grid of V_2 is driven to cutoff, a steady state is reached because any further decrease in grid voltage cannot increase plate voltage.

Tube V_1 continues to conduct and V_2 remains cut off in the absence of any external triggering voltage. If a narrow positive-going pulse is applied at the trigger input, it has little effect on V_1 which is already conducting heavily. If the triggering pulse is sufficiently large, however, it raises the grid voltage of V_2 above cutoff, and this tube conducts.

When V_2 conducts, its plate voltage drops. This causes a negative increase in the grid voltage of V_1 which, in turn, results in a reduction of plate current and an increase in plate voltage. This positive voltage increase is coupled to the grid of V_2, and the entire cycle is repeated. A new steady state is quickly reached, with V_1 cut off and V_2 conducting.

In a similar manner, the application of another positive-triggering pulse causes V_1 to conduct and V_2 to be cut off. Each time a positive triggering pulse of sufficient amplitude is applied, the condition of the circuit is reversed.

Now suppose a negative triggering pulse is applied to the circuit. With V_1 conducting and V_2 cut off, it has no effect on V_2, but the grid voltage and plate current of V_1 are reduced. This, in turn, applies a positive voltage to the grid of V_2. If the triggering pulse is made large enough to raise the grid voltage of V_2 above cutoff, this tube starts to conduct and the condition of the circuit is rapidly reversed. Thus, a negative triggering pulse of sufficient amplitude also causes a transition from one steady state to the other.

The operation of the transistor circuit, Fig. 13-1(b), is similar to that described for the vacuum-tube arrangement. Physically, the resistances are of lower values than those used in the tube circuit, since the transistors exhibit relatively high conductance.

During saturation, the collector dissipation of the conducting transistor is low, since the collector voltage is often less than one volt; also most of the supply voltage, V_{cc}, is available as output. Although collector dissipation increases greatly during switching from *on* to *off*, this switching action occurs so rapidly that the transistor is not likely to be damaged.

13.2 Self-biasing

The vacuum-tube binary is made self-biasing by using the circuit arrangement of Fig. 13-2(a). The bias voltage developed across R_k is applied to both tubes. Since the cathode current of one tube decreases as that of the other increases, there is little voltage variation across R_k and a bypass capacitor is not needed.

Capacitors C_1 and C_2 are called *speed-up* (also commutating) capacitors; they have a value of about 50 $\mu\mu$f each. Their function is to ensure that voltage changes at the plate of either tube are transmitted rapidly to the grid of the opposite tube. Because these capacitors cannot charge or discharge instantaneously, rapidly changing plate voltages are accompanied by nearly equal changes of grid

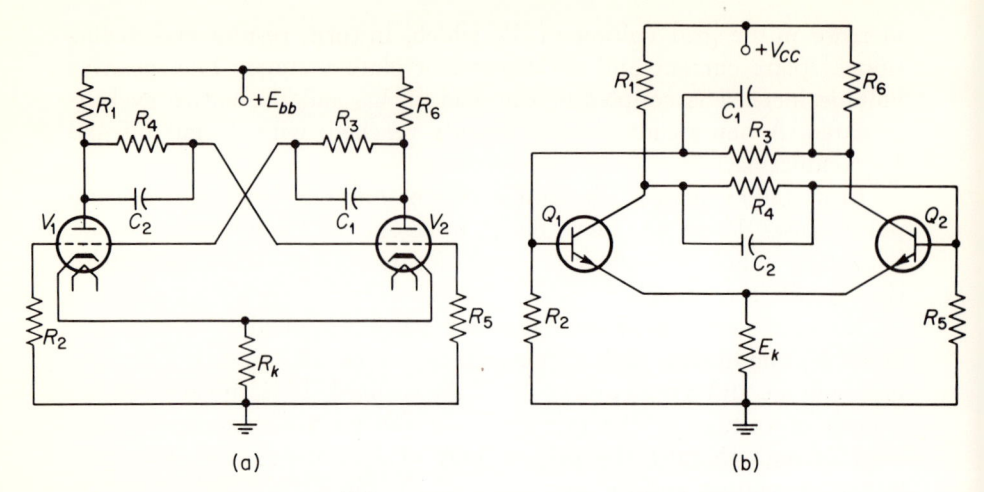

Figure 13-2 Self-biased binary (a) vacuum-tube circuit and (b) transistor circuit.

voltage. If these capacitors are omitted, it is possible that the binary will not respond to trigger pulses of short duration.

Suppose that a short trigger pulse is applied to the grid of V_1 and drives that tube beyond cutoff. Voltage e_{p1} rises toward E_{bb} at a rate determined by R_1 and the output capacitance of the tube. This increase is applied to the grid of V_2 through the resistive network R_4R_5, which introduces a delay. Voltage e_{g2} is further delayed by the charging of the V_2 input capacitance through a resistance equal to R_4 and R_5 in parallel. If the sum of both delays is too long, the trigger pulse applied to V_1 may end before e_{g2} is raised above cutoff. Under these conditions, the circuit would not respond to the trigger pulse.

To ensure circuit response, e_{g2} must be raised far enough beyond cutoff so that the voltage drop across R_6, resulting from the conduction of V_2, is sufficient to keep V_1 cut off in the absence of the input pulse. As e_{g2} rises above cutoff and V_2 begins to amplify, the input capacitance increases because of the Miller effect, and the time required for this capacitance to charge increases accordingly. It is this charging time that is of greatest importance. The addition of capacitors C_1 and C_2 introduces compensation, such that $R_4C_2 = R_5C_{i1}$, and $R_3C_1 = R_2C_{i2}$. In each case, C_i refers to the effective input capacitance of the tube.

The shortest time between triggering pulses for which the binary operates reliably is called the *resolution time*. To minimize this time, stray capacitances are reduced as much as possible. The values of

Sec. 13.3   Triggering vacuum-tube bistable multivibrators   255

the plate load and the coupling-network resistors should also be reduced to improve the rise time of the waveform at the plates and to reduce the recharging time of the speed-up capacitors.

A transistor counterpart of the tube circuit is shown in Fig. 13-2(b); its operation is basically the same as described above.

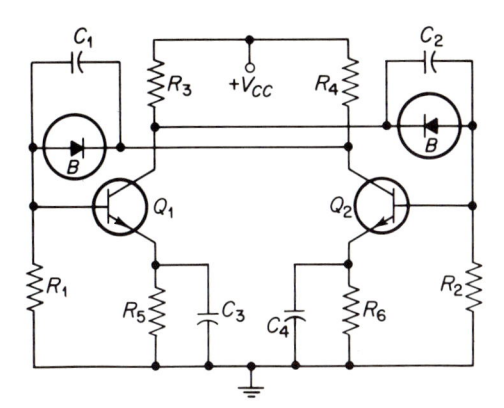

Figure 13-3 Using breakdown diodes as the coupling elements to improve switching time and static stability.

The arrangement shown in Fig. 13-3 is often used in transistor binary circuits[1], and occasionally in vacuum-tube circuits, to improve switching time. Breakdown diodes are used in place of the coupling resistors. It has been shown that a breakdown diode, at reverse voltages above the breakdown value, acts like a constant-voltage source in series with a low resistance. The breakdown diodes are selected to have a breakdown voltage that is equal to the required difference between the direct voltages of the collector and base. Because of the relatively low variational resistance of the diodes, almost the entire change of collector voltage is applied to the base.

Since the diodes maintain the voltage across the coupling capacitors practically constant, the capacitors do not shorten the switching time; this function is performed by the emitter bypass capacitors, C_3 and C_4. For reliable switching, the time constants R_5C_3 and R_6C_4 are made large in comparison with the duration of the switching pulse.

13.3 Triggering vacuum-tube bistable multivibrators[2]

Triggering pulses are usually applied through suitable devices rather than directly. Ideally, a triggering device should act as a switch which closes when the trigger is applied and then opens to disconnect the circuit from the trigger source. The multivibrator is then unable to follow the reversing tendency of the trigger, which is now changing in the wrong direction. In addition, the trigger source does not load the multivibrator.

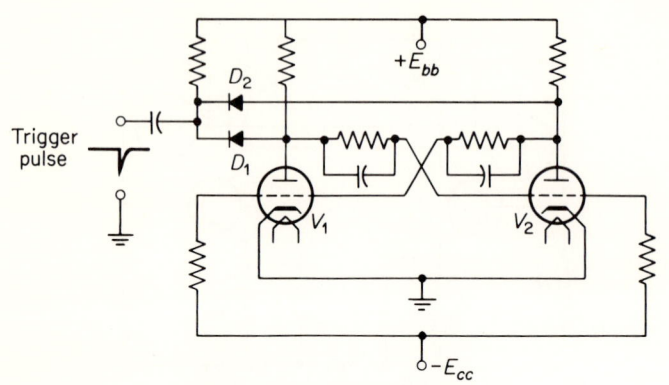

Figure 13-4 Diode triggering.

A satisfactory method of obtaining the desired isolation is by means of a series-connected diode. A crystal diode is preferred to a thermionic type because the crystal reduces the stray capacitance. By inserting either of these diode devices in the proper direction, triggers of either polarity can be applied to the circuit. An example of the use of crystal diodes for this purpose is shown in Fig. 13-4. Recovery time is also reduced by the *plate-catching* action of the diode. (It catches the plate at a voltage lower than the voltage E_{bb} to which the plate returns if the diode is not used.)

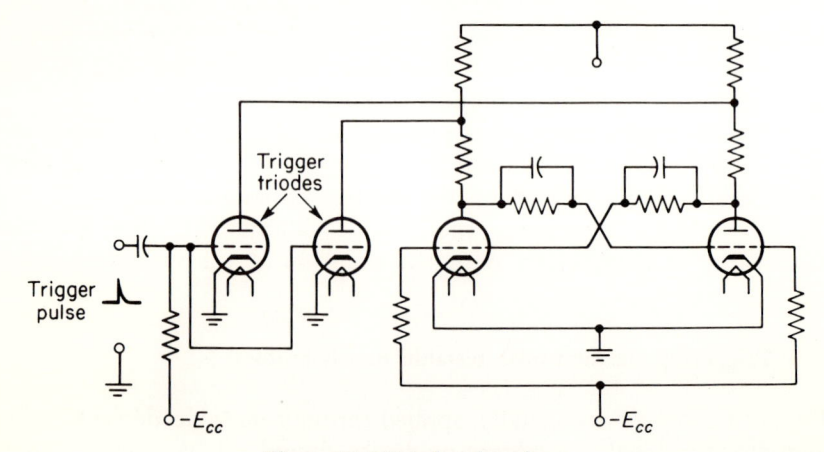

Figure 13-5 Triode triggering.

Sec. 13.4 *Triggering transistor bistable multivibrators* 257

Triggering pulses may also be applied by using triodes (instead of diodes) arranged as shown in Fig. 13-5. The triggering triodes are normally biased beyond cutoff and do not load the multivibrator. When positive triggering pulses are applied, the tubes momentarily conduct. The normal reversal in voltage between grid and plate causes a negative trigger pulse to be applied to the multivibrator.

The triggering pulses can also be applied directly through a coupling capacitor as in Fig. 13-1, but precautions must be taken to prevent the trigger pulse from being differentiated, in which case the overshoot turns the multivibrator off again just as soon as it begins to trigger. If this arrangement is used, a large-value capacitor is best (to provide a long time constant) and it should be connected to a point, such as a grid or cathode, that does not change rapidly during the switching action.

Figure 13-6 shows an example of a binary with positive triggering pulses injected at both cathodes across a small common-cathode resistor. The input signal is derived from a low-impedance source.

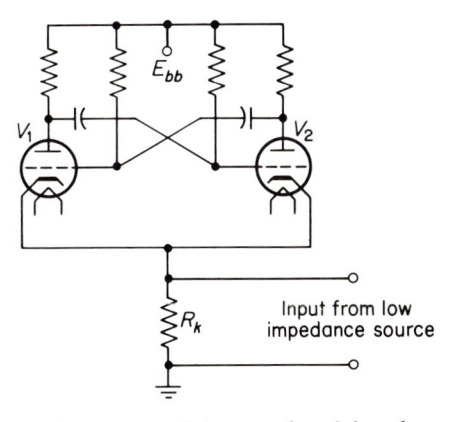

Figure 13-6 Trigger pulses injected at cathode across a small common resistor R_k.

13.4 Triggering transistor bistable multivibrators

In Fig. 13-7, separate inputs are available at the bases of Q_1 and Q_2. Assuming that Q_1 is initially *off*, a transition can be produced by applying either a negative trigger to the base of Q_1 or a positive trigger to the base of Q_2. The circuit can then be restored to its initial condition by the application of a triggering positive pulse to

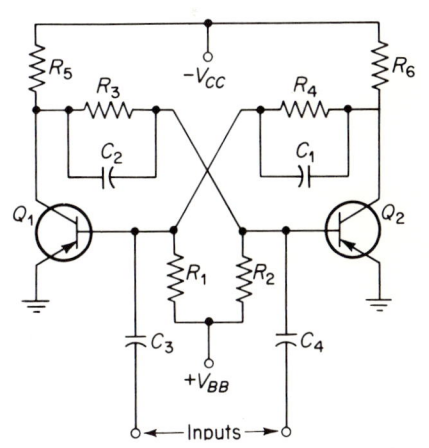

Figure 13-7 Transistor binary with separate base inputs through capacitors C3 and C4.

the base of Q_1 or a negative trigger to the base of Q_2. Capacitors C_1 and C_2 are speed-up capacitors.

In some applications, it is desirable to trigger the flip-flop *on* and *off* repeatedly with input pulses of the same polarity. If the pulse were applied simultaneously to both transistors, switching would be delayed. Assume, for example, that a negative pulse is applied to a bistable MV using PNP transistors. The tendency is to drive the *off* transistor into conduction and the *on* transistor into saturation. Thus the feedthrough from the collector of the *off* transistor must overcome both the initial bias of the *on* transistor and the additional bias provided by the input pulse. This necessitates increased signal voltage, and the transition process is slowed. The rise and fall times of the output signal are increased, thereby decreasing the pulse repetition rate (prr) of the circuit.

This unfavorable condition can be avoided by the use of *pulse-steering diodes*[3]. Either negative or positive pulse steering can be used in a bistable MV circuit. Again assuming the use of PNP transistors, a negative pulse-steering circuit is illustrated in Fig. 13-8(a).

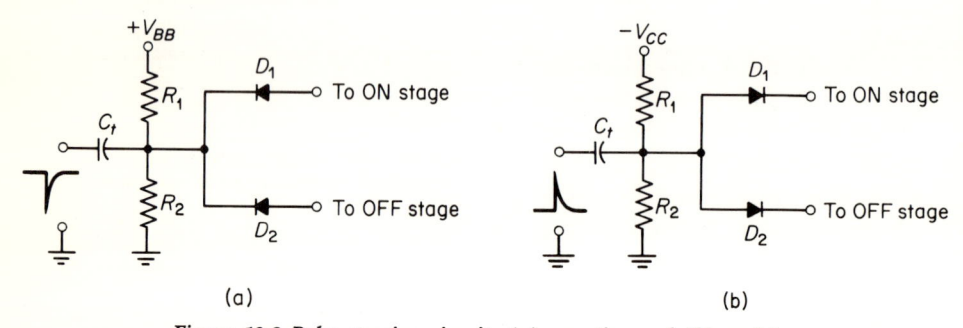

Figure 13-8 Pulse steering circuits (a) negative and (b) positive.

The cathodes of D_1 and D_2 are at a positive potential with respect to ground as a result of the voltage divider action of resistors R_1 and R_2, which are connected between $+V_{BB}$ and ground. Diode D_1 is reverse-biased, and diode D_2 is forward-biased. When a negative trigger pulse is applied through capacitor C_t, the reverse-biased diode D_1 keeps the pulse from being applied to the base of the *off* transistor; but the forward-biased diode D_2 permits the pulse to be applied to the base of the *on* transistor and the next trigger pulse passes through D_2 to that stage and causes a second circuit transition back to the original steady state. A negative-going trigger pulse has no effect on the circuit since it increases the reverse bias on the diodes.

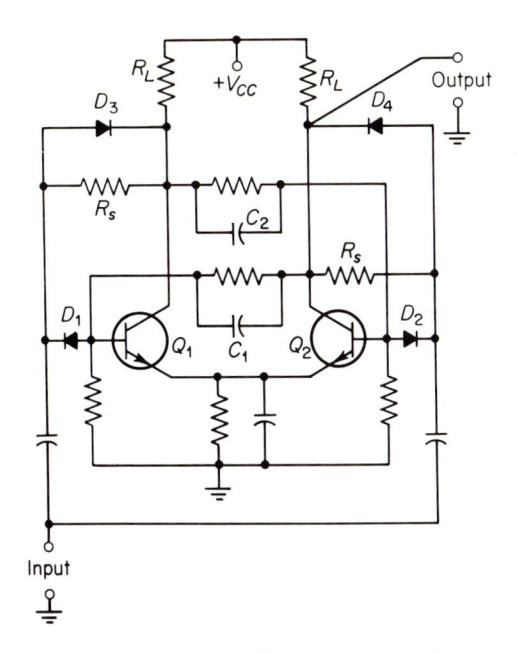

Sec. 13.4   *Triggering transistor bistable multivibrators*   259

Figure 13-9 Transistor flip-flop using speed-up capacitors and steering diodes.

Figure 13-9 shows a flip-flop using speed-up capacitors and steering circuits. The steering circuits consist of the diodes D_1, D_2, D_3, and D_4, which are connected to the transistor bases and to the two resistors R_s. The diodes connected across R_s are required for the circuit to respond at the maximum repetition rate. If several flip-flops are cascaded to form a *binary counter* (to be studied later), only the first and possibly the second stage require diodes across R_s.

Negative base triggering is used for all flip-flops; that is, for NPN transistors, negative pulses are applied to the base and used to turn the *on* transistor *off*. A flip-flop usually must drive a number of circuits; therefore, its output must be sufficient to handle the load. When flip-flops are cascaded, the output voltage or current must be greater than the input voltage or current amplitude required for triggering the next stage. The input pulse width and amplitude are also important and will usually determine the input and cross-coupling capacitor values.

The speed-up capacitors are selected after designing the basic flip-flop. They should have the smallest value which will give the fastest operation for all units at about 50°C. The value of the input capacitor may vary in size from about 100$\mu\mu$f to as high as 3000$\mu\mu$f.

The value of R_s in Fig. 13-9 may be approximated by using

$$R_s C_i = \frac{1}{f_{max}}$$

where R_s = the steering resistor
C_i = the trigger input capacitor
f_{max} = the maximum repetition rate of the input pulses

Good results should be obtained if $3R_L < R_s \leq 10R_L$.[4]

13.5 Nonsaturated flip-flops[3]

In all of the circuits studied thus far, the transistor is assumed to switch between cutoff and saturation. These circuits generally are not as fast as flip-flops using transistors that do not saturate, because the storage time of minority carriers in a saturated transistor adds delay time. With a given dissipation rating for a transistor, a saturated flip-flop can switch more current than one using a nonsaturated transistor. Also, for a given current level, there is less average dissipation in the saturated flip-flop because the two operating points have low dissipation; the *on* transistor is at low voltage and high current, while the *off* transistor is at high voltage and low current. The power loss at both of the steady-state points is small, whereas the power loss incurred during the transient period in switching from one state to the other may be negligible at low frequencies but can be significant at high frequencies. On the other hand, the *on* point of a nonsaturated transistor is a point of high dissipation and represents a large amount of power. Therefore, the switched current may be kept lower.

A flip-flop using nonsaturated transistors can switch the same amount of load power at a faster rate than a saturated flip-flop. Since the transistor is never saturated, storage time is low; however, there is more lost power. Lower power efficiency is the result of a demand for speed.

One method of achieving nonsaturated operation is to hold the transistors out of saturation with a breakdown (reference) diode and a clamping diode. A typical circuit arrangement is shown in Fig. 13-10. When transistor Q_1 is *on*, diode D_1 conducts and clamps the collector voltage above the base voltage by the reference voltage of D_{B1}, thus keeping the transistor out of saturation. Diode D_{B1} is passing more than enough current to control the collector current and voltage. When transistor Q_1 is *off*, diode D_1 does not conduct and

Sec. 13.5   Nonsaturated flip-flops

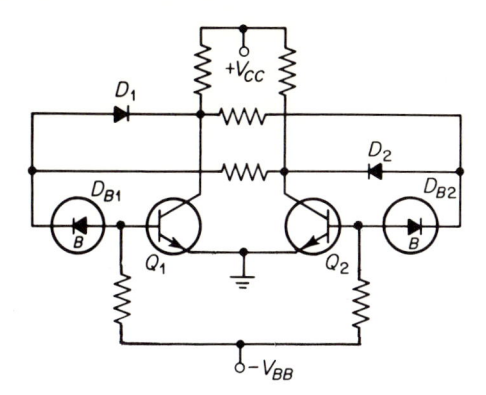

Figure 13-10 A simple circuit arrangement for nonsaturated operation.

D_{B1} still maintains a reference voltage but now passes only a very small current.

Coupling capacitors, input capacitors, and steering circuits may be added to the circuit of Fig. 13-10. Such an arrangement is shown in Fig. 13-11. Note, however, that the input steering diode D_s connects to the junction of D_1 and D_{B1} of Fig. 13-11, rather than to the base of the transistor. Smaller trigger voltages are required with this connection.

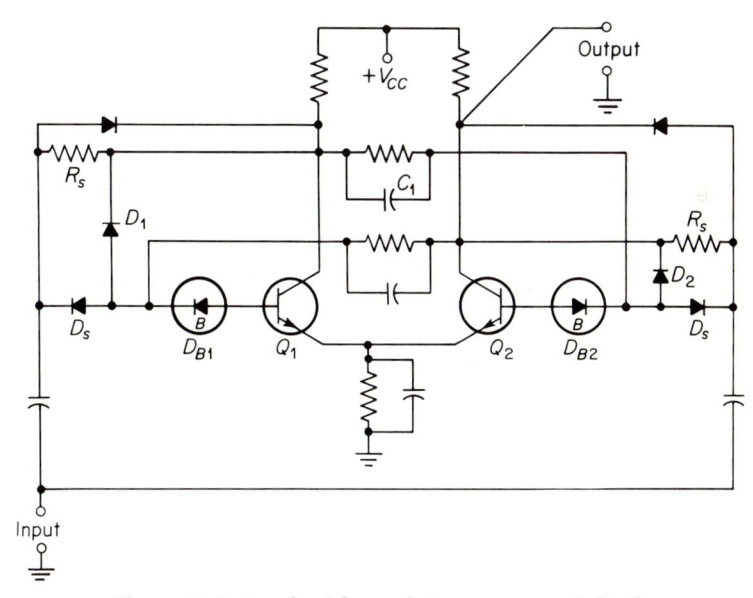

Figure 13-11 A refined form of the nonsaturated flip-flop.

13.6 Cathode-coupled binary[5]

A cathode-coupled binary, frequently called the *Schmitt trigger* circuit after its inventor, is shown in Fig. 13-12. Notice that one grid-to-plate coupling network has been replaced by common-cathode coupling across the unbypassed resistor R_k from which the circuit derives its name. Two stable states occur: one with V_2 conducting and V_1 cut off as the result of bias developed across R_k; and the other with V_1 conducting and V_2 cut off. The switching action is initiated either by the use of a trigger pulse or by raising or lowering the grid of V_1. The latter method is used in Fig. 13-12.

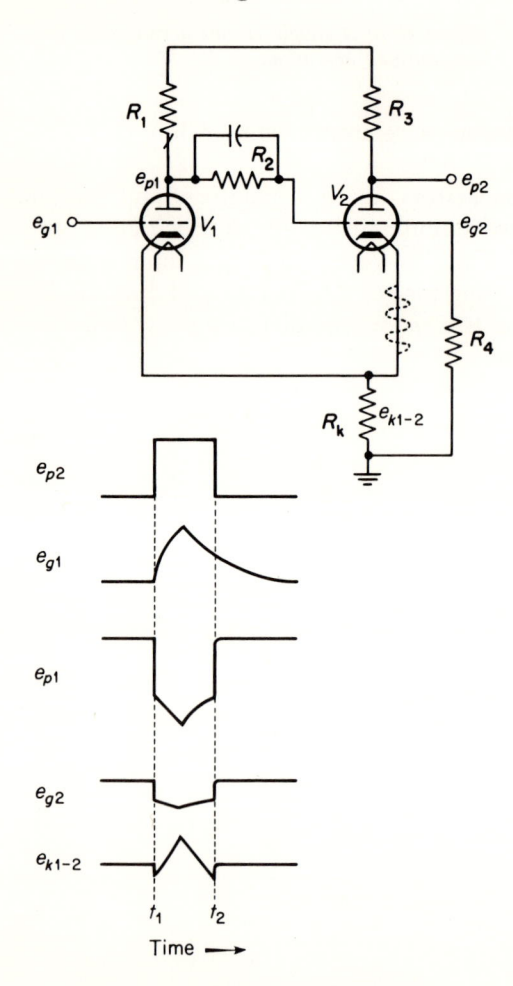

Figure 13-12 Cathode-coupled binary.

Sec. 13.6   Cathode-coupled binary   263

Initially, V_2 is conducting and V_1 is cut off. As the grid of V_1 is raised, a point is reached that causes the tube to conduct. The resulting voltage drop across R_1 causes the plate voltage of V_1 to decrease, and this decrease is transmitted to the grid of V_2 by the R_2R_4 voltage-divider network. The negative voltage applied to V_2 causes a decrease in current through R_k. This lowering of bias causes a further increase in the conduction of V_1. Very rapidly, the switching action occurs, making V_2 conduct and V_1 cut off. On the trailing edge of e_{g1}, the reverse action occurs, but at a lower voltage. This difference in critical switching voltage is the result of a form of *hysteresis* which occurs when the loop gain is greater than unity. Hysteresis may be reduced by a resistor in series with the cathode of V_2, of the proper value to limit loop gain to unity. The addition of this resistor in the circuit of Fig. 13-12 is indicated in dashed form. The output voltage waveform, e_{p2}, is a square wave whose duration is equal to the time spent by e_{g1} between its two critical voltage values. It should be noted that, for a triangular or sinusoidal input, the output is a square wave. For this reason the Schmitt trigger is sometimes used as a squaring circuit.

There are several advantages in taking the output from the plate of V_2. Because the stray capacitance from this point to ground is small, the speed of the circuit is improved. In addition, any type of load can be applied without seriously affecting the performance of the rest of the circuit. Further improvement may be achieved by making V_2 a heavy-current tube, reducing R_3, and returning the cathode of V_2 to a point only part way up on R_k.

A transistor version of the Schmitt trigger circuit is shown in Fig. 13-13(a). In the stable-state condition, if Q_1 is *off*, $v_{c1} = V_{cc}$. This negative voltage is coupled to the base of Q_2, and $v_{b2} = (V_{EE} - v_{R2})$.

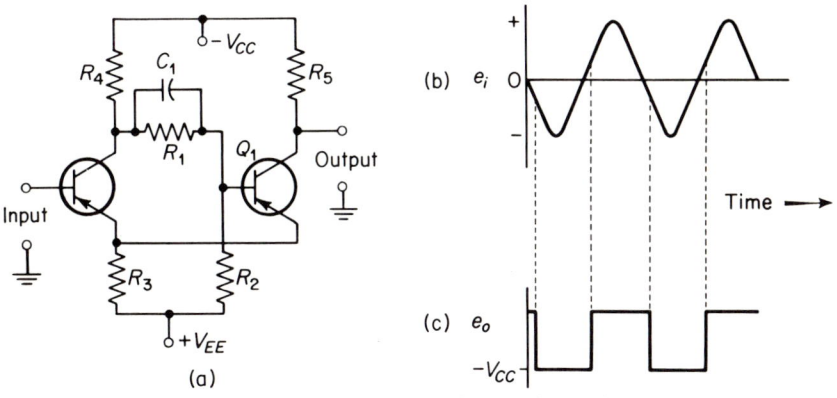

Figure 13-13 A transistor Schmitt trigger circuit.

Emitter current from Q_2, through resistor R_3 to the positive terminal of battery V_{EE}, maintains the emitter of Q_1 at a negative potential. The reverse bias developed between the emitter and base of Q_1 keeps that stage *off*. The high negative voltage at the base of Q_2 forward-biases the base-to-emitter junction and causes Q_2 to operate in the saturation region.

If a negative-going signal, Fig. 13-13(b), of sufficient amplitude to overcome the reverse bias, is applied to the base of Q_1, it turns that stage *on*. The collector voltage of Q_1 then decreases, and the change is coupled to the base of Q_2. This causes the emitter current of Q_2 to decrease; it also lowers the potential across R_3, which in turn causes the collector current of Q_1 to increase even more. As a result of the regenerative circuit action, in a very short time Q_1 is full on (saturated), and Q_2 is full off. The output voltage taken from the collector of Q_2 is a maximum negative voltage; that is, $e_{c2} = -V_{cc}$, as shown in Fig. 13-13(c).

The new steady-state condition continues until the input signal becomes positive-going. At some low value, about one volt, it is of sufficient amplitude to bring Q_1 out of saturation. At that point, the now familiar switching action is initiated; and in a very short time, the circuit has reverted to the original condition with Q_1 full off and Q_2 full on. The output voltage is essentially equal to zero.

13.7 Direct-coupled bistable multivibrator[6,7]

The direct-coupled bistable MV is shown in Fig. 13-14. Obviously, the arrangement is not suitable for use with vacuum tubes, since both tubes would remain on.

To show that this flip-flop has two distinct stable states, temporarily disregard the presence of Q_1. Then the circuit consists simply of Q_2 with its emitter grounded and its base and collector returned to V_{cc} through R_1 and R_2, respectively. The base current in Q_2 is equal to $V_{cc} R_1$, and the components are selected to ensure that this is more than adequate to drive Q_2 into saturation. Typically, the base-to-emitter voltage of Q_2 is of the order of a few tenths of a volt, and the collector-to-emitter voltage is practically zero.

If we now consider the presence of Q_1, it is seen that the base-to-emitter voltage of Q_1 is equal to the collector-to-emitter voltage of Q_2; and the base-to-collector voltage of Q_1 is equal to the base-to-emitter voltage of Q_2. Under these conditions, the collector and emitter currents of Q_1 are very small, so that Q_1 is very nearly at cutoff.

Sec. 13.8 Bistable MV applications

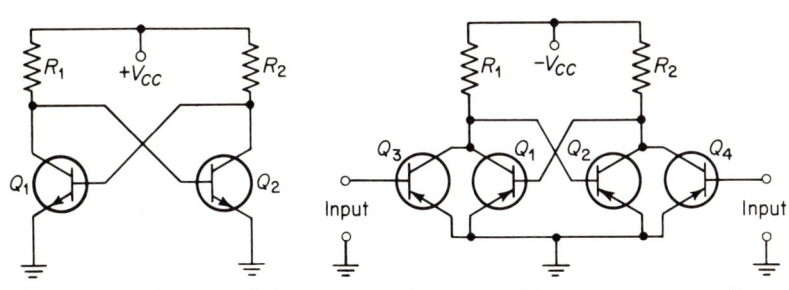

Figure 13-14 Direct-coupled bistable multivibrator.

Figure 13-15 Usual arrangement of direct-coupled MV.

The usual arrangement for a direct-coupled multivibrator is shown in Fig. 13-15. Transistors Q_3 and Q_4 are used only to provide input control of the conducting state of the circuit. The load for each stage of the binary consists of R_1 and R_2 in parallel with the input resistance of the other transistor.

Assume initially that Q_1 is *off* and that Q_2 is *on*. Circuit triggering is accomplished by effectively grounding the collector of the *off* transistor. When a negative-going trigger pulse is applied at the input terminals of Q_3, that stage conducts through R_1. The collector potential of Q_1 and the base potential of Q_2 rise toward the saturation voltage of Q_3. The base potential of Q_2, although slightly negative with respect to ground, cuts off Q_2, and the Q_2 collector voltage increases negatively, driving Q_1 farther into conduction. The positive feedback between collectors and bases of the transistors results in the rapid change of state of the circuit. Circuit equilibrium is reached; Q_1 is full on and Q_2 is full off.

When a negative pulse is applied to the input terminals of Q_4, that stage conducts through R_2. This initiates switching, and very soon Q_2 is full on and Q_1 is full off.

13.8 Bistable MV applications

A very important application of the flip-flop is the basic *scale-of-two* circuit shown in Fig. 13-16, which produces an outgoing pulse for every second input pulse. In Fig. 13-16, the neon lamp connected across the plate load will be lit if V_2 is conducting, owing to the voltage drop across R_7.

A positive-going input trigger pulse drives the cathode of V_2 positive because of increased current through V_1. This reduces the plate current of V_2, decreases the bias on V_1, etc., until V_1 is full

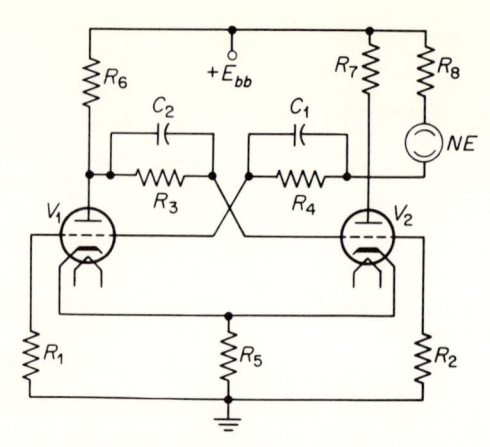

Figure 13-16 Scale-of-two counter.

on and V_2 is full off. The neon bulb is extinguished in the new steady state. The next trigger pulse results in the flip-flop reverting to its original state, and the neon bulb glows again. Thus, for each second input pulse, V_2 is on and the neon bulb is also on. Thus, scale-of-two counting is achieved. If two such flip-flops are connected in cascade, the first circuit registers every second input pulse, and the second circuit registers every fourth input pulse. This indicates that if several flip-flops are connected in cascade, each subsequent scale-of-two would register the second, fourth, eighth, sixteenth, thirty-second, sixty-fourth, etc., pulse. Consequently, flip-flops are the heart of the scalers found in many modern instrumentation systems.

Flip-flops can also be used to actuate relays in industrial control circuit applications.[8] A simplified arrangement is shown in Fig. 13-17, with V_1 and V_2 representing the two stages of the flip-flop. A normally open relay, Rel 1, has a solenoid in the plate circuit of V_1, and a normally closed relay, Rel 2, has a solenoid in the plate circuit of V_2. With V_1 off and V_2 on, Rel 1 is in its normal de-energized position (open); but Rel 2 is not, since current through its solenoid, Sol 2, pulls the relay open.

When a positive-going pulse is applied to the grid of V_1, a transition to the second steady state occurs, with V_1 on and V_2 off. Current through solenoid Sol 1 then causes Rel 1 to close. The absence of current through Sol 2 also causes Rel 2 to *close*.

Thus, by using one normally open relay and one normally closed relay, both are caused to perform the same function (close) when the trigger pulse is applied. Successive pulses applied to the flip-flop alternately close or open circuit terminals 1, 2, 3, 4.

Exercises 267

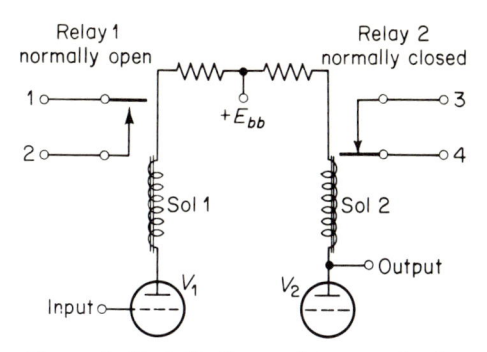

Figure 13-17 A flip-flop used to control relay operation.

This circuit uses relays that are sensitive to small changes of current. These, in turn, can be used to open or close larger relays where high power is to be handled. The output can either be applied to a following flip-flop or used to trigger devices such as *thyratrons*.

REFERENCES

1. Linvil, J. F., "Nonsaturating Pulse Circuits Using Two Junction Transistors," *Proc. IRE*, July 1955, p. 826.
2. Chance, B., et al., *Waveforms*. New York: McGraw-Hill Book Co., Inc., 1949.
3. "Silicon Transistor Flip-Flop Circuits," *Application Notes*, Staff of Texas Instruments, Inc. June 1960.
4. Hull, James E., "Flip-Flop Circuits Using Saturated Transistors," *Elec. Ind.*, Sept. & Oct. 1959.
5. Schmitt, O. H., "A Thermionic Trigger," *J. Sci. Inst.*, Jan. 1938.
6. Beter, R. H., W. E. Bradley, R. B. Brown, and M. Rubinoff, "Surface Barrier Transistor Switching Circuits," Philco Corp. Publication, Phila., Pa.
7. Millman, J., and H. Taub, *Pulse and Digital Circuits*. New York: McGraw-Hill Book Co., Inc., 1956.
8. Mandl, Matthew, *Industrial Control Electronics*. Englewood Cliffs, N. J.: Prentice-Hall, Inc., 1961.

EXERCISES

13-1 Draw the basic circuit and explain the operation of a vacuum-tube binary.

13-2 What is the purpose of capacitors C_1 and C_2 in Fig. 13-2? Explain how they achieve this purpose and what may happen if they are omitted.

13-3 What steps can be taken to improve the resolution time of a binary?

13-4 Explain how breakdown diodes may be used to improve binary switching time, and draw a transistor binary circuit which includes these devices.

13-5 Draw the circuit of a binary triggered through a series-connected triode, and state why this method of triggering is an improvement over direct triggering.

13-6 If triggering pulses of the same polarity were applied simultaneously to both transistors in a flip-flop, what effect would this have on switching time? Why? Draw and explain a circuit that is suitable for the application of such triggering pulses.

13-7 Why are saturated flip-flops not as fast as nonsaturated circuits?

13-8 What are the advantages and disadvantages of saturated operation? Nonsaturated operation?

13-9 Draw and explain the circuit of a nonsaturated transistor flip-flop.

13-10 In the Schmitt trigger circuit of Fig. 13-12, how is hysteresis reduced?

13-11 Show that the direct-coupled flip-flop has two distinct stable states.

13-12 Draw the circuit of a scale-of-two counter and explain its operation.

13-13 Draw the transistor version of the Schmitt trigger circuit and include pertinent waveforms.

13-14 What is meant by the "plate catching" action of a diode?

13-15 In the cathode-coupled binary, what is the advantage of taking the output from the plate of V_2?

13-16 Why is a crystal preferred in place of a thermionic diode in Fig. 13-4?

14

MONOSTABLE
MULTIVIBRATORS

A monostable multivibrator is named for its self-restoring action: it is a multivibrator with only one permanently stable state. A correctly applied triggering pulse may produce a reversal of the stable state, but the circuit returns spontaneously to its original permanently stable condition. The temporary-stable state exists for only a finite period of time. Each time a triggering pulse is applied, the circuit first switches to the quasi-steady condition, and then, after a finite period of time, reverts to its original permanently stable condition where it remains until another pulse is applied.

Other names commonly used to describe the same circuit are *one-shot*, *driven*, and *triggered* multivibrator.

14.1 Operation of the vacuum-tube monostable MV circuit

The circuit of a vacuum-tube monostable MV is shown in Fig. 14-1. In the absence of a triggering pulse, the steady state exists. Tube V_1 is cut off by the high negative bias applied through resistor R_1, and V_2 is conducting because there is no current through R_2 and its grid is at ground potential. As

270 Monostable multivibrators   Chap. 14

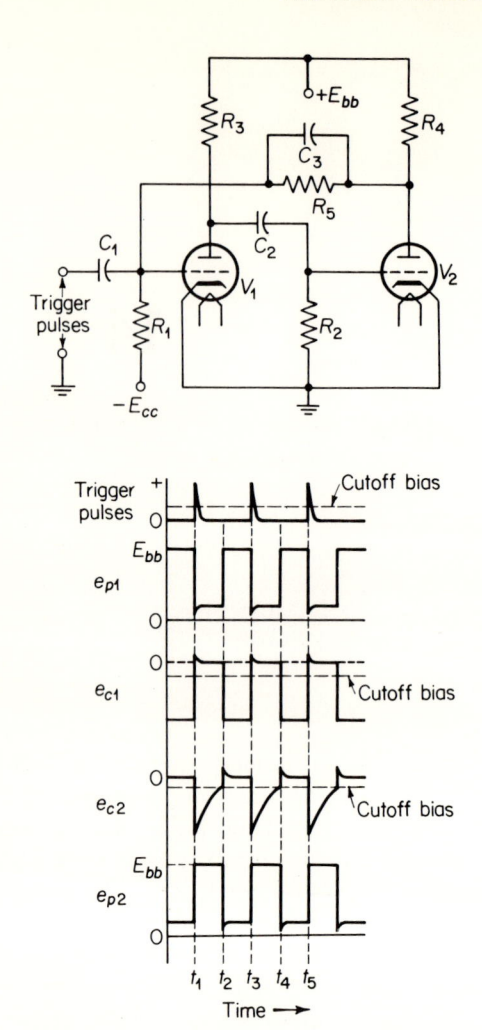

Figure 14-1 A monostable MV requiring positive triggering pulses and pertinent waveforms.

shown on the accompanying waveforms, a minimum positive potential from the plate of V_2 is applied to the grid of V_1 through resistor R_5. The fixed negative bias, $-E_{cc}$, is much larger than this small positive voltage, and the grid of V_1 is held well beyond cutoff. The plate voltage of V_1 is at a maximum, and capacitor C_2 is charged to this high value through R_3 and the low grid-to-cathode resistance of V_2.

When a positive triggering pulse of sufficient amplitude is applied to the grid of V_1, at time t_1, its grid voltage increases above cutoff, as shown by waveform e_{c_1}, and the tube conducts. This lowers the

Sec. 14.1 *Operation of the vacuum-tube monostable MV circuit* 271

plate voltage, e_{p_1}, and applies a negative voltage, e_{c_2}, to the grid-cathode circuit of V_2 across resistor R_2, since the charge on capacitor C_2 cannot change instantaneously. Capacitor C_2, which has been charged to the value of supply voltage E_{bb}, starts to discharge through R_2 and the low plate resistance of V_1 to the lower value of the V_1 plate voltage. This negative voltage drives V_2 beyond cutoff.

With V_2 cut off, its plate voltage is at a maximum positive value, which is large enough to maintain the grid voltage of V_1 above cutoff after the triggering pulse decays. When V_1 conducts and V_2 is cut off, the quasi-steady state exists.

Capacitor C_2 discharges exponentially. After a certain period of time, its current decreases to the point where the negative voltage developed across R_2 is not large enough to bias V_2 beyond cutoff and it starts to conduct (at time t_2 in Fig. 14-1). The resulting decrease in the V_1 plate voltage drives the grid of V_2 in the negative direction. The plate current of V_1 then decreases and its plate voltage increases. Capacitor C_2 charges quickly, e_{c_2} going slightly positive and then dropping exponentially to zero as C_2 assumes maximum charge. This further increases the plate current of V_2, drives the grid voltage of V_1 beyond cutoff, and produces the steady equilibrium state.

The output voltage taken from the plate of V_2 is normally low; it increases to a maximum value for a period of time equal to the cut-off period of V_2. This is shown in waveform e_{b2}. The output voltage returns to a minimum value when the tube conducts, and it stays at this level until the next triggering pulse is applied to the circuit. A single triggering pulse, therefore, produces one complete cycle. The frequency of the circuit, is, obviously, equal to the trigger pulse frequency. Note that the switching action could also be initiated by applying a negative-going pulse to the plate of V_1.

The duration of the output pulses depends on the maximum negative value of e_{c2}, the voltage cut-off point, and the time constant of the discharge circuit. The maximum negative value of e_{c2} in turn depends on the tube characteristics and on load resistor R_3. The greater the increase in plate current and the value of R_3, the larger will be the maximum negative grid voltage, which means a longer duration. The pulse duration is also increased for a larger time constant, which is primarily a function of R_2C_2. In summary, pulse duration is dependent upon the value of load resistance, the tube characteristics, E_{bb}, and the R_2C_2 time constant.

A modification of Fig. 14-1 is shown in Fig. 14-2. Grid resistor R_2 is returned to E_{bb} rather than to the cathode. In the steady-state condition, V_1 is *off* and V_2 is *on*, because of the positive bias applied

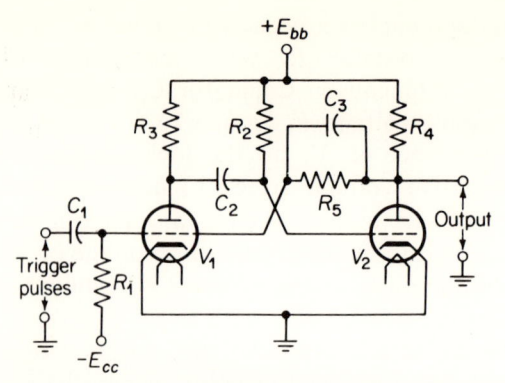

Figure 14-2 A modification of the MV shown in Fig. 14-1 with R_2 returned to E_{bb}.

through R_2. The grid of V_2 does not reach a high value of bias because, during periods of grid current, the grid-to-cathode voltage depends upon the voltage divider action of the small grid-to-cathode resistance and the high value of resistance R_2.

The waveforms produced by this circuit are similar to those shown in Fig. 14-1. The pulse duration depends primarily on the R_2C_2 time constant, and the frequency of the circuit is the same as the frequency of the triggering pulse.

There is an important difference in the R_2C_2 discharge curves of the last two multivibrators discussed. These curves are shown in Fig. 14-3. The R_2C_2 discharge curve of the first circuit, which has its grid returned to cathode, is shown in (a). The curve approaches ground potential until cutoff is reached. The tube then starts to conduct, and the capacitor stops discharging. A large portion of the original charge on C_2 is lost before cutoff is reached, and the grid voltage changes gradually in the vicinity of cutoff. Because a slight change in tube characteristics causes an appreciable change in the time at which cutoff is reached, such a gradual approach does not provide good frequency stability or accurate timing. This condition is often referred to as jitter.

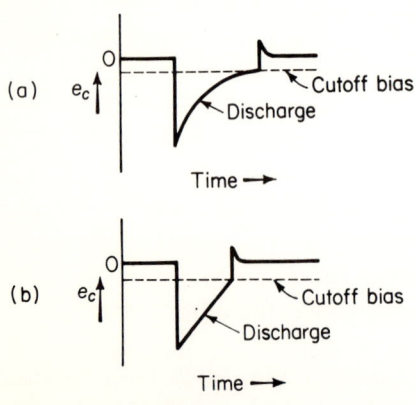

Figure 14-3 Discharge curves (a) cathode return (b) E_{bb} return.

In the discharge curve for the positive grid-return circuit, Fig. 14-3(b), capacitor C_2 loses a smaller portion of its total charge by the time cutoff is reached. The curve is relatively linear and varies more rapidly. Any slight change in tube characteristics causes a negligible change in the time required to reach cutoff, and circuit stability is noticeably improved.

14.2 Vacuum-tube monostable cathode-coupled multivibrator

The circuit diagram of a monostable cathode-coupled multivibrator is shown in Fig. 14-4. The steady state exists when V_1 is *off* and V_2 *on*. In the absence of a triggering pulse, V_2 conducts; and, because there is no grid current through resistor R_2, its grid is at cathode potential. The grid of V_1 is negative with respect to its cathode because R_1 is returned directly to ground and the cathode is at some potential above ground as the result of the V_2 plate current through resistor R_k. Voltage e_k is sufficient to bias V_1 beyond cutoff.

If a positive pulse of sufficient amplitude is applied, the grid voltage of V_1 rises above cutoff and drives the tube into conduction. The resulting decrease of e_{p1} is coupled to the grid of V_2 through capacitor C_2, and decreases the V_2 plate current. The voltage developed across R_k is also reduced, which further increases the plate current of V_1. The action is cumulative, and V_2 is quickly cut off. The quasi-steady state of the circuit then exists. After capacitor C_2 has discharged sufficiently to raise e_{c2} above cutoff, the circuit reverts to the steady state.

The steady state initially exists for the various waveforms of Fig. 14-4. Plate voltage e_{p1} is maximum, and grid voltage e_{c1} is beyond cutoff. Plate voltage e_{p2} is minimum, and grid voltage e_{c2} is equal to the cathode potential, which is at its maximum value.

A positive trigger pulse is applied to the circuit at time t_1. This causes the grid voltage of V_1 to rise above cutoff; the tube conducts, and plate voltage e_{p1} decreases, driving the grid of V_2 in the negative direction. The cycle is now initiated. Within a very short time, e_{p1} drops to its minimum value, e_{c2} is driven beyond the cut-off point, e_{p2} increases its maximum, and e_k decreases to its minimum.

At time t_2, e_{c2} rises above cutoff. Tube V_2 now conducts, and the increased current through R_k develops sufficient voltage to drive the grid of V_1 beyond cutoff. Capacitor C_2 charges through resistor R_3, supply voltage E_{bb}, resistor R_k, and the cathode-to-grid circuit of V_1.

The frequency of the pulse output is the same as the frequency of the triggering pulse, provided that the circuit is permitted to

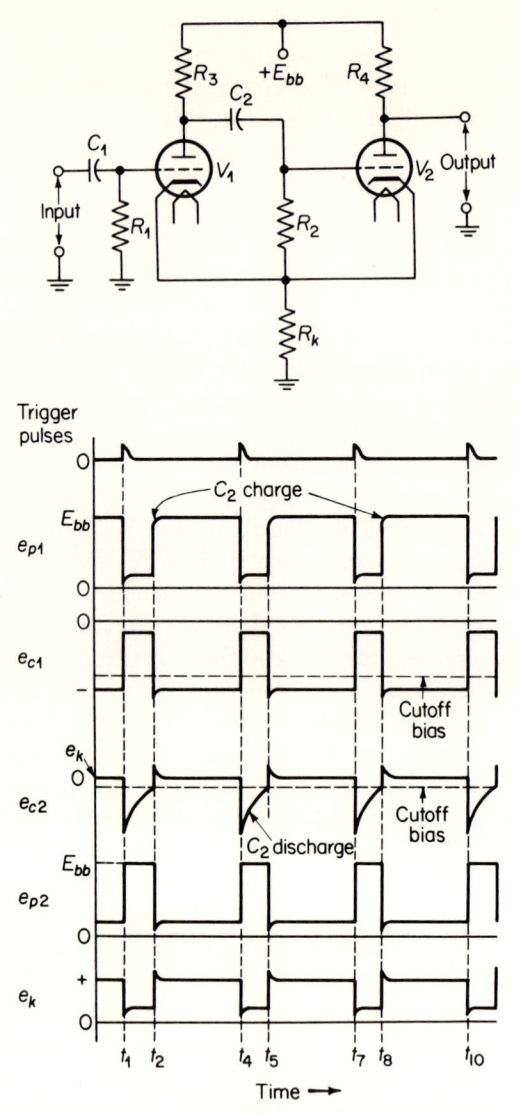

Figure 14-4 Monostable cathode-coupled MV.

recover to the steady-state condition after each triggering pulse. The output pulse duration depends primarily on the time constant of the discharge circuit.

The same circuit, but with grid resistor R_2 returned to E_{bb} instead of to the cathode, is shown in Fig. 14-5. In the steady state, V_1 is *off* and V_2 is *on*. The negative bias applied to both tubes, as a result

Sec. 14.3   Other vacuum-tube circuit modifications

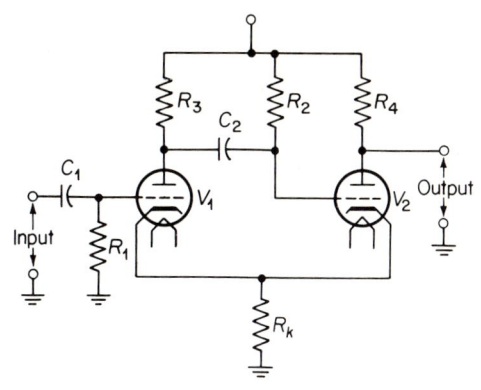

Figure 14-5 Improvement of frequency stability by using positive grid return.

of the voltage developed across resistor R_k, is sufficient to drive V_1 *off*; but the positive bias applied to the grid of V_2 through resistor R_2 prevents the simultaneous cutoff of V_2.

When a positive pulse of sufficient amplitude to overcome the bias is applied to its grid, V_1 conducts. Plate voltage e_{p1} drops, and this decrease is coupled to the grid of V_2 through capacitor C_2. The voltage drop across R_k decreases, and switching action begins. The quasi-steady state is quickly reached and continues until C_2 discharges enough for grid voltage e_{c_2} to rise above cutoff. The circuit then returns rapidly to the steady-state condition.

14.3 Other vacuum-tube circuit modifications

In the circuits of Figs. 14-2 and 14-5 (where grid resistor R_2 is returned to E_{bb}), if R_2 is large in comparison with the resistance of R_3 in parallel with the plate resistance of V_1, the time for which V_2 remains *off* is approximately equal to

$$t_o = R_2 C_2 \ln \frac{a+b}{a+1/\mu} \tag{14-1}$$

where t_o = the time V_1 is *off*
 a = the ratio of grid supply voltage of V_2 to E_{bb}
 b = the ratio of the negative swing of e_{c_2} to E_{bb}

When R_2 is connected to $+E_{bb}$ as in both figures, $a = 1$. An approximate value for b is found by load line analysis, with the approximate load resistance R_3 being used to determine the slope of the load line when $R_2 \gg R_3$.

Equation 14-1 indicates that the length of the output pulse may be changed by varying R_2, C_2, or a. Large changes of pulse length in the circuits mentioned are usually made by varying C_2, and continuous adjustment is made by varying R_2. To produce very short output pulses, C_2 can be made small; but this reduction is limited by the input capacitance of V_1. If C_2 is made too small, no switching action can occur, and the circuit becomes completely stable. Any reduction in the size of R_2 is limited by the steady-state grid current of V_2 and by the reduction of b in Eq. 14-1 when R_2 approaches R_3.

The maximum pulse length is limited primarily by capacitor C_2 leakage, which reduces the effective values of resistor R_2. Except at low values of E_{bb}, the maximum value of R_1 may be limited by the maximum grid circuit resistance of V_2 that can be used without danger of cumulative increase of grid current. This value is specified by the tube manufacturer. The maximum pulse-repetition rate is dependent upon the time required for C_2 to recharge after a pulse is terminated; it may be improved by using small C_2 and R_3.

As the *on* time of V_1 is reduced, the ratio of its transition time to conduction time is likewise reduced. When these two times are of the same order of magnitude, the rise and fall times of the output pulse are an appreciable fraction of the pulse duration, and the pulse is of the general form shown in Fig. 14-6. To reduce the

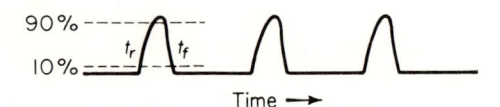

Figure 14-6 *General form of output pulse when t_r and t_f are of the same order of magnitude and constitute an appreciable fraction of the pulse duration.*

transition time, tubes having high transconductance and low interelectrode capacitance should be used. *Plate-catching diodes* also help considerably and, in the cathode coupled circuit, the use of a bypass capacitor across resistor R_k is also helpful. The best value of this bypass is usually between 50 and 200 $\mu\mu$f; but care should be taken not to make it too large, as too great a value causes rounding of the leading edge of the pulse and decreases the maximum pulse-repetition frequency.

The use of a triggering diode, as shown in Fig. 14-7, speeds up switching. As previously mentioned, a negative triggering pulse is used to reduce the triggering voltage requirements.

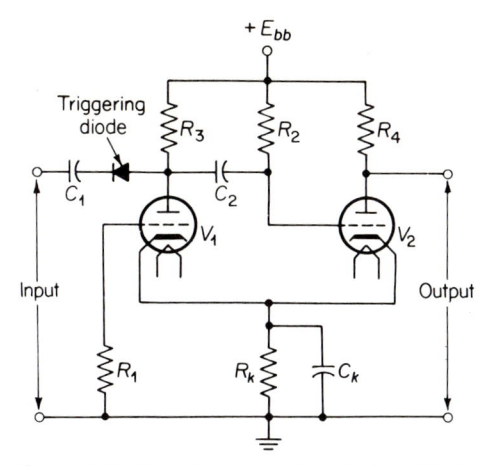

Sec. 14.3   Other vacuum-tube circuit modifications 277

Figure 14-7 Use of a triggering diode with the monostable MV.

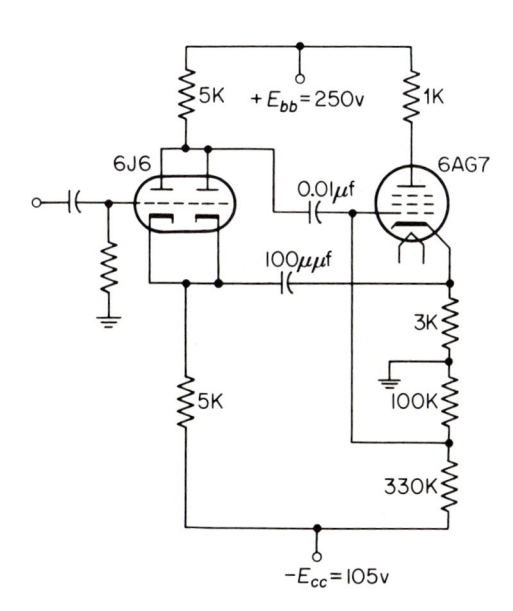

Figure 14-8 Cathode-coupled MV for generating short waveforms (from Waveforms, McGraw-Hill Book Co., 1949).

Variation of the output voltage during or following the pulse is caused by the capacitance of the external load shunted across the output terminals. To minimize the effect of load capacitance, a cathode-follower stage may be inserted between the load and the output of the monostable circuit, as shown in Fig. 14-8.[1] This circuit produces pulse lengths as small as 0.1 μsec that are practically independent of load resistance.

14.4 Transistor monostable MV

A transistor monostable MV and the waveforms at various electrodes are illustrated in Fig. 14-9. From what has been said previously, the reader should have no difficulty in understanding the circuit

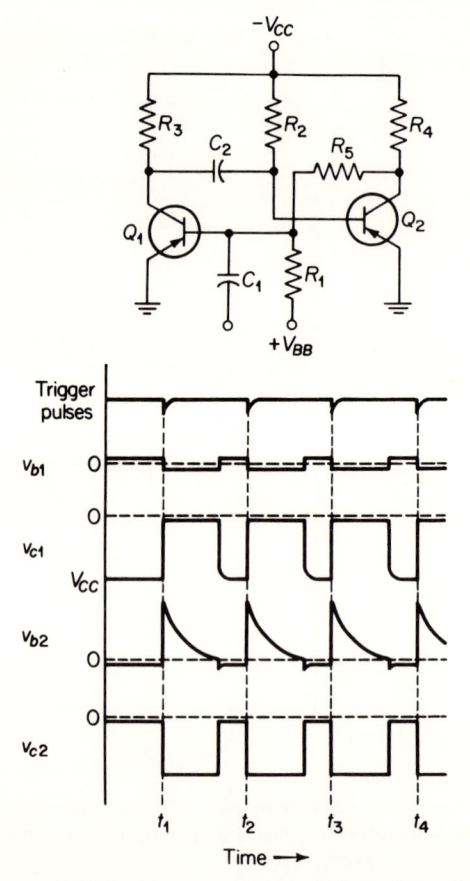

Figure 14-9 Transistor monostable MV and waveforms.

Sec. 14.5 Modifications of the basic circuit 279

operation. Briefly, in the steady state, Q_1 is *off* and Q_2 is *on*. The application of a negative-going trigger pulse of sufficient amplitude to the base of Q_1 turns Q_1 *on* and Q_2 *off*. The circuit remains in the quasi-stable state until C_2 discharges sufficiently to turn Q_2 back *on* and Q_1 *off*.

Simplified equivalent circuits, resembling those used to demonstrate the action of the astable MV in Chap. 12, can also be drawn for the monostable MV, as shown in Fig. 14-10.[2] In part (a), Q_1 is *off* and Q_2 is *on*, and capacitor C_2 charges through resistor R_2. The base, B_1, and collector, K_1, of transistor Q_1 are at ground potential. After the transition occurs, the simplified equivalent circuit is changed to that shown in Fig. 14-10(b). The quasi-steady state lasts until C_2 loses its charge and B_1 goes slightly positive.

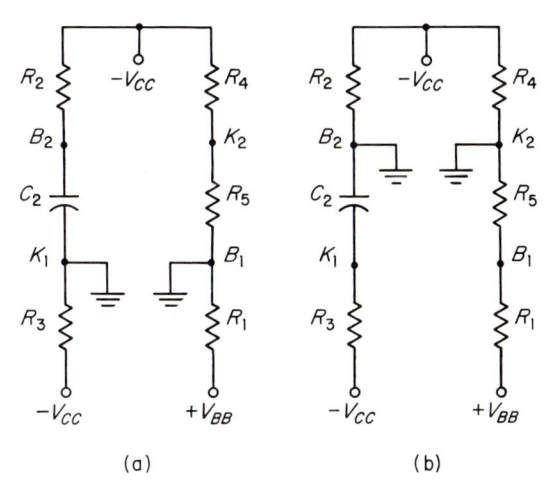

Figure 14-10 Simplified equivalent circuits of transistor monostable MV.

14.5 Modifications of the basic circuit

Several useful arrangements of the transistor monostable MV are shown in (a),[3] (b)[4], and (c)[5] of Fig. 14-11. In (a), transistor Q_3 is used for the application of triggering pulses. The circuit in (b) decreases the transition time as a result of the direct coupling between the collector of Q_1 and the base of Q_2. In (c), the base of transistor Q_1 is connected directly to the collector of Q_2, even though the emitter of Q_2 is grounded. This arrangement is possible only with certain *alloy junction* transistors, which cut off when the base is

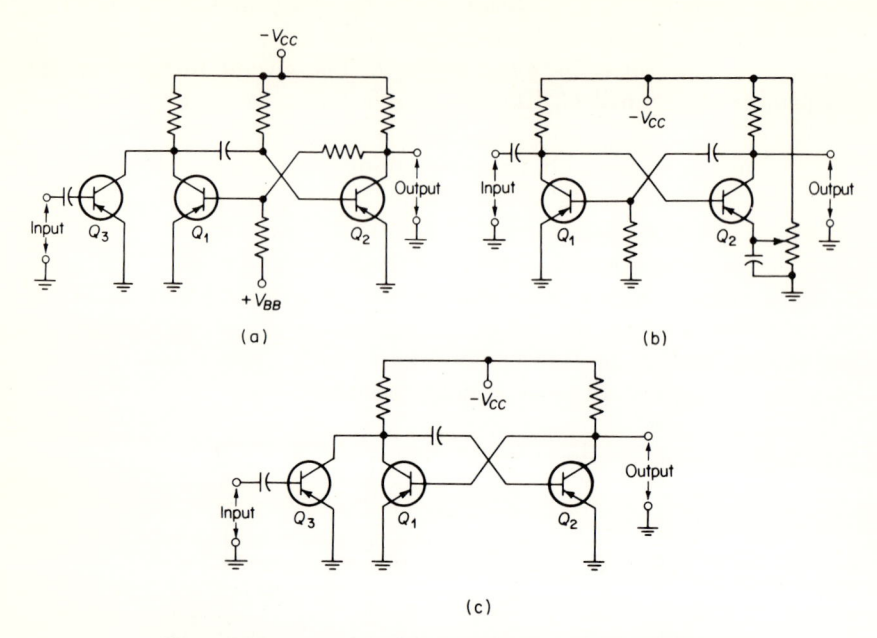

Figure 14-11 *Several useful arrangements of the transistor monostable MV.*

still forward-biased with respect to the emitter. Transistor Q_3, in Fig. 14-11(c), is used for the application of triggering pulses.

Figure 14-12 illustrates an unusual monostable circuit in which the duration of the output pulse is determined by the L/R time constant of inductor L_1 and transistor Q_1 instead of by the RC constant of a resistor-capacitor circuit. Since L_1 and R_1 connect the bases of transistors Q_1 and Q_2, respectively, to $-V_{cc}$, both transistors begin to conduct when the circuit is first switched on. Since the collector currents are d-c, however, a very small direct voltage appears across L_1, and a high negative potential, practically equal to $-V_{cc}$, is applied to the base of Q_2. Consequently, that stage conducts heavily, and the voltage drop across R_1 reduces the collector voltage of Q_2 and the base voltage of Q_1 to approximately ground potential. This causes Q_1 to be cut off, and the circuit is in its stable state.

A positive trigger pulse of sufficient amplitude applied to the input terminals at time t_1 drives Q_2 *off*. The collector voltage of Q_2 then swings negative, producing the leading edge of the output pulse, e_o, and driving Q_1 into conduction. Because of the reactance of series inductor L_1, the collector current of Q_1 cannot rise to a maximum

Sec. 14.5   Modifications of the basic circuit   281

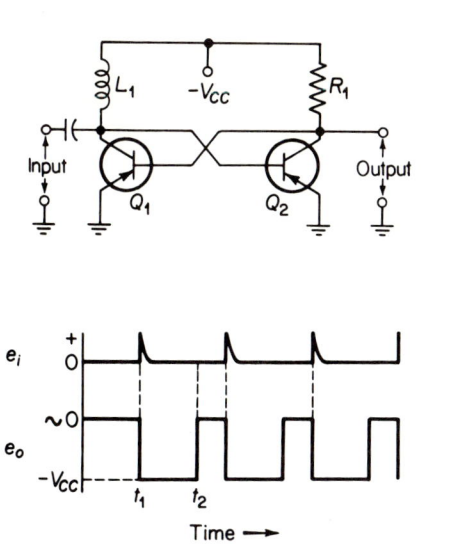

Figure 14-12 A monostable MV with duration of the output pulse determined by L/R time constant.

immediately. Instead, it increases exponentially at a rate determined by the time constant L/R, where L is the inductance of coil L_1 and R is the sum of the resistances of transistor Q_1 and the coil winding. As the collector current rises, the expanding lines of magnetic flux cut the turns of L_1 and induce a back emf. This voltage opposes $-V_{CC}$ and, therefore, holds the base of Q_2 at a very low negative potential, keeping Q_2 in cutoff. Thus, the collector voltage of Q_2 remains constant at a low negative value, forming the flat portion of the output pulse during the interval of time required for the Q_1 collector current to build up to a maximum.

When the collector current of Q_1 approaches saturation, or the maximum level permitted by the circuit values, it is no longer changing rapidly and the slowly expanding flux cannot induce sufficient back emf across L_1 to hold Q_2 in cutoff. Stage Q_2 then begins to conduct, and its collector voltage swings in the positive direction, ending the output pulse and driving Q_1 into cutoff. If desired, this circuit may be triggered through a third transistor.

An unusual and interesting variation of the monostable multivibrator in which both transistors are off during the timing cycle, to reduce the effects of transistor variations on timing accuracy, is shown in Fig. 14-13.[6] This circuit utilizes the *complementary symmetry* properties of transistors, and both stages are initially *on*. When a negative-going triggering pulse is applied through diode D_1, it causes

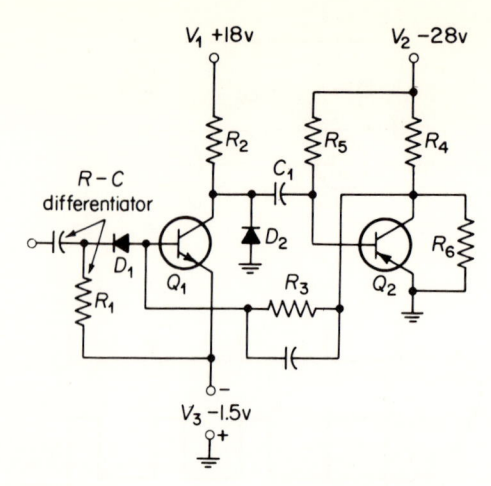

Figure 14-13 Use of complementary transistors to reduce recovery time of monostable MV.

the collector voltage of Q_1 to go slightly less than supply voltage V_1. This change appears also at the base of Q_2 and turns that stage *off*. Capacitor C_1 discharges toward supply voltage V_2 through resistors R_2 and R_5.

When the base voltage of Q_2 goes negative, Q_2 conducts and turns Q_1 on. Capacitor C_1 now discharges through the saturation resistance of Q_2. When C_1 is fully discharged, Q_1 and Q_2 are *on* and ready for another input trigger to start the circuit recycling. *Frequency division* is accomplished by controlling the length of time that Q_1 and Q_2 are *off*, so that a specified number of input triggers can be accepted without causing circuit action.

The use of complementary transistors, in place of similar conductivity transistors, provides a higher charge-to-discharge ratio of the timing capacitor, in addition to the advantages previously noted. A high charge-to-discharge ratio is desirable because unreliable operation can occur if the input trigger is coincident with the discharge of C_1.

In a conventional monostable circuit, the timing capacitor discharges through a collector load resistor of one transistor and the input resistance of the other, whereas in the circuit described here, the capacitor discharges through the saturation resistance of one transistor and an input resistance of the other. Since the sum of the latter resistance is low, an increase of five or more in the charge-to-discharge ratio can be obtained with this circuit.

A circuit that reduces transistor time by preventing saturation is shown in Fig. 14-14.[7] Breakdown diodes D_1 and D_2 are used to prevent the fall of v_{c1} to zero and the rise of v_{c1} sufficiently high to drive Q_2 into saturation. Diode D_3, in combination with D_1 and D_2, also ensures that the minimum and maximum values of the output waveform are independent of collector characteristics. Diode D_4 is the triggering diode.

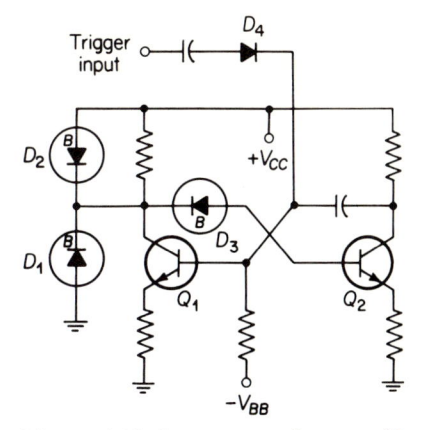

Figure 14-14 A non-saturated monostable MV.

14.6 Applications

A monostable MV is often used to delay the application of a pulse to a circuit by a time equal to the MV duration time. A simple circuit arrangement for accomplishing this is shown in Fig. 14-15. When switch S_1 is in position 1, the positive-going input pulse is applied directly to the grid of amplifier V_3. When the switch is thrown to position 2, however, the output of the MV is applied to V_3. The amount of delay can be varied by methods previously discussed.

The monostable MV is also frequently used for *pulse equalization*. Suppose that pulses are being counted by a circuit (counters are discussed in a later chapter) that is critical with respect to the size and shape of the input pulses. The pulses to be counted, however, are of many sizes and shapes, and thus some means of correction is needed before they are applied to the counter.

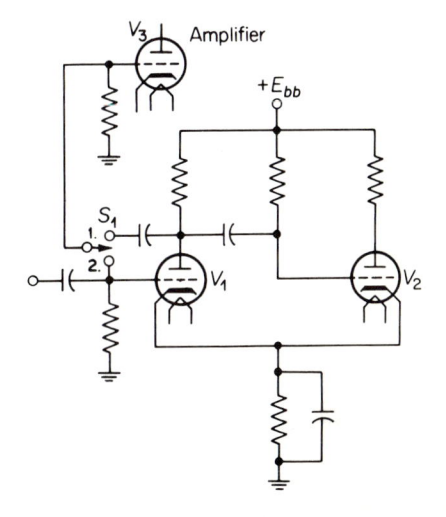

Figure 14-15 Simple delay circuit.

The irregular pulses are simply used to trigger a monostable MV, and the uniform output pulses of the MV may then be counted.

In this process, the input pulse repetition rate must, of course, be within the capabilities of the MV; that is, the MV must revert fully to the steady state before the next pulse is applied. Also, the rise time of the pulse must be quite uniform so that the MV is triggered at the same time after the start of each pulse.

REFERENCES

1. Chance, Britton, et, al., *Waveforms*. New York: McGraw-Hill Book Co., 1949.
2. Neeteson, P. A., *Junction Transistors in Pulse Circuits*. New York: The Macmillan Co., 1959.
3. Prugh, T. A., "Junction Transistor Switching Circuits," *Electronics*, Jan. 1955.
4. Sulzer, P. G., "Junction Transistor Circuit Applications," *Electronics*, Aug. 1953.
5. Beter, R. H., W. E. Bradley, and M. Rubinoff, "Directly Coupled Transistor Circuits," *Electronics*, June 1955.
6. Aronson, A. I., and C. F. Chong, "Monovibrator Has Fast Recovery Time," *Electronics*, Dec. 1957.
7. Linvil, J. L., "Non-Saturating Pulse Circuits Using Two Junction Transistors," *Proc. I R E*, July 1955.

EXERCISES

14-1 Draw the circuit and explain the operation of a vacuum-tube monostable MV. Include pertinent voltage waveforms.

14-2 Draw the circuit of a vacuum-tube monostable cathode-coupled MV and show the voltage waveforms at the grid and plate of each stage and across resistor R_k.

14-3 State the mathematical relationship for determining the *off* time of V_2 in the circuits of Figs. 14-2 and 14-5. Does this equation give any indication of how the length of the output pulse may be changed? Explain.

14-4 Draw and explain the operation of a transistor monostable MV in which both transistors are cut off during the timing cycle.

14-5 Draw the circuit of a nonsaturated transistor monostable MV.

14-6 Explain the operation of the circuit shown in Fig. 14-12.

14-7 Draw a circuit showing how a monostable MV may be used to delay the application of a pulse to a circuit by a time equal to the MV duration time.

14-8 What factors determine the duration of the output pulses in the vacuum-tube monostable MV?

Exercises

14-9 What steps may be taken to reduce the transition time of a vacuum-tube monostable MV?

14-10 Draw the equivalent circuits of a monostable transistor MV. Label each.

14-11 Can the circuit arrangement of Fig. 14-11(c) be used with any type of transistor? Explain your answer.

14-12 Why is a high charge-to-discharge ratio desirable in the circuit of Fig. 14-13? How is this higher ratio achieved?

15

BLOCKING OSCILLATORS

Blocking Oscillators are single tube or transistor transformer-coupled feedback oscillators, Fig. 15-1, in which plate or collector current is permitted to flow for one-half of a cycle, after which cut-off bias is imposed upon the grid or base to prevent further oscillation.

15.1 General[1]

Most blocking oscillators are designed to produce nearly rectangular pulses of plate or collector voltage or current, that are made possible by the special characteristics of the feedback transformer. In the ordinary feedback oscillator used to produce sinusoids, either the grid (base) circuit, plate (collector) circuit, or both are tuned to resonance at the frequency of operation. In the blocking oscillator, the *resonant period* of the transformer and its associated stray capacitance is short compared with the duration of the pulse desired in the output, and its Q (figure-of-merit) is kept as low as possible. This results in a rapid rise in plate current when regeneration is initiated. The duration of the pulse is limited by the low-frequency response of the transformer.

Sec. 15.1   General

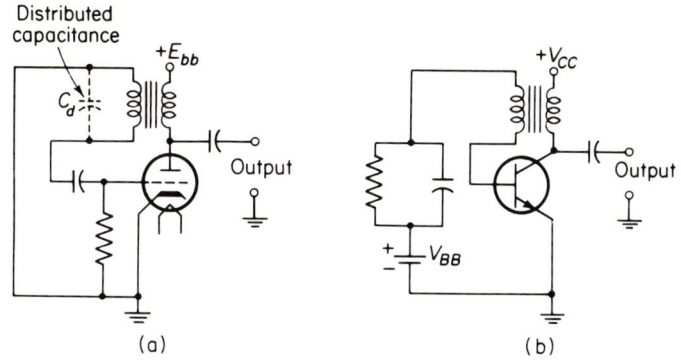

Figure 15-1 (a) Vacuum tube and (b) transistor forms of the blocking oscillator.

The low-impedance feedback path and the very short time lag in the transformer permit the generation of rapidly rising and falling pulse currents of very large magnitude. For this reason, the ratio of the pulse length to repetition period must be kept small, in order that the dissipation rating of the tube or transistor is not exceeded.

Blocking oscillators are used as low-impedance pulse generators for triggering and switching where large currents are desired and the tolerances in pulse width, shape, and amplitude are sufficiently broad to permit their use. It must be emphasized that the blocking oscillator is not a precision device, and the characteristics of its output are influenced markedly by the d-c operating potentials, the residual magnetism in the transformer core, and the age and condition of the tube or transistor. Like the gas-tube pulse generator (discussed in a later chapter) but unlike the multivibrator in which one tube must always be conducting and wasting power, it has the tremendous advantage of drawing plate or collector current only during the pulse. Unlike the gas-tube generator which must deionize after each pulse, the recovery time of a blocking oscillator can be made very short.

After the pulse has terminated, the grid or base is well into the cut-off region because of the action of the R-C network. The grid or base voltage then rises exponentially. If it is permitted to rise to a potential at which the blocking oscillator will fire, a new pulse is generated and the cycle is repeated. The circuit is then functioning as an astable blocking oscillator and is equivalent to the astable multivibrator. If the grid or base voltage is not permitted to rise to the firing potential, a monostable circuit results. An external trigger must be applied to cause the tube or transistor to fire. The monostable blocking oscillator is usually referred to as a *triggered* blocking oscillator and is equivalent to the monostable MV.

Bistable blocking oscillators cannot be made, since feedback is provided by the transformer and positive grid voltage cannot be maintained unless current is changing in the transformer.

A blocking oscillator is used to perform three general types of functions. First, it may be employed to generate a pulse of very short duration when a slowly varying trigger voltage is applied. (Appreciable jitter of the output pulse may occur if the input waveform rises very slowly.) Second, a blocking oscillator may be used to generate a pulse of appreciable peak power and specified shape. It is most economical since plate or collector current occurs only during the pulse. Third, a blocking oscillator may be used as a low-impedance switch. The circuit being switched is usually connected in the self-biasing circuit of the blocking oscillator. The switching action is used in frequency dividers and step counters (discussed later) to discharge the timing or counting circuits. Circuits have also been developed in which the plate (collector) or cathode (emitter) circuits are used as low-impedance switches. They are used in free-running time-bases to replace a gas-tube switch. The blocking oscillator is frequently used to furnish low-impedance pulses to actuate external diode or triode switches for modulation or demodulation, or for the initiation of waveforms. It is very suitable for applying a switching voltage between ungrounded points, since a transformer would have to be employed for isolation even if another type of circuit were used to generate the pulse.

15.2 Pulse transformers[2,3]

Because of its importance not only in the blocking oscillator, but also in many other circuits used to transmit pulses, considerable attention must be given to the *pulse transformer*. First, let us review briefly general magnetics theory.

When an electric current flows in a conductor, a magnetic potential gradient is established. *Magnetomotive force* (mmf) is nearly analogous to the *electromotive force* (emf) that causes current to flow in an electric circuit. Magnetomotive forces have their source in electric currents which commonly link the magnetic circuit in the form of a coil. When current, I, flows in each of the linking turns, N, of a coil, the effective mmf expressed in *gilberts* is:

$$\mathcal{H} = 0.4\pi NI \qquad (15\text{-}1)$$

Sec. 15.2   Pulse transformers   289

The magnetic potential gradient is

$$H = \frac{\mathcal{F}}{l} \text{ gilbert per centimeter} \qquad (15\text{-}2)$$

This quantity, more commonly known as magnetizing force, has been given the name *oersted*.

Magnetizing force causes a magnetic field to surround the conductor. This field of force can be conveniently considered as consisting of a number of closed loops or lines. The strength of the field depends upon the density of the lines. The unit of magnetic flux density is the *gauss*. One gauss is equal to one line per square centimeter.

If the current-carrying conductor is coiled around a magnetic material, flux lines tend to exist in this material rather than in surrounding non-magnetic materials. *Permeability* (μ) of the magnetic material is a measure of the preference of the flux for magnetic over nonmagnetic materials. For example, suppose a coil carrying a certain current produces a flux density of 1 gauss in air at its center. If replacing the air core with a closed loop of magnetic material increased the flux density to 2000 gauss, the permeability of the material is 2000. The fundamental expression that relates these parameters is

$$B = \mu H \qquad (15\text{-}3)$$

Of greatest interest to the transformer designer are materials which have permeabilities greater than one. These include the magnetic ceramics, called *ferrites*, with permeabilities up to 5000; *silicon steels* with permeabilities in the range of 10,000 to 20,000; and the *nickel irons* in which permeabilities of up to one million have been realized. In working with these materials, one soon realizes that permeability may vary directly and sometimes inversely with the flux density. For example, a certain ferrite material may have a permeability of 800 at a low level (say 10 gauss) and 3000 at a high level (say 2000 gauss). As the ferrite approaches saturation, say 5000 gauss, the permeability decreases to say, 1500. Temperature also affects the permeability of magnetic materials to a marked extent.

The effects of these nonlinear permeability changes can be reduced by the inclusion of an air gap (or other nonmagnetic material) in the magnetic circuit. The action of the gap may be understood if one thinks of the magnetic circuit in terms of a series electrical circuit analog. The gap is equivalent to a large-value fixed resistor, whereas the rest of the core may be represented by a small-value variable resistor. If the values of the resistors are in the ratio of

10 : 1, a 300 per cent change in the small resistor (core permeability) causes only a 30 per cent change in the current (flux density) in the analog. Inclusion of an air gap in a pulse transformer has a second beneficial effect that will be made apparent shortly.

Another important characteristic of magnetic materials is the phenomenon of saturation. In the example cited above, if the magnetizing force (current) is increased, a flux density will be reached where further increases in magnetizing force will produce no significant increase in flux. This flux density is termed the *saturation flux density*, B_{SAT}, for the material.

Materials used at their saturation levels undergo great reductions in permeability. Saturation levels vary greatly. Most of the magnetic ferrites saturate at 3000 to 5000 gauss, nickel iron at 6000 to 8000 gauss, and silicon steels at 12,000 to 14,000 gauss. Because of their low saturation levels and their sensitivity to mechanical shock, high-permeability nickel-irons are generally used in special low-level applications such as digital computer circuits. Silicon steel and ferrites are most often used in pulse applications.

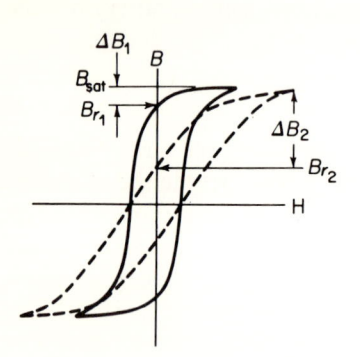

Figure 15-2 Hysteresis loop.

If the relation between B and H is plotted, the familiar hysteresis (B-H) loop for the material is obtained. The loop shown in solid lines in Fig. 15-2 gives this relation for a *toroidal* (gapless) sample of a typical magnetic material used in pulse transformers. Note that in the absence of magnetizing force ($H = 0$) there is a residual (remnant) flux, B_r, left in the core. The remnant flux in the core of a transformer that must pass unidirectional pulses presents a problem because it reduces the available flux change to ΔB_1. For this reason, most pulse transformer cores are gapped slightly. The dashed B-H loop in Fig. 15-2 is for the same core with a gap in the magnetic circuit. Notice that the remnant flux has been reduced to B_{r2}, and the available flux swing increased to ΔB_2.

Magnetic material loss characteristics are an important consideration in the choice of material for a pulse transformer core. Core loss manifests itself in the form of core heating. This heating, in addition to the winding copper losses, determines the temperature rise of the transformer, and thus is a determining factor in the life of the transformer.

Sec. 15.2 Pulse transformers

Compared to the ferrites, silicon steel has relatively high eddy-current losses. Losses in silicon steel cores can be lowered by reducing the thickness of the metal used. One mil is about the thinnest silicon-steel tape commercially available.

Permeability and core losses during pulse operation can most conveniently be related to the rate of change of flux in the core. One-mil and two-mil oriented-grain silicon steels can be operated at rates of change of flux up to 8 and 16 kilogauss per microsecond, respectively, without excessive eddy-current loss. Under these conditions, effective pulse permeabilities of 5000 are often obtained. For a given rate of change of flux, pulse loss in one-mil steel tape is about 25 per cent of that in two-mil tape. Ferrites, on the other hand, because of their low losses, may be operated at higher rates of change of flux. Twenty kilogauss per microsecond is not uncommon.

To summarize: The specific requirements of any application will always dictate the core material to be used. In general, two-mil steel is used for pulse lengths greater than 1 μsec, one-mil steel for 0.1 μsec and longer, ferrite below 0.25 μsec.

Pulse transformers differ from conventional types primarily in their capacity to pass a broad band of frequencies with minimum attenuation and phase distortion. They are usually more difficult to specify, since the impedances they work into are not always easily determined and since their required bandwidth depends upon the pulse rise time, droop, fall time, and backswing. (Backswing is defined as the amplitude of the first maximum excursion in the negative amplitude direction after the trailing edge of the pulse, expressed as a percentage of the 100 per cent amplitude.) See Fig. 15-3 for a typical transformer output pulse.

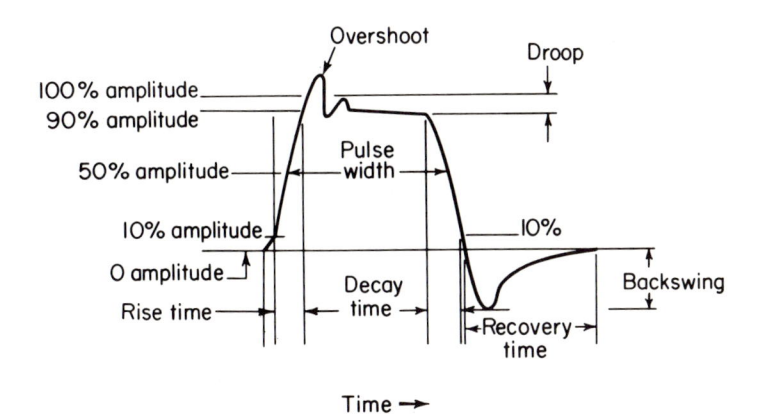

Figure 15-3 A typical transformer output pulse.

Pulse width is a term used in pulse transformer specifications to indicate the widest pulse a transformer can handle without introducing more than 10 per cent *droop*. As shown in Fig. 15-3, it is measured between the 50 per cent pulse amplitude points in accordance with generally accepted practice.

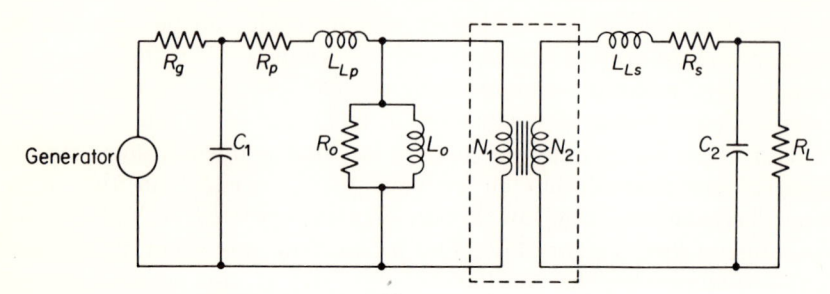

Figure 15-4 Equivalent transformer circuit.

The influence of each transformer parameter on the waveform may best be understood by considering the complete equivalent circuit of a transformer shown in Fig. 15-4. In this figure the parameters are defined as follows

R_g = generator resistance
R_p = resistance of the primary winding
R_c = core loss (negligible for ferrite)
R_s = resistance of the secondary winding
R_L = load resistance
C_1, C_2 = primary and secondary equivalent lumped capacitances
L_{1p}, L_{1s} = leakage inductance of primary and secondary, respectively
L_0 = primary inductance
$N_1 : N_2 = 1 : n$ = turns ratio of an ideal transformer

A pulse such as that described in Fig. 15-3 has many frequency components. The high-frequency components are responsible for the leading edge (rise time), while the flatness of the top and pulse width are functions of lower frequencies.

It is possible to clarify the understanding of pulse transformer operations by considering the response of the circuit during build-up time, the pulse duration, and the decay period.

Since the rise time is a function of the high-frequency response, we shall first evolve the high-frequency circuit, which is shown in Fig. 15-5. This circuit was obtained by first transferring all the

Sec. 15.2   Pulse transformers

elements on the right side of the ideal transformer to the left side by multiplying each impedance on the right by $1/n^2$. Since the ideal transformer provides perfect coupling, it can be ignored. In ferrite core transformers, R_0 and L_0 are both negligible. If a step-up transformer is considered, C_1 is small compared to $n^2 C_2$. (If the transformer is step-down, the equivalent circuit is altered in form but the results are not changed appreciably.) The leakage inductances are then combined to give a single value, L_1. Capacitance C_d represents the total distributed capacitance. From Fig. 15-5, it is readily seen that the important factors in considering rise time are the leakage inductance and distributed capacitance.

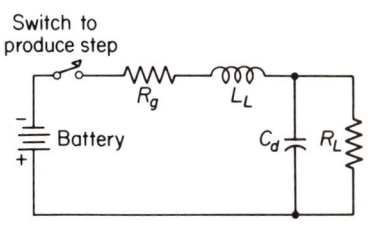

Figure 15-5 High-frequency circuit.

$$\text{Rise time} = K\sqrt{L_1 C_d} \qquad (15\text{-}4)$$

The value of R_L must be such that the value of the time constant $L_1/(R_g + R_L)$ is short. Too small a value of R_L, with respect to the leakage inductance, will cause poor rise time; higher values will improve rise time, within limits.

The first cycle of the oscillation which occurs at the top of the output pulse, in response to a square-wave input pulse, is called an overshoot, and its amplitude is expressed as a percentage. The figure given for it in a specification is a measure of the amount by which the overshoot will exceed the 100 per cent amplitude value. Overshoot may be reduced by introducing losses in the transformer itself, to damp the oscillation. These losses, however, have a tendency to increase the rise time. For good high-frequency response and, therefore, fast rise time, pulse transformers are generally designed to have about 5 to 10 per cent overshoot.

The pulse duration is determined by the low-frequency response of the transformer. Referring again to Fig. 15-5, all capacitances in the circuit may be considered negligible. The leakage inductance is small compared to L_0 and may be eliminated. If the primary and secondary resistances are again lumped into R_L, the equivalent circuit will be that of Fig. 15-6.

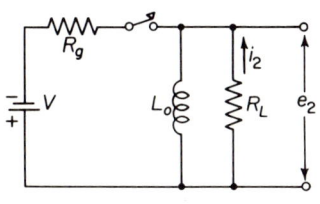

Figure 15-6 Low-frequency equivalent circuit.

During this phase of the pulse, we should consider that the rise time portion of the pulse is complete. Were the voltage across R_L to remain constant, the top of the output pulse would have no slope associated with it. The top would have the same degree of flatness as that of the ideal square wave which produced it. If it could be shown that the rate-of change of current through the primary inductance were a constant (thus inducing a constant voltage), this condition would be satisfied. The response of this current to a step voltage, V, can be determined mathematically. It can be shown that the current i_2 flowing through the resistance R_L as a function of time is given by

$$i_2(t) = \frac{V}{R_g + R_L} \epsilon^{-Rt/L_0} \tag{15-5}$$

where $R = \dfrac{R_g R_L}{R_g + R_L}$, and $e_2(t)$, the voltage across R_L, as a function of time is given by

$$e_2(t) = \frac{VR_L}{R_g + R_L} \epsilon^{(-Rt/L_0)} \tag{15-6}$$

The voltage $e_2(t)$ is also the voltage across the primary inductance. This is related to the primary inductance by the expression

$$e_2(t) = L_0 \frac{d(iL_0)}{dt} \tag{15-7}$$

Then

$$\frac{d(iL_0)}{dt} = \frac{K\epsilon^{(-Rt/L_0)}}{L_0} \tag{15-8}$$

where $e_2(t) = K\epsilon^{(-Rt/L_0)}$.

Thus, when t equals zero (at the beginning of the pulse duration period) the slope (droop) of the pulse will be zero. It is apparent from the equations, however, that the rate of change of current is no longer a constant, and that the pulse amplitude does decrease with time. Ideal square input pulses of different pulse widths, fed into the same coupling transformer, will appear as output pulses with varying degrees of droop. The greater the input pulse width, the greater will be the output pulse droop, for any transformer.

During the fall time of the pulse, the equivalent circuit is shown in Fig. 15-7.

As indicated in Fig. 15-7, the decay time response depends upon the open circuit inductance, the distributed capacitance, and the load resistance. Capacitance C_d represents the entire shunt capacitance referred to the primary. Since there is no independent source of voltage in this circuit, its behavior must be contained in the initial conditions. Assume that during the pulse-duration part of the cycle, just prior to the time when the input pulse returns to zero, there is

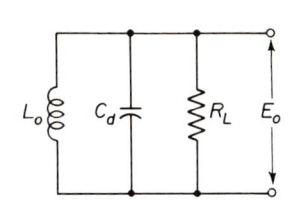

Figure 15-7 Parameters important during pulse trailing edge.

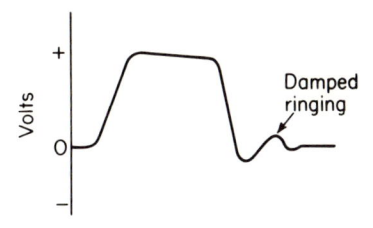

Figure 15-8 Damped ringing effect caused by continued interchange of energy between L_o and C_d in Fig. 15-7.

some energy stored in the inductor L_o. When the input pulse returns to zero, the inductance must discharge this energy into the load. Since the discharge path through C_d and R_L is not a short circuit, the discharge takes a finite time. This is the *decay time*.

During the decay time, C_d has become charged and must now discharge through L_o and R_L, causing the voltage E_o to swing negative. If the value of R_L is relatively high while the other circuit losses are relatively low, it is possible for L_o and C_d to continue to interchange energies and produce a *damped ringing effect* about the zero line as shown in Fig. 15-8.

15.3 The vacuum-tube type of blocking oscillator

The basic circuit of a vacuum-tube type of blocking oscillator is shown in Fig. 15-9. The transformer primary is in the plate circuit, and the transformer secondary is coupled into the grid circuit in such a manner that increasing plate current causes a positive voltage to be coupled to the grid. The positive voltage causes a further increase in plate current, and a cumulative action occurs. The regenerative action continues until plate current saturation is reached. Thus, a decreasing plate current through the transformer primary produces a negative grid voltage. The voltage developed depends on the rate-of-change of the plate current. The dots at opposite ends of the transformer windings indicate similar polarities.

The distributed capacitance, C_d, indicated by dashed lines in Fig. 15-9, makes the transformer act like a resonant circuit. The voltage developed across the secondary charges this shunt capacitance. When the voltage decreases, the circuit oscillates. The grid then swings from a positive voltage to a highly negative voltage. During the positive half-cycle, capacitor C_1 charges to the value of the positive grid voltage through the low grid-to-cathode resistance of the tube.

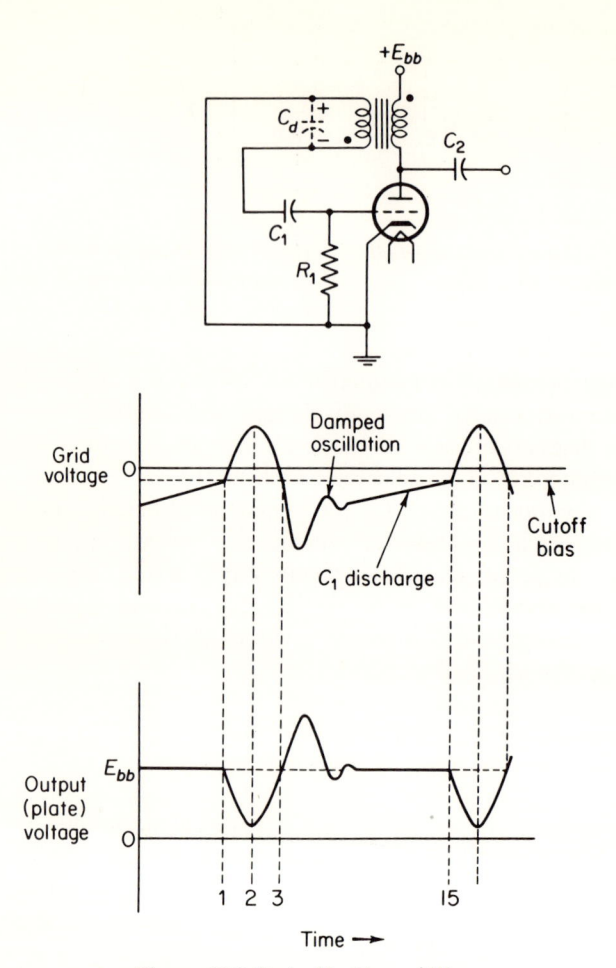

Figure 15-9 Basic blocking oscillator.

On the negative half-cycle, the full negative voltage is applied to the grid because the charge on the capacitor cannot change instantaneously, and the tube is driven beyond cutoff. Capacitor C_1 must now discharge through grid resistor R_1, because the negative voltage makes the grid-to-cathode resistance very high.

The duration of cut-off time depends primarily on the R_1C_1 time constant. When C_1 discharges sufficiently for the grid voltage to reach cutoff, conduction starts. The increasing plate current causes an increasing grid voltage and the cycle described is repeated.

The desired turns ratio between primary and secondary is a function of the tube type and the choice between maximum plate current

Sec. 15.3   *The vacuum-tube type of blocking oscillator*

and maximum plate-voltage swing. Maximum plate current is obtained when the plate resistance reflected to the grid circuit is equal to the grid resistance at the top of the pulse. Turns ratios from 1:1 to 3:1 are commonly used.

Assume that operating potentials are applied to the circuit at time $t = 1$ on the plate and grid voltage waveforms. Rising plate current in the transformer primary produces a positive voltage at the grid and a further increase in plate current. The circuit is obviously regenerative.

Between $t = 1$ and $t = 2$, the plate voltage drops and the grid voltage rises abruptly. Capacitor C_1 charges to the value of the maximum grid voltage through the low-resistance grid-to-cathode path.

Plate current saturation occurs at $t = 2$. The nonlinearity of the tube limits regeneration, and further increases in the grid voltage cannot produce an increase in the plate current. For a brief instant, the plate current does not change, and C_1 begins to discharge through R_1. This produces a negative voltage on the grid and a decrease in the plate current. Regeneration, in the reverse direction, is caused by these two actions. The negative voltage at the grid end of the transformer is coupled to the grid through C_1, and the grid voltage drops abruptly.

The grid voltage is driven beyond cutoff by the oscillating action of the resonant circuit, and the plate voltage reaches the value of E_{bb}. Capacitor C_1, which has been charged to the maximum positive grid voltage, now starts to discharge. A large negative voltage appears across R_1; this voltage gradually drops toward zero as the capacitor discharges.

At $t = 3$, oscillations continue in the tuned circuit formed by the transformer windings. As noted in the previous section, special precautions, such as a low Q, are taken in the physical construction of the transformer to damp these oscillations quickly. Another factor that contributes to quick damping is that the tube is cut off and does not supply energy to the oscillatory circuit. If the damping is inadequate, the amplitude of undesired oscillations may be sufficient to start the regenerative action at a point earlier than desired, by driving the grid voltage above cutoff. The blocking oscillator would then behave more like a generator of a distorted sinusoidal waveform than a generator of separated pulses. Adequate damping is assumed in the waveforms shown.

By making the R_1C_1 time constant greater than the period of oscillation, the elapsed time between output pulses is much larger than the pulse duration. The grid voltage then increases slowly in accordance with the capacitor discharge curve. Plate voltage rises to a single peak, or a series of sinusoidal oscillations, then drops to the level of supply voltage E_{bb} where it stays for the remainder of the cutoff period.

At time $t = 15$ in Fig. 15-9, the grid voltage rises to cutoff and the tube again conducts. The increase in the plate current starts the regenerative process. The grid voltage rises rapidly to a positive value, and the plate voltage decreases to its minimum value. Capacitor C_1 charges to the maximum grid voltage through the low-resistance grid-to-cathode path, and the blocking oscillator cycle is produced again.

15.4 Shape of the output pulse[1]

Damping of undesired oscillations in the plate voltage waveform is

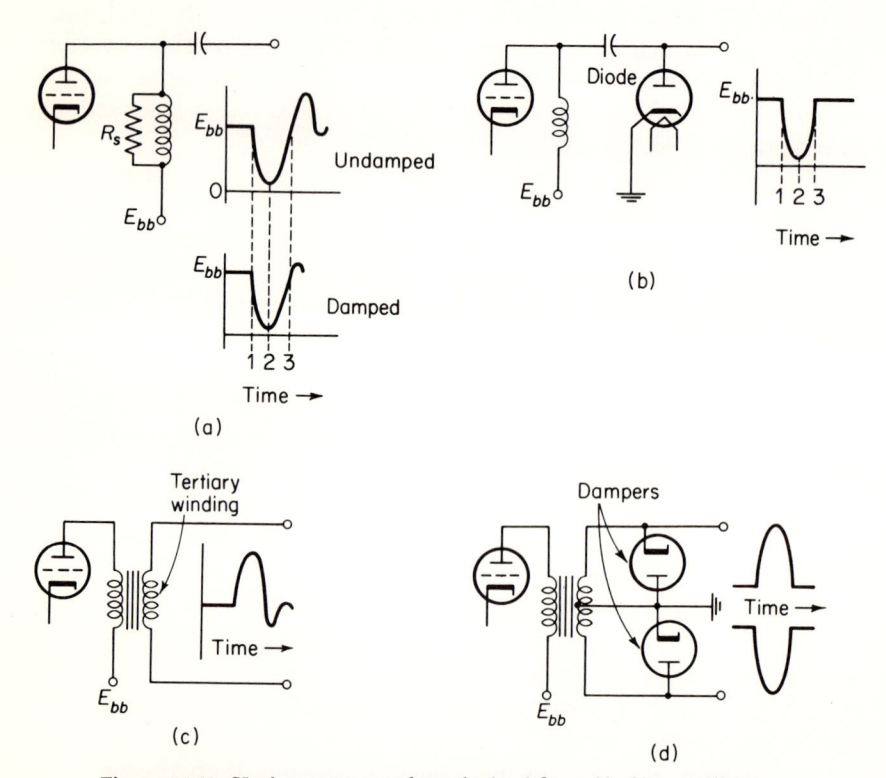

Figure 15-10 Various output pulses obtained from blocking oscillator: (a) shunt damping resistor, (b) damping diode, (c) tertiary winding, (d) a center-tapped tertiary winding and dampers.

Sec. 15.4   *Shape of the output pulse*   299

accomplished by the use of a shunting resistor, R_s, across the transformer primary, as shown in Fig. 15-10(a). Compare the damped and undamped output pulses. The overshoot is completely eliminated by using a *damping diode* as shown in Fig. 15-10(b).

The phase of the output pulse is reversed 180° by using a tertiary winding on the transformer as shown in Fig. 15-10(c). The undershoot is attenuated by using a damping resistor or diode. If the tertiary winding is center-tapped as shown in Fig. 15-10(d), pulses

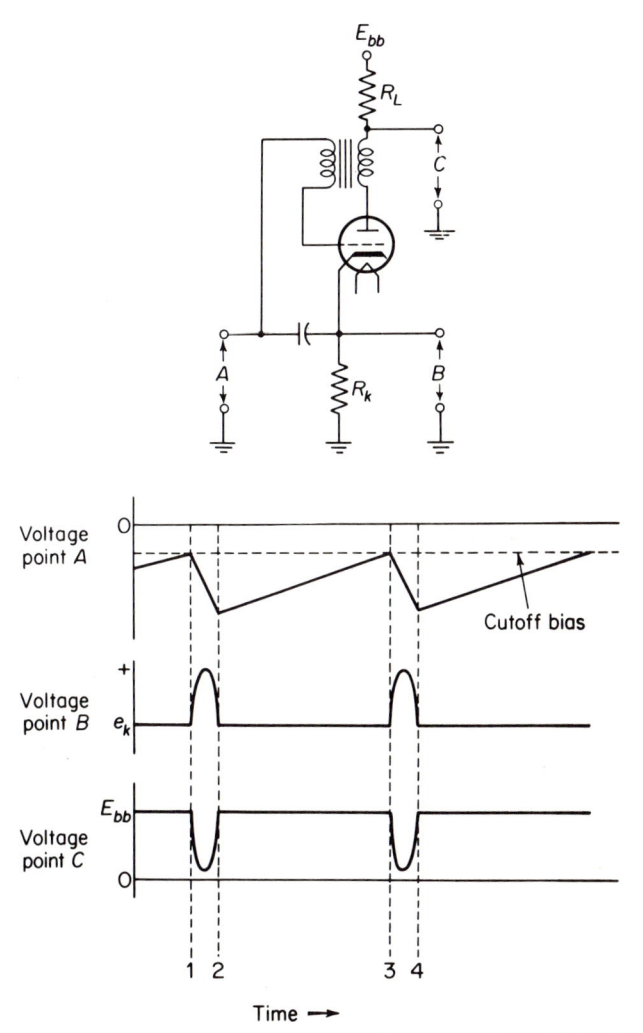

Figure 15-11 Modified blocking oscillator provides additional output waveforms with respect to ground.

of both polarities with respect to ground are obtained. The damping diodes eliminate undershoot and overshoot.

If the circuit is modified to include cathode and plate load resistors, as shown in Fig. 15-11, additional output waveforms are obtained with respect to ground. At point A, the waveform is essentially a sweep-type voltage. Excessive loading is avoided at this point to maintain the correct pulse shape and period. At point B, the waveform consists of a positive pulse with no undershoot and at a very low output impedance. At point C, a negative pulse with no overshoot is obtained.

15.5 Frequency and duration of output

The time required for the oscillation is almost completely independent of recovery time. The duration of the pulse is equal to the time required to generate one half-cycle of oscillation. The period of oscillation is equal to $2\pi\sqrt{LC}$. If L is the effective inductance of the transformer and C is the shunt capacitance, the duration of the pulse is determined by substitution in the above expression. Recovery time is proportional to the R_1C_1 time constant, the maximum negative voltage reached, and the grid cut-off level. The total period is the duration of the pulse plus the recovery time. The frequency of operation is equal to the reciprocal of the total period. Thus, pulse frequency varies inversely as transformer inductance, shunt capacitance, R_1C_1 time constant, maximum negative grid voltage, and grid cut-off voltage. The two most important factors are the R_1C_1 time constant and the maximum negative grid voltage.

As in multivibrator circuits, if the grid resistor is returned to a positive voltage instead of to ground, frequency stability is improved because of greater linearity in the discharge curve.

The circuit recovery time depends on the R_1C_1 time constant, the maximum negative grid voltage, the value of E_{bb}, and the grid cut-off voltage.

The repetition of the astable blocking oscillator will readily synchronize in harmonic relation with the frequency of an injected voltage, just as in the case of the multivibrator. The mechanism involved in such synchronization is essentially the same as that previously discussed; that is, the injected voltage controls the instant at which the pulse is initiated.

15.6 The vacuum-tube monostable blocking oscillator

When a pulse is required to overcome an initially present bias, the circuit is operating in the monostable mode (see Fig. 15-12).

In the steady state, the tube is normally cut off by the high negative bias applied through R_1. An output pulse occurs only when a positive triggering pulse of sufficient amplitude to overcome the bias is applied to the input terminals.

At time $t = 1$, the required triggering pulse is applied. The plate current increases and drives the grid further in the positive direction. Because the circuit is regenerative, the grid voltage remains above

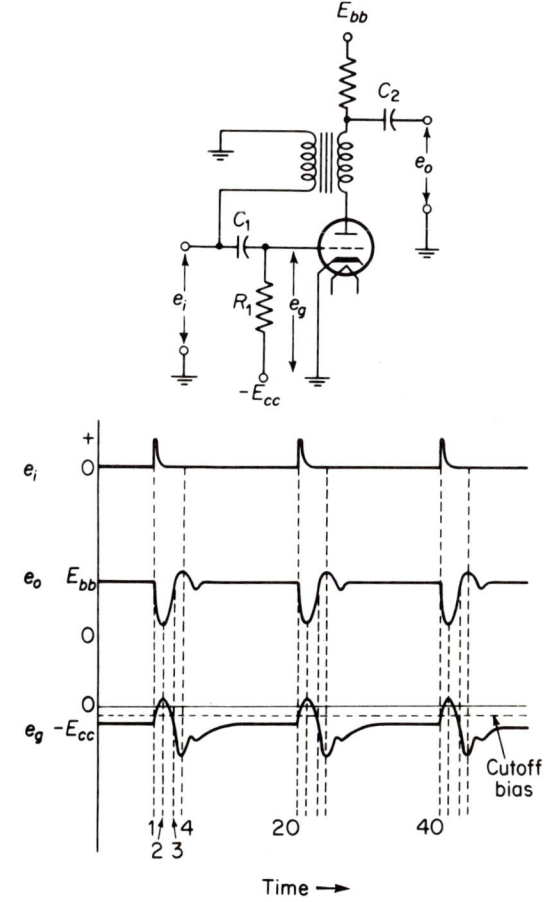

Figure 15-12 Monostable blocking oscillator.

cutoff after the short-duration trigger pulse decays. The pulse is needed only to initiate the action. Plate-current saturation is quickly reached. During the plate-current build-up period, capacitor C_1 charges to the positive grid voltage through the low-resistance grid-to-cathode path, and the plate voltage decays toward the minimum value.

At $t = 2$, the grid voltage is at its maximum positive value and the plate voltage is minimum. After these peak values are reached, a switching action begins in the reverse direction because the plate current decreases. The grid voltage then decreases and the plate voltage increases. This action, in conjunction with that of the resonant circuit, quickly drives the grid voltage beyond the value of $-E_{cc}$. Capacitor C_1 starts to discharge through R_1 toward $-E_{cc}$, and the plate voltage rises toward E_{bb}.

At $t = 3$, the grid voltage reaches the cut-off level and the plate voltage equals E_{bb}. The grid voltage continues to swing negative to some maximum value. It then increases exponentially toward the fixed bias level. The plate voltage rises above the value of E_{bb}, and then returns to that level and remains there during cutoff. The steady-state conditions continue until another positive triggering pulse is applied to the input terminals at $t = 20$.

If sufficient time is allowed for the grid to return to the fixed bias potential between successive triggering pulses, the frequency of oscillation equals the frequency of the triggering pulses. The duration of the output pulse depends mainly on the transformer resonant circuit. The R_1C_1 time constant is not very critical and serves only to determine the time required for the grid to return to the fixed bias level.

Negative triggering pulses can be used in this circuit only if they are applied to the plate circuit, because of the polarity reversing action of the transformer. If the amplitude of the inverted pulse is sufficient to raise the grid voltage above cutoff, conduction occurs and the cycle already described begins.

15.7 Transistor blocking oscillators[3]

Qualitatively, the operation of the transistor blocking oscillator circuit is nearly identical to that of a circuit employing a vacuum tube.

To avoid needless repetition, a monostable type is used here to describe the operation, but the reader should experience no difficulty in understanding astable operation. The circuit is shown in Fig. 15-13.

It is assumed that initially the emitter is biased at zero volts with respect to the base, and the transistor is cut off. A negative trigger pulse applied to the base causes the emitter to be biased in the forward, low-resistance direction and produces emitter current. The resulting collector current, in passing through the collector winding of the transformer, induces a voltage in the emitter winding in such a direction as to increase the forward bias of the emitter. When the current gain around this feedback loop reaches unity, the action becomes regenerative. The trigger pulse is no longer required, and the transistor switches into its saturated region. At this instant, the transformer magnetizing inductance is charging at a nearly constant rate, and a nearly constant voltage is induced. This forms the flat top of the output pulse waveform.

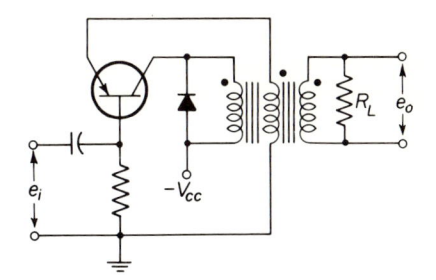

Figure 15-13 Transistor monostable blocking oscillator.

Eventually, the collector is not able to sustain the reflected load current plus the constantly increasing magnetizing current. When magnetizing current can no longer increase, the induced voltage drops to zero and the transistor is forced out of its saturation region.

The energy stored in the transformer primary must now be dissipated in the load. As the magnetizing current decreases, there is a reversal in the sign of the induced voltage. In the vacuum-tube blocking oscillator circuit, this backswing need not be a matter of particular concern. In a transistor circuit, however, the maximum allowable reverse collector voltage is severely limited. A diode should be shunted across the collector winding as a precaution against damage to the transistor. This is shown in Fig. 15-13.

15.8 Triggering methods[1]

The method selected for insertion of the triggering pulse in blocking oscillators should not only produce the best operation of the blocking oscillator but also the least interference with the trigger source. The rise time and amplitude of the triggering pulse determines the time interval between its application and the time at which the blocking oscillator pulse nears its maximum amplitude. In some applications, this interval must be minimized, and a pulse having fast rise time and large amplitude is needed.

The shape of the output pulse is also largely determined by the character of the triggering pulse. Due to the low impedance of the transformer, the gain around the feedback loop is not very large even before heavy grid current is drawn by the tube, and the rise in potential follows a positive exponential curve until limiting occurs. With a slow trigger, the exponential starts in a region of small slope. With a fast trigger (rapid rise time and large amplitude), however, the exponential starts in a region of high slope, and the leading edge of the output pulse is much steeper.

There are two basic methods of triggering a tube-type blocking oscillator. One method, called *parallel triggering*, requires a constant current source in parallel with some winding of the transformer. The other method, called *series triggering*, requires a constant voltage source in series with the grid-cathode loop.

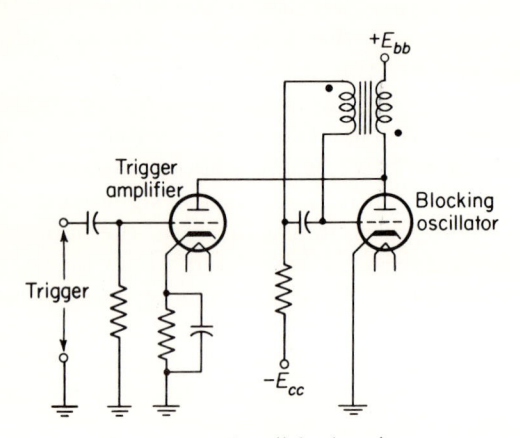

Figure 15-14 Parallel triggering.

Parallel triggering is illustrated in Fig. 15-14. Although the ideal constant-current generator is approximated by means of a pentode, a triode can be used in most cases since the minimum r_p near zero bias is usually large compared with the impedance of the blocking oscillator and transformer. Sharp negative triggers in the plate circuit of a triggering tube are obtained more easily than positive triggers. For this reason, a negative trigger is usually applied to the plate of the blocking oscillator, as indicated in Fig. 15-14, and this pulse appears as a positive trigger at the grid due to the inverting action of the transformer. The trigger must have a sufficiently steep edge to be effectively reproduced by the transformer.

A trigger of long duration will be partially differentiated since the transformer will not pass the low-frequency components.

Parallel triggering, using a separate triggering tube, has several

Sec. 15.8 Triggering methods

advantages including: (a) minimum interference with the blocking oscillator; (b) practically no effect on the triggering source by the oscillator; (c) time delay between the trigger and output pulse is reduced since the gain of the trigger tube increases both the slope and amplitude of the triggering pulse; and (d) the amplitude of the trigger pulse may be readily controlled by the trigger amplifier.

The disadvantages of the circuit of Fig. 15-14 are the need for an additional tube and a trigger with fairly large initial slope.

Series triggering is illustrated in Fig. 15-15. The constant-voltage source is approximated by the cathode-follower.

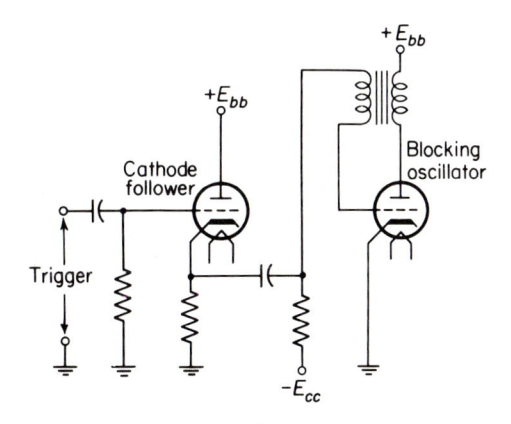

Figure 15-15 Series triggering.

A serious disadvantage of this circuit is the interaction by the blocking oscillator on the triggering source, which is made to draw grid current. Also, power is wasted in the feedback circuit since the triggering impedance is in series with the grid circuit.

Two advantages of series triggering are: (a) a very slow trigger can be used since it will not be appreciably differentiated with consequent loss of amplitude; and (b) if a very fast trigger is used, there is a minimum delay between the trigger and the output pulse since it appears directly in the grid-cathode circuit of full amplitude.

Another method of series triggering is illustrated in Fig. 15-16. Resistor R_i is small compared with R_g. Since these resistors can be very large, the source impedance can be moderately high without much loss of trigger voltage. When the oscillator tube starts to conduct, a negative pulse appears in series with capacitor C. The diode conducts and furnishes a very low impedance path for the grid current while the oscillator continues its cycle in the normal manner.

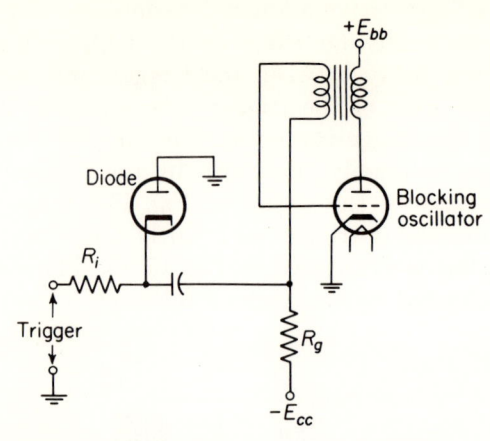

Figure 15-16 Another form of series triggering.

Several direct methods of triggering blocking oscillators are shown in Fig. 15-17. Since no triggering amplifier is used, the triggers must be of large amplitude and interaction must be relatively unimportant.

Parallel triggers, having the polarities indicated, may be applied through resistors or small capacitors at points A, B, C, and D in Fig. 15-17. Series triggering may be applied at point E.

The trigger pulse requirements of a transistor blocking oscillator are similar to those in its vacuum-tube equivalent. The output shape is not independent of the trigger pulse characteristics. Two types of input pulses are generally satisfactory. The first type is a negative

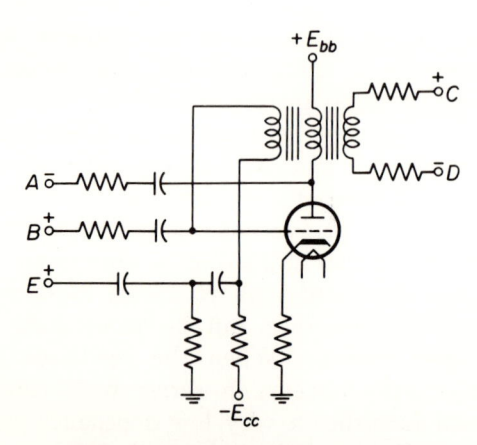

Figure 15-17 Several direct methods of triggering a blocking oscillator.

Sec. 15.8   *Triggering methods*   307

spike of short duration similar to that which might be obtained with R-C differentiation of a pulse with a fast rise time. The second type is a pulse as wide as or wider than the output pulse, with a rather slow rise time. In all cases, the trigger pulse should have a width less than 5 per cent or greater than 100 per cent of the output pulse width. Otherwise, the positive-going fall time of the trigger pulse may cause the emitter to turn off, resulting in a premature termination of the output pulse.

When fast triggers are used, the rise time of the output pulse is a function of both the input trigger rise time and its amplitude. In general, the greater the triggering amplitude, the faster will be the output rise time. Too high a triggering voltage causes excessive overshoot. Trigger pulse amplitudes from five to ten times the minimum requirement will cause overshoot of about 10 per cent.

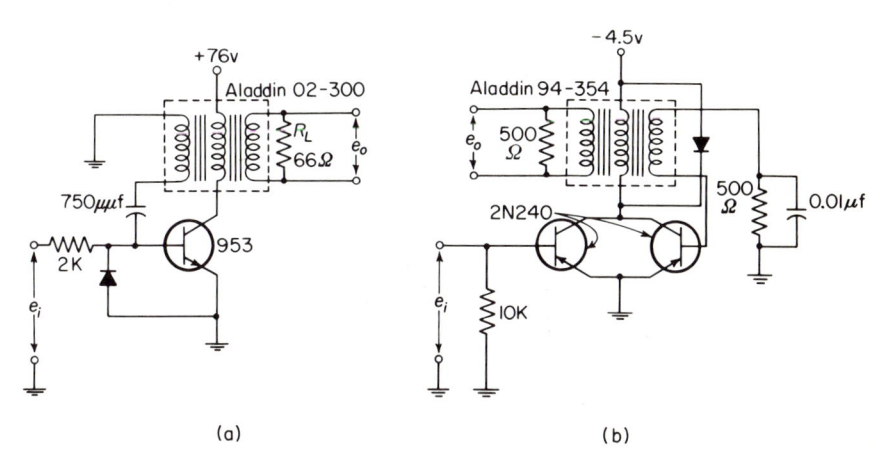

Figure 15-18 Typical examples of (a) series and (b) parallel triggering applied to transistor blocking oscillators.

Typical examples of series and parallel triggering applied to transistor circuits are illustrated in Figs. 15-18(a) and (b), respectively. Using the components illustrated, these two circuits exhibit the following input and output characteristics. For the circuit in (a), the input pulse characteristics are: amplitude, 10 v; rise time, 1 μsec; width, 1 μsec; repetition rate, 0 to 50 kc. The output pulse characteristics are: amplitude, 10 v $\pm$ 1 v; rise time, 0.35 μsec maximum; pulse width, 1.75 μsec $\pm$ 0.25 μsec measured at 10 per cent amplitude; droop, 0; backswing, 15 per cent maximum; overshoot, 5 per cent maximum; repetition rate, same as input.

For the circuit in (b), the input characteristics are: amplitude, 2 to 10 v; width, spike; rise time 0.02 to 1.0 μsec; repetition rate, 100 kc maximum. The output characteristics are: amplitude, 4.5 v ± 0.5 v; width, 2.0 μsec ± 20 per cent; droop, 10 per cent maximum for 0.02 μsec trigger and 0.5 μsec maximum for 1.0 μsec trigger.

15.9 Applications

Blocking oscillators are commonly used as frequency dividers. Effective division by three is achieved by having the blocking oscillator synchronized by a sync signal whose frequency is three times that of the blocking oscillator. One output pulse is obtained for every three applied to the grid.

There is a practical limit to the amount of frequency division that can be realized. For example, assume that the blocking oscillator is synchronized to every tenth pulse, as shown in Fig. 15-19. The

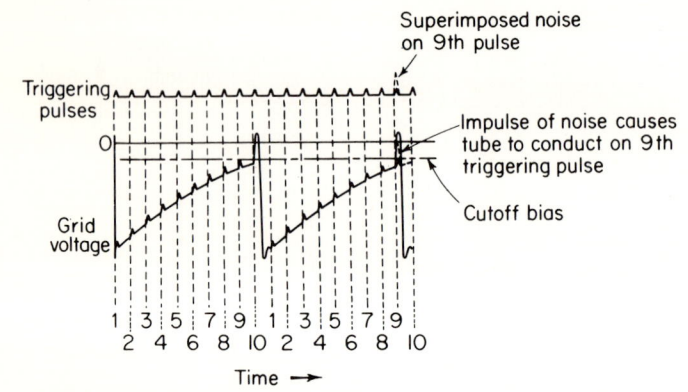

Figure 15-19 An illustration of the practical limit of frequency division using a blocking oscillator.

ninth pulse drives the grid voltage almost above cutoff, which would cause conduction. Any small change in the capacitor discharge time, or even some small noise impulses during the ninth pulse, would be sufficient to cause synchronization at the ninth rather than the tenth pulse. Under these conditions, a frequency division of 9, rather than 10, can easily occur. As frequency-division ratios increase, this effect is more likely to occur. For smaller frequency-division ratios, relatively large deviations from normal would be required to cause incorrect operation. Therefore, circuit stability is much greater at the lower frequency-division ratios.

Sec. 15.9   Applications   309

More than one blocking oscillator is generally used for high frequency-division ratios. For example, suppose we wanted to divide by 25. This could be accomplished easily by using two blocking oscillators, each providing division by 5, with the output of the first oscillator being used as the input to the second.

Frequency-divider circuits also find common use in pulse counters. The limitations of mechanical systems are thereby eliminated.

A basic diode counter is shown in Fig. 15-20. This circuit counts the rate at which positive pulses are applied to its input. Appropriate circuits are used to provide input pulses of uniform amplitude and width. On the positive pulses, D_2 conducts, and D_1 is cut off. Current through R_1 develops voltage across this resistor. It also charges capacitor C_1 and causes the side connected to the diode to be negative with respect to ground. Between pulses, D_2 is cut off, but D_1 conducts because of the negative voltage applied to its cathode by C_1. The capacitor rapidly discharges through this low-resistance path.

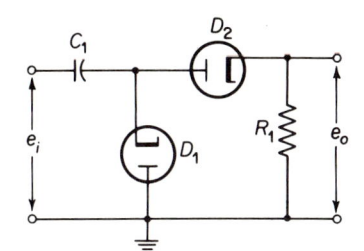

Figure 15-20 Basic diode counter.

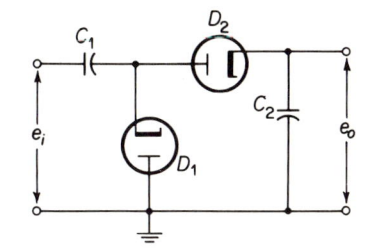

Figure 15-21 Step-by-step counter.

Capacitor C_1 charges and discharges for each cycle. Current flows through R_1 during each positive pulse. The average current through R_1 for a period of one second is equal to the current drawn during each cycle multiplied by the pulse-repetition rate (prf). Therefore, as the prf increases, average current increases. By placing an average-reading d-c milliammeter in series with the resistor and calibrating the meter in terms of prf, an accurate count of the number of pulses per second can be obtained. This circuit does not provide a total pulse count. Negative input pulses may be used by reversing the connections to D_1 and D_2.

A step-by-step counter, shown in Fig. 15-21, is used to obtain a total pulse count. The only difference between this circuit and the basic counter of Fig. 15-20 is the use of a capacitor, C_2, in place of the resistor. Capacitor C_2 has a much greater capacitance than C_1.

Diode D_2 conducts when a positive pulse is applied to the input, and both capacitors charge. Capacitors C_1 and C_2 form a voltage-divider network, with the input voltage division inversely proportional to their capacitance. For illustration purposes, we assume an input voltage of 10 v, with 9 v (90 per cent) appearing across C_1 and 1 v (10 per cent) across C_2. (Any voltage drop across V_2 is disregarded for practical purposes.)

Between pulses, C_1 discharges through D_1. Capacitor C_2 maintains its charge, because D_2 is not conducting and presents a practically infinite impedance.

On the next positive half-cycle, only a 9 v potential difference exists between the two capacitors, since the 1 v maintained across C_1 is in opposition to the input voltage. With the same percentage distribution of input voltage between the two capacitors as noted above, 8.1 v appear across C_1, and 0.9 v across C_2. Note that the total voltage across C_2 is now the sum of the voltages from each pulse and equals 1.9 v.

After each positive pulse, the voltage across C_2 increases another 10 per cent of the net voltage applied to the two capacitors. Thus, by placing a voltmeter across the capacitor, it is possible to determine the number of pulses applied to the circuit. In the example given, a reading of 1 v would correspond to 1 pulse, 1.9 v to 2 pulses, etc. There is, of course, a practical limit to the number of pulses that can be counted, because the maximum voltage that capacitor C_2 can charge to is 10 v.

A blocking-oscillator counter is shown in the diagram of Fig. 15-22. A step-by-step counter applies triggering pulses to this circuit. In the absence of input pulses, the blocking oscillator produces one oscillation when the plate-supply voltage is first supplied. Since C_2 has no discharge path, it maintains a negative voltage across the

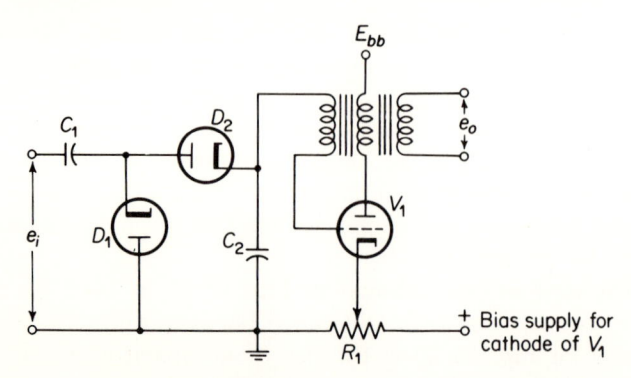

Figure 15-22 Step-by-step counter used to trigger a blocking oscillator.

Sec. 15.9   Applications

blocking-oscillator input terminals of sufficient amplitude to keep the tube biased beyond cutoff, as shown at $t = 0$ in Fig. 15-23.

As pulses are applied to the input of the counter circuit, the voltage across C_2 increases. In the example shown, the grid voltage reaches cutoff after five pulses ($t = 5$). This causes the blocking oscillator to produce another oscillation. During this time, C_2 discharges through the low-resistance grid-to-cathode path of the tube, and then returns to a highly negative voltage value by the grid oscillation. The circuit has now returned to its initial condition. After five more pulses, another pulse appears at the output of the blocking oscillator. Therefore, the circuit acts as a frequency divider providing one output pulse for every five pulses applied.

The number of pulses required to return the grid to cutoff depends on the grid bias, the peak-input voltage, and the capacitor ratio. The grid bias can be adjusted by varying R_1. The greater the step increase in voltage caused by the pulse making the grid voltage exceed cutoff and the previous step, the greater will be the stability of the circuit. Thus, stability is greatest for a low-pulse input-to-output ratio.

A number of blocking-oscillator counters can be connected in series to obtain any desired frequency division. Again, the degree of frequency division depends on the maximum pulse-repetition rate and the maximum counting speed of the ultimate recording device.

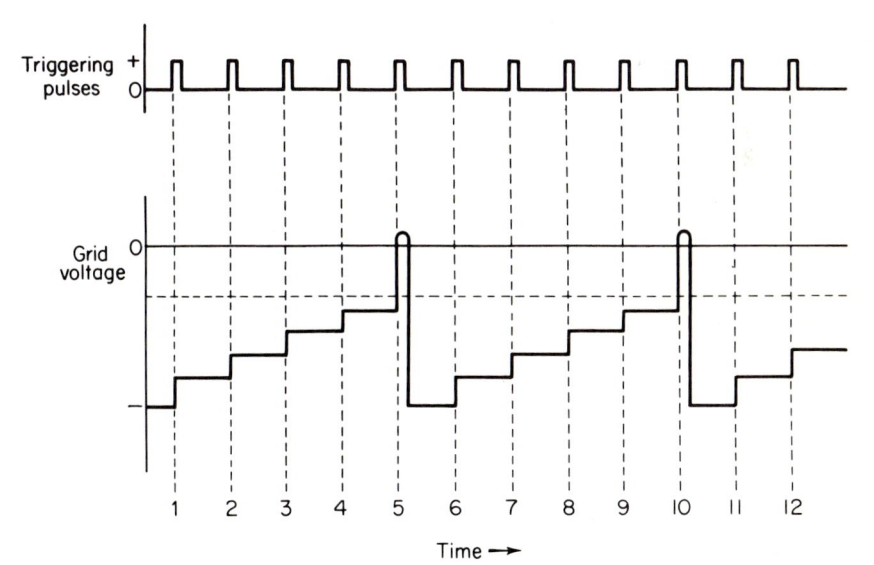

Figure 15-23 Every fifth triggering pulse drives the blocking oscillator on and produces an output pulse.

The blocking-oscillator counter has a number of advantages over the counters described previously. For example, the number of pulses applied to each blocking oscillator can be readily determined by connecting a voltmeter, calibrated directly in terms of the number of pulses, across capacitor C_2. Another advantage is that the pulse-repetition rate need not be constant. Circuit operation is not affected even if 5 pulses are applied in the first second and 50 in the next second.

REFERENCES

1. Chance, B., et. al., *Waveforms*. New York: McGraw-Hill Book Co., Inc., 1949.
2. *Catalog and Application Manual of Pulse Transformers*, Catalog 202, Pulse Engineering Inc., Santa Clara, Calif., 1959–1960.
3. Engineering Bulletin No. 55-1, *Pulse Transformer Encyclopedia*, Aladdin Electronics, a Division of Aladdin Industries, Inc., Nashville, Tenn., May 1958.

EXERCISES

15-1 How does the Q of the pulse transformer affect the output waveform of a blocking oscillator?

15-2 What factor limits the duration of the output pulse of a blocking oscillator?

15-3 If the ratio of pulse length to pulse-repetition period is large, what effect may it have on the tube or transistor used in a blocking oscillator?

15-4 What factors affect the characteristics of the output pulse produced by a blocking oscillator?

15-5 Is it possible to have a bistable blocking oscillator? Explain your answer.

15-6 Why are most pulse transformers gapped slightly?

15-7 Define the terms "backswing" and "pulse width."

15-8 On what factors does the decay-time response of a pulse transformer depend?

15-9 Draw the basic circuit and waveforms for an astable blocking oscillator.

15-10 Draw the circuits showing two methods of damping.

15-11 Draw the circuit and waveforms of a vacuum-tube monostable blocking oscillator.

15-12 In a transistor blocking oscillator, what precaution is taken to prevent damage to the transistor?

Exercises

15-13 Draw circuits illustrating parallel triggering of vacuum-tube and transistor versions of the blocking oscillator.

15-14 Draw circuits illustrating series triggering of vacuum-tube and transistor versions of the blocking oscillator.

15-15 What two types of triggering pulses are generally satisfactory for triggering a transistor blocking oscillator?

15-16 Draw the circuit of a blocking oscillator used as a frequency divider, and include waveforms clarifying the operation.

16

TIME-BASE
GENERATORS

When a cathode-ray tube (CRT) is used to display information, a voltage or current that varies linearly with time is applied to one of two or more sets of deflection plates or coils. This voltage or current, which is called a sweep, causes the electron beam to move horizontally across the face of the CRT. To generate a sweep waveform, use is made of a time-base generator. In general, if the display device employs an electrostatic deflection system, a voltage time-base generator is used. With electromagnetic deflection, a current time-base generator is employed.

The linear time-base waveform is usually a sawtooth wave, as shown in Fig. 16-1. The time required (t_0 to t_1) for the ramp portion of the sawtooth to rise from its initial value to a maximum (and move the beam across the screen from a reference position) is called

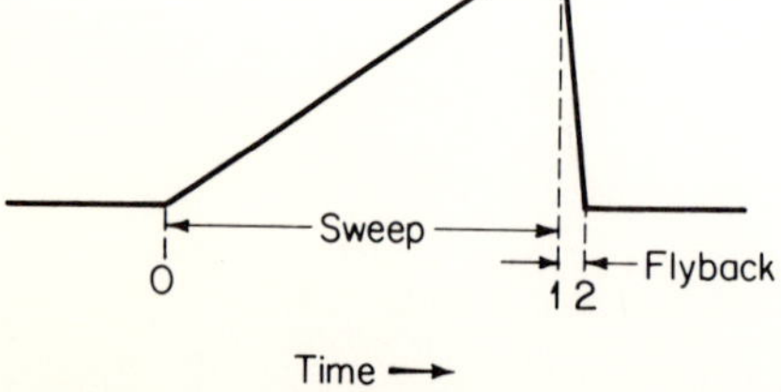

Figure 16-1 Sawtooth time-base waveform.

314

Chap. 16   Time-base generators   315

the *sweep time*. The time required (t_1 to t_2) for the waveform amplitude to drop to its initial value is called the *flyback time*.

The sawtooth waveform is generally required to start from the same position, move a given distance, and then return to the point of origin. The sweep waveform must, therefore, always start at the same minimum amplitude and rise to some maximum amplitude. Clamping circuits, described in Chap. 11, are used for this purpose.

A voltage or current time-base generator may be either free-running or triggered. If it is free-running, the operation of other circuits is synchronized with its operation. If it is triggered, a synchronizing pulse from a stable pulse generator is applied to initiate the sweep.

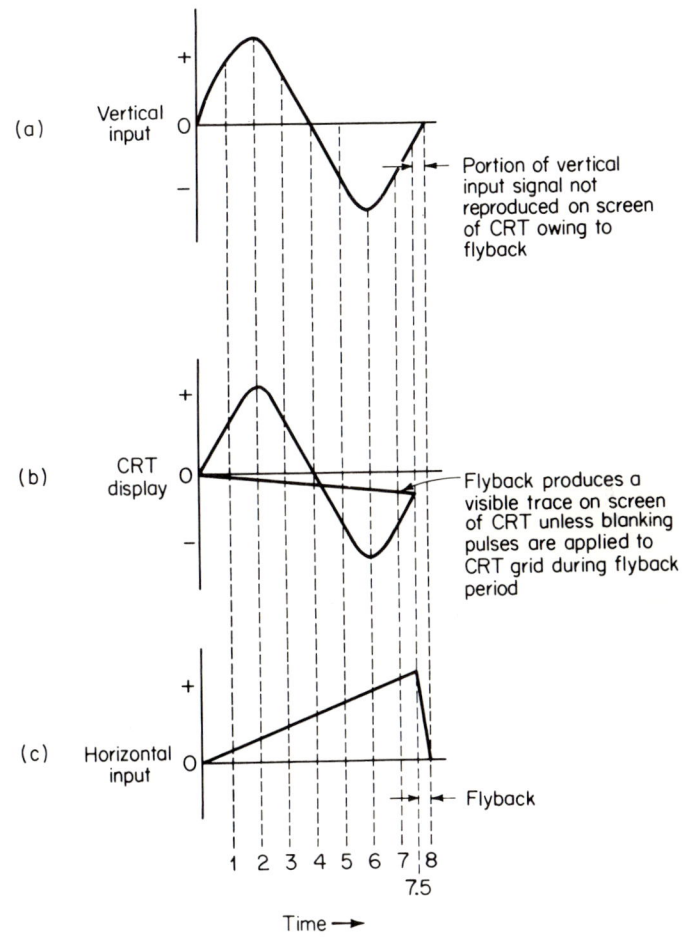

Figure 16-2 Illustration of flyback.

Because the flyback time can never be reduced to zero, the return trace is visible during this time, as illustrated in Fig. 16-2(b). The sine wave (a) is applied to the vertical deflection plates of a CRT, and the sawtooth wave (c) to the horizontal deflection plates. During the time interval from $t = 0$ to $t = 7.5$, the input waveform is faithfully reproduced on the screen of the CRT. The sawtooth flyback begins at $t = 7.5$, however, and the electron beam moves across the screen from right to left back to its original starting position where it begins to trace the next cycle. In so doing, it fails to reproduce the final segment of the sine wave input and also leaves a visible *retrace* which may cause confusion in the case where more complex information is presented. To avoid this possibility, the retrace is eliminated by using a blanking pulse to bias the CRT grid so that it cuts off the electron beam during the flyback time.

16.1 A neon sawtooth generator (relaxation oscillator)

The neon sawtooth generator (relaxation oscillator), shown in Fig. 16-3, is the simplest type of voltage time-base generator. When

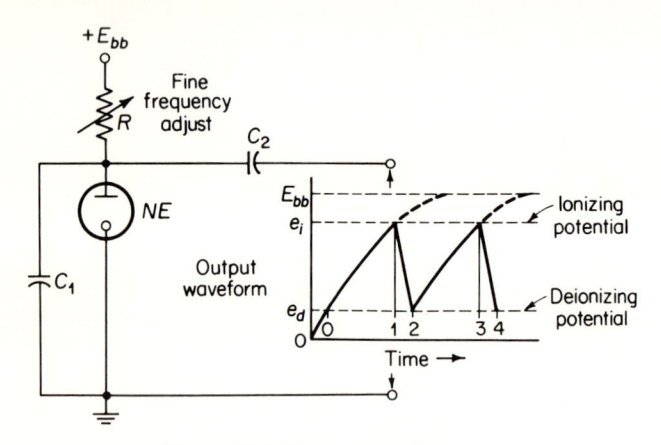

Figure 16-3 Neon sawtooth generator.

voltage E_{bb} is applied, capacitor C_1 starts to charge exponentially through resistor R. The voltage across the neon gas tube is the same as the voltage across C_1 because they are in parallel. The gas tube acts as an open switch until the voltage across it reaches a critical value, e_i, called the *firing* or *ionizing potential*. At this point, the molecules of gas in the tube ionize and gaseous conduction occurs, offering a low-resistance path through which C_1 rapidly

Sec. 16.1   A neon sawtooth generator

discharges. Conduction continues until the voltage across the tube and capacitor reaches e_d, the *deionizing potential* of the tube as shown in Fig. 16-3. When conduction stops, the gas tube again acts as an open switch, and capacitor C_1 begins to recharge through R once more.

The output waveform, as shown in Fig. 16-3, is a positive-going sawtooth. If the firing potential of the neon tube is a small fraction of the supply voltage, the initial rise of the waveform is nearly linear.

The frequency of the sawtooth output is determined by the RC_1 time constant, which can be set in this circuit by varying R. The smaller the RC_1 product, the higher the frequency of the sweep voltage will be. But as R decreases, the ramp voltage becomes less linear. The upper frequency limit of the circuit, however, is determined by the deionization time of the neon tube.

Because the firing and deionization potentials of the tube are not affected by the frequency of operation, the output always remains between these two levels. By using several capacitors of different value for C_1 with a switching arrangement, it is possible to obtain different sweep frequency ranges. The capacitor switch acts as a coarse adjustment and the variable resistor as a fine adjustment for each range.

The application of a neon lamp as a time-base generator is limited. The most serious disadvantage of this device is that the peak-to-peak value of the output voltage is determined by the firing and deionization potentials. Typical values are of the order of 10 v to 15 v. Also, because operation is fairly high up on the exponential curve, excessive curvature is likely. This causes distortion of the time base and, consequently, of the information presented on the CRT screen. Further, the firing and deionization potentials of a neon tube are not particularly stable. Finally, when synchronization must be used

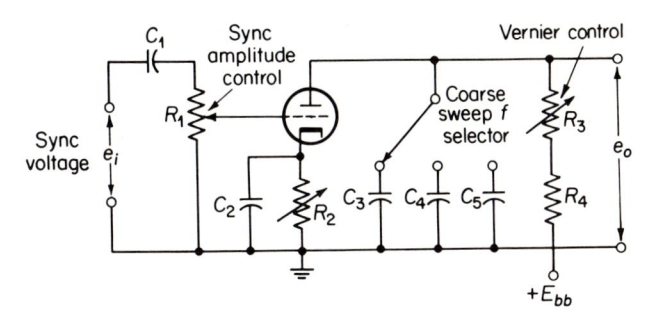

Figure 16-4 Thyratron sweep circuit.

to lock the sweep voltage to the input pattern, the gas diode is difficult to synchronize over a wide frequency range. For these reasons, thyratron tubes are used rather than gas diodes.

16.2 Thyratron sweep generator

A simple thyratron sweep circuit is shown in Fig. 16-4. The firing potential of the thyratron and, therefore, the amplitude of the output waveform, is determined by the direct bias voltage, which is adjusted by resistor R_2. Any variation of R_2, however, changes the frequency of operation, and the vernier control, resistor R_3, must be varied to compensate for this effect. It should be noted that any change in the value of R_3 by itself has no effect on the amplitude of the output waveform but does affect the frequency because it changes the RC time constant.

Resistor R_4 is used to limit the plate current once ionization occurs. When the tube fires, the grid loses all control. The resistance of the tube drops to a low value, and the only factor limiting the plate current is the externally connected resistor. Without this resistor, plate dissipation would be excessive and the thyratron would be destroyed.

As in the case of the simple neon-tube circuit, the higher the plate supply voltage and the smaller the percentage of the total charging time taken by the sweep, the more linear the output voltage waveform will be.

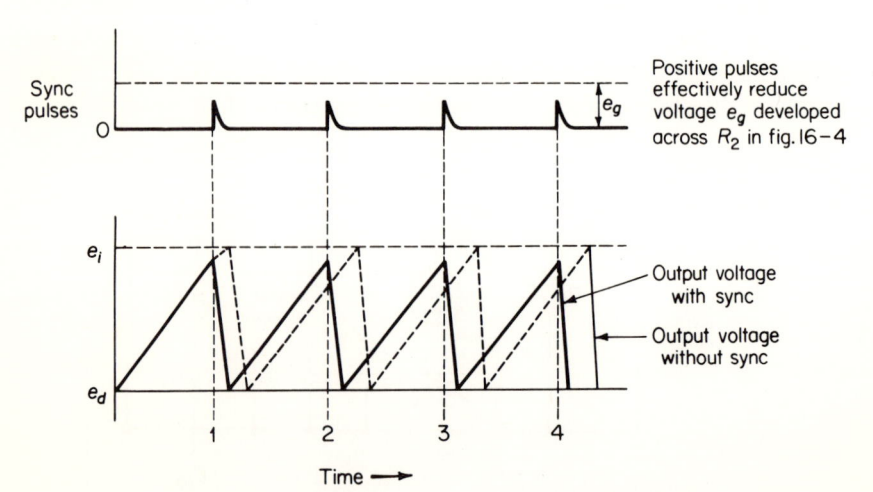

Figure 16-5 Synchronization with positive-going pulses.

Sec. 16.3   Improving linearity of the thyratron time-base generator   319

Variations in circuit constants, gas temperature, ionizing potential, and load all affect the frequency of the thyratron time-base generator. To prevent frequency instability, circuits of this type are, therefore, usually synchronized either by triggering pulses or by a sine wave of constant frequency. The frequency of the sync signal is usually a little above that of the generator.

The action of the sync signal is illustrated in Fig. 16-5. The positive-going sync pulses are applied to the thyratron control grid. The output waveform without sync applied is indicated by the dashed sawtooth. Just before the thyratron firing potential is reached, and while the output waveform is rising toward its maximum amplitude, the positive-going sync pulse effectively reduces the bias voltage, e_g, developed across R_2 in Fig. 16-4, and the thyratron fires sooner than it would otherwise. The output waveform with sync is represented by the solid sawtooth waveform in Fig. 16-5.

16.3 Improving linearity of the thyratron time-base generator[1]

The circuit shown in Fig. 16-6 acts to improve the linearity of the output voltage waveform. If a capacitor is charged by a constant

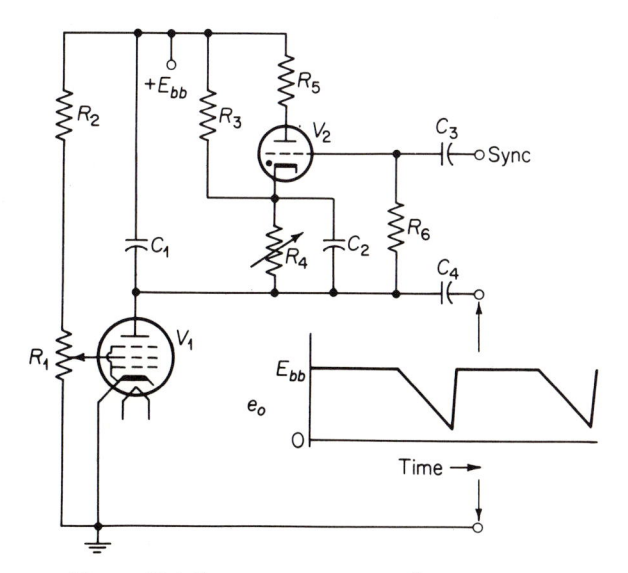

Figure 16-6 Constant-current pentode sweep generator.

current, the voltage across it rises linearly. A pentode is used in practice as the constant-current source because its plate current is relatively independent of changes in the plate voltage when a sufficiently high plate voltage is used.

Capacitor C_1 is initially uncharged, and the output voltage is practically equal to E_{bb}. The capacitor then charges rapidly through pentode V_1 in a linear manner, causing a similar decrease in the output voltage. Because the thyratron, V_2, is connected across the capacitor, it conducts when the voltage on the capacitor reaches the firing level. The capacitor then discharges through V_2 at a rate determined by the value of R_5.

The output voltage waveform is a negative-going sawtooth with a linear sweep. Its frequency can be varied by adjusting R_1 because this resistor controls the plate current of V_1 and, therefore, the rate of charge of capacitor C_1. The amplitude of e_o is determined by the setting of the thyratron grid-bias resistor R_4. A synchronizing pulse can be applied to the grid of V_2 in the manner previously described. Various frequency ranges can be provided by using different values for capacitor C_1 with a switching arrangement similar to that shown in Fig. 16-4. Resistor R_1 then acts as the fine-frequency control for each range.

Thyratron sweep circuits are limited to repetition rates up to

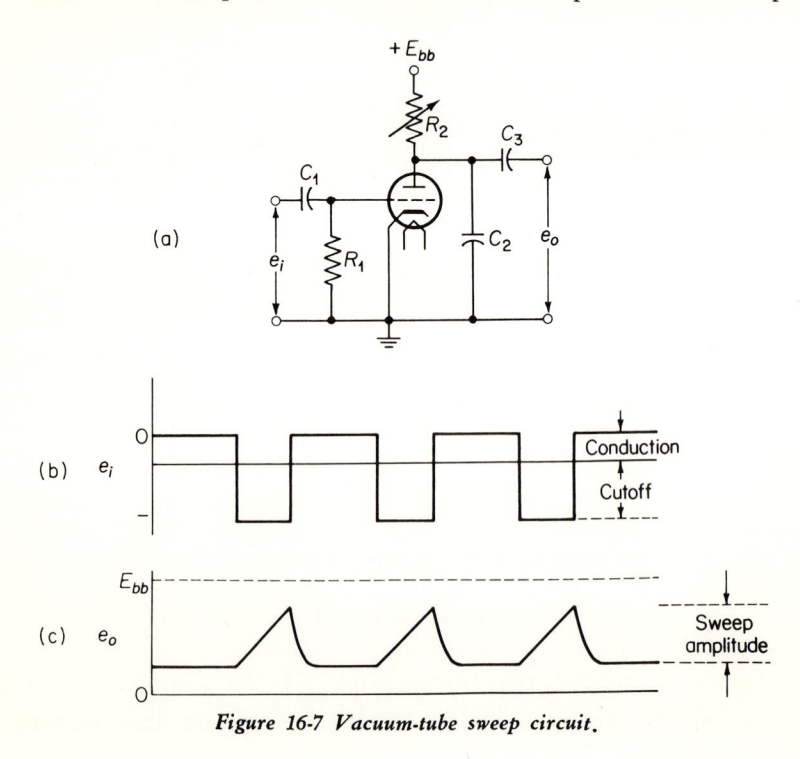

Figure 16-7 Vacuum-tube sweep circuit.

Sec. 16.5 *Improving linearity of the vacuum-tube sweep circuit* 321

50 kc chiefly because of the time required for the thyratron to deionize.

16.4 Vacuum-tube and transistor sweep circuits

A sweep circuit using a triode vacuum tube is shown in Fig. 16-7(a). In the absence of an input signal, the grid is at zero potential. The tube conducts heavily, causing a large voltage drop across adjustable plate resistor R_2. Both the plate voltage and the voltage across C_2 are at a low constant value.

The application of the negative-going input pulse waveform shown in Fig. 16-7(b) drives the grid beyond cutoff, and the plate current ceases. The plate voltage, as shown in (c), then rises toward the value of E_{bb} at a rate determined by the R_2C_2 time constant. When voltage e_i returns to zero and cutoff is effectively removed, the grid again returns to zero potential and the tube conducts, rapidly discharging capacitor C_2. Note that the frequency of the output waveform is controlled by the input pulse. The pulse duration of voltage e_i also controls the amplitude of e_o for a given R_2C_2 time constant. The circuit is adjusted so that only a small portion of the exponential charging curves is used for best linearity of e_o. Since the sweep voltage is triggered (driven) by the input waveform, this type of circuit is usually called a triggered (driven) sweep circuit. To improve the frequency response of this circuit, high g_m pentodes are used in lieu of triodes.

A transistor version of the vacuum-tube sweep circuit is shown in Fig. 16-8. The operation is essentially the same as that of the vacuum-tube circuit and requires no additional comment. By using the proper type of transistor, however, either a positive-going or a negative-going sweep can be generated across capacitor C_2.

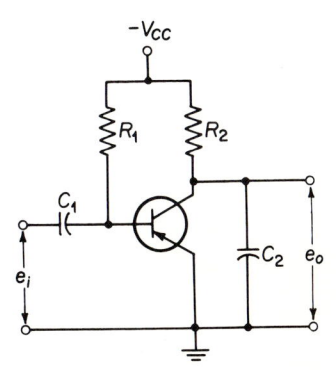

Figure 16-8 Transistor sweep circuit.

16.5 Improving linearity of the vacuum-tube sweep circuit[1]

The linearity of the circuit just discussed is considerably improved by using a *bootstrap* arrangement with feedback to the sweep generator. A circuit of this type is shown in Fig. 16-9. Tube V_1 is the

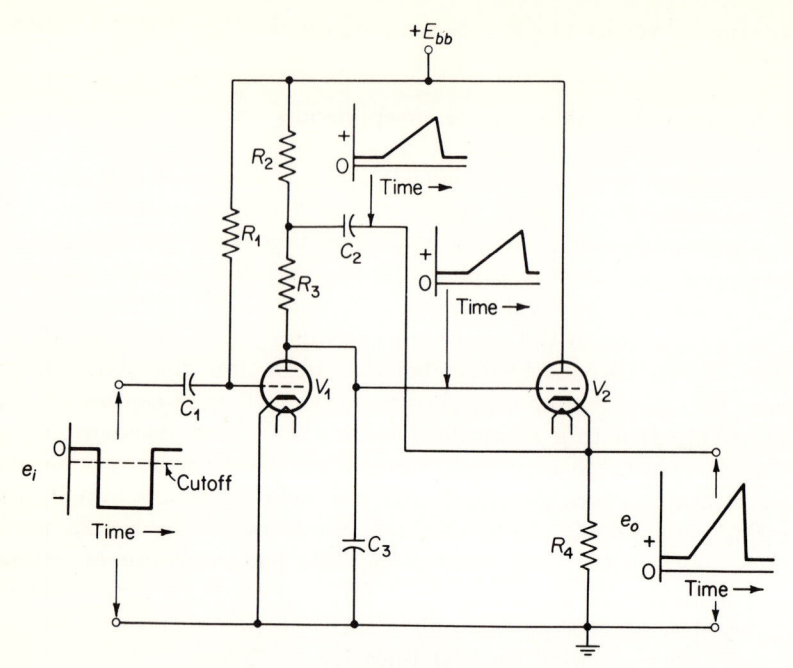

Figure 16-9 The bootstrap sweep circuit.

triggered sweep generator. It normally conducts heavily because the control grid is returned to E_{bb} through R_1. As a result, the plate voltage and the voltage across capacitor C_3 are low. The grid of cathode-follower V_2 is directly coupled to the plate of V_1. The plate current through V_2 produces a voltage drop across R_4 that makes the cathode positive with respect to the grid. Feedback is applied from the cathode of V_2 through C_2 to the junction of resistors R_2 and R_3 in the plate circuit of V_1.

When the negative-going square-wave voltage is applied to the input of V_1, it drives the grid beyond cutoff and the plate current ceases. Capacitor C_3 then begins to charge through R_2 and R_3 in the usual manner. The positive-going sawtooth voltage across R_3 is directly coupled to the grid of V_2. As the grid voltage of V_2 rises, its cathode potential also rises. This rising voltage is fed back to the junction of R_2 and R_3 through capacitor C_2. Thus, the voltage at this junction rises as the sweep voltage increases. The name *bootstrap* is derived from this action, which effectively increases the supply voltage to charging capacitor C_3 as the sweep voltage increases. Because the output of the cathode-follower is almost equal in amplitude to its grid voltage, the upper end of R_3 rises in potential almost

Sec. 16.5   *Improving linearity of the vacuum-tube sweep circuit*

as rapidly as the lower end. This produces a constant current through R_3, and the sweep capacitor charges at a uniform rate. The cathode-follower serves only to improve the linearity of the output sweep voltage.

The operation of this circuit has some limitations. Some loss actually occurs between the input and output of the cathode-follower. This means that the feedback voltage to R_3 is always less than the rise in the plate voltage of V_1. In addition, capacitor C_2 must charge slightly during the sweep so that the rise in potential on the terminal connected to R_3 is less than the rise on the opposite terminal. If C_2 is made large, this voltage loss is minimized. Another disadvantage is that R_2 acts as a load on the feedback network in parallel with R_3 and the charging circuit. The loading effect is reduced if R_2 is made large in comparison to R_3. For the best results, however, resistor R_2 may be replaced by a diode connected as shown in Fig. 16–10.

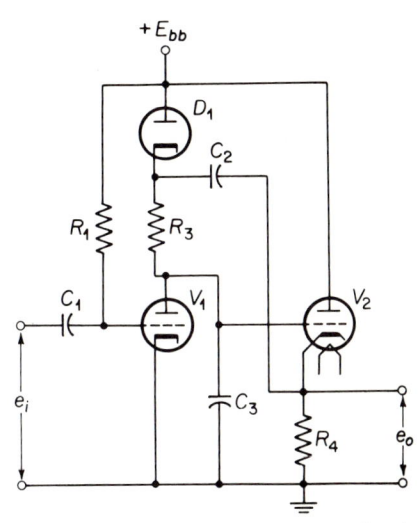

Figure 16-10 *The bootstrap sweep circuit using a diode for improved linearity.*

The triggered sweep-generator tube V_1 and the diode D_1 conduct simultaneously. Because the forward impedance of the diode is very small, the voltage at the upper end of R_3 is nearly equal to E_{bb}. When the negative-going square-wave pulse is applied to the input of V_1, it drives the grid beyond cutoff as before, and capacitor C_3 starts to charge through resistor R_3. The increasing voltage across C_3 is applied to the grid of V_2, causing an increase in the plate current and a greater voltage drop across R_4. The rise in voltage is returned through the large capacitance of C_2 to the upper end of R_3. The upper end of R_3 rises above the value of E_{bb} and cuts off the diode. Because the diode then acts as an infinite impedance, the voltage at the top of R_3 can rise to a considerably greater value than E_{bb}. This action improves the linearity of the sweep voltage and also increases its amplitude with a given supply voltage.

A transistor counterpart of the circuit of Fig. 16–10 is shown in

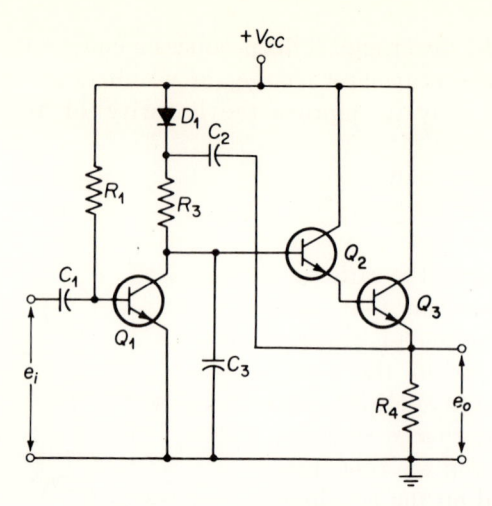

Figure 16-11 Transistor version of bootstrap sweep generator.

Fig. 16-11[2]. The negative-going trigger pulse cuts off Q_1, and C_3 charges through R_3. This increases the forward bias on Q_2 and, therefore, the voltage drop across R_4. This increase in turn is applied to the cathode of D_1 and cuts off the diode. Capacitor C_2 then discharges through R_3 and the base of Q_2. The reason for using compound-connected transistors instead of a single transistor is to make the base current small.

Another method of improving the linearity of a sweep waveform

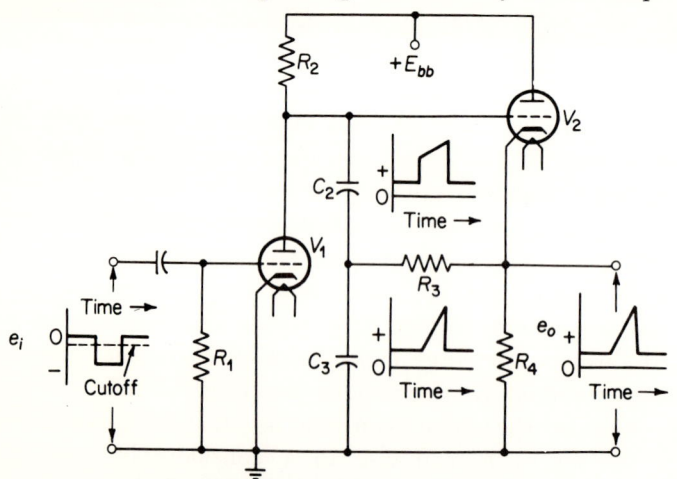

Figure 16-12 Improvement of linearity by using an integrating network.

Sec. 16.6   The Miller sweep circuit

is by the use of an integrator circuit as shown in Fig. 16-12[1]. Sweep voltage is developed across two series-connected capacitors, C_2 and C_3. Resistor R_3 and capacitor C_3 form an integrating network. The voltage waveforms across C_2, C_3, and R_4 are shown. The voltage across C_2 rises rapidly at first, then at a decreasing rate as the charging current through R_2 decreases exponentially. The resulting waveform, of course, is not linear. The rising voltage at the grid provides a rising voltage at the cathode of V_2 which is applied to the integrating network. The voltage developed across C_3 rises slowly at first, then more rapidly as the applied voltage increases. The combined effect of these two rising voltages is applied to the grid of V_2. If the value of R_2 is correctly chosen, the nonlinearity of the sweep generator is offset by the integrating effect of R_3 and C_3, and a linear sweep voltage results.

Obviously, the merit of this integrator circuit is that the feedback from the cathode of V_2 to the plate of V_1, when combined with the integration effect, provides a sweep voltage of good linearity and large amplitude without the use of an excessively high value of supply voltage.

16.6 The Miller sweep circuit[3]

Another type of sweep circuit that will improve linearity of the sweep is the so-called Miller integrator or sweep. Before this particular sweep circuit can be understood, it is necessary to take another look at tube interelectrode capacitances. In a triode, the capacitances of concern are those existing from grid-to-plate, grid-to-cathode, and plate-to-cathode, as shown in Fig. 16-13.

At high frequencies, these capacitances, however small, may tend to resonate with stray circuit inductances, causing instability and distortion. As their reactances are low at high frequencies, they shunt the input and output circuits, reducing gain and causing regenerative feedback.

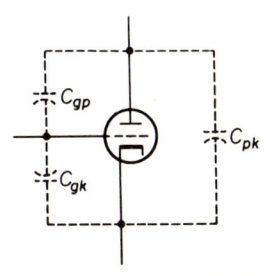

Figure 16-13 Illustration of the principal interelectrode capacitances of a triode vacuum-tube.

The interelectrode capacitances C_{gk} and C_{gp} are effectively in parallel with respect to the grid. Momentarily disregarding the plate voltage, the current in the grid-cathode circuit through C_{gk} and C_{gp} is, by Ohm's law

$$I = \frac{E_g}{X_{C_{gk}}} + \frac{E_g}{X_{C_{gp}}}$$

$$= \frac{E_g}{1/\omega C_{gk}} + \frac{E_g}{1/\omega C_{gp}}$$

$$= E_g \omega C_{gk} + E_g \omega C_{gp}$$

$$= E_g \omega (C_{gk} + C_{gp}) \qquad (16\text{-}1)$$

This expression does not, however, consider the amplified voltage present at the plate as a result of the stage gain, A. The voltage across C_{gp} is not just E_g, but $(E_g + E_g A)$, because the amplified plate voltage is E_g multiplied by the stage gain.

This means that the current through C_{gp} is not merely $E_g/X_{C_{gp}}$, as shown in the first line of Eq. 16-1, but $(E_g + E_g A)/X_{C_{gp}}$. Comparing this expression with the expression previously derived, which did not consider A, it is apparent that the effective value of C_{gp} has become $(C_{gp} + AC_{gp})$, or $C_{gp}(1 + A)$. This increase in C_{gp} resulting from stage gain is termed the *Miller effect*.

With the Miller effect present, and assuming no phase difference between the resistive and actual load voltages, the capacitances shunting the grid and cathode are

$$C_{gk} + C_{gp} + AC_{gp} = C_{gk} + C_{gp}(1 + A) \qquad (16\text{-}2)$$

If a phase difference does exist between the resistive and actual load voltages, the expression becomes

$$C_{gk} + C_{gp}(1 + A \cos \theta) \qquad (16\text{-}3)$$

where θ is the phase angle.

It has been shown that the effective value of C_{gp} increases by the factor $(1 + A)$ when a tube conducts. If a physical capacitor of value C is connected from grid to plate, its effective value becomes $C(1 + A)$ during conduction as a result of the Miller effect. This increase in capacitance makes it possible to produce a substantially linear sweep of considerable amplitude without using a high voltage to charge the time-base capacitor.

Use is made of the Miller effect in the Miller sweep circuit of Fig. 16-14. In the steady-state condition, the suppressor grid is maintained below cutoff by the biasing voltage $-E_{\text{sup}}$, and there is no plate current. The plate voltage is, therefore, equal to E_{bb}, and sweep capacitor C_2 is charged to almost this value. Because the control grid is returned to E_{bb} through resistor R_2, grid current exists and develops a large voltage drop that maintains the grid slightly positive with respect to the cathode. The screen potential is also low owing to screen current.

Sec. 16.6 The Miller sweep circuit   327

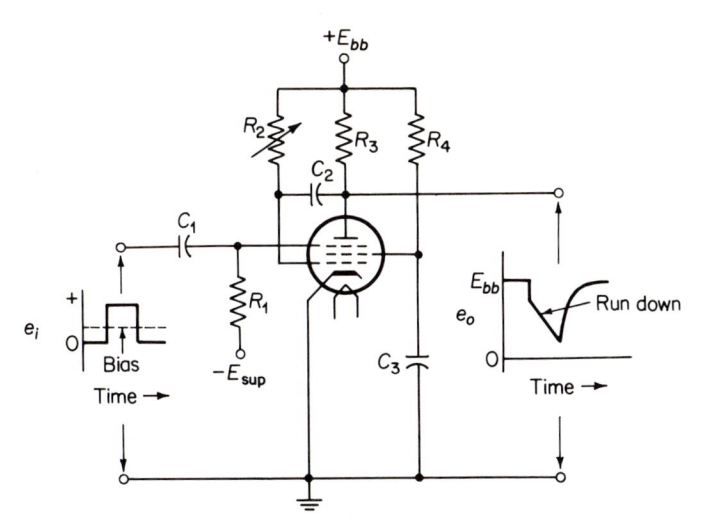

Figure 16-14 Miller-type linear sweep generator.

When the positive-going input trigger voltage is applied to the suppressor, it overcomes the negative bias and produces plate current through load resistor R_3 and a drop in plate voltage. The negative-going plate voltage is fed back to the control grid through C_2 and drives the grid negative to a point just above cutoff.

Capacitor C_2 then begins to discharge through the tube and through resistor R_2. The grid voltage rises toward E_{bb} at a rate determined by the R_2C_2 time constant. As the grid voltage rises, the plate current increases and the plate voltage decreases correspondingly. The decreasing plate voltage is fed back to the control grid through C_2, and tends to oppose the rise that caused it. As a result of this action, the rise of grid voltage is not exponential, as might be expected from the normal discharge of a capacitor through a resistor, but rather, a slower and much more linear rise. As the grid voltage rises linearly, the plate voltage decreases in the same manner. The linear decrease in plate voltage is called the *run-down.*

When the trigger input voltage drops to its initial value, the suppressor grid is again driven beyond cutoff and plate current ceases. Grid current again occurs, and capacitor C_2 charges rapidly through the plate-load resistor and the grid-to-cathode path of the tube. The circuit then remains in the quiescent condition, with the plate voltage rising exponentially to E_{bb}, until the next square-wave pulse is applied to the suppressor grid. The operation described is then repeated.

16.7 The monostable screen-coupled phantastron[1]

The positive pulse applied externally to the suppressor grid in the Miller sweep generator can be injected internally from the screen. Because the internally injected pulse is used, the circuit requires only a positive triggering pulse on the suppressor, or a negative triggering pulse on the plate, to initiate the Miller sweep action. The circuit then becomes known as a *monostable screen-coupled phantastron*. The original explanation of circuit operation was so unbelievably fantastic that the name *phantastron* was conceived.

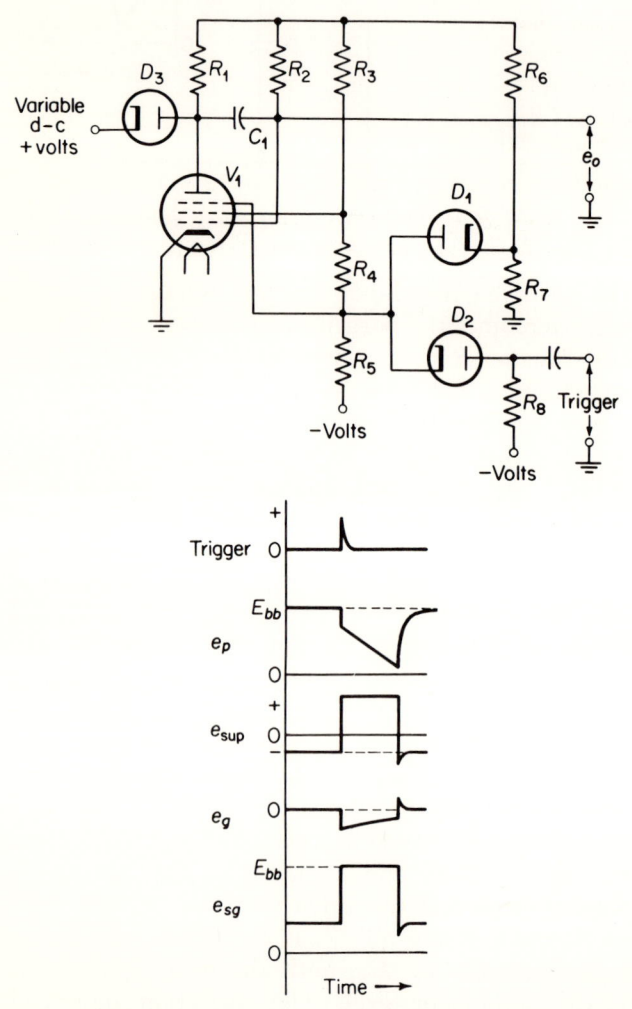

Figure 16-15 Monostable screen coupled phantastron and waveforms.

Sec. 16.7 *The monostable screen-coupled phantastron* 329

The circuit is shown in Fig. 16-15. During quiescent operation, there is no V_1 plate current as the result of a negative bias applied to the suppressor grid from the voltage-divider network consisting of resistors R_3, R_4, and R_5. The entire tube current passes to the screen grid. When a positive trigger is applied to the suppressor through diode D_2 it overcomes the negative bias, and the resulting current through load resistor R_1 causes the plate voltage to drop. The grid voltage drops by the same amount because of the coupling across capacitor C_1. The decrease in screen current causes its voltage to rise. Because the suppressor grid obtains its voltage from the same voltage divider network as the screen, suppressor voltage also rises. The action is regenerative. When the plate voltage of V_1 reaches its minimum value, capacitor C_1 must rise toward E_{bb} at a rate determined by the $R_1 C_1$ time constant because the plate current continues. The increase in the grid voltage, resulting from the charging of capacitor C_1, causes the cathode and screen currents to increase and the screen voltage to decrease. An increase in the screen current means that the plate current must decrease and the plate voltage must increase. The voltage increase is again coupled back to the control grid. Because of this regenerative action, the entire current rapidly switches to the screen, the plate current is cut off, and the suppressor grid is driven negative.

Diode D_1, shown connected to the suppressor, acts as a clipper to ensure that the suppressor voltage does not go too high. The suppressor is then unable to "stick" as a result of secondary emission, but the plate can still be turned full on during the Miller run-down. Another advantage resulting from the use of D_1 is a quicker transition from the quiescent to the on position because the total movement of the suppressor is reduced. Diode D_2 is used as a trigger-injecting diode. It is ideal for this purpose, because it acts as a switch that closes during the initial voltage step of the trigger and then opens to disconnect the circuit from the trigger source. Any loading from the trigger source is prevented. Diode D_3 makes possible linear control of the rectangle and triangle length by controlling the initial level of plate voltage, that is, the level at which the run-down begins. A diode used in this

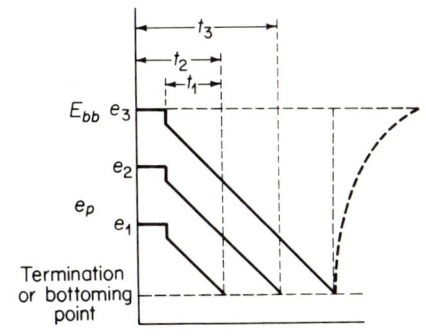

Figure 16-16 Production of an adjustable time delay.

manner is often called a *plate-catching* diode.

The method used to provide an adjustable time delay for the circuit of Fig. 16-15 is to vary the *d-c* voltage level at which the plate is found at the instant of applying the trigger. Note, in Fig. 16-16, that the slope of the run-down is obviously constant, as is the termination or bottoming level of the plate voltage. Thus, any attempt to vary recovery by adjusting the plate voltage also affects the duration of the run-down.

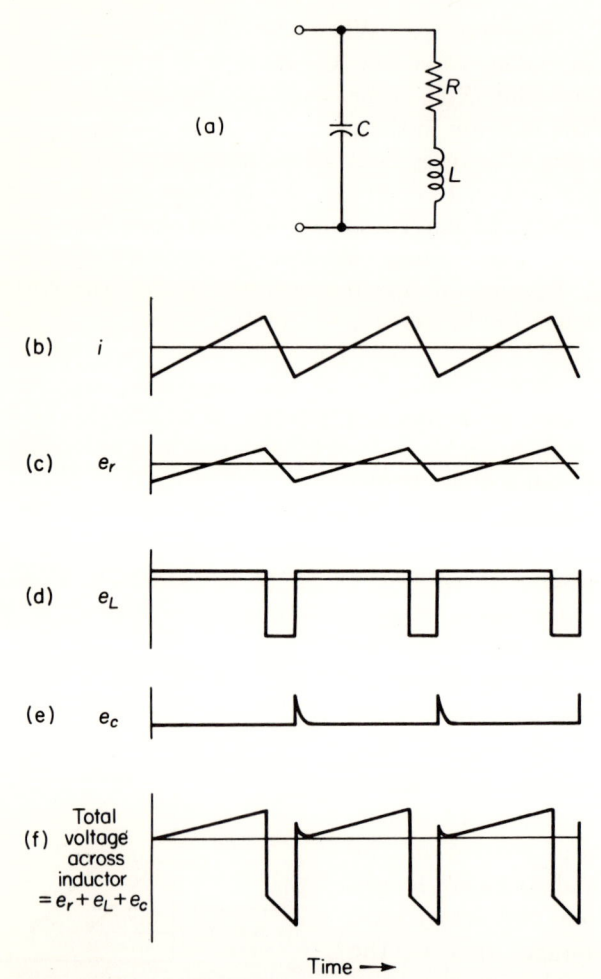

Figure 16-17 (a) Equivalent circuit of an inductor and (b–f) formation of a trapezoidal waveform.

16.8 Current time-base generators

The sweep circuits considered thus far are collectively known as voltage time-base generators and are designed for use with display tubes employing electrostatic horizontal deflection of the electron beam. For electromagnetic deflection, a current time-base generator is required, since deflection is proportional to the flux density of the magnetic field created by the deflection coil. Assuming a uniform flux field, the beam will be deflected by an amount which is proportional to the current generating the flux. Thus, the linearity of the sawtooth of current through the coil determines the linearity of the beam deflection.

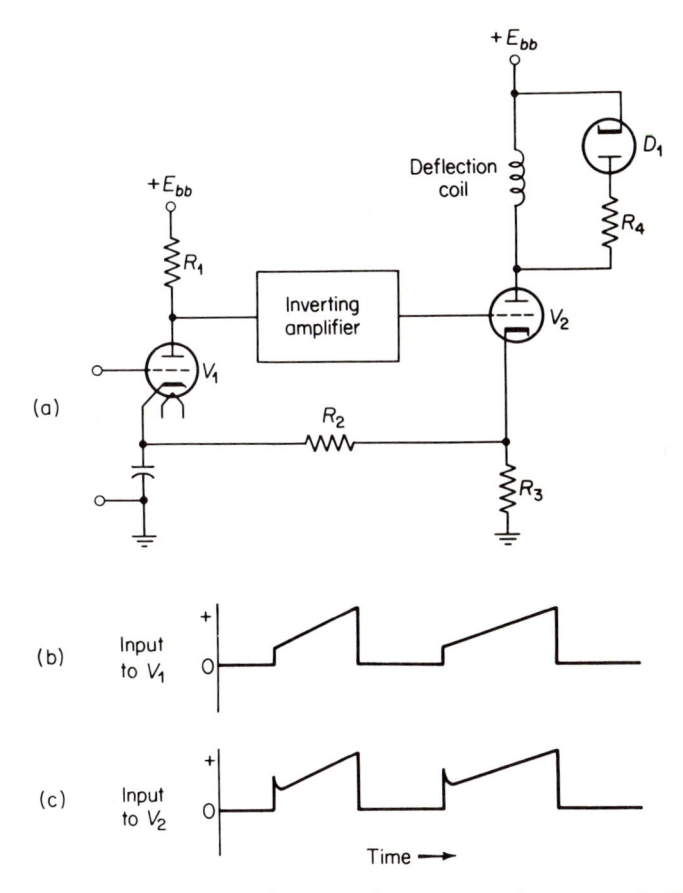

Figure 16-18 Achievement of linear deflection in an electromagnetic CRT.

All practical inductors contain some resistance and capacitance in addition to inductance. We can, therefore, represent an inductor by the equivalent circuit shown in Fig. 16-17(a). If a sawtooth current, as in (b), is now applied to the inductor, the resistive component produces a sawtooth voltage waveform, as in (c); the inductive component produces a pulsed waveform, as in (d); and the capacitive component produces a spike, as in (e). The total voltage across the inductor has a trapezoidal waveform with a spike on the leading edge, as in (f), which is formed graphically by adding the waveforms of (c), (d), and (e).

If, then, a voltage wave of the shape shown in Fig. 16-17(f) is applied to the inductor, a sawtooth wave of current, as in (b), passes through the coil. If this sawtooth is linear, and if the inductor is the deflection coil of an electromagnetically deflected CRT, the electron beam will be deflected linearly across the tube face.

The simplified circuit of Fig. 16-18(a) may be used for linear deflection in an electromagnetic CRT. Stage V_2 is normally biased *off* to prevent current through coil L between sweeps. Thus the input sawtooth voltage must have a step as shown in (b) of the figure. Current feedback is applied to V_1, and a small capacitor, C, is placed across the feedback network to attenuate the high-frequency components of the feedback waveform. In this manner, the leading edge of the trapezoidal waveform applied to V_2, in part (c) of Fig. 16-17, is emphasized to create the desired spike.

Diode D_1 and resistor R_4 act as a damping circuit. When the V_1 plate current ceases, the energy stored by coil L is returned to the circuit, and the tank circuit formed by coil L, its distributed capacitance, and stray capacitance, is shock-excited into oscillation. The first negative half-cycle of oscillation makes the plate of D_1 positive with respect to its cathode, and the diode conducts. The power dissipated by R_4 rapidly damps out subsequent oscillation. If the damping network is not used, oscillations cause distortion of the succeeding time-base sweeps.

REFERENCES

1. Chance, B., et al., *Waveforms.* New York: McGraw-Hill Book Co., Inc., 1949.
2. Reich, Herbert J., *Functional Circuits and Oscillators.* Princeton, N. J.: D. Van Nostrand Co., Inc., 1961.
3. MIT Radar School Staff, *Principles of Radar*, 2nd. ed. New York: McGraw-Hill Book Co., Inc., 1946.

EXERCISES

16-1 What are the disadvantages of a neon sawtooth generator?

16-2 Draw the circuit and waveforms of a current time-base generator.

16-3 Explain the Miller effect.

16-4 Explain the operation of a trapezoidal waveform generator.

16-5 Draw the circuit and waveforms of a simple bootstrap sweep generator.

16-6 Draw the circuit and waveforms of a constant-current pentode sweep generator.

16-7 Explain how the integrating network in Fig. 16-12 helps to improve sweep linearity.

16-8 What disadvantages are associated with the thyratron sawtooth generator?

16-9 Draw the circuit and waveforms of a Miller linear-sweep generator.

16-10 Draw the circuit and waveforms of a monostable screen-coupled phantastron.

16-11 Why is a current-limiting resistor needed in a thyratron sweep-generator circuit?

16-12 What disadvantages are associated with the bootstrap sweep generator of Fig. 16-9?

16-13 Explain the purpose of each of the three diodes used in the circuit of Fig. 16-15.

17

TRANSMISSION GATES

A gate is a circuit having a single output and a multiplicity of inputs so designed that the output is available (appears) only when a certain definite set of input conditions is met.

An *ideal transmission gate* produces an output identical to the input during a selected time interval (gate open) and zero output (gate closed) at all other times. In practice, the idealized transmission gate is not realized; but, within limits, this is of relative unimportance so long as an output is produced at the correct time. Transmission gates are also referred to by names descriptive of their functions such as, *coincidence, time selector, linear gate,* and *separator* circuits.

17.1 Unidirectional diode gate

The unidirectional diode gate shown in Fig. 17-1 passes signals of positive polarity only. In the absence of a gating signal, a battery, of voltage E, makes the anode of diode D_1 negative with respect to its cathode, and conduction cannot occur as long as the amplitude of the input pulse train, e_s, does

Sec. 17.1 Unidirectional diode gate   335

not exceed the value E. For the sake of illustration, let us assume that E is $-6\,\text{v}$ and e_s is $+2\,\text{v}$.

From time t_0 to t_5, diode D_1 cannot conduct and e_o is zero. At time t_4, however, a positive-going gating signal, E_g, equal in amplitude to the negative bias voltage, is also applied to the circuit. The gating

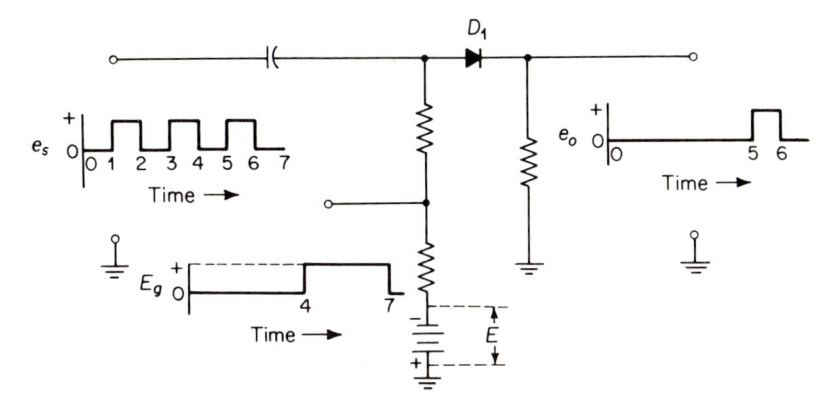

Figure 17-1 A unidirectional gate that passes signals of positive polarity only.

voltage, which has an amplitude equal to E but is opposite in polarity, effectively cancels the bias so that the anode of D_1 is at zero potential with respect to its cathode. Signal voltage e_s then causes D_1 to conduct during time t_5 to t_6, and the positive-going output signal, identical to e_s, appears across load resistor R_L.

At time t_7, the gating signal E_g drops to zero, and the initial circuit conditions apply. It is apparent that an output signal is produced only during the application of a gating signal: in this instance, between time t_4 and time t_7. To ensure proper operation, the width of the gating signal is larger than that portion of e_s to be passed to the output.

The action described above is idealized in that perfectly shaped waveforms are depicted and the output voltage, e_o, is zero at all times except during the application of the gating pulse. From previous discussions, it is apparent that this is not the case in practice. A minimum capacitance is always present and tends to round the corners of the waveforms. In addition, e_o cannot exactly equal e_s in amplitude because the resistance of the diode during periods of conduction is not zero, but some low value.

The circuit of Fig. 17-1 is sometimes modified for use as a *threshold gate*. To illustrate, let us suppose that the amplitude of the gating pulse is only $+5\,\text{v}$ compared to the $-6\,\text{v}$ of the battery, E.

Under these conditions, the plate of D_1 is still 1 v negative with respect to its cathode. Only when the amplitude of e_s exceeds $+1$ v can the diode conduct and produce an output across R_L. The similarity to a clipper is easily recognized. This arrangement is useful in removing spurious or small, low level noise signals in the output.

It is also interesting to observe the circuit action when the amplitude of the gating signal is greater than bias voltage E. For example, suppose that the gating signal has an amplitude of $+8$v. This makes the plate of the diode 2 v positive with respect to its cathode, and the resulting conduction produces an output of $+2$ v amplitude, as shown in Fig. 17-2. When a signal having an amplitude of, say, 3 v, arrives, conduction increases and produces a signal of $+3$ v on top of the $+2$ v signal produced by a portion of the gating pulse.

That portion of the gate appearing in the output is called a *pedestal*. In applications where a pedestal is undesirable, the bias and gate amplitudes must be adjusted carefully.

If it is desired to pass signals of negative polarity only, the circuit of Fig. 17-2 is rearranged by reversing the biasing battery and diode and by using negative gating and input signal pulses.

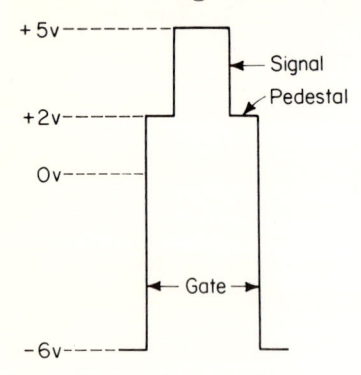

Figure 17-2 A formation of a pedestal when the amplitude of the gate exceeds the bias applied to diode D_1.

17.2 The coincidence gate

A form of unidirectional gate which requires the simultaneous application of multiple gating pulses to produce an output is shown in Fig. 17-3. Because the time at which the gating pulses are applied must coincide, the circuit is usually called a *coincidence gate*. If the three gating pulses are not in coincidence, diode D_4 cannot conduct because its plate is held negative with respect to its cathode by any one of the three parallel diode gates. If all three gating pulses are in coincidence, the negative bias developed by the bias battery on the plate of D_4 is effectively removed, conduction occurs, and an output signal is produced across R_L.

Suppose for example, that a gating pulse is not present at gate input 1. The negative-bias voltage applied to the cathode of D_1 causes that diode to conduct, producing a current through the series circuit

Sec. 17.2   The coincidence gate

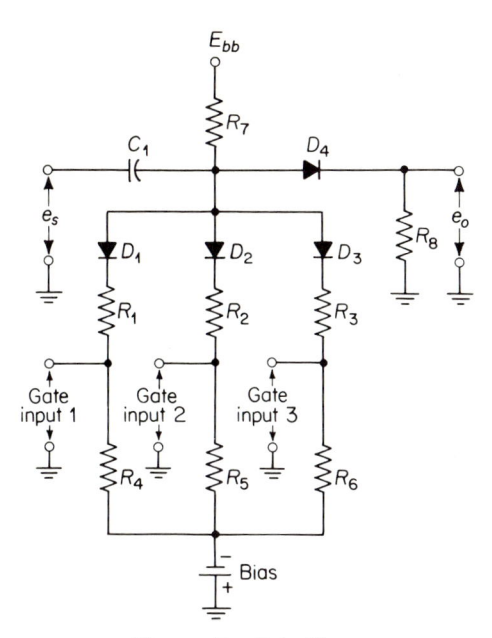

Figure 17-3 Coincidence gate.

composed of R_4, R_1, D_1, and R_7. The value of R_7 is very large compared to $R_4 + R_1 + R_f$ (the forward resistance of D_1). Under these conditions, most of the voltage between ground and E_{bb} appears across R_7. Because the anode of D_4 is connected to the bottom (negative end) of R_7, it cannot conduct; and no output signal can appear across R_L unless the amplitude of e_s is sufficient to overcome the bias on D_4. A similar action occurs if gating pulses are absent from gate input 2 or 3.

When all gating pulses are applied in coincidence, the bias battery is effectively removed from the circuit. The backward resistance of diodes D_1, D_2, and D_3 becomes very large compared to R_7 and, for practical purposes, the total supply voltage is dropped by the diodes, removing the negative bias from the plate of D_4. A positive-going input signal then causes D_4 to conduct, and produces an output signal across R_L that is a reasonable reproduction of e_s. When any or all of the gating pulses are removed, e_o again drops to zero.

In Fig. 17-3, resistors R_1, R_2, and R_3 serve to isolate the input signal from the gate inputs. Resistors R_4, R_5, and R_6 develop the gating-pulse voltages and isolate the gating terminals from the bias battery.

17.3 Bidirectional gates[1,2]

The bidirectional gating circuit shown in Fig. 17-4 can select input signals of opposite or identical polarity; i.e., negative and/or positive. In the absence of gating voltages, diodes D_1 and D_2 conduct; and, for practical purposes, e_o is zero since the output is effectively short-circuited by the diodes. The application of gating signals serves to make the diodes nonconducting and allows the input signal to appear at the output.

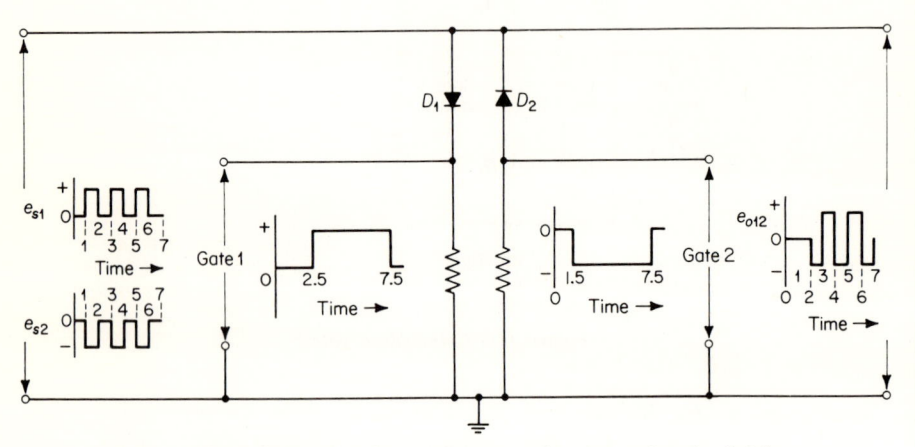

Figure 17-4 A bidirectional gate that can select input signals of either or both polarities.

With e_{s1} applied to the input terminals, suppose that a positive-going gating pulse is applied to gate input 1 at time $t_{2.5}$. The first pulse, at t_1 to t_2 of the input pulse train e_{s1}, produces no output for the reasons discussed above. Before the second input pulse occurs, however, the cathode of D_1 is made positive with respect to its plate as a result of the gating signal. The shunting diode can no longer conduct. With the input-output line now open instead of short-circuited, the second input pulse appears across the output terminals as e_{o1}. As long as the gating signal is present, the output signal is a reasonable reproduction of the input.

With e_{s2} applied to the input terminals, and a negative gating pulse applied to gate input 2, an output signal e_{o2} is produced. The action is identical to that already described. The combined output signal is labeled e_{o12}.

Another form of bidirectional gate using two diodes is shown in Fig. 17-5. In the absence of gating signals, e_o is practically equal to e_s. The application of gating signals removes either or both positive-

Sec. 17.3   *Bidirectional gates*   339

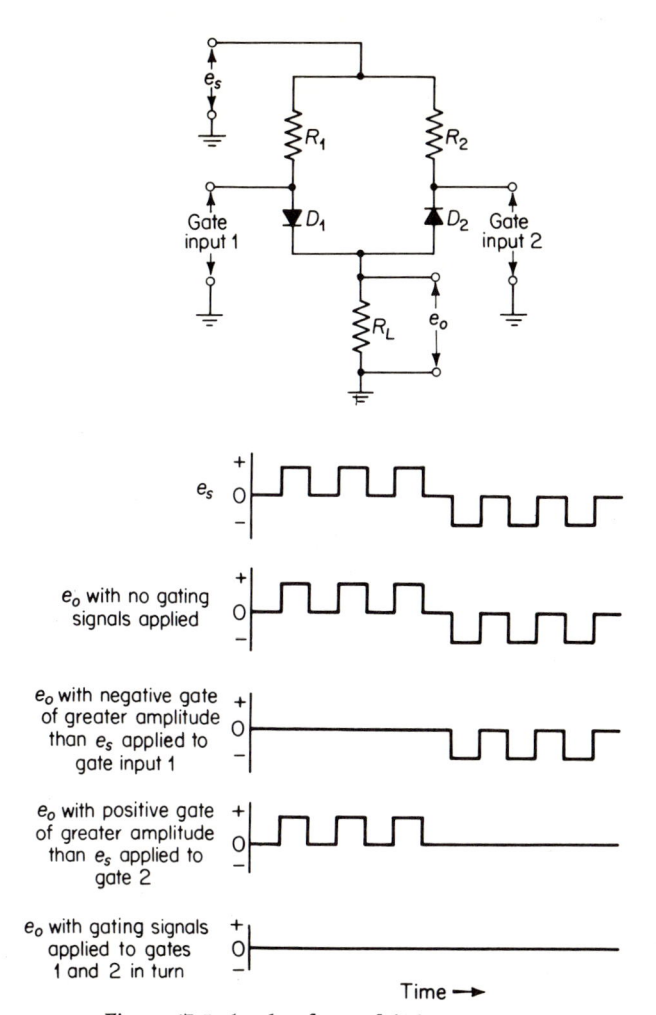

Figure 17-5 Another form of bidirectional gate.

going or negative-going input polarities of e_s.

Suppose that the input signal consists first of a positive-pulse train, then a negative-pulse train. The positive-going signal makes the anode of D_1 positive with respect to its cathode. Current passes from ground, through R_L, D_1, R_1, and the input circuit, back to ground, producing a positive-going output signal of slightly less amplitude than e_s owing to the small drop across isolating resistor R_1. A similar action occurs for the negative input, with current passing through R_L, D_2, R_2, and the input circuit, back to ground.

If a negative gating pulse is applied to gate input 1, of greater amplitude than the pulses of the positive input signal, conduction through D_1 cannot occur and the output drops to zero. Similarly, if a positive gating pulse is applied to gate input 2, of greater amplitude than the pulses of the negative input signal, D_2 cannot conduct and e_o drops to zero.

Obviously, each gating pulse controls the time intervals during which output signals of a desired polarity can appear across R_L.

17.4 Four-diode gate

The two-diode bidirectional gate of Fig. 17-4 has some disadvantages: (1) the output signal is attenuated as a result of the voltage drop across the isolating resistors, and (2) there is some signal leakage to the output in the absence of gating pulses, because of the capacitance of the diodes. Considerable improvement is achieved by using four diodes and arranging the circuit as shown in Fig. 17-6. Symmetrical gating voltages are applied to the gate inputs; that is, with reference to the zero axis, the maximum positive and negative amplitudes of the gating voltages are equal. Two additional d-c voltages, $+E$ and $-E$, of equal amplitude but opposite polarity, are also connected as shown.

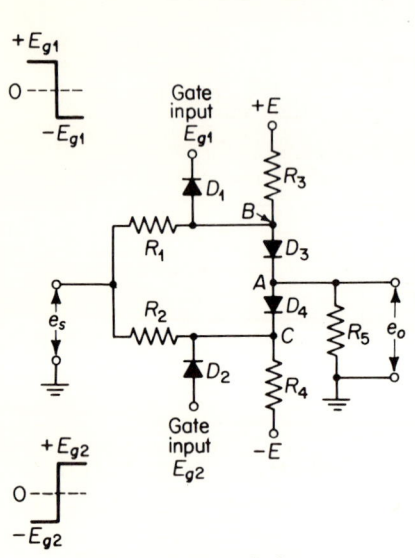

Figure 17-6 Four-diode gate.

The d-c voltages cause current to pass from $-E$ through R_4, D_4, D_3, and R_3 to $+E$. Because the circuit is balanced, the net voltage appearing at point A is zero, and no output appears across R_5.

When gate signal $-E_g$ is applied to gate input 1, and signal $+E_g$ to gate input 2, diodes D_1 and D_2 are back-biased and cannot conduct. If an input signal, e_s, is now applied, it passes through R_1 and R_2 to diodes D_3 and D_4. If e_s is positive-going, it increases the conduction of D_3; if it is negative-going, it increases the conduction of D_4. In either event, the net voltage at point A is no longer zero. A current is, therefore, produced through R_L and develops an output voltage that is reasonably similar to e_s.

Reversing the polarities of the gating signals permits D_1 and D_2

Sec. 17.5   Six-diode gate   341

to conduct, clamping points B and C to the gating voltage potentials. This back-biases D_3 and D_4, effectively isolating the output circuit from the input, and e_o is zero. Input signals of either polarity serve only to increase the degree of isolation. A positive input signal causes greater conduction through D_1. This current passes through R_1, making point B and, therefore, the anode of D_3, even more negative than the gating signal alone. Similarly, a negative input signal increases the conduction of D_2 and increases the back-bias on D_4.

It is also important to note that when D_1 and D_2 are cut off, the gating signal, regardless of its amplitude, has no effect on the output circuit, and the formation of a pedestal, as previously described, cannot occur.

Although the four-diode gate is an improvement over the two-diode circuit, the input signal is still attenuated somewhat because of the drop across R_1 and R_2 when the gate is open, that is, when D_1 and D_2 are cut off.

17.5 Six-diode gate

The six-diode gate of Fig. 17-7 causes practically no attenuation of e_s when the gate is open and only negligible feed-through when the gate is closed. The only difference between this circuit and the four-diode gate is that diodes D_5 and D_6 replace resistors R_1 and R_2.

With the gating signal applied to D_1 at $+E_g$, and that applied to D_2 at $-E_g$, input signals are transmitted to the output in the same manner as previously described. When e_s is positive, D_6 conducts; and when e_s is negative, D_5 conducts. Because the resistance of the diodes is very low during conduction, e_s suffers practically no attenuation in passing to the output circuit.

When the polarity of the gating signals is reversed, points B and C are again clamped to the gating potential, cutting off D_3, D_4, D_5, and D_6. Notice that the anodes of D_3 and D_5 are connected to point B, and the

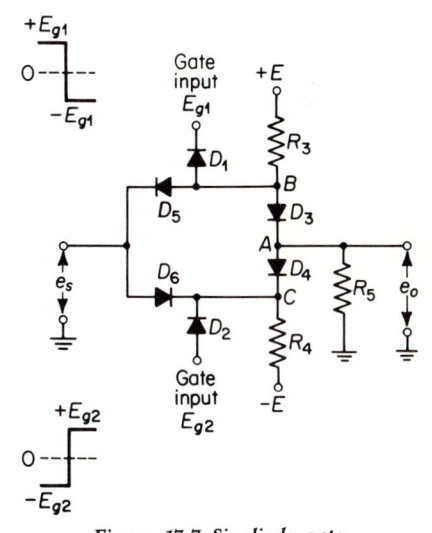

Figure 17-7 Six-diode gate.

cathodes of D_4 and D_6 to point C. The back-bias on D_5 and D_6 effectively decouples the input and output circuits. Any input signal that might pass through the capacitances of D_5 and D_6 still has to pass through the capacitances of D_3 and D_6; it then produces a negligible output across R_L.

17.6 A simple triode gate[3]

It has been shown previously that triodes may perform some function of diodes with the added advantage of grid control. A simple form of triode gate is shown in Fig. 17-8. An a-c signal voltage is applied

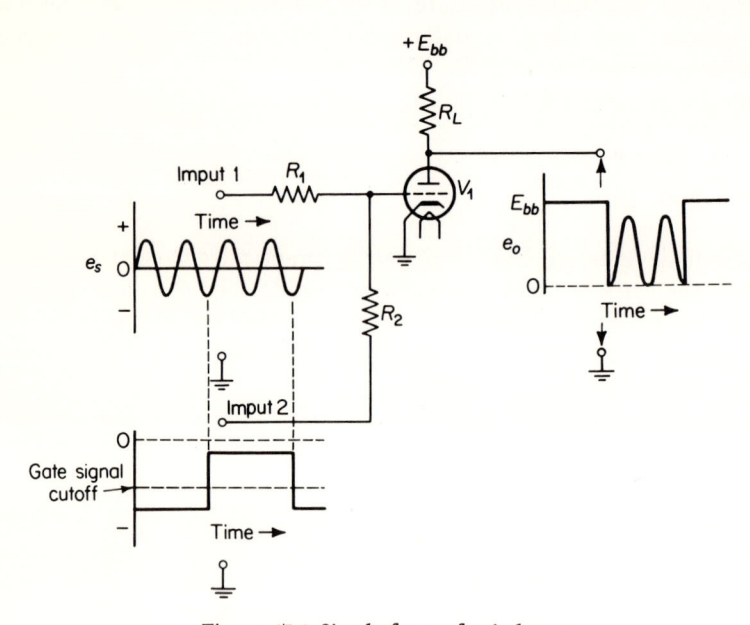

Figure 17-8 Simple form of triode gate.

to the control grid through input 1, while a gating signal is also applied to the grid through input 2.

In the absence of a gating pulse, the grid of V_1 is biased beyond cutoff by the steady negative voltage applied at input 2. The amplitude of e_s is not sufficient to overcome this bias. Under these conditions, there is no voltage drop across R_L, and the potential at the output terminals is practically equal to E_{bb}.

The application of a gating pulse shifts the operating point of V_1 to the approximate midpoint of the characteristic curve so that

Sec. 17.7 A two-triode gate   343

the tube operates as a Class A amplifier. An amplified and inverted replica of e_s then appears across R_L. Notice that this waveform is superimposed on an inverted pedestal as a result of decreasing plate voltage when V_1 conducts.

When the gating pulse is removed, the circuit is restored to its original condition. A disadvantage of this circuit is the small degree of isolation provided between the signal and the gating input terminals by resistors R_1 and R_2.

17.7 A two-triode gate

Isolation is improved by using the circuit arrangement shown in

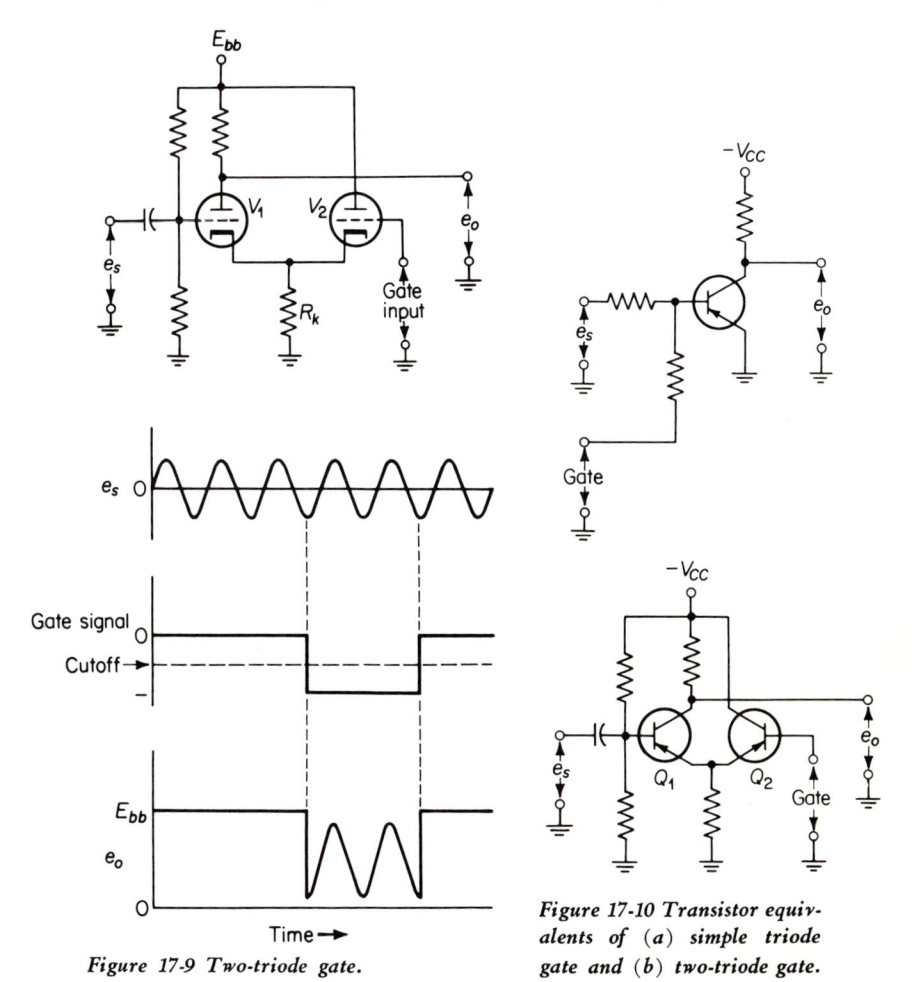

Figure 17-9 Two-triode gate.

Figure 17-10 Transistor equivalents of (a) simple triode gate and (b) two-triode gate.

Fig. 17-9. In the absence of a gating signal, V_2 conducts heavily, and its plate current through R_k produces a voltage drop that makes the cathode of V_1 sufficiently positive with respect to its grid to produce plate-current cutoff. Under these conditions, the output voltage is at a steady value practically equal to E_{bb}.

A negative gating signal drives V_2 beyond cutoff and permits V_1 to conduct. The plate current of V_1 produces at the cathode a positive voltage that would ordinarily limit plate current to a very low value. The grid of V_1, however, is returned to E_{bb} through a dropping resistor. The resulting positive voltage overcomes part of the bias developed across R_k, and the operating point of the tube is again at the approximate midpoint of the characteristic curve.

Under these conditions, signal voltage e_s appears in an amplified and inverted form across R_L. Again, the output signal appears to be superimposed on an inverted pedestal.

Transistor equivalents of the simple triode and two-triode gates are shown in Fig. 17-10(a) and (b).

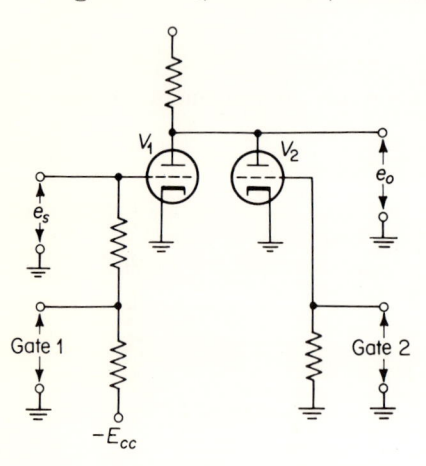

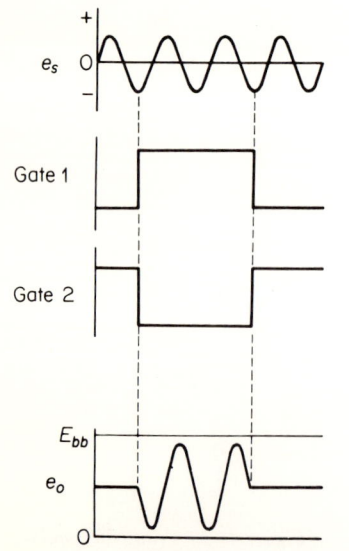

Figure 17-11 *Triode gate in which V_2 acts to remove pedestal from e_0.*

17.8 Removal of pedestal

In the circuit of Fig. 17-11, V_1 acts as a gating tube and V_2 acts to remove any pedestal from the output waveform. Gating signals of opposite polarity are applied to gate inputs 1 and 2. In the absence of signal input, the gating voltages hold the plate output voltage constant whether the gate is open or closed.

A negative gating signal at gate input 1 cuts off V_1. At the same time, a positive gating signal at gate input 2 permits V_2 to conduct. This current through R_L drops the output voltage to a level below E_{bb}.

The gate is opened by applying a positive gating signal to gate input 1 as shown in Fig. 17-11. At the same time, the gate input to V_2 becomes negative and causes plate-current cutoff. The circuit is arranged so that V_1 and V_2 pass equal plate current during their periods of conduction. The voltage drop across R_L, therefore, remains constant regardless of which tube conducts. Because the opening of the gate does not produce a change in the output voltage level, no pedestal appears in the output.

17.9 Pentode gates

The additional grids of a pentode provide better isolation between the signal and gate inputs. The gate is closed when the plate current is cut off by a negative bias voltage applied to the suppressor grid, and the output is almost equal to E_{bb}. The gate is opened by applying a positive pulse to the suppressor grid. Input signals then appear in amplified and inverted form in the output.

Control-grid bias depends on the nature of the input signal. If positive pulses only are to be passed during the gating interval, the bias is set beyond the cutoff point. If pulses of both polarity are to be passed, the bias is set for Class A operation.

17.10 Multicoincidence pentode gate

A typical pentode gate is shown in Fig. 17-12. The gate is closed as a result of negative bias voltages applied to each grid. When no positive-going pulse is present at either gate, the bias applied to that grid is sufficient to prevent plate current and no output is produced. When positive pulses are present at each grid to overcome the biasing voltages, the tube conducts and an amplified and inverted version of signal input appears in the output.

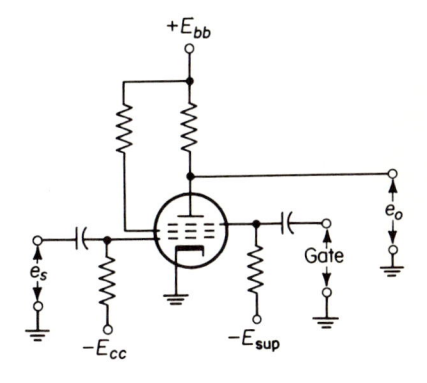

Figure 17-12 Pentode gate.

17.11 Gate circuit applications

Gating circuits are used in a wide variety of applications. Perhaps one of the most common uses is in electronic counters, and this subject is discussed at considerable length in the following chapter.

REFERENCES

1. Millman, J., and T. H. Puckett, "Accurate Linear Bidirectional Gates," *Proc. IRE*, Jan. 1955.
2. Millman, J., and Herbert Taub, *Pulse and Digital Circuits*. New York: McGraw-Hill Book Co., Inc., 1956.
3. Chance, B., et al., *Waveforms*. New York: McGraw-Hill Book Co., Inc., 1949.

EXERCISES

17-1 Explain the operation of a unidirectional diode gate.

17-2 What causes a pedestal in the output waveform of the circuit shown in Fig. 17-1?

17-3 How can the pedestal in the output waveform be removed?

17-4 What is the basic difference between a unidirectional and a bidirectional gate?

15-5 What are the disadvantages of the circuit shown in Fig. 17-4?

17-6 Draw the circuit of a four-diode gate and explain its operation.

17-7 What is the major advantage of the six-diode gate as compared to the four-diode gate.

17-8 Draw a simple triode gate and explain its operation.

17-9 How does the circuit of Fig. 17-8 operate?

17-10 What is the advantage of using a pentode gate as compared to the diode or triode types?

18

COUNTERS

Counters are finding ever-increasing application. For example, they are used in nucleonics, medical research, chemical and petroleum processing, digital computers, electronic data processing systems, and radar.

Many of the commercially available counters may be used for making a number of different measurements. For example, the instrument illustated in the photograph of Fig. 18-1 can measure frequencies to 10.1 megacycles and display their readings in digital form on an eight-place indicating system. In addition to making direct frequency measurements, this counter can also measure period, frequency ratio, and total events. The counter of Fig. 18-1 can also count random events such as are encountered in nuclear work.

The first part of this chapter is concerned with the techniques that may be used in the construction of a counter for a specific purpose. Following this discussion, a typical high-quality counter, the Hewlett-Packard Model 524D, is described. Although the trend in counters is toward complete transistorization, vacuum-tube instruments, such as the 524D, will be in use for many years to come. The principles in-

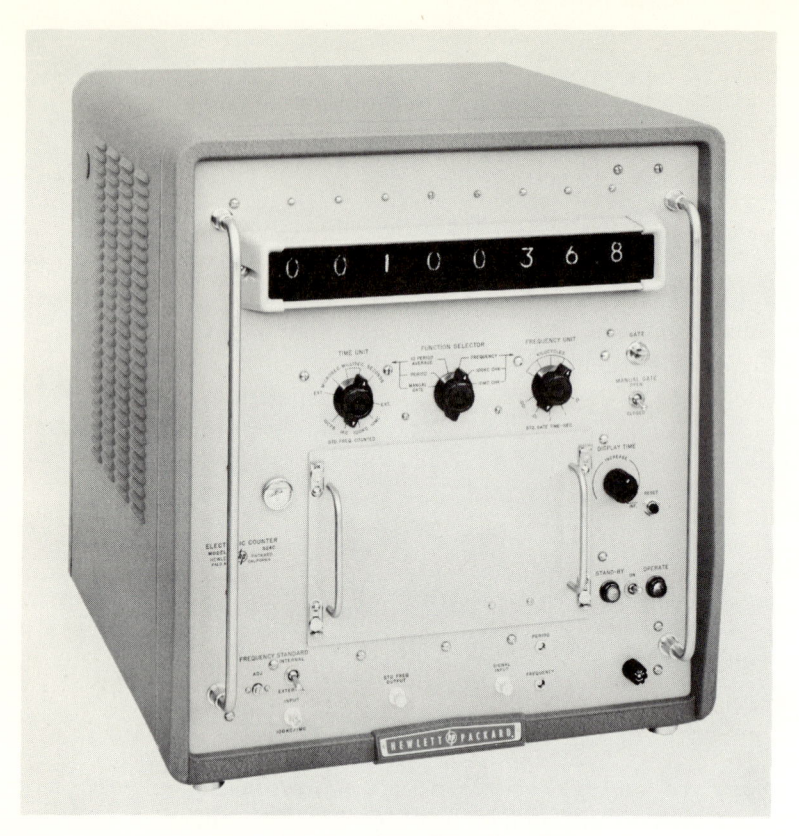

Figure 18-1 A modern electronic counter.

volved, rather than the type of amplifying device used, should be the major concern of the reader.

It is also to be expected that the indicating devices described herein will undergo change within a relatively short period of time. Again, however, these devices are indicative of current practice, and an understanding of their operation should prove beneficial to the practicing engineer and technician.

18.1 Counting to the base 2

A binary used as a scale-of-two counter has already been discussed. Because this circuit has two stable states and remains in either until the application of a proper triggering pulse, it is convenient to designate these states by means of the symbols 0 and 1. The selection of either symbol to designate either state is arbitrary,

Sec. 18.1   *Counting to the base 2*   349

although 0 most often represents *off* and 1 most often represents *on*. This convention is used here. Also, when a binary is said to be in the *one* state, this will be understood to mean that the input stage is *on* (conducting) and the output stage is *off* (nonconducting). Similarly, any statement to the effect that that binary is in the *zero* state will mean that the input stage is *off* (nonconducting) and the output stage is *on* (conducting).

Two cascaded binaries are illustrated in Fig. 18-2(a). Assuming that both binaries are initially in the *zero* state, it will be shown that this circuit counts by fours when a negative-going pulse train, Fig. 18-2(b), is applied to point A in Fig. 18-2(a).

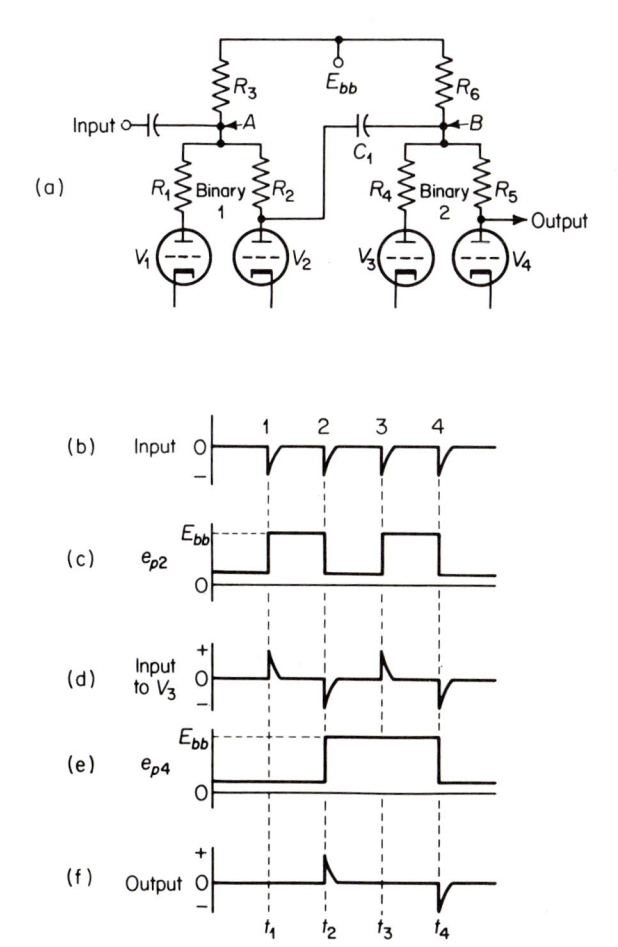

Figure 18-2 Cascaded binaries that count by four.

Since binary 1 is in the *zero* state, V_1 is *off* and V_2 is *on*. Under these conditions, V_1 effectively represents a very high resistance and V_2 a very low resistance. Stage V_1 and resistor R_1, and stage V_2 and resistor R_2, act as voltage-divider networks for the applied pulse. From what has been said, it is obvious that a negligible portion of the input pulse appears at the plate of V_2, but practically a full-amplitude pulse appears at the plate of V_1. This pulse is applied to the grid of V_2 to initiate the switching action, and very rapidly the binary is in the *one* state with V_1 *on* and V_2 *off*.

With V_2 *off*, its plate voltage, e_{p2}, rises to the value of E_{bb} as shown in Fig. 18-2(c), and then remains steady until the second pulse is applied. The positive step in e_{p2} is applied to point B of binary 2. The R-C coupling network between the two binaries, consisting of C_1 and the input resistance of binary 2, has a very short time constant compared to the positive step duration and, therefore, acts as a differentiator and produces the first positive-going narrow pulse shown in Fig. 18-2(d).

With binary 2 in the *zero* state, this positive-going pulse has no effect and does not produce a switching action. The plate voltage, e_{p4}, of stage V_4, therefore, remains at a constant low level as shown in Fig. 18-2(e).

This can be understood by recalling that with binary 2 in the *zero* state, V_3 is *off* and V_4 is *on*. Again because of voltage-divider action, the positive input appears only at the plate of V_3. It is then coupled to the grid of V_4, but produces no effect on e_{p4} since V_4 is already fully conducting. Of course, there is no output from V_4 at this time, as indicated by the waveform shown in Fig. 18-2(f). All of the actions described thus far occur at time t_1 on the common time axis for the waveforms of Fig. 18-2.

At time t_2, the second negative-going input pulse is applied to point A of binary 1. This again initiates a switching action, making binary 1 revert to its original *zero* state with V_1 *off* and V_2 *on*.

When V_2 switches *on*, its plate voltage again drops to some low value and a negative-going pulse is produced across the R-C coupling network between the binaries. This negative-going pulse does cause switching in binary 2. Thus, voltage e_{p4} rises to the value of E_{bb}, and a positive-going pulse appears at the output.

Referring to the various waveforms, notice that the application of two negative pulses to binary 1 still has not produced a negative pulse at the output of binary 2, even though binary 1 has changed state twice and binary 2 has changed state once.

At time t_3, the third negative-going pulse is applied to point A

Sec. 18.1   Counting to the base 2   351

of binary 1. This produces another change of state, driving V_1 *on* and V_2 *off*. The positive step in e_{p2} again produces a positive-going pulse at point B. This pulse has no effect on binary 2, for the reasons previously explained. Thus e_{p4} and the output of binary 2 remain unchanged.

At time t_4, the fourth negative-going pulse is applied to binary 1 and produces a switching action. With V_2 driven *on*, its plate voltage drops to a low value, and a negative-going pulse is applied to point B of binary 2. This pulse causes binary 2 to change to the *zero* state, with V_3 *off* and V_4 *on*. When e_{p4} drops to a low value, it produces a negative-going pulse at the output of binary 2. Thus, the fourth negative-going input pulse produces the first negative-going output pulse. The series of operations described is repeated as long as the pulse train is applied to the input. Thus, it can be said

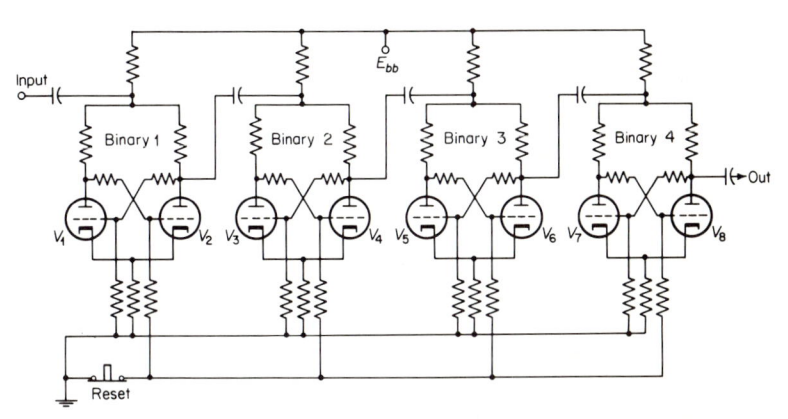

Figure 18-3 Scale-of-sixteen counter with reset switch for setting initial condition of all binaries.

that the circuit counts by fours. We may also think of the circuit as dividing by four, since for 16 input pulses, it would produce 4 output pulses.

To establish the initial condition prevailing in a *binary chain* (a group of cascaded binaries), use is made of a *reset switch*, as illustrated in Fig. 18-3, where four binaries are arranged to form a scale-of-sixteen counter. With the reset switch closed, all of the grid leak resistors are connected to ground. When the switch is opened, however, bias is removed from the output stage of each binary, causing the output stages to conduct. This places all the binaries in the *zero* state. Thus, the counter can be reset to the *zero* state at any time by momentarily depressing the reset switch.

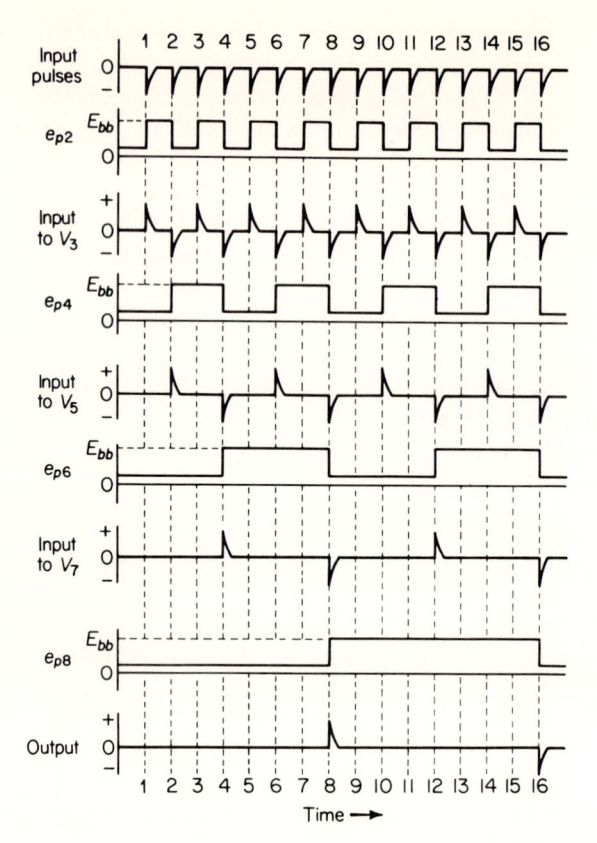

Figure 18-4 Waveforms for scale-of-sixteen counter.

The waveforms resulting from the application of a negative pulse train to the circuit of Fig. 18-3, assuming that the counter is initially reset to the *zero* state, are shown in Fig. 18-4. A negative pulse appears at the output of each binary whenever the output stage reverts from the *one* state back to the *zero* state. Thus a negative pulse is applied to the input of binary 2 on every second input pulse, to binary 3 on every fourth input pulse, and to binary 4 on every eighth input pulse. For every sixteen input pulses, one negative pulse appears in the output. Thus, the circuit is a scale-of-sixteen counter.

It is apparent from the discussion thus far that the output of each binary represents a power of 2; that is, the output of the first binary represents 2^1, the output of the second binary represents 2^2, the third binary, 2^3, and so on. Thus, it is possible to count to any scale that is a power 2 simply by adding the required number of binaries. The binary counter, in effect, performs binary addition.

Sec. 18.1   Counting to the base 2   353

To make practical use of the counter in Fig. 18-3, some form of *visual readout* device is required. (Various kinds of readout devices are discussed later.) A simple but effective arrangement is shown in Fig. 18-5. Instead of taking the output from the plate of V_8, as in Fig. 18-3, it is taken from the plate of V_7. Since binary 4 is driven to the *zero* state by the sixteenth pulse applied to the counter input, V_7 is switched *off* and its plate voltage rises to the value of E_{bb}. A positive-going pulse is, therefore, applied to the grid of V_9, which is normally biased beyond cutoff by bias voltage $-E_{cc}$. The amplitude of the applied pulse is sufficient to bring V_9 momentarily out of cutoff.

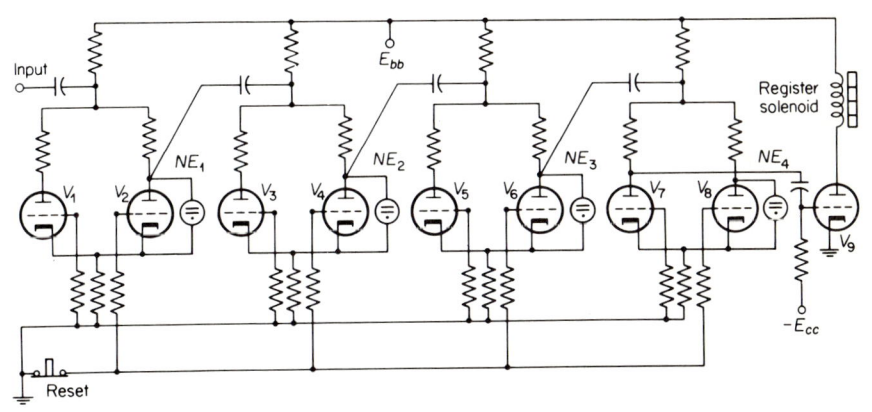

Figure 18-5 Scale-of-sixteen counter with a reset switch and provision for exact visual readout.

An electromechanical register is connected between the plate of V_9 and E_{bb}. When V_9 conducts, the register produces a numerical indication in one of the windows. For the first nine pulses received by V_9, a corresponding numeral appears in the left window. On the tenth pulse the reading in the left window becomes 1, and a zero appears in the next window to the right; and so on. The maximum register count is 9999. On the 10,000th pulse, the register is cleared and the count begins once more as described above.

Since one pulse triggers V_9 *on* for every sixteen input pulses applied to the counter, the register does not provide an exact count. The neon bulbs connected between the plate and cathode of each binary output stage provide the means of obtaining an exact count.

Each light is *on* only when its associated binary is in the *one* state with the input stage conducting and the output stage cut off. Under these conditions, the plate voltage of the output stage is equal to E_{bb} and causes the neon lamp to ionize. When the circuit switches to the *zero* state, the plate voltage of the output stage drops and the neon lamp deionizes.

Now, during the time that the lamps are *on*, if the lamp in binary 1 represents 1, the lamp in binary 2 represents 2, the lamp in binary 3 represents 4, and the lamp in binary 4 represents 8; either alone or in combination with each other, the lamps can represent any value between 1 and 15, since $1 + 2 + 4 + 8 = 15$.

Thus, if the register reads 20, and the first and third lights are lit, the exact count is 325, since $1 + 4 + (16 \times 20) = 325$. If the register reads 920, and the second and third lights are lit, the exact count is 14,726 since $2 + 4 + (16 \times 920) = 14,726$. Whenever the reset switch is depressed, all binaries are reset to the *zero* state. At that time, the register is also cleared.

Counters of this type are entirely satisfactory for many applications. For example, if a photoelectric cell and its exciter lamp are set on opposite sides of a conveyor belt, each time an article passes the lamp, the photocell produces a current through a suitable load. This current causes a voltage which can be amplified and shaped to produce the input pulse train to the counter. Thus, an exact count of the number of articles passing the photocell is automatically available at any time. Information of this type is invaluable for proper production records, inventory control, and cost accounting.

18.2 Counting to a base other than 2

Very often it is either desirable or necessary to count to some base other than 2. As long as this base is some power of 2, all that we need do is to cascade the appropriate number of binaries. Thus, if we wanted a scale-of-thirty-two counter we could use five binaries; for a scale-of-sixty-four counter we could use six binaries; and so on. It is obvious, however, that simple cascaded binaries are not suited to counting to some other base, such as 3 or 10, for example.

18.3 Counting to a scale of 3[1]

A circuit can be arranged so that it has three stable states and produces one output pulse for every three input pulses. Such a circuit is shown in Fig. 18-6; it is a modification of the Eccles-Jordan circuit. Although this circuit has found only limited application, it is interesting and merits study.

Sec. 18.3 Counting to a scale of 3

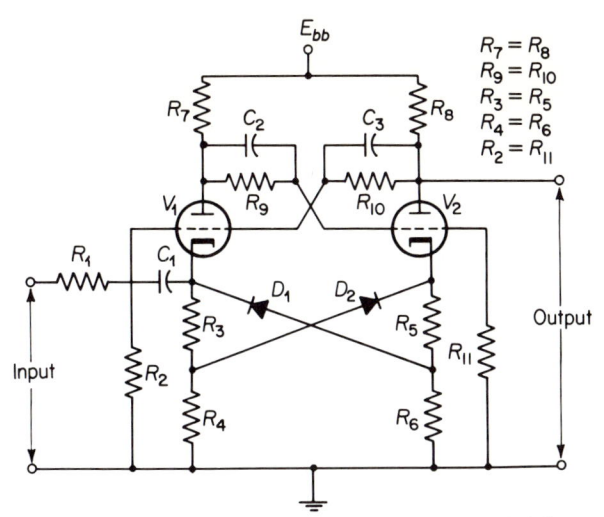

Figure 18-6 Scale-of-three counter that is a modified form of the Eccles-Jorden circuit.

The values of resistors R_3, R_4, R_5, and R_6 are such as to prevent either D_1 or D_2 from conducting when the plate currents of V_1 and V_2 approach equality. With the two diodes *off*, negative feedback increases and reduces the loop amplification below unity; at the same time, positive feedback decreases. Under these conditions the circuit becomes stable.

If the plate current of V_1 is increased and that of V_2 is correspondingly decreased, a switching action occurs. The increase in current through R_3 and R_4 makes the anode of D_2 swing in the positive direction, while the decrease in plate current through R_5 and R_6 makes the cathode of D_2 swing in the negative direction. Both actions serve to forward-bias D_2 and it conducts, causing a portion of the V_1 plate current to be diverted from R_4 to R_5 and R_6. The resulting decrease in the V_1 cathode potential and increase in the V_2 cathode potential (both with respect to ground) causes a further unbalance. The action is cumulative, rapidly driving V_1 *full on* and V_2 *full off*. This also is a stable state. Similarly, a stable state is produced when the plate current of V_1 decreases and the plate current of V_2 increases. Diode D_1 then conducts, and V_1 is rapidly driven to *full off* while V_2 is turned *full on*. Thus, the circuit of Fig. 18-6 has three stable states.

If a negative-going pulse train of sufficient amplitude is applied to the cathode of V_1 in Fig. 18-6 through resistor R_1 and capacitor C_1, the three stable states occur in sequence.

To illustrate, assume that initially V_1 is *off* and V_2 is *on* with D_1 conducting. The first pulse raises V_1 out of cutoff and it conducts, thereby initiating the switching action. Provided that the duration of the triggering pulse is very short compared to the switching time, D_1 is cut off when the plate currents approach equality, and the circuit becomes stable with the two plate currents equal.

The second triggering pulse upsets the plate current balance, making the V_1 plate current the larger, and diode D_2 starts to conduct. Very rapidly, V_2 is *off* and V_1 is *on*. In this state, the V_1 grid current maintains the grid of that stage at approximately zero voltage with respect to its cathode.

The third triggering pulse drives the cathode V_1 negative with respect to its grid. The resulting grid current raises the voltage across C_3 by an amount approximately equal to the input pulse amplitude. This additional voltage decays sufficiently slowly after the end of the triggering pulse to prevent a stable condition when the two plate currents are equal, and the switching action continues until the circuit reverts to its original condition, with V_1 *off* and V_2 *on*.

Because an appreciable current difference is required to make the diodes conduct, a triggering pulse of large amplitude is required. Also, as previously noted, the duration of the triggering pulse must be short compared to the circuit transition time, if the equal-current state is to be achieved.

Two stages of the modified Eccles-Jordan circuit are cascaded in Fig. 18-7. Because each stage produces one output pulse for every three input pulses, such arrangements are sometimes called "ternary" counters. With two stages cascaded, the count on the mechanical register will advance 1 for every ninth pulse applied to the input, since two stages count to the second power of three; that is, $3^2 = 9$.

The capacitors shown connected between the cathodes of V_2 and V_4 and ground are used to reduce the fall time of waveforms e_{p2} and e_{p4} in Figs. 18-7(d) and (h) when the ternary reverts to the original *zero* state; that is, with the input stage *off* and the output stage *on*. This produces a sharp output pulse to trigger the following stage. The diode clipper is used to remove the positive pulses appearing at the output of the differentiator. See waveforms (e) and (f) in Fig. 18-7.

An exact count can be obtained in the circuit of Fig. 18-7(a) by connecting neon lamps across the plate load resistor of each stage. Alone or in combination, these lamps can then represent any value between 1 and 9 according to the table of Fig. 18-8. Thus if lamps

Sec. 18.3   Counting to a scale of 3   357

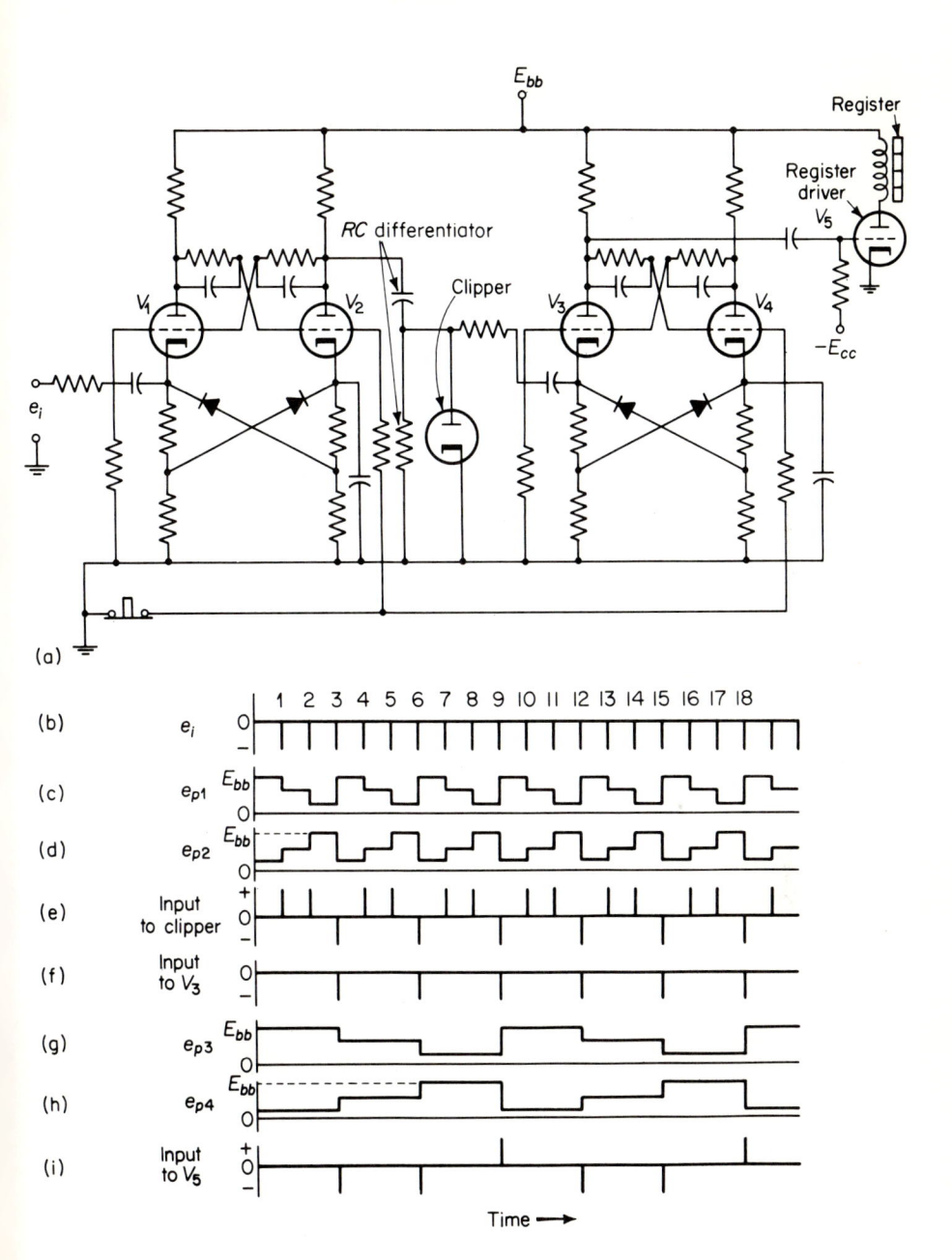

Figure 18-7 Cascaded ternary scale-of-nine counter.

1 and 3 are on, the count is exactly nine times that indicated on the electromechanical register. For any other single light, combination of lights, or absence of any light, the count is equal to 9 times the amount indicated on the register *plus* the number of pulses indicated in the table of Fig. 18-8. The lights must ionize, of course, when the plate voltage is equal to E_{bb} and must deionize before the plate voltage drops to the intermediate level occurring when the plate currents are equal.

Pulse	Lamp(s) On
1	3
2	2 and 3
3	1
4	None
5	2
6	1 and 4
7	4
8	2 and 4
9	1 and 3

Figure 18-8 Table indicating pulse count between 1 and 9.

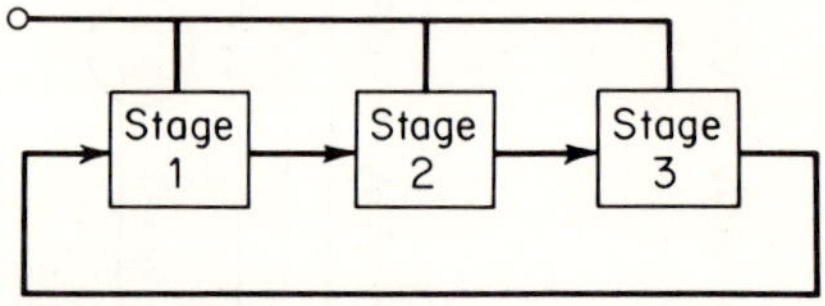

Figure 18-9 Block diagram of a three stage ring counter.

Another scheme for counting by three's is shown in the block diagram of Fig. 18-9[3]. Because the output of each stage provides the input to the next, this circuit is sometimes called a *ring counter*. With no signal applied to the input, one tube is *on* and all the others are *off*. When a trigger pulse is applied to all the tubes simultaneously, the *on* tube is turned *off* and the next tube in the series-connected ring is turned *on*. The succeeding pulse turns this latter tube *off* and turns the next tube *on*. Thus, if three stages are connected in a ring, any one tube responds to every third pulse, and a frequency division of three is achieved. One of the tubes operates a counting relay which, therefore, responds to every third pulse. Although three stages are shown in Fig. 18-9 to illustrate counting by three's, a ring counter may use up to about five stages and provide counting to the base corresponding to the number of stages used. Beyond this number, however, the circuit becomes uncommonly complicated and the distributed capacitance to ground becomes a limiting factor. Although certain modifications can be made to overcome these difficulties, the necessary circuit complexity makes the use of such arrangements undesirable.

A three-stage vacuum-tube ring counter is shown in Fig. 18-10. Notice that the plate of each stage is resistively coupled to the grids of the other two stages. As a result of this coupling, the low plate

Sec. 18.3   Counting to a scale of 3

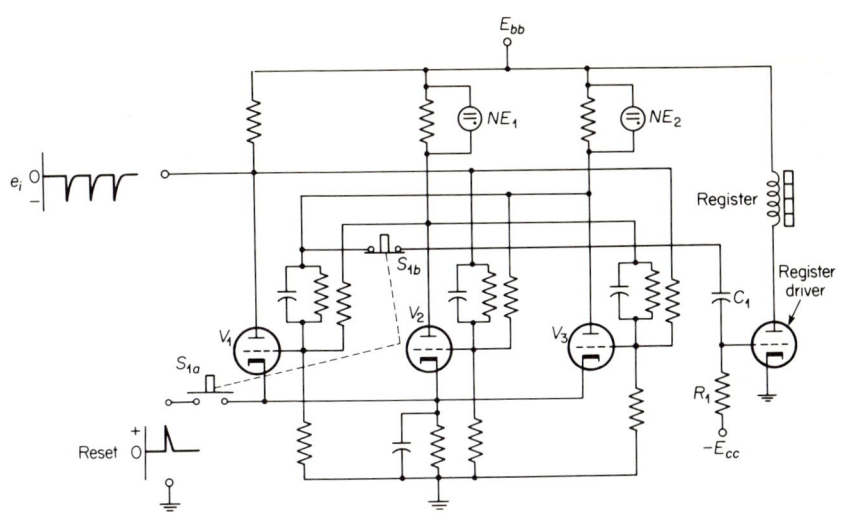

Figure 18-10 *Vacuum-tube ring counter. Every third input pulse produces one output pulse to activate register.*

voltage of a single *on* stage will prevent the other stages from conducting. Notice also that from the output of any stage, a speed-up capacitor is used with the coupling resistor to only one of the other two stages. It will be shown that, with a given stage *on*, the stage switched *on* by the next triggering pulse is the one having the speed-up capacitor in its coupling network.

Initial conditions are set by momentarily closing the reset switch. This applies a positive pulse to the grid of V_1, turning that stage *on* and turning V_2 and V_3 *off*. The reset pulse may be obtained from either an external or internal source. For example, it could easily be provided internally from a free-running MV by a differentiator and a negative clipper.

The first negative input pulse is applied to all three grids through the plate load resistors. Because V_1 is the only *on* stage, the negative pulse drives V_1 to cutoff but has no effect on V_2 and V_3. When V_1 stops conducting, its plate voltage rises to the value of E_{bb} and is coupled to the grids of both V_2 and V_3. Owing to the presence of the speed-up capacitor, the grid of V_2 feels the increase in the V_1 plate voltage before the grid of V_3. When V_2 conducts, the decrease in the plate voltage is coupled to the grids of V_3 and V_1, and effectively offsets the positive increase applied to V_3. Of course, the actions described occur almost instantaneously. Stage V_3 is, therefore, maintained in the *off* state. Thus, after the application of one input pulse,

V_2 is *on*, and V_1 and V_3 are *off*. With V_2 conducting, the voltage drop across its plate load resistor ionizes neon bulb NE_1, providing an exact count.

When the second input pulse is received, the same actions described above reoccur, but now V_3 is turned *on*, and V_1 and V_2 are *off*. The third input pulse turns V_1 *on* and turns V_2 and V_3 *off*, completing the ring. No neon bulb is used across the plate load resistor of V_1 since it is *on* for every third pulse only, and the exact count is indicated on the register.

Notice that the reset switch is a double-pole master switch with one normally open and one normally closed position. When S_{1a} is closed, S_{1b} is opened so that the initial set pulse is not applied to the register driver. When S_{1a} is opened, S_{1b} is closed, and every third input pulse produces an output from the plate of V_1 that is differentiated by C_1 and R_1 and turns the register driver *on* for a brief interval. The negative spike of the differentiated waveform has no effect on the register driver since it serves only to increase the cut-off bias.

Although a binary is inherently a scale-of-two device, it may be used in a ring-connected ternary counter. Such a circuit is shown in Fig. 18-11. The plate of each binary output stage is coupled to the grid of the succeeding binary. Initially, binary 1 is driven into the *zero* state (V_1 *off* and V_2 *on*) by the application to its grid of a negative reset pulse of sufficient amplitude. With V_2 *on*, V_4 and V_6 are *off* because of the coupling arrangement. Thus, with no input

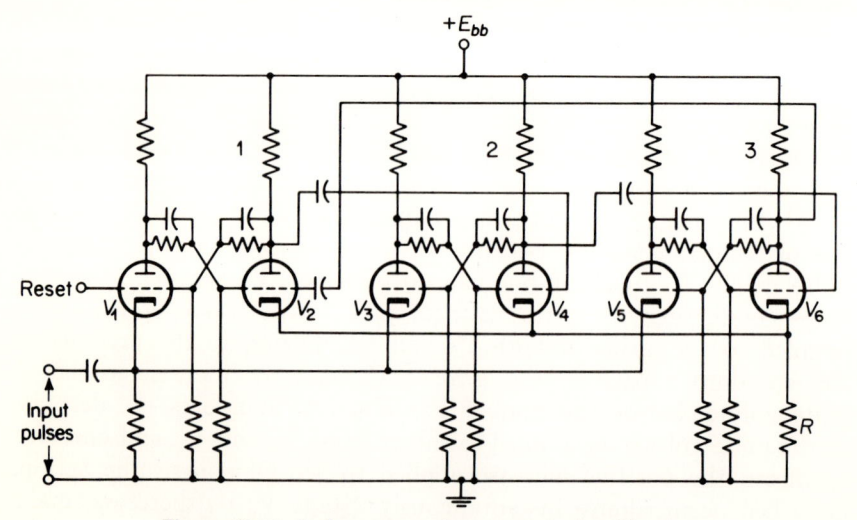

Figure 18-11 A three stage ring counter using binaries.

Sec. 18.4 Decade counters

signal applied, binary 1 is in the *zero* state, while binaries 2 and 3 are in the *one* state. With the assumption that all input pulses are of sufficient amplitude and of short duration compared to the binary switching time, circuit operation can now be described.

The first input pulse, applied simultaneously to the cathodes of V_1, V_3, and V_5, turns V_1 *on*; but it has no effect on V_3 or V_5 since they are already *on*. Binary 1, therefore, switches to the *one* state, and the plate voltage of V_2 rises to the value of E_{bb}. The increase in the V_2 plate voltage is coupled to the grid of V_4, turning that stage *on* and switching binary 2 to the *zero* state. The low plate voltage of V_4 coupled to the grid of V_6 is not sufficient to bring that stage out of cutoff, so binary 3 remains in the *one* state. Thus, after the first input pulse, binary 2 is in the *zero* state, while binaries 1 and 3 are in the *one* state.

Application of the second input pulse turns V_3 *on*, but has no effect on V_1 or V_5. Thus, binary 2 switches to the *one* state, and the plate voltage of V_4 rises to the value of E_{bb}. This increase is coupled to the grid of V_6, turning that stage *on*, and causing binary 3 to switch to the *zero* state. Thus, after the second input pulse, binary 3 is in the *zero* state, while binaries 1 and 2 are in the *one* state.

From the above, it is obvious that the application of the third input pulse causes a switching action in binaries 3 and 1, but has no effect on binary 2. Thus, after the third pulse, binary 1 is in the *zero* state while binaries 2 and 3 are in the *one* state. Switching around the ring has been completed. The continued application of input pulses causes the cycle of operation described to be repeated.

18.4 Decade counters[3,4,5]

A *decade counter* is one that produces a single output pulse for every ten input pulses. Since our numbering system uses the base 10, the decade counter is, by far, the most prominent.

Although we can construct a decade counter by using a single binary in combination with a ring-of-five counter, the circuitry is unnecessarily complex. A much more satisfactory arrangement is to use a scale-of-sixteen counter with feedback circuits that insert extra pulses to make the four binary stages go through a full cycle with only ten input pulses.

One method of accomplishing this is shown in the diagram of Fig. 18-12; this arrangement is sometimes called the *simultaneous feedback* method. The feedback extends from the plate of V_7 to the

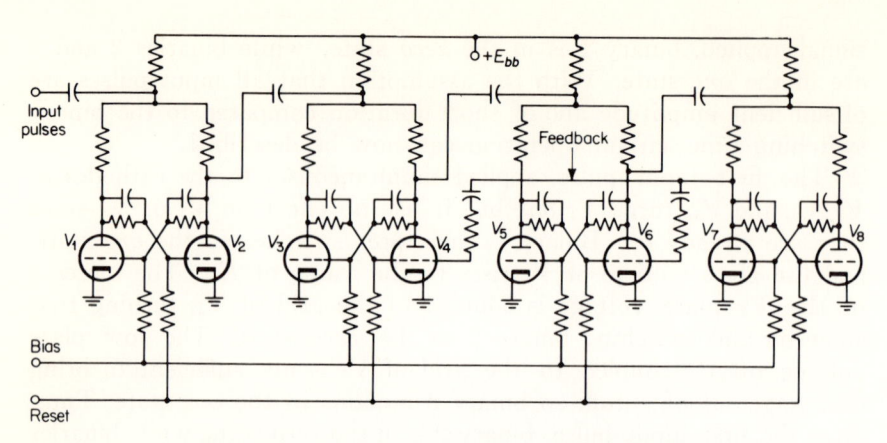

Figure 18-12 Simultaneous feedback used to convert a scale-of-sixteen counter to a decade counter.

grids of V_4 and V_6. Momentarily disregarding the feedback, the plate voltages of each binary input stage (V_1, V_3, V_5, and V_7) would appear as shown in Fig. 18-13(a). Note in particular that after the eighth input pulse, binaries 1, 2, and 3 are in the *one* state while binary 4 is in the *zero* state for the first time.

Now refer to the waveforms shown in Fig. 18-13(b). The switching of binary 4 to the *zero* state, upon the application of the eighth input pulse, causes the plate voltage of V_7 to rise to the value of E_{bb}. This increase is fed back to the grids of V_4 and V_6 in binaries 2 and 3, respectively, which have just completed switching to the *one* state, and causes them to switch again, this time back to the *zero* state. Thus, binaries 2, 3, and 4 are all in the *zero* state. Referring to Fig. 18-13(b), it is seen that, in the scale-of-sixteen counter without feedback, this condition does not occur until the fourteenth input pulse. The feedback has, therefore, advanced the counter by 6, that is, from eight to fourteen. The ninth input pulse sets the binaries to a state equivalent to fifteen, and the tenth recycles all the binaries, resulting in a true decade count.

Another method of converting a scale-of-sixteen counter to a decade counter is shown in Fig. 18-14.

In this arrangement, two separate R-C feedback networks are used, one from the plate of V_7 to the grid of V_6, and the other from the plate of V_5 to the grid of V_4.

Without feedback, binary 2 would switch to the *zero* state (V_3 *off*, V_4 *on*) on the fourth input pulse, and would remain there until the sixth input pulse. Now refer to Fig. 18-15(c). On the fourth input pulse, binary 3 reverts to the *one* state, so that the plate voltage

Sec. 18.4   Decade counters

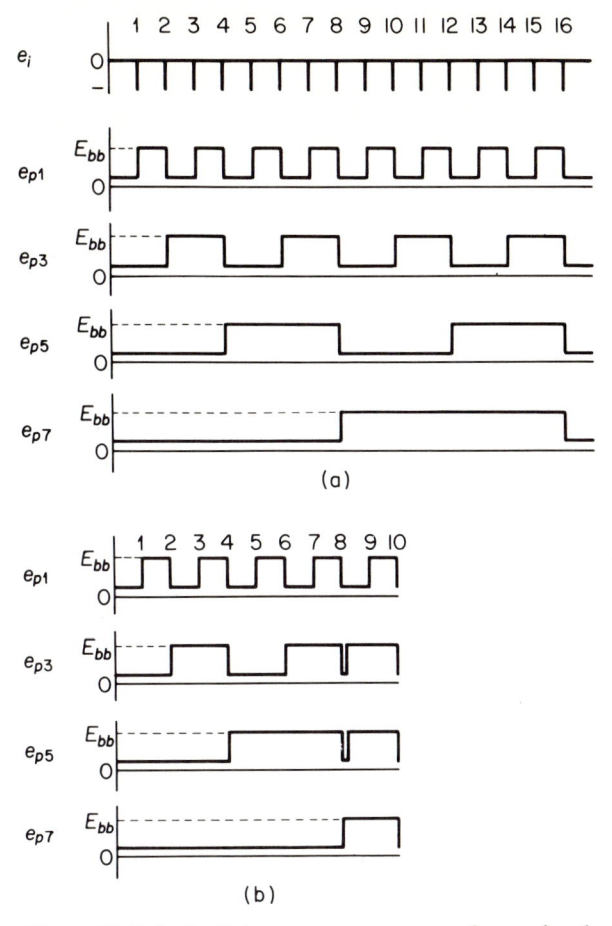

Figure 18-13 Scale-of-sixteen counter converted to a decade counter by using feedback (a) waveforms at plates of binary stages without feedback (b) with feedback.

of V_5 drops to a low value. This decrease is fed back to the grid of V_4 and switches that stage *off* within a very short time after it had switched *on*, and its plate voltage rises to the level of E_{bb}. Since, without feedback, e_{p4} would not rise to E_{bb} until the sixth pulse, the feedback has advanced the count by 2: from four to six.

On the sixth input pulse, binary 4 switches to the *one* state. The resulting low plate voltage of V_7 is fed back to the grid of V_6, driving that stage *off*. Notice, in Fig. 18-15(d), that just prior to the application of the feedback voltage to its grid (that is, upon application of the sixth input pulse), V_6 had switched *on*. The *off* condition of V_6 would not normally occur until the application of the

364   *Counters*   *Chap. 18*

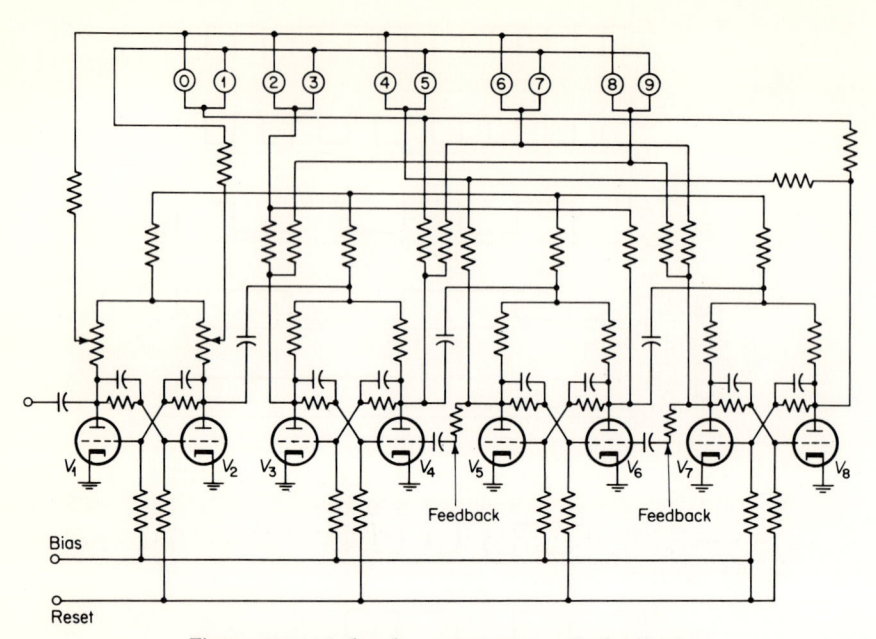

Figure 18-14 A decade counter using split feedback with provision for exact readout.

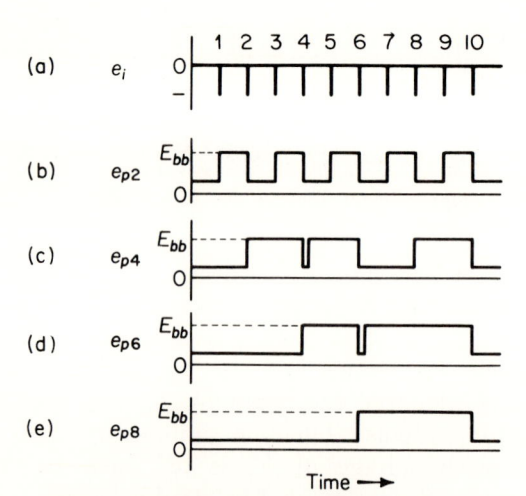

Figure 18-15 Waveforms at plates of output stages of binaries of Fig. 18-14.

twelfth input pulse. Thus, the count is now advanced by 6.

The seventh input pulse sets the decades to a state equivalent to thirteen, the eighth input pulse to the equivalent of fourteen, the ninth input pulse to the equivalent of fifteen, and the tenth input pulse to the equivalent of sixteen. At this point, the counter recycles and the counting process is again repeated as described above.

Referring back to Fig. 18-14, the neon lamps are arranged so that only one lamp lights for any combination of conducting tubes. One side of each lamp is connected in five pairs, namely 0-1, 2-3, 4-5, 6-7, 8-9; and the other side in two groups of five each: 0-2-4-6-8 and 1-3-5-7-9. Selection of the odd or even group is made by binary 1, the selected group receiving the full positive plate potential, while the other group receives one-half. The selection of the group of two's (pairs) is accomplished by a combination of two plate voltages received from the plates of other binaries, the selected group receiving the full negative potential. Only one group of two's (only one pair) receives that potential for any combination of binary positions.

18.5 Preset counters

A *preset counter* is one that produces an output pulse after receiving a number of input pulses determined by the setting of a *preset switch* mounted on the front panel. The versatility of preset counters accounts for their widespread use in such applications as the packaging of aspirins, vitamin capsules, machine screws, etc.

Preset counting techniques can be applied either to pure cascaded binaries or to decade counters. In the pure binaries, an output pulse may be produced at any setting from one to the maximum rate of the binary chain. In the decade counters, the output pulse may be produced at any setting between one and ten.

A simple preset counter using four cascaded binaries is shown in Fig. 18-16. The grids of all stages are connected to a four-deck, sixteen-position, ganged rotary switch. The decks are identified as S_{1a}, S_{1b}, S_{1c} and S_{1d}. The rotating arm of each deck is connected to a positive voltage source whenever the preset relay in the plate of driver V_9 closes. The grid of any stage in contact with the rotating arm, when the preset relay closes, is positively biased, turning that stage *on*. Thus, instead of having all binaries start counting in the *zero* state, the count may be effectively advanced to any number between 1 and 16 depending on the setting of the preset switch.

In Fig. 18-16, the counter is preset to produce one output pulse

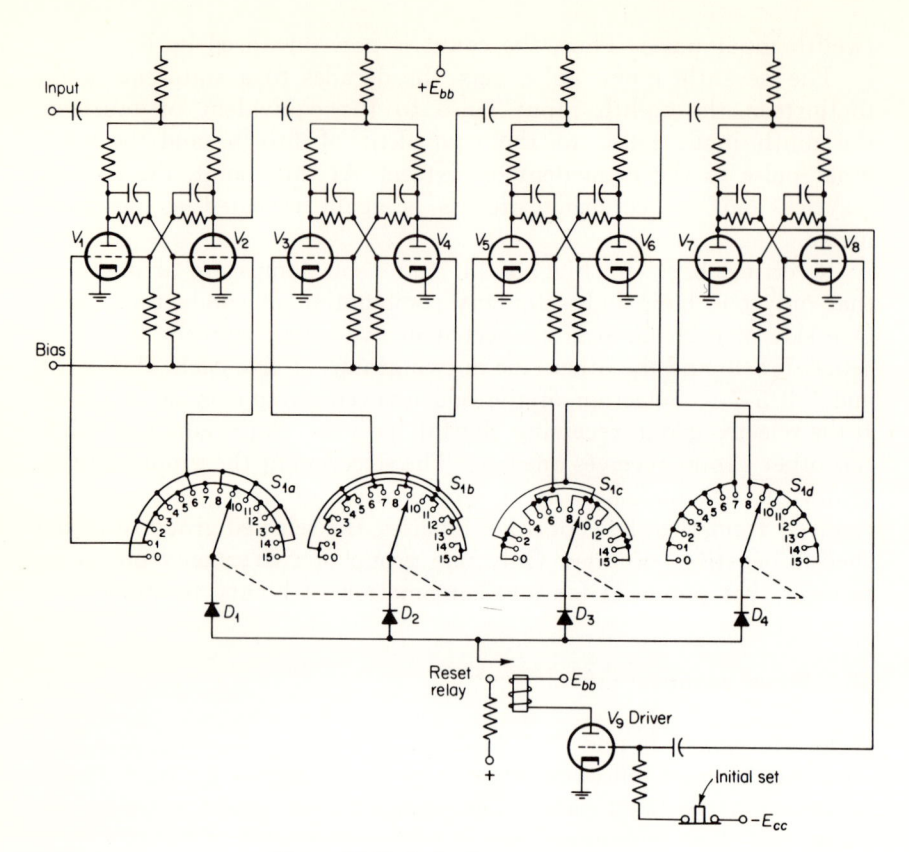

Figure 18-16 A simple preset counter.

for every six input pulses by presetting the binaries to the states corresponding to nine input pulses. This may be verified by referring back to the waveforms of Fig. 18-4 which show that, on the ninth input pulse, binaries 1 and 4 are in the *one* state while binaries 2 and 3 are in the *zero* state.

As a result of the presetting, all binaries in Fig. 18-16 are in the *zero* state upon receipt of the sixth input pulse. The positive step voltage then produced at the plate of V_7 is differentiated and used to turn *on* momentarily the *preset relay driver* V_9 which is normally biased beyond cutoff by voltage $-E_{cc}$. This action energizes the preset relay, applying a positive voltage to the grids of stages V_1 and V_7. Under these conditions, the binaries are again preset to those states corresponding to the ninth input pulse, and the counting-by-six process is repeated as often as desired.

Diodes D_1 through D_4 in Fig. 18-16 are isolation diodes used to

Sec. 18.5   *Preset counters*   367

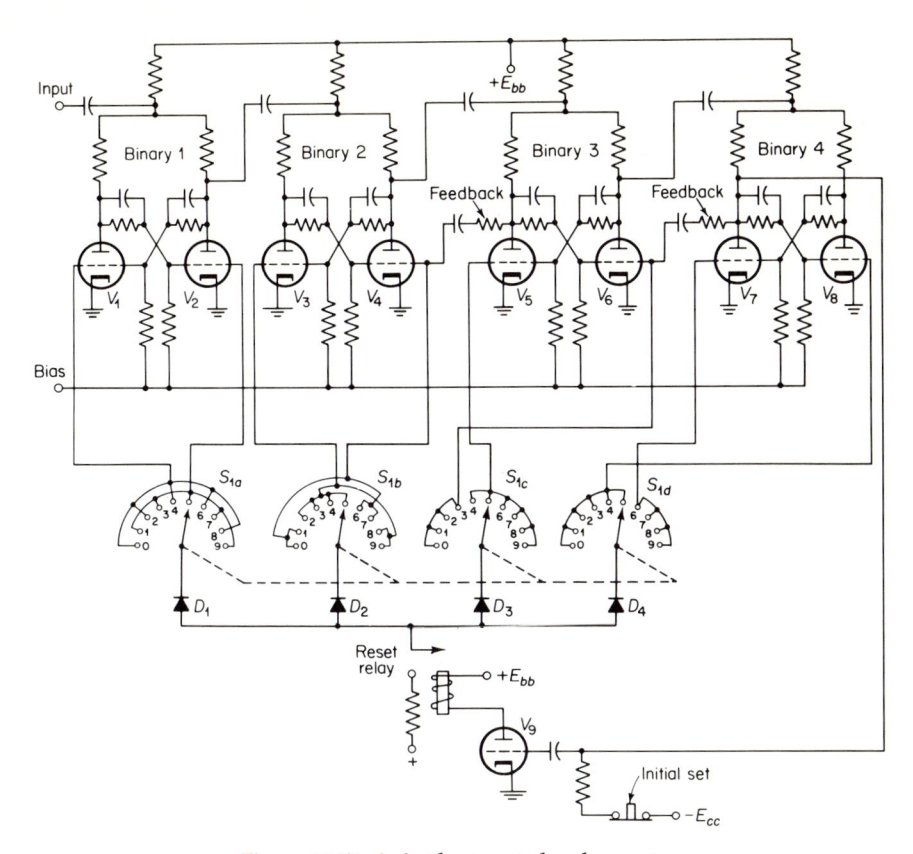

Figure 18-17   A simple preset decade counter.

prevent interaction between the various grids. Notice that in this particular circuit, the front panel switch is set to the complement of the desired count with respect to 16. If more stages were added, the preset switch is set to the complement of the desired count with respect to the maximum ratio. For example, if five binaries are used and it is desired to count by sixes, the *preset* switch would be set to 26, which is the complement of 6 with respect to 32. In preset decade counters (discussed below), the preset switch is set to the complement of the desired count with respect to ten.

A simple preset decade counter in which the feedback is between binaries 4 and 3 and binaries 3 and 2, is shown in Fig. 18-17. With the preset switch in position 5, an output pulse is produced on every fifth input pulse. Referring back to Fig. 18-15, it is seen that the fifth input pulse places binaries 1, 2, and 3 in the *one* state and binary 4 in the *zero* state. Since the switch setting in Fig. 18-17 turns *on* stages V_1, V_3, and V_5, and turns *off* stage V_7, the binaries are in corresponding states.

By using a neon indicator arrangement, the preset count may be visually indicated.

If we go one step further and cascade a number of decades, a counter having a maximum ratio of 100, 1000, 10,000, or even higher, may be achieved. Such arrangements are commonly used in practice with the neon readout of the first decade representing units, the output of the second decade representing tens, the output of the third, hundreds, and so on. Each decade is constructed as a modular plug-in unit, and a single preset driver and relay permit repetition of the preset count. This is illustrated in the block diagram of Fig. 18-18, which shows a preset counter having a maximum ratio of 10,000. With the preset switches in the position shown, an output pulse is produced for every 9750 input pulses.

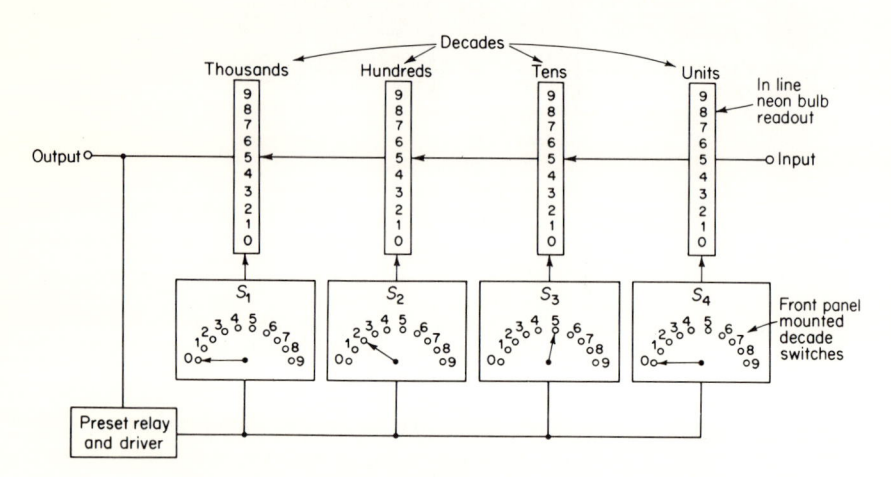

Figure 18-18 Cascaded decades used in a preset counter having a maximum ratio of 10,000.

18.6 Gate and count types of instruments[6]

For time interval and frequency measurements, gating circuits are used in conjunction with the counter. Two high-quality instruments of this type are the Hewlett-Packard Electronic Counters, Model 524C and Model 524D, which are identical except for the readout. The 524C, shown in the photo of Fig. 18-19(a) uses eight in-line digital display tubes behind a light filter, and can blank the readout during blank time. The 524D, shown in Fig. 18-19(b) uses eight

Sec. 18.6   Gate and count types of instruments   369

(a)

(b)

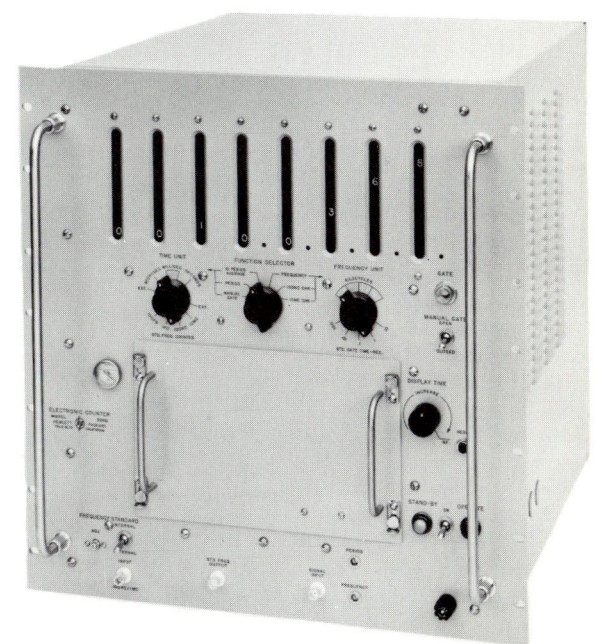

Figure 18-19 Hewlett-Packard Electronic Counters Models (a) 524C and (b) 524.

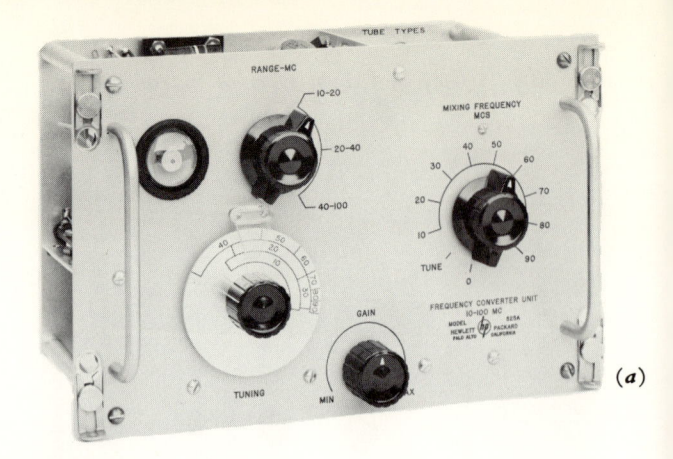

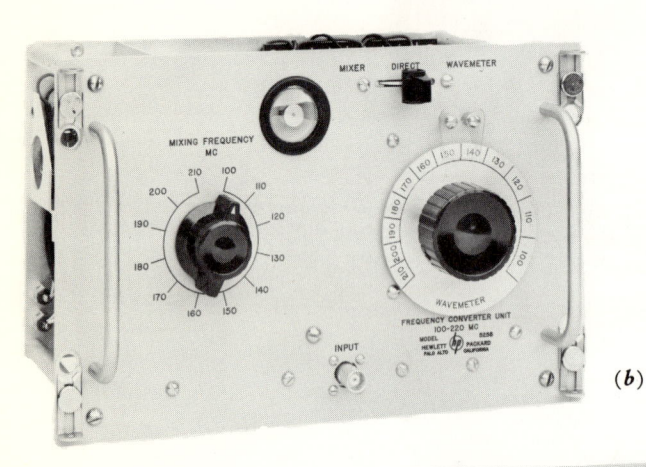

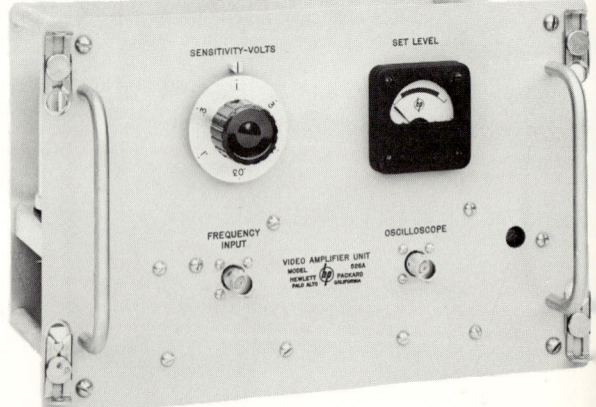

Figure 18-20 Plug-in accessories for Hewlett-Packard Electronic Counter Model 524D. (a) *Model frequency converter to extend frequency range from 10.1 mc to 100 mc.* (b) *Model 525B frequency converter—unit enables 524D to measure frequencies between 100 mc and 220 mc.* (c) *Model 526A video amplifier—unit increases sensitivity of 524D to 10 mv in the range from 10 cps to 10.1 mc.* (d) *Model 526B time intervals from 1 microsecond to 10^7 seconds.* (e) *Model 526C period multiplier extends 524D period measurement range up to 10,000 periods of unknown frequency.* (f) *A separate instrument, the model 540B transfer oscillator extends the range of the 524D equipped with a 525B frequency converter to about 12,000 mc.*

(d)

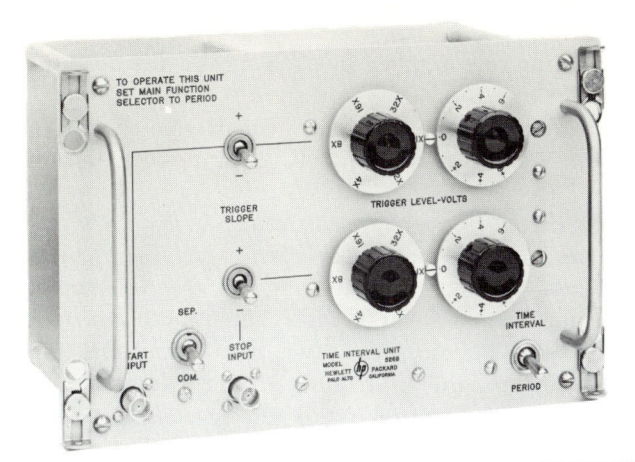

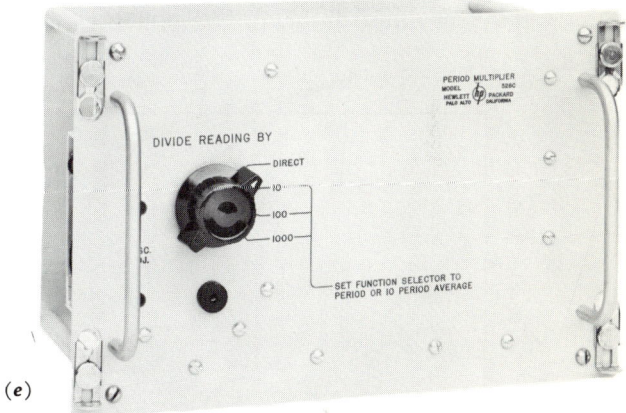

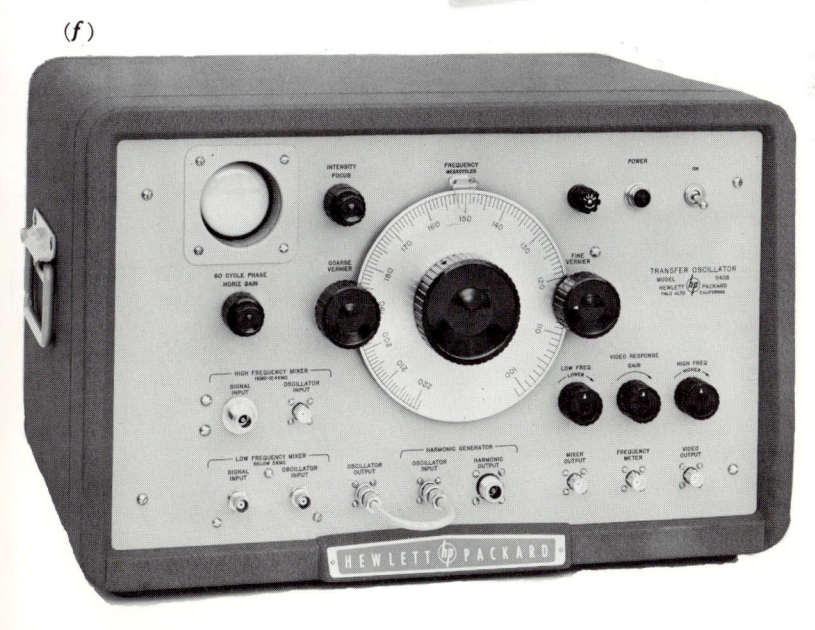

(e)

(f)

numbered masks, illuminating one number in each mask; the readout is visible at all times.

The text and illustrations that follow refer primarily to the 524D. They apply equally well, however, to the 524C except where specifically noted.

In addition to making direct frequency measurements, the Hewlett-Packard counters can measure the period, the frequency ratio, total events, and random events. To increase the range of measurements, five accessory plug-in units and a separate instrument, the *transfer oscillator*, are available. The plug-in units and their functions are shown in Figs. 18-20(a) through(f). The operating controls and terminals of the 524D are identified in Fig. 18-21.

Fig. 18-22 shows the basic circuit arrangement of the 524D. For frequency measurement, the input signal is fed through a signal gate

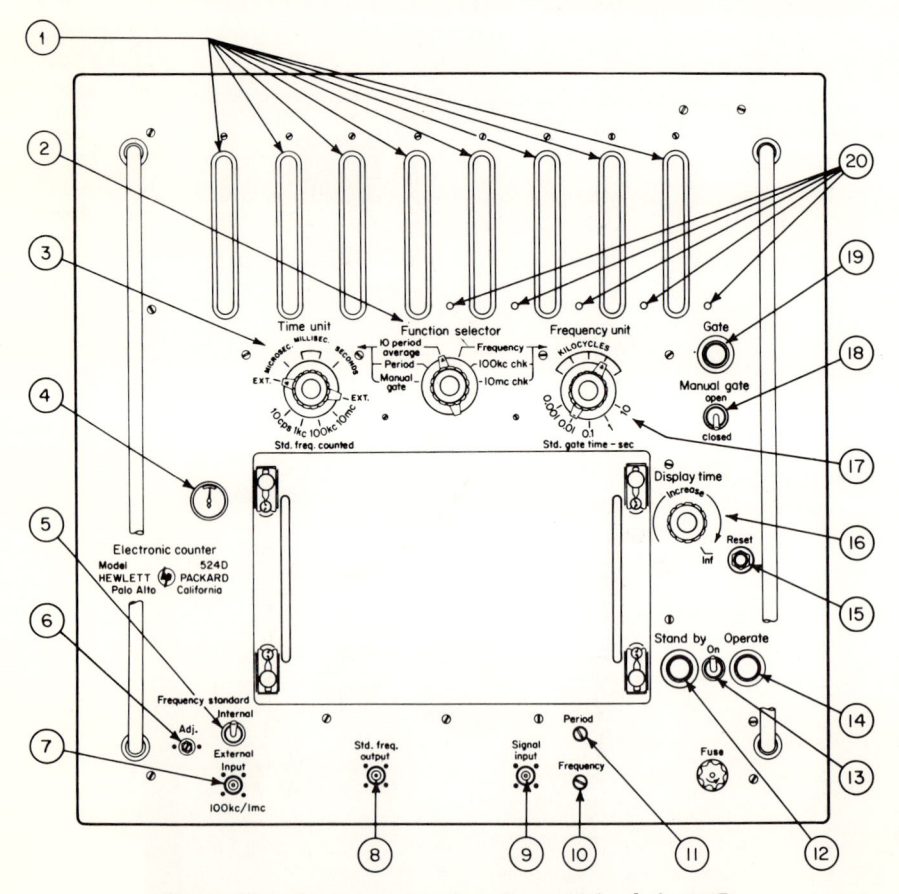

Figure 18-21 Operating centrals and terminals of the 524D.

Sec. 18.6   *Gate and count types of instruments*   373

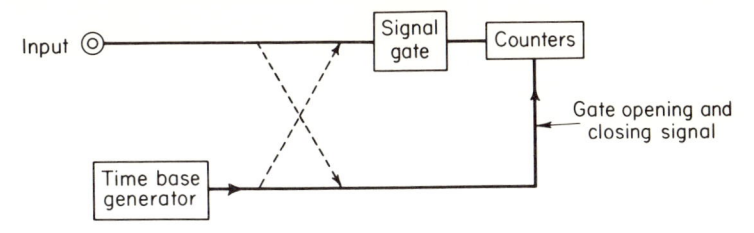

Figure 18-22 Basic circuit arrangement of 524D.

to a series of visual-indicating digital-type counters. A precision time interval obtained from the time-base generator section opens and closes the signal gate for an extremely accurate period of time, for example, 1 second. The counters count the number of pulses entering through the gate during the one second interval and then display the total. The answer is read directly as the number of kilocycles occurring during the one-second interval. The period of time during which the signal gate remains open is set by the *frequency unit switch* (Fig. 18-21, callout 17). For each position of this switch, the illuminated decimal point is automatically positioned so that the answer is always read directly in kilocycles per second. The answer is automatically displayed for a period of time determined by the setting of the display time control on the front panel or by the gate time, whichever is greater.

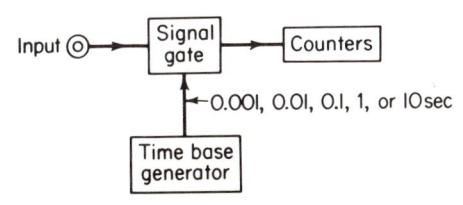

Figure 18-23 Block diagram of 524D for frequency measurements below 10 mc.

To measure a period or time interval, the 524D reverses the application of the two signals, as shown by the dotted lines in Fig. 18-22. The period or time interval to be measured is connected in such a manner as to open and close the signal gate while one of the standard frequencies from the time-base section is passed through the signal gate to the counters. When measuring a period, one cycle of the incoming signal opens the gate, and the next cycle closes it. The number of cycles of the standard frequency of the time base that occurred during the period are then indicated on the counters. The standard frequencies obtained from the time base have been selected so that the answer to the measured period will always be displayed in direct-reading units of time: seconds, milliseconds, or microseconds.

Provision is also made in the circuit to permit measurement of the average of ten periods of the unknown frequency. Higher accuracy can then be obtained than with period measurements.

The basic 524D measures frequencies up to 10.1 mc. The block diagram of Fig. 18-23 shows the circuit arrangement of the basic counter when measuring frequencies in this range. To measure frequencies up to 220 megacycles, one of the two frequency converter units is required. This arrangement is shown in the block diagram of Fig. 18-24. The 525A frequency converter is used for frequency measurements up to 100 mc., and the 525B frequency converter is used between 100 and 220 mc. In both frequency converters, the

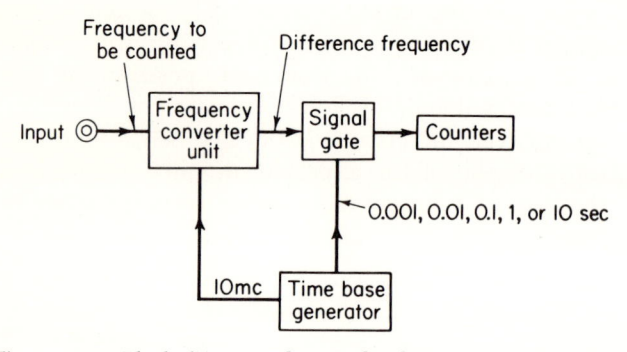

Figure 18-24 Block diagram of 524D for frequencies above 10 mc.

input signal is mixed with a harmonic of 10 mc so that the difference between the signal and the harmonic is not more than 10.1 mc. The difference frequency is counted and displayed by the counter. To determine the signal frequency, the count displayed is added to the known 10 mc harmonic.

Both frequency converters have tuning systems to indicate the correct mixing frequency. If the mixing frequency is within 1 mc of the unknown frequency, however, there is a possibility of ambiguity; that is, the problem arises of whether to add or subtract the displayed reading from the mixing frequencies to determine the unknown frequency. When making the final measurement, a mixing frequency which is at least 100 kc away from the unknown is selected.

This counter is also arranged so that it can measure directly the period of the frequency to be measured. When set for period measurement, the time-base circuit and the signal-input circuit are inter-

Sec. 18.6   Gate and count types of instruments

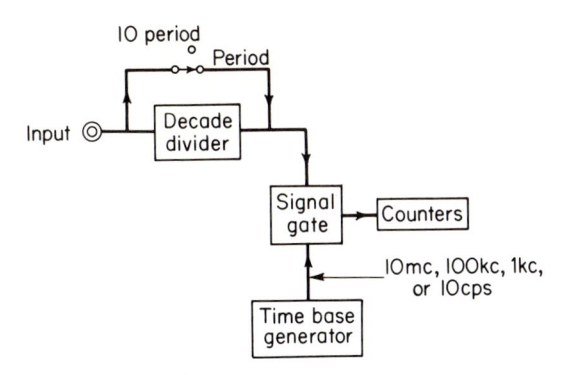

Figure 18-25 Block diagram of 524D for period measurements.

changed from their frequency measurement positions, as shown in the block diagram of Fig. 18-25. With the circuit so connected, the counters count the output of the time base for the period of the unknown input signal. Thus, the standard frequency generated in the time base is used as a unit of time to measure the unknown period in terms of microseconds, milliseconds, or seconds.

The accuracy of period measurement is largely determined by the accuracy with which triggering occurs at the same point in consecutive cycles of signal voltages having a slow rate-of-rise. Fig. 18-26 shows the error in triggering for a single-period measurement of a sine wave. Note that as the single-to-noise ratio improves, the triggering accuracy also improves. Averaged over ten periods, the single-period error is reduced by a factor of ten. By using the period multiplier unit, the error may be reduced an additional factor of ten for each factor of ten the measurement is extended. The accuracy of triggering is considerably improved when the waveforms being measured have a fast rise-time. For example, a significant reduction in error is achieved by applying square waves instead of sine waves to the input.

To follow the slowest-changing waveforms, the period measurement input circuits are direct-coupled and are adjusted to trigger at the zero-volt crossing of a negative-going voltage. Thus, any d-c component in the input signal will shift the triggering level so that the maximum slope no longer occurs at the zero-volt level, resulting in a loss of accuracy. A large d-c component may result in no triggering at all.

The 524D may also be used to measure the ratio of two frequencies. The higher frequency is passed through the signal gate to the counters and is counted for a period of time determined by

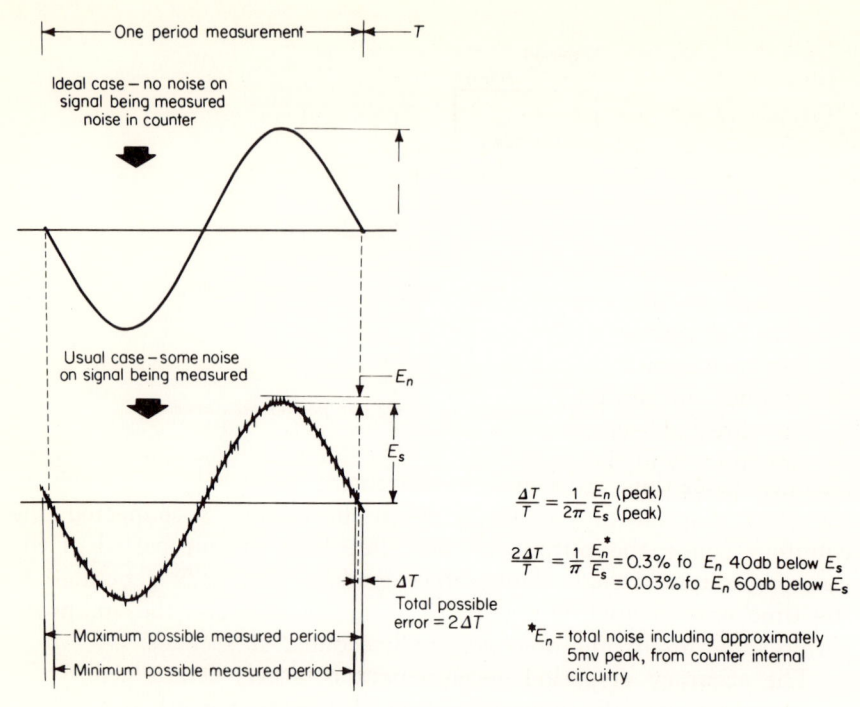

Figure 18-26 Possible errors in period measurement of sine waves.

$$\frac{\Delta T}{T} = \frac{1}{2\pi} \frac{E_n \text{ (peak)}}{E_s \text{ (peak)}}$$

$$\frac{2\Delta T}{T} = \frac{1}{\pi} \frac{E_n^*}{E_s} = 0.3\% \text{ fo } E_n \text{ 40db below } E_s$$
$$= 0.03\% \text{ fo } E_n \text{ 60db below } E_s$$

*E_n = total noise including approximately 5mv peak, from counter internal circuitry

either one period or ten periods of the lower frequency, which controls the opening and closing of the gate. See Fig. 18-27.

The accuracy of ratio measurement is determined by the same factors as period measurement accuracy: consistency of triggering by the lower input frequency, and the inherent error of ±1 count of the higher frequency.

To make interval measurements, the time-interval unit must be

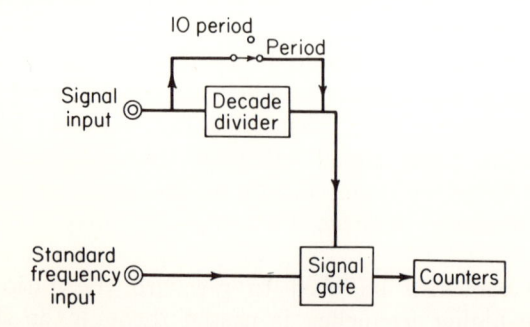

Figure 18-27 Block diagram of 524D for ratio measurements.

Sec. 18.6 Gate and count types of instruments

installed. Time-interval measurements are similar to period measurements except that the points on the signal waveforms, or the waveforms at which the measurement starts and stops, are adjustable. The adjustable threshold feature allows measurements from one part of a waveform to another part of the same waveform, or the use of separate waveforms as start and stop signals.

As in the case of period measurement, the input signals control the opening and closing of the gate while the standard frequencies are passed to the counters, as indicated in Fig. 18-28. Thus, the accurate frequencies generated in the time base are used as units of time to measure the unknown interval in terms of microseconds, milliseconds, or seconds.

The threshold-selecting controls adjust the start and stop channels so that they will be actuated only by signals of predetermined polarity, amplitude, and slope. Time-interval measurements begin

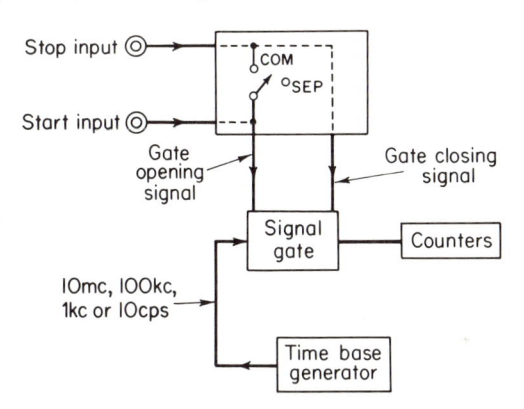

Figure 18-28 Block diagram of 524D for time-interval measurements.

when the "start" signal crosses the selected start threshold value in the selected direction, and end when the "stop" signal crosses the selected threshold value in the selected direction. The threshold controls are only approximately calibrated, and in some applications special precautions must be taken to obtain the interval desired.

The 524D may also be used as a high-speed totalizer, capable of counting at a maximum rate of 10.1 million events per second. Fig. 18-29 shows the basic circuit.

A block diagram of the time-base section of the Model 524D is shown in Fig. 18-30 (see back-flap envelope). A 1-mc oscillator V901 is the time-base reference. Amplifier V902 drives an AGC (automatic gain control) circuit, CR901 and CR902, and one input grid of mixer

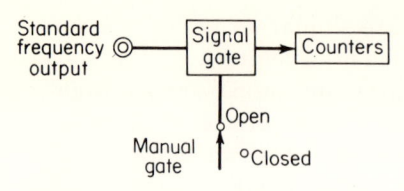

Figure 18-29 Block diagram of 524C/D for totalizing.

V903. The AGC circuit varies the oscillator bias to maintain the oscillator output at a constant amplitude. Mixer V903 mixes the 1-mc output of the amplifier and the 900-kc output of an X9 filter, T901 and C912, to produce a 100-kc signal. This signal drives blocking oscillator V236A. The blocking oscillator pulses the X9 filter, which rings at 900 kc.

The mixer, blocking oscillator, and X9 filter form a regenerative divider, which is not self-starting. To start it, a relaxation oscillator, VR1001 and VR1002, periodically triggers the blocking oscillator, and the blocking oscillator pulses the X9 filter. After a few tries, the regenerative divider starts, and rectifier V235A cuts off the relaxation oscillator. The rectifier also controls the voltage applied to a hold-off circuit, VR201 and VR202, in the gate section, discussed shortly.

An external 1-mc or 100-kc signal may be used as the time-base reference. If the external is 100 kc, the mixer operates as an amplifier, and the X9 filter is not active.

The mixer provides a 100-kc check signal and drives a 10-mc multiplier. The 10-mc multiplier provides a 100-kc check signal and drives the harmonic generators in the frequency converter plug-in units. The output of the blocking oscillator passes through buffer V236B to five phantastron frequency-dividers, V235B and V241 through V247. The phantastrons provide the signals used by the gate circuits to time the signal gate during frequency measurements. (More is said about the phantastron circuit later.)

Four frequencies are available for measuring period and time. The available frequencies are 10 cps, 1 kc, 100 kc, and 10 mc. An external frequency connected to the "Std Freq Out" (Standard Frequency Output) connector can also be used.

The gate section of the 524D is shown schematically in Fig. 18-31 (see back-flap envelope). The "Function Selector" switch determines the type of operation. All signals to be counted are applied to trigger unit Z201, a stabilized amplifier followed by a Schmitt trigger. The trigger unit shapes the signal to provide the fast-rise, constant-amplitude waveform required to drive the "units" counter. Signal gate V202 opens and closes the signal path.

Gate-timing signals for frequency and period measurements are applied to amplitude discriminator Z202. Like trigger unit Z201, the

Sec. 18.6   Gate and count types of instruments

amplitude discriminator is a stabilized amplifier followed by a Schmitt trigger; it provides a fast-rise, constant-amplitude signal to drive the circuit that follows: decade divider Z203 or the gate binary, V206 and V207.

Decade divider Z203 divides its input by ten. The divider consists of four cascaded binaries, with feedback from the fourth binary to the second and third binaries to provide the input-output ratio of 10:1.

The gate binary, V206 and V207, controls signal gate V202 through cathode-follower buffer amplifier V208A, delay line DL201 (delay lines are covered in a later chapter), and gate control V203. In the 524C, the gate binary also controls readout blanking circuit V224, which blanks the readout while the signal gate is open. When a "start" pulse causes the gate binary to open the signal gate, the gate binary triggers blocking oscillator V208B through the cathode-follower buffer amplifier. The blocking oscillator fires the reset thyratron to generate the positive pulse which sets the counter units to zero. The display-time thyratron triggers display-time control V204, a Schmitt trigger, which closes diode gate CR205. The diode gate remains closed for a time determined by the front-panel "Display Time" control. A "stop" pulse causes the gate binary to close the signal gate, but succeeding "start" pulses are blocked until display-time control V204 reopens the diode gate. Hold-off control VR201 and VR202 also close the diode gate if there is no signal out of mixer V903 in the time-base section.

The "Reset" switch provides manual reset control. When pressed, the reset switch fires reset thyratron V209, and when released, it sets the decade divider to a count of eight. If the "Display Time" control is set to INF, the reset switch also causes display-time control V204 to open the diode gate. When switched from one setting to another, the frequency unit switch (in the time-base section) sets the decade divider to a count of eight and sets the gate binary to its gate-closed position.

A "Manual Gate" switch controls the signal gate when the function selector is set to "Manual Gate". If set to "Open", the "Manual Gate" switch prevents readout blanking in the 524C regardless of the "Function Selector" setting.

The Model 524D units counter, Fig. 18-32 (see back-flap envelope), has two sections: a counting section and a readout section. The counting section consists of four binaries (V210–V213 and V215–V218), a coincidence network (CR231–CR233), and feedback gate V214. The readout section consists of four readout amplifiers (V1200–V1203) and the readout itself.

380   *Counters*   *Chap. 18*

The binaries of the counting section have two stable conditions (either half conducting and the other half cut off) and remain in either condition until actuated by a trigger pulse. The trigger pulse causes the triggered binary to reverse its condition; the previously conducting half cuts off and the previously cut-off half conducts. The binaries quickly change from one condition to the other. Thus, the output of one half is a positive voltage step, and the output of the other half is a negative voltage step. Circuit components between the 1st and 2nd binaries, 2nd and 3rd binaries, and 1st binary and feedback gate differentiate the voltage steps into pulses. However, except for the positive reset pulse, only negative pulses can trigger the binaries.

The reset pulse from reset thyratron V209 in the gate section sets the counting section to the *zero* state. Fig. 18-33 shows the output waveforms of one half of each binary, the coincidence net-

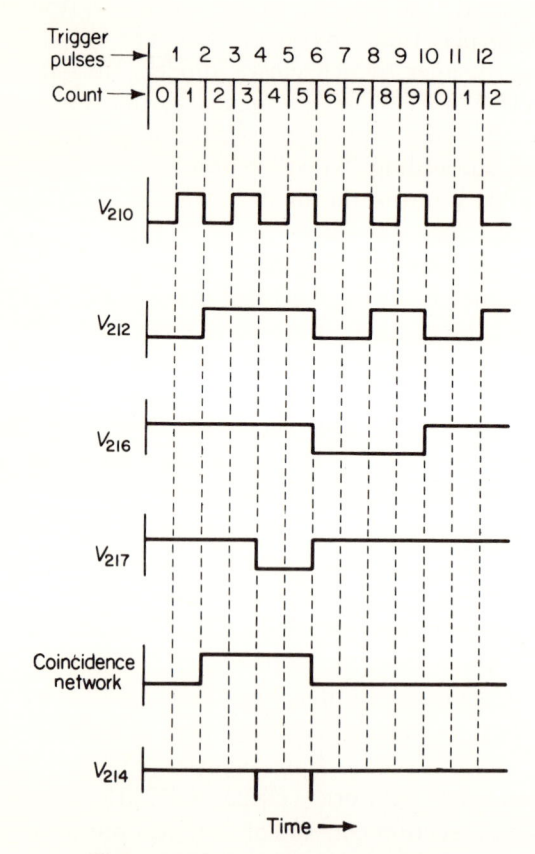

Figure 18-33 Unit counter waveforms.

Sec. 18.6   *Gate and count types of instruments*   381

work, and the feedback gate for a succession of negative pulses from the signal gate. The coincidence network opens feedback gate V214 only when both V212 and V216 are nonconducting. Starting from the reset position, the counting section operates as shown in the table of Fig. 18-34.

The binaries position the four readout amplifiers, which drive the readout. In the 524D, the readout consists of ten neon lamps behind a numbered mask; in the 524C the readout consists of ten neon lamps, ten semiconductors, and a digital display tube. (Special counter tubes are discussed in Sec. 8.7.) A sum of the voltages from three different readout binary halves is applied to each neon lamp. For a given number, only one of the ten possible sums is large enough to light a lamp. In the 524D, the lighted lamp illuminates a number on the mask. In the 524C, the lighted lamp illuminates a photoconductor which causes one number of the digital display tube to light. In either case, the number lighted corresponds to the number stored in the positions of the four binaries.

In concluding the study of the 524D, a closer look at the phantastron frequency-divider of the time-base section is interesting. One such divider is shown in Fig. 18-35. The heart of this phantastron is the 6AS6/5725, a special tube with two control grids, G1 and G3. Grid G1 controls the total cathode current, and G3 controls the division of current between G2 and the plate. Initially, G1 and the cathode are about 25 volts positive, and G3 is sufficiently negative with respect to the cathode to hold the plate current to zero. The cathode current flows mainly to G2; some of it, however, flows to G1.

A negative trigger applied to the input is coupled through C1, V1, and C2 to G1 of V2. The negative pulse on G1 reduces the cathode current and thus reduces the cathode potential. If the pulse is large enough, the potential difference between G3 and the cathode decreases to the point where plate current flows. When the plate currents starts flowing, the plate potential decreases; this decrease is coupled through C2 to G1, further decreasing the cathode potential and increasing the plate current. Grid G1 quickly reaches a voltage level where any further increase also increases the plate current; here the G1 potential stops dropping. As C2 discharges, the G1 potential rises, increasing the cathode current. Most of the increasing cathode current goes to the plate, further decreasing the plate potential. The plate signal, coupled to G1, slows the rise of G1 caused by the discharge of C2. Eventually, the plate potential approaches the cathode potential, and the plate current stops in-

Number	Operation
1	1st binary: negative pulse from signal gate causes V210 to cut off and V211 to conduct.
2	1st binary: negative pulse from signal gate causes V210 to conduct and V211 to cut off. 2nd binary: negative pulse from V210 causes V212 to cut off and V213 to conduct. Coincidence network: coincidence network opens feedback gate, for both V212 and V216 are cut off.
3	1st binary: negative pulse from signal gate causes V210 to cut off and V211 to conduct.
4	1st binary: negative pulse from signal gate causes V210 to conduct and V211 to cut off. Feedback gate: feedback gate amplifies and inverts positive pulse from V211. 4th binary: negative pulse from feedback gate causes V217 to conduct and V218 to cut off. 2nd and 3rd binaries: feedback from V217 keeps V212 and V216 cut off, V213 and V215 conducting. Negative pulse from V210 to 2nd binary has no effect.
5	1st binary: negative pulse from signal gate causes V210 to cut off and V211 to conduct.
6	1st binary: negative pulse from signal gate causes V210 to conduct and V211 to cut off. Feedback gate: feedback gate amplifies and inverts positive pulse from V211.

Number	Operation
	4th binary: negative pulse from feedback gate causes V217 to cut off and V218 to conduct. When V217 is cut off, it removes the feedback from the 2nd and 3rd binaries. 2nd binary: negative pulse from V210 causes V212 to conduct and V213 to cut off. 3rd binary: negative pulse from V212 causes V215 to cut off and V216 to conduct. Coincidence network: coincidence network closes feedback gate.
7	1st binary: negative pulse from signal gate causes V210 to cut off and V211 to conduct.
8	1st binary: negative pulse from signal gate causes V210 to conduct and V211 to cut off. 2nd binary: negative pulse from V210 causes V212 to cut off and V213 to conduct.
9	1st binary: negative pulse from signal gate causes V210 to cut off and V211 to conduct.
10	1st binary: negative pulse from signal gate causes V210 to conduct and V211 to cut off. 2nd binary: negative pulse from V210 causes V212 to conduct and V213 to cut off. 3rd binary: negative pulse from V212 causes V215 to conduct and V216 to cut off. The counting section is again in its zero or reset position.

Figure 18-34 Operation of counting section.

Sec. 18.6   *Gate and count types of instruments*   383

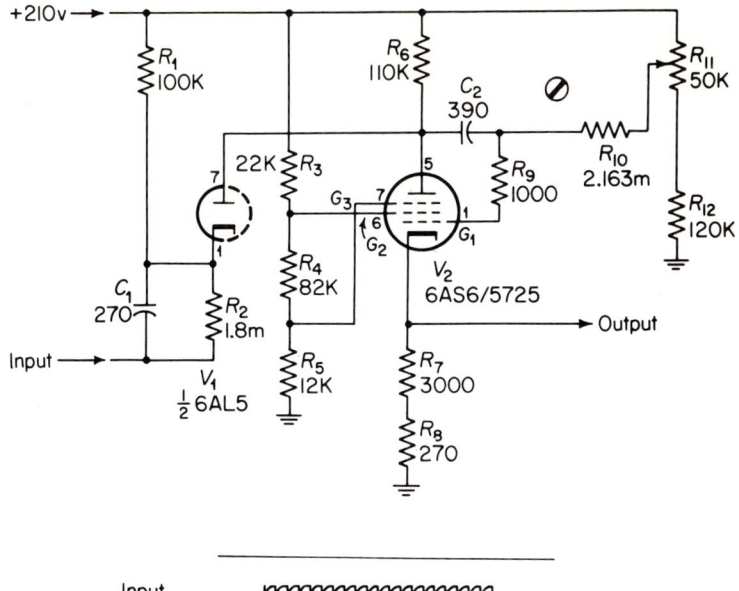

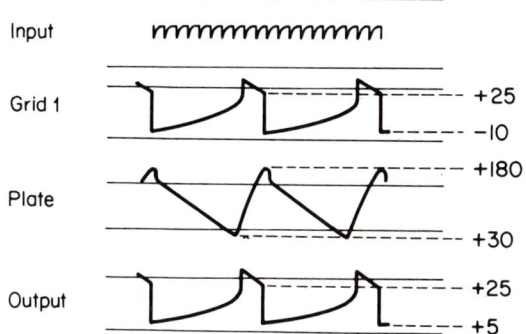

Figure 18-35 Phantastron and waveforms.

creasing. The new G1 potential increases more rapidly. The cathode potential follows the G1 potential and becomes sufficiently positive with respect to the G3 potential to cut off the plate current. As C2 recharges, the circuit returns to the initial condition.

The cathode of V1 is biased very close to the supply voltage; thus, V1 is open and blocks the input signal as long as the plate voltage of V2 is below its initial value.

The time required for the phantastron to complete a cycle is determined primarily by the values of C2 and R10. Potentiometer R11 permits adjustment of the required time over narrow limits. In the 524D, each phantastron is adjusted to respond to every tenth input trigger, to provide the required 10:1 division ratio.

18.7 Readout indicators: the Nixie and the Pixie

This section describes two special readout indicators that are in current use. These are the *Pixie* and the *Nixie* (registered trademarks), both manufactured by the Burroughs Corporation. Again, the reader should remember that rapid changes in devices of this kind are to be expected.

The Pixie, Fig. 18-36, is a gas-filled tube containing ten cold cathodes located under a disk-shaped anode. The numerals 0 through 9 are perforated in the anode. When a voltage of sufficient amplitude to produce ionization is applied between the anode and any one cathode, the resulting glow lights the corresponding numerical perforation. As the voltage shifts from one cathode to the other, the

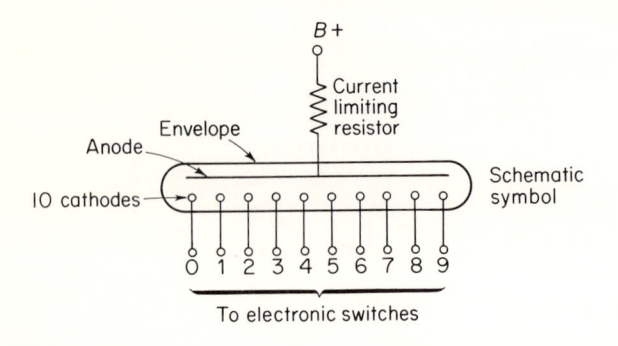

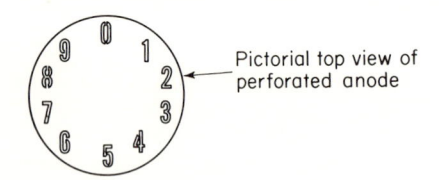

Figure 18-36 Pixie (Burroughs) indicator.

numerals are lighted in succession.

The Nixie, Fig. 18-37, is also a gas-filled tube, containing a common anode and ten cold cathodes, each shaped like one of the numerals 0 to 9. When a voltage of sufficient amplitude to produce ionization is applied between the anode and any one cathode, the ionization glow surrounds the cathode and provides visual readout. The circuit and the voltages are adjusted in such a way that only one numeral is ionized at any one time.

Sec. 18.8   *The Beam-X Switch*   385

Figure 18-37 Nixie readout tube.

18.8 The Beam-X Switch[1]

The *Beam-X Switch* (registered trade name), manufactured by the Burroughs Corporation, is a special device which combines vacuum and magnetic techniques. The theory of operation is first discussed, followed by a typical application of the Beam-X switch as a decade counter. It is believed that this device will assume increasing importance and, therefore, warrants a rather thorough discussion.

The Beam-X switch consists of ten identical sections (arrays) arranged symmetrically about a central cathode, as shown in Fig. 18-38. Each array comprises a spade element for beam-forming and locking; a target element to provide constant-current output; and a switching grid which is used to switch a beam from position to

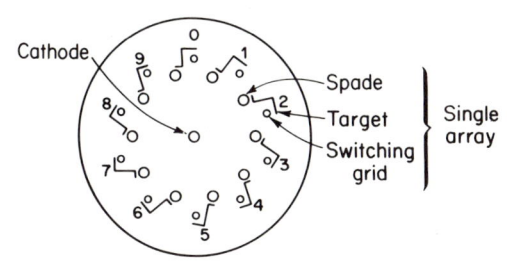

Figure 18-38 Symbolic cross section of the Burroughs Beam-X switch.

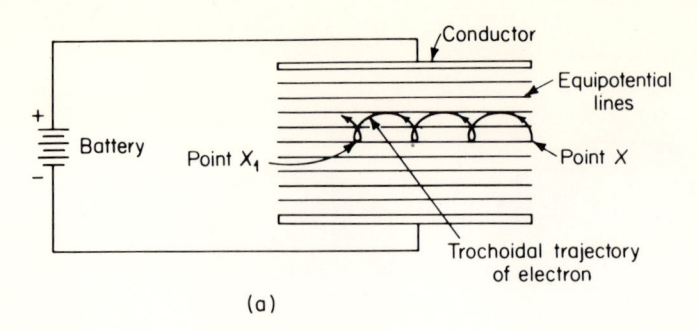

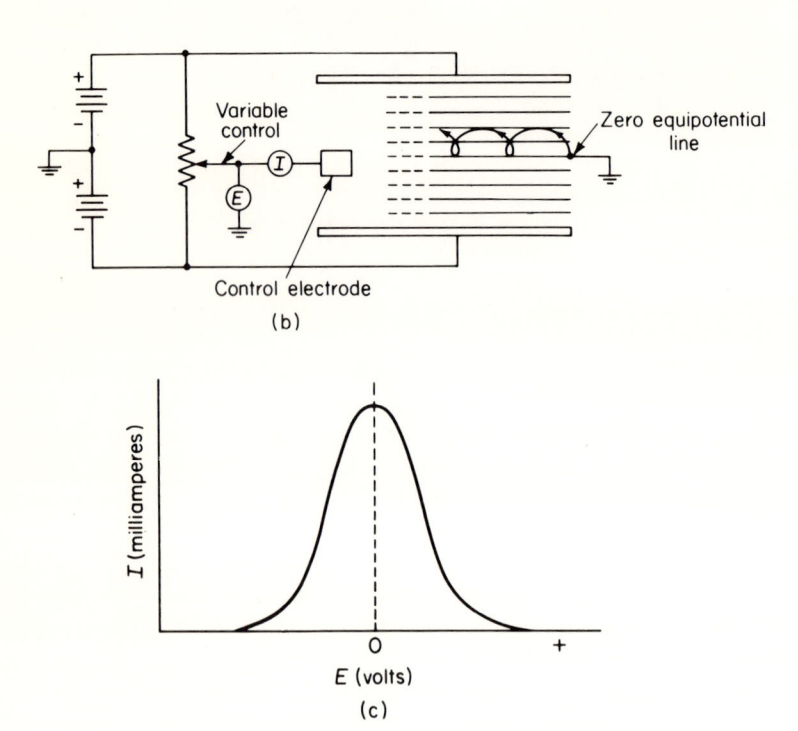

Figure 18-39 Parallel conductors to (a) opposite terminals of battery (b) electron source introduced at zero equipotential line and (c) characteristic curve.

position. The spade element is a small rod magnet that provides a magnetic field and also serves an electrical function.

The Beam-X switch utilizes crossed electric and magnetic fields in its operation; therefore, in order to clarify its operation, it may be beneficial to review briefly the behavior of an electron in crossed electric and magnetic fields.

Sec. 18.8   The Beam-X Switch

Fig. 18-39(a) shows two parallel conductors connected to opposite polarity terminals of a battery. The resulting equipotential lines (plots of a given voltage existing in the gradient between conductors) are shown parallel to the conductors. If an electron with no initial velocity is positioned on one of these lines (for example, at point X, as shown), an orthogonal magnetic field would have no influence on it. The electric field, however, would cause the electron to move toward the positive electrode. The resulting magnetic field developed by and surrounding the moving electron would strengthen the permanent magnetic field on one side of the electron and weaken it on the other side, thus causing the electron to turn and be diverted back toward an equipotential line. The direction of the turn would be dependent upon the polarities of the magnetic and electric fields. Upon turning, the electron loses its motion and stops at the equipotential line. Without motion, the magnetic field loses its influence, and the electron again begins moving toward the positive electrode. This cycle is repeated. The result is an electron drifting along equal potential lines (from X to X_1) in a trochoidal trajectory. The magnitude of each step is determined by the relative field strengths of the crossed electric and magnetic fields.

In Fig. 18-39(b), a source of electrons has been introduced at the zero-volt equipotential line. In addition, a third control electrode has been placed between the two parallel conductors. With a stream of electrons drifting along the zero-volt equipotential line, the E-I characteristic of the third electrode can be plotted. When the third electrode is highly positive, the zero-volt equipotential line and the electron beam which flows along that line are deflected below the third electrode. Thus, when it is highly positive, the third electrode receives no electron current. In the same manner, when the third electrode is very negative, the zero-volt equipotential line and the electron beam are deflected above it, and again the third electrode receives no electron current. It follows then, that, as the potential of the third electrode is varied from positive to negative, the beam of electrons following the zero-volt equipotential line is deflected across the third electrode. The resulting volt-milliampere curve of the third electrode is shown in Fig. 18-39(c).

Figure 18-40(a) shows a symbolic cross section of the Beam-X switch. The volt-ampere characteristic curve that results when the tube is connected as a diode (that is, when all electrodes except the cathode are connected to one power supply), is shown in Fig. 18-40(b). Figures 18-40(a) and (b) illustrate the cut-off condition that exists when B+ is initially applied to the Beam-X switch. In

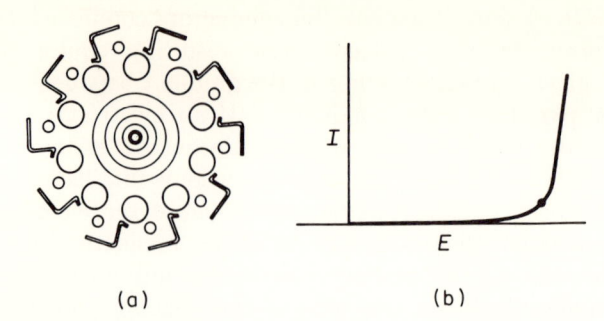

Figure 18-40 Equipotential distribution—no beam formation.

this condition, all electrons emanating from the cathode remain in the vicinity of the cathode, forming a virtual cathode. The diameter of the virtual cathode is determined by the relative field strengths of the electric and magnetic fields. With the magnetic field fixed, an increase in the electrode supply voltage will increase the velocity of the electrons, causing them to reach an equipotential line of higher voltage. The switch remains in the cut-off condition until the virtual cathode strikes the anodes. This is the maximum cut-off voltage. In normal applications, the spade electrode voltage is kept well below the maximum cut-off voltage.

In Fig. 18-41(a), the equipotential lines are shown with all the electrodes at B+ with the exception of one spade which is at or near the cathode potential. This represents the beam-formed-and-locked condition, with the beam formed to the low-potential spade. All of the electrons arriving at an equipotential line higher in potential than the saddle point (the equipotential line cross-over) constitute the beam current. Fig. 18-41(b) shows the static E-I characteristic curve of one spade taken with all the other electrodes

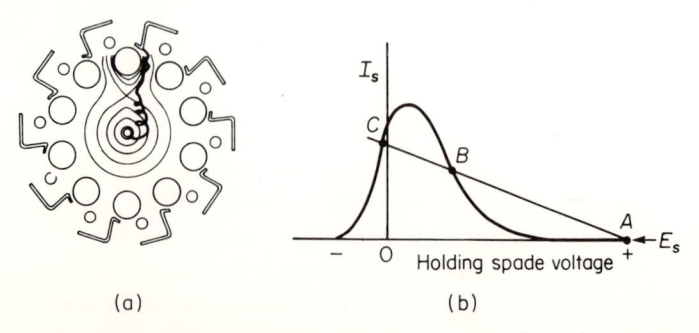

Figure 18-41 Equipotential distribution (a) beam formed and (b) static spade characteristic.

Sec. 18.8   The Beam-X Switch   389

at the spade supply voltage. Note the similarity between Figs. 18-39(c) and 18-41(b). This static characteristic curve contains a negative resistance slope which, when intersected with suitable load lines, exhibits a bistable state.

The resistance load line (typical 130 K) shown in Fig. 18-41(b) intersects the curve at points A, B, and C. Point A corresponds to the potential of the spades which are of the spade bus potential. Point B is a regenerative point having positive feedback; it is unstable, since any variation in either the current or the voltage will cause a change of the other in such a way as to increase the original change. Point C is a degenerative point having negative feedback; it is stable, since any variation in either the voltage or the current will cause a change in the other parameter cancelling the original change. Thus, with a proper resistor in series with each spade, lowering the potential of one spade to slightly below point B will result in the regenerative action that forms and locks the electron beam to the spade at point C. Approximately 85 per cent of the beam current is passed on to the associated target output electrode, where it is available to perform useful work without influencing the beam forming and locking characteristics.

The conducting target exhibits a pentode-like constant-current output that is influenced by: (1) the spade bus voltage; (2) the value of the spade load resistor; and (3) the *on* target voltage. Fig. 18-42 is a plot of I_s (the holding spade current) and I_t (the con-

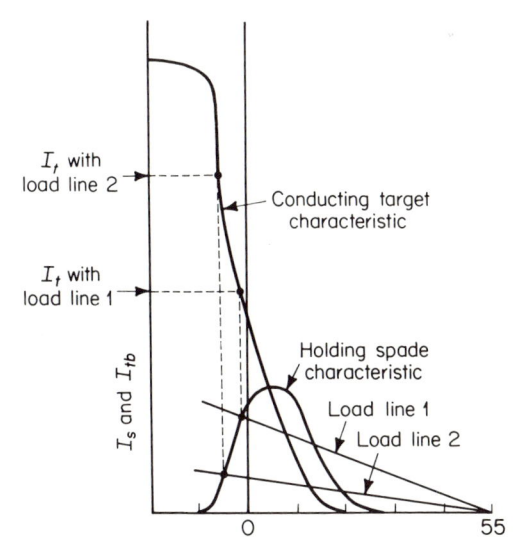

Figure 18-42 Typical spade-target characteristic, beam formed.

ducting target current) as a function of E_{sb} (the holding spade voltage) with E_s (the spade bus voltage) and E_t (the target bus voltage) held constant. The I_t-E_s curve indicates that the conducting target current is increased as the holding spade voltage is lowered in potential. This is the result, in part, of two factors: (1) the holding spade does not require as much current; and (2) the equipotential lines are altered so that the saddle point shown in Fig. 18–41(a) is moved closer to the cathode with a resulting increase in beam current. With a fixed spade bus voltage, a larger holding spade resistor will make the holding spade more negative and thus increase the *on* target current.

Fig. 18–43 shows a family of curves of I_t (the conducting target current) versus E_t (the conducting target voltage), with E_t equal to E_s. Two important factors are pointed out: (1) the output current is a function of the spade bus voltage; and (2) the output

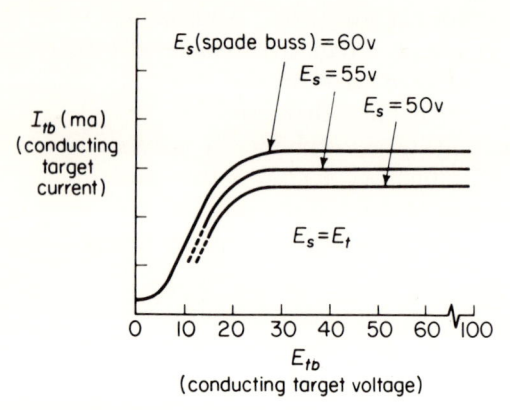

Figure 18-43 Target output current.

current is constant over a wide range of E_{tb}. With suitable R_t and E_t, large output voltage swings may be obtained.

To alleviate any stringent requirements on the input switching pulse, the switching grids are connected in two odd and even groups of five each. With this odd and even connection, applying a switching pulse to one set of grids will cause the beam to switch to one position. Since the beam switches from one position to the next in millimicroseconds, it follows that the grid potential in the next holding position must be at or near the grid bias voltage; otherwise, the beam will switch more than one position. Figs. 18–44(a) and (b) explain the switching action of the grids. The dotted curve in Fig. 18–44(a) shows the static spade characteristic curve. The

Sec. 18.8   *The Beam-X Switch*

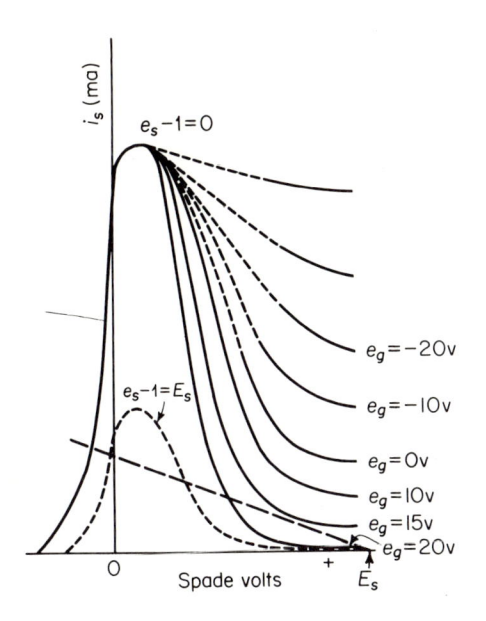

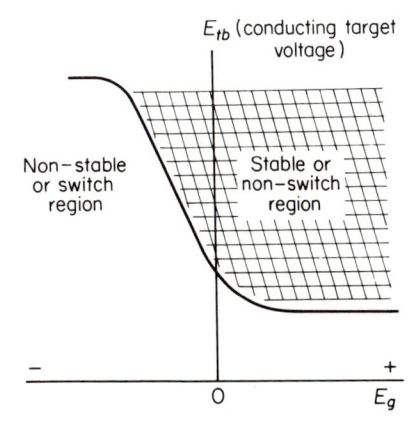

Figure 18-44 (a) Spade current vs switching input (b) switching voltage vs target voltage.

larger (dynamic) curve is a plot of the leading spade current versus the leading spade voltage obtained when the holding spade is held at the cathode potential. This simulates the condition that exists during the switching interval. In actual operation, the dynamic characteristic curve exists only during the switching transition and becomes the static characteristic as the voltage of the spade from which the beam has switched recovers through its R-C time constant from point C to point A in Fig. 18-41(b). The influence of a negative switching pulse on the dynamic curve is shown by the family of dotted additions to the dynamic curve.

A negative voltage applied to the switching grid in the holding position alters the equipotential line configuration in the area between the holding spade and the leading spade, thus allowing the higher potential electrons to strike the leading spade. These electrons lower the potential of the leading spade and alter the equipotential lines, thereby allowing additional electrons to flow to the leading spade.

When sufficient electron current flows into the leading spade resistor, lowering its voltage to below point B in Fig. 18-41(b), the beam will immediately switch into this position. As shown in Fig. 18-44(a), a larger negative switching pulse will divert more current to the leading spade and will result in a faster switching transition and higher maximum frequency. Fig. 18-44(b) illustrates the influence of the conducting target voltage on the grid switching voltage. The more positive the conducting target, the closer the electron beam is to the holding spade. As a result, the switching grid associated with this position must be made more negative to assure switching. If the conducting target voltage is allowed to fall too close to the cathode voltage, the beam will automatically switch regardless of the voltage of the associated grid. E_t and R_t are usually selected so that E_{tb} is approximately equal to E_s.

Figure 18-45 shows various methods of applying pulses to the switching grids. Figure 18-45(a) illustrates the use of a SPDT switch applying a negative voltage alternately to the even and odd grids, causing the beam to be switched sequentially. The electronic equivalent of the SPDT switch is the flip-flop, which supplies negative pulses alternately to the even and odd grids in turn, advancing the beam sequentially as shown in Fig. 18-45(b). The diodes are used for d-c restoration at frequencies above which the R-C coupling time constant does not differentiate the output of the flip-flop. The diodes are used at high frequencies to take full advantage of the negative edge of the output of the flip-flop. Figures 18-45(c) and (d)

Sec. 18.8   The Beam-X Switch

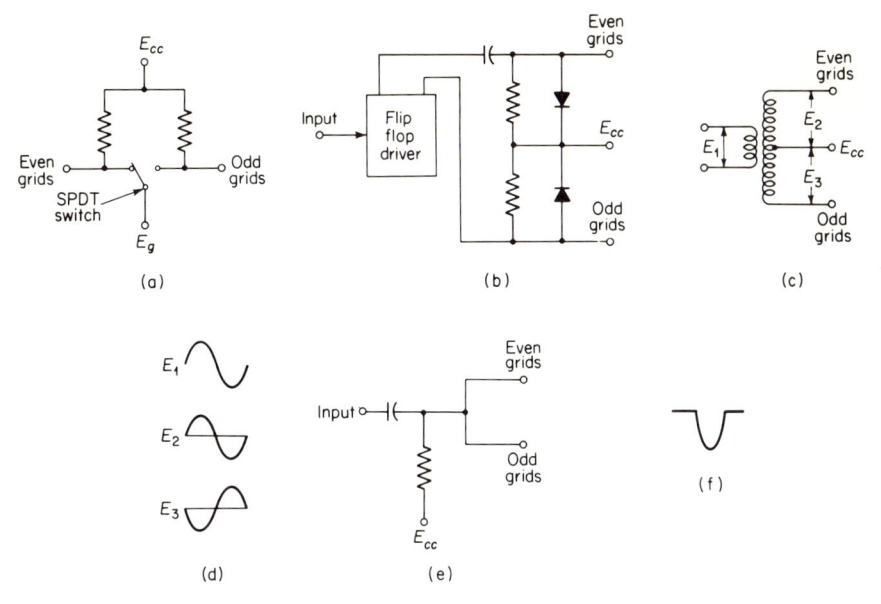

Figure 18-45 Typical switching inputs.

show a push-pull transformer method of driving the Beam-X switch. As shown, each set of grids is driven negatively during one input cycle, thus advancing the beam two positions. A factor to be considered when using this drive method is the necessity of obtaining the required amplitude as measured from one side of the secondary to the center tap, which is normally used as the grid-bias return. Only one-half of the peak-to-peak amplitude (from center tap to one side) is effective in switching the beam. Thus, only one-quarter of the peak-to-peak amplitude of the entire secondary winding is effective. Figures 18-45(e) and (f) illustrate single-ended drive. Using this technique, the odd and even grids are connected together and are driven by a *quantized* pulse, that is, a pulse containing a certain amount of energy. The energy contained in the quantized pulse is very important and somewhat critical. A pulse of too little energy would not switch the beam, whereas a pulse containing too much energy would switch the beam more than a single position. This method of drive is not recommended at frequencies above 10 kc. At lower frequencies, the switching time may be slowed by adding capacitance to the spade circuitry. This makes the energy requirement of the pulse less dependent upon the variation of characteristics between individual positions within a tube and from tube to tube.

When supply voltages are applied to the Beam-X switch circuit, the tube will assume one of its stable states known as cutoff. In this state, no current flows to any of the targets, and grid driving pulses, if applied, have no effect. The tube cannot perform any useful function until an electron beam is established in one of its ten positions. This operation is called *zero-setting, resetting*, or *beam-setting*.

If the tube is initially in the cutoff state, the beam can be established in any of the ten positions by merely lowering the potential of the spade associated with the selected position.

It is often necessary to reset the beam to a certain position when the beam has already been formed in some other position. An example of this is the zero-setting of counters. Zero-setting an existing beam in the Beam-X switch entails more than merely lowering the potential of the zero spade; it involves first clearing the beam and then re-establishing it in the desired position. Although two successive operations appear necessary, zero-setting can be accomplished by a single contact closure or by a single pulse.

Clearing the electron beam in the tube is accomplished by reducing the spade bus potential, E_s, to zero. This can be done either by lowering the spade supply potential to the cathode potential, or by raising the cathode potential to equal the spade supply potential.

In an electronic reset circuit, it may be difficult to keep the amplitude of the clear pulse equal to E_s; a maximum increase of 100 per cent in the clear pulse amplitude is permissible in such cases without imparing reliability. The minimum duration of the clear pulse should be somewhat greater than the spade R-C time constant. Normally, it is greater than 2.0 μsec, although with more complex circuitry this time can be reduced to less than 1.0 μsec.

After the tube is cleared, a beam may again be formed to any of the ten positions by lowering the potential of the spade associated with the selected position. For reliable setting of the tube, the spade potential, E_s, should be lowered by an amount at least equal to E_s but no more than 1.5 times E_s. If the spade is not reduced to a sufficiently negative value, erratic resetting or no resetting at all will result; whereas, if the spade is lowered to an excessively negative value, the beam may skip to a different position.

The minimum time required to set the beam is very short and is limited primarily by the associated circuitry. Setting of the beam generally is accomplished in a much shorter period of time than it takes to clear the beam.

It is important to note, however, that when setting the beam in the Beam-X switch, current is permitted to flow from the space to

Sec. 18.8   The Beam-X Switch   395

the reset pulse source only. If any flow of current in the reverse direction existed, it would be impossible to establish the beam. The reason for this requirement will become evident when the examples of Figs. 18-46, 18-47, and 18-48 are considered.

Most of the Beam-X switch circuits are employed in conjunction with a flip-flop driver circuit. Since the beam in the Beam-X switch may have to be reset from either an even or an odd position, it becomes necessary to reset the driver flip-flop to the even state whenever the Beam-X switch is zero-set. To avoid instability, the flip-flop should already be in the even state by the time the beam is formed in the Beam-X switch. For this reason, the leading edge of the zero-set pulse that clears the Beam-X switch should be employed to set the flip-flop. By the time the beam has been set in the Beam-X switch, the flip-flop has already reached equilibrium in the even state.

The circuit in Fig. 18-46 illustrates a method of zero-setting the Beam-X switch by the use of a switch contact closure. The relative magnitudes of the component values for this circuit are also indicated.

Since the ratio of R_3/R_2 is rather large, the charge on capacitor

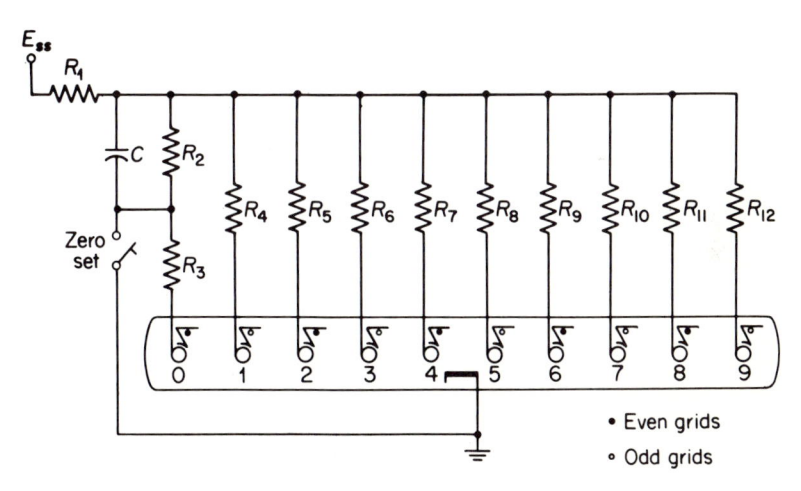

$R_4, R_5, \ldots R_{12} = R_s$ (spade load resistance)
$R_2 + R_3 = R_s$
$R_3/R_2 \approx 10$
$R_2/R_1 \approx \frac{3}{2}$
$R_2/C > 100 \mu sec$

Figure 18-46 *Manual zero-setting and relative magnitudes of the components.*

C will be negligible if the beam is in the zero position, or the charge will be zero if the beam is in any of the other positions. At the instant the zero-set push-button is closed, the potential of the junction of R_2 and R_3, and also the spade bus potential, drop to zero. This clears the tube. Immediately after the closing of the switch, capacitor C begins to charge through R_2. Capacitor C charges toward an equilibrium potential of somewhat greater than one-half of normal E_s. At this equilibrium potential, the beam normally does not form. If the switch is now opened, the spade bus voltage tends to assume its normal operating potential, E_s. Because of the charge on capacitor C, the zero spade is held negative with respect to the spade bus voltage immediately after the opening of the switch, and the beam forms to the low-potential spade. The time constant, R_2C, is made rather large to overcome the effects of switch contact bounce.

If the capacitor C is considered to be the reset generator, then (to illustrate the statement made previously) current flow, during the setting of the beam, will exist in one direction only, flowing from the junction of R_2 and R_3 towards the lower terminal of C.

Figure 18-47 illustrates an electronic reset circuit requiring a

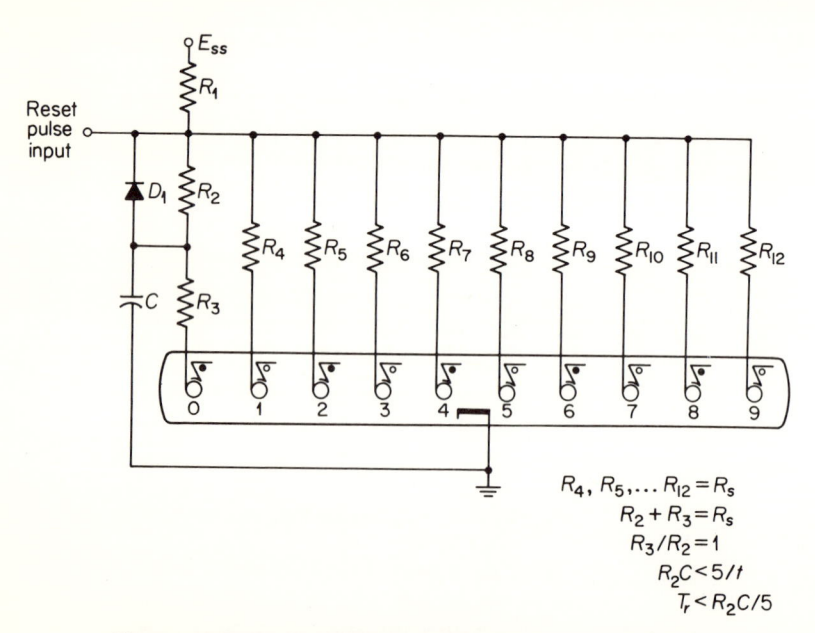

Figure 18-47 An electronic reset with relative magnitudes of components, where T_r is the positive-going rise time of input pulse and f is the counting frequency.

negative pulse. While the tube is stepping sequentially, capacitor C will have negligible effect on the performance of the circuit. When the beam is to be reset to zero, a negative pulse, as indicated in Fig. 18-47, is applied to the spade bus. This negative pulse lowers the spade bus potential to zero, thereby clearing the tube. The same pulse also discharges capacitor C completely through the forward impedance of diode D_1. When the duration of the reset pulse is over, the spade bus potential recovers to its normal value, E_s, in a short time equal to T_r. Since the time constant of R_2C is several times larger than time T_r, the junction of R_2 and R_3 and the zero spade will remain near cathode potential for some time after the remaining nine spades have recovered to their normal potential, E_s. As a result, the electron beam forms to the negative spade.

Note that diode D_1 conducts only during the "clearing interval," and that beam setting is accomplished by the capacitor-charging current flowing through R_2 which also reverse-biases D_1.

It is essential that the minimum duration of the reset pulse be at least 2.0 μsec. If this pulse were of shorter duration, when the beam was to be reset to zero from the number one spade, the latter might not completely recover to its normal potential, E_s, at the time the beam is already established on the zero spade. This condition will result in the beam advancing to the still somewhat lower potential number one spade. The reset pulse should not have any positive overshoot on its trailing edge since such an increase in the spade voltage, E_s, would cause instability or skipping. The value of resistor R_1 is not critical. It should be selected to provide a sufficiently high impedance for the reset pulse source to drop the available B+ potential to the proper operating spade bus voltage, E_s. The resistor R_1 may form the plate load resistance of a tube or the collector load resistance of a transistor, which in turn produces the reset pulse. If the spade bus potential is supplied directly through the secondary of the reset pulse transformer, resistor R_1 can be entirely eliminated. The increased switching requirements resulting from the use of R_1 should be taken into consideration in the design of the circuit.

With the proper selection of component values and reset pulse dimensions, the circuit of Fig. 18-47 is ready to accept switching pulses within two to five microseconds after the beam has been reset in the tube.

The circuit of Fig. 18-48 is another form of electronic reset requiring a positive pulse at the cathode of the tube. The relative magnitudes of the component values are again given.

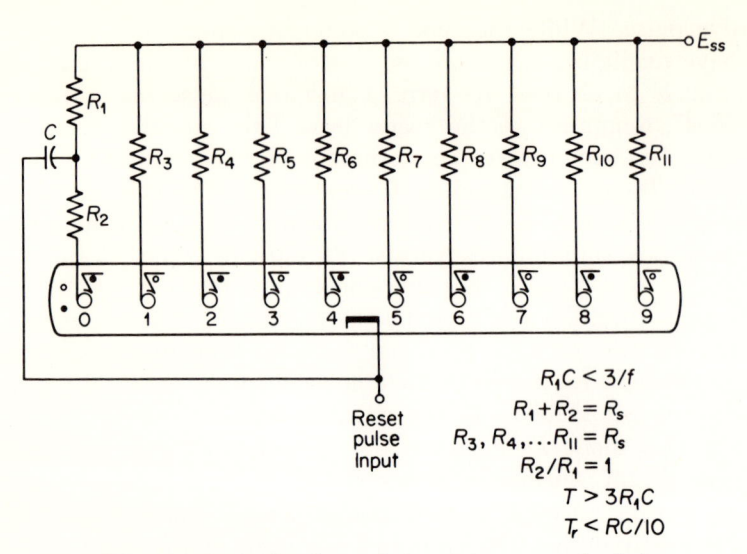

$R_1 C < 3/f$
$R_1 + R_2 = R_s$
$R_3, R_4, \ldots R_{11} = R_s$
$R_2/R_1 = 1$
$T > 3 R_1 C$
$T_r < RC/10$

Figure 18-48 Another form of electronic reset and relative magnitudes of component parts.

The positive pulse, applied to the cathode of the Beam-X switch, clears the beam; it also discharges capacitor C through resistor R_1. At the end of the period T_r, the negative-going edge of the reset pulse is coupled through capacitor C and appears across R_1. At the end of the period T_r, the cathode is again at its normal operating d-c level; the junction of resistors R_1 and R_2 is also approximately at cathode potential; and the beam forms to the zero spade.

Capacitor C could also be connected directly to the spade; however, the arrangement shown in Fig. 18-48 reduces the undesirable capacitive effect for high-speed operation. The trailing edge of the reset pulse should be sufficiently fast to couple through capacitor C with negligible loss in amplitude. Although the minimum duration of the reset pulse is generally on the order of 10 to 15 μsec, the circuit is ready to accept switching input pulses within one microsecond after the termination of the reset pulse.

This circuit is especially adaptable to transistor circuitry because the reset pulse can be supplied by a normally saturated NPN transistor operated as a common-emitter amplifier. If the cathode of the Beam-X switch is connected to the collector of the transistor, then reverse-biasing the base, for the period T_r, will produce the desired reset pulse.

Figure 18-49 shows a typical ten-position circuit for the Beam-X switch containing the voltages, components, and auxiliary circuitry

Sec. 18.8   The Beam-X Switch

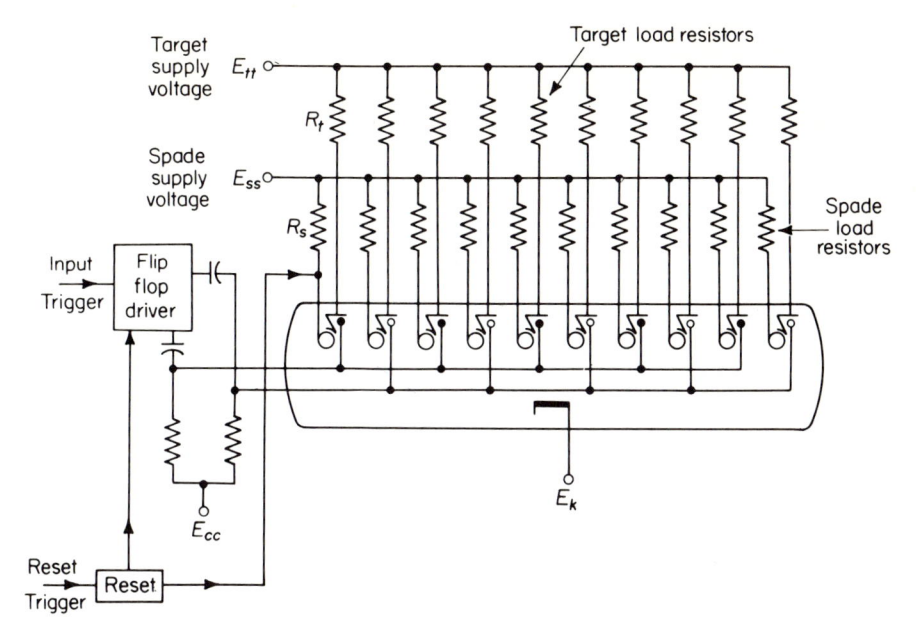

Figure 18-49 Basic ten-position Beam-X switching circuit.

required for decimal operation. A single B+ supply can normally be used.

Separate supplies have been designated for explanatory purposes. As an example, the spade bus of one type of Beam-X switch may be operated at any voltage from 15 v to 75 v by employing spade load resistors of appropriate values. The spade bus voltage generally determines the magnitude of the target constant-current output. This output current can range from 300 microamperes to as much as 4 milliamperes. Other types may be operated at higher voltages with correspondingly higher currents obtainable.

As in any constant-current device, the target supply voltage requirements are extremely flexible. When the electron beam is formed to any one of the ten target outputs, there is negligible current to any other target. The constant-current output, in conjunction with the target supply voltage and load resistance, can be used to obtain output voltage steps from zero to 200 v. However, the voltage and resistance should be properly selected to satisfy design parameters affecting the conducting target voltage switching input requirements and maximum wattage considerations. The target resistors can be replaced with other loads including relay coils, transistors, gas discharge devices, and even large capacitances as long as the target load line and its effect on switching requirements are properly compensated.

The grid-bias voltage is also a function of the spade supply voltage; it is somewhat positive with respect to the cathode, thereby assuring a stable static beam. The grid-switching drive voltage requires a negative input whose magnitude is determined primarily by the spade bus voltage and its load line, and secondarily by the target voltage and load line and the maximum frequency. A rule of thumb is that the amplitude of the negative grid-driving signal should be approximately equal to the spade bus voltage. When operating at extremely high frequency and/or providing large target outputs, a larger input drive pulse is recommended.

The preferred method of driving the Beam-X switch is by means of a flip-flop. The negative-going pulse from the alternate outputs of the flip-flop are coupled to the even and odd grids of the Beam-X switch. Therefore, for each input pulse, the beam within the Beam-X switch is advanced one, and only one, position. Other types of drive circuits, including mechanical switches, push-pull transformers, and discrete single-ended pulsing, have been used to drive the Beam-X switch, as previously explained.

When the d-c supply voltage is first applied to the Beam-X switch circuit, the tube normally remains in its cleared or "cut-off" state. In order to form the beam to the zero position initially, the spade must be lowered to below the first intersection of the load line with the static spade characteristic curve. At the same time, the flip-flop must be set to the appropriate state so that the first input pulse that triggers that flip-flop produces a negative pulse from the flip-flop to the even grids to advance the beam from zero position.

In order to reset the beam to zero from any position, the tube must first be cleared, as the beam will not step backwards within the tube. After clearing, the reset pulse may be applied to any spade, thereby resetting the tube to that position. As mentioned earlier, in order not to lose a count, the flip-flop must be set to the appropriate state whenever resetting or presetting the Beam-X switch. The functions of clearing and resetting may be done simultaneously with a single pulse at a single point as shown in Fig. 18-49.

Figure 18-50 shows one of the major applications of the Beam-X switch, namely, decade counting. In decade counting, the input trigger pulses are applied to the driver of the units decade, advancing the beam in the Beam-X switch one position for each trigger pulse. Each time the beam steps from position nine to position zero, an output "carry" pulse complements the flip-flop driver of the ten's

Sec. 18.8   *The Beam-X Switch*   401

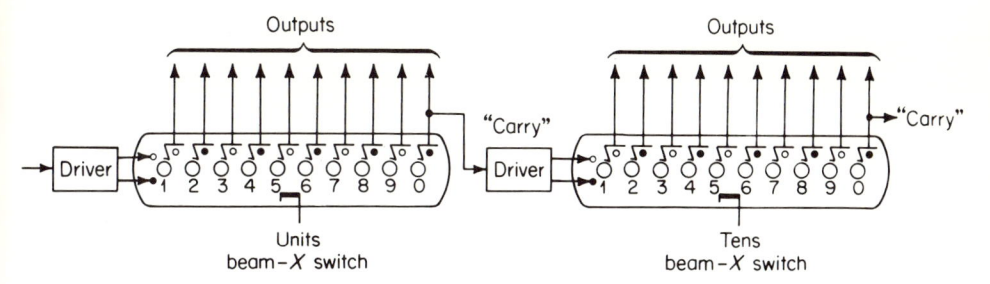

Figure 18-50 Decade counting.

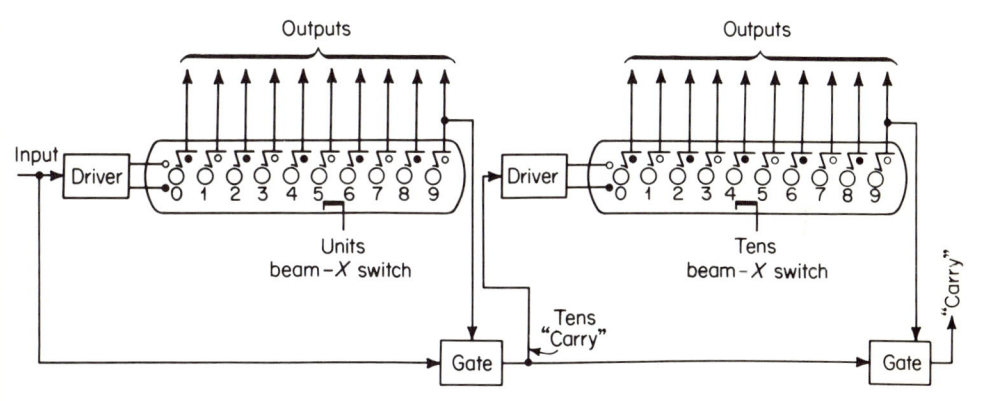

Figure 18-51 Decade counting with no cascade delay.

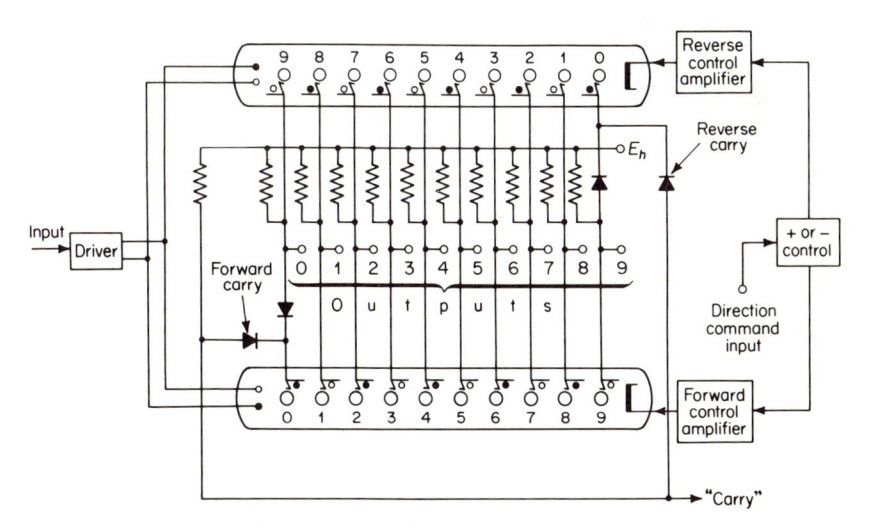

Figure 18-52 A bidirectional counter.

decade, advancing the count in this decade. A "carry" output could be provided from each succeeding decade, in order that many decades of counting may be performed. In addition to the cascade or "carry" pulse, Fig. 18-50 also illustrates that decimal outputs are available from each Beam-X switch. These outputs are directly applicable to the operation of such devices as the Nixie indicator tubes.

Figure 18-51 illustrates one method of eliminating accumulated cascade delay time in decade counting. The output from the number nine position of the units Beam-X switch will open a gate so that the tenth input trigger not only advances the units decade from nine to zero, but also advances the ten's decade to its next position. This technique can be utilized for any number of decades.

Figure 18-52 shows how two Beam-X switches can be interconnected to perform as a single decade bi-directional counter. The tubes are so connected that, for a given mode of count, one Beam-X switch is active and is the master for the other, which is in the cut-off state. One Beam-X switch counts in the forward direction, the other in the reverse direction; thus, by controlling which tube is the master, bi-directional counting is performed. The use of common spade and target load resistors simplifies circuitry. A method of obtaining the "carry" pulse from the active Beam-X switch for cascading decades is shown. Therefore, when counting up, the "carry" pulse is produced as the count advances from 9 to 0; when counting down, the "carry" pulse is produced as the count descends from 0 to 9.

Figure 18-53 shows the utilization of the decimal outputs of Beam-X switches for preset counting applications. The particular configuration shown will provide an output after from 1 to 99 input pulses as determined by the setting of the preset switches. Dual, triple, or multiple preset outputs can be obtained. In addition, the output may be used to trigger the reset generator to make a preset/

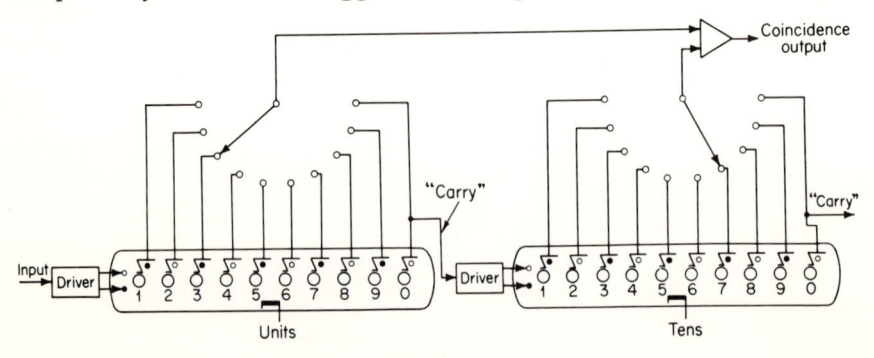

Figure 18-53 Preset counter.

reset counter. Another method of building preset counters, using the Beam-X switch, is to select the spade to which the beam will be formed in each decade when initially setting the tube or tubes.

18.9 Transistor counters[8]

Transistor counter circuitry is generally similar to that already described for vacuum tubes. To round out our discussion, however, we will consider two transistorized counters as well as a special type of decade indicator that operates directly off transistors.

A block diagram of the over-all system to be described is shown in Fig. 18-54. Counting is accomplished by using four binary stages per decade, with appropriate feedback to produce a decimal count.

The resetting circuit sets the counter to the complement of the number to be counted. For example, if the number to be counted is 3, the decade is preset to the seventh state so that three input pulses set the decade to the tenth state and produce the desired

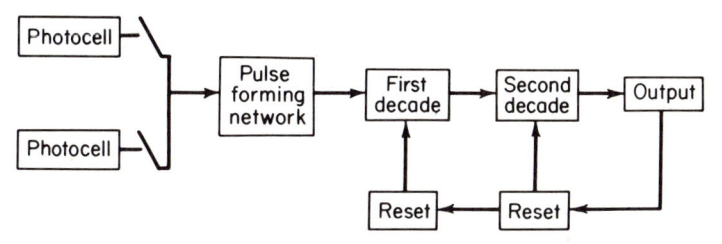

Figure 18-54 Overall block diagram.

output. Because this counter uses more than one decade, it is necessary that the reset circuit be disabled during the counting period. For example, the procedure used to count the number 33, which requires two decades, is as follows. The states of both decades are preset to the complement of the number to be counted — the units decade to the seventh state, and the ten's decade to the sixth state. After three input pulses, the units decade transmits a pulse to the ten's decade. Because the reset circuit is disabled, the units decade now requires ten input pulses to produce additional outputs. Thus, 33 input pulses are required to trigger the ten's decade to the tenth state, and produce the desired output.

The output stage of the counter is a bistable stage capable of driving a relay. It is also possible to use a monostable output stage, in which case the counter will return to the counting made after some predetermined interval of time.

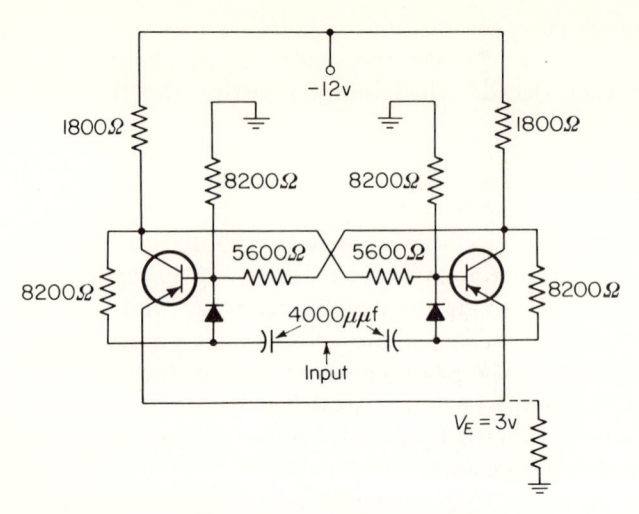

Figure 18-55 Flip-flop circuit.

A Schmitt circuit (pulse-forming network) driven by an amplified signal from a photocell, a microswitch, or an electrical signal, provides the input signal to the decades.

The binaries used in the circuit of Fig. 18-55 are designed so that the *on* transistor is saturated, the *off* transistor is reverse-biased by some positive voltage, and the collector current of the *on* stages does not exceed the maximum specified.

The binaries are triggered by a positive-going signal which is gated to the base of the *on* transistor.

Each decade uses simultaneous feedback to obtain a decimal (base of 10) count from the normal scale-of-sixteen circuit.

For reliable resetting, a voltage which is positive with respect to the emitters is applied to the bases of the transistors of the binaries to be reset. This circuit is shown in Fig. 18-56. Under full-load conditions, this resetting voltage must be more positive than the voltages of all the emitters in the three decades. The nominal voltage on the emitters is −3 v, and ground potential is needed to insure reliable operation at high temperatures. For these reasons, a reset voltage of −1 v is used because it represents a voltage greater than −3 v and less than ground (0 v).

The −1 v reset voltage is obtained from the tapped emitter resistor of the output stage. Because the ratio of current required by the output stage, to current required by the reset circuit, is very large, the use of this tapped resistor causes a negligible reaction. The output of the reset stage, therefore, swings from nearly −1 v

Sec. 18.9   Transistor counters

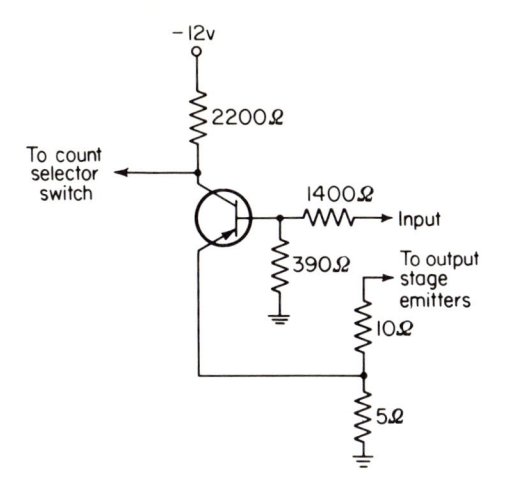

Figure 18-56 Reset circuit.

(-1 v plus V_{CE} of the stage) to -12 v. This output is fed directly to the anodes of the reset diodes, and the cathodes of these diodes are, in turn, connected directly to the appropriate base through the count-selector switch sw_4, as shown in Fig. 18-57 (see back flap envelope).

The reset circuit, triggered by the output stage, is turned on during the reset period. The reset circuit, in turn, applies a -1 v potential to the bases of the transistors to be reset, and turns them off. When counting is resumed, the reset circuit is turned off by the output stage. The reset circuit is then disabled from the decades by means of a reverse bias of 9 v applied to the reset diodes.

This particular equipment was designed to provide input signals to the counter by means of two methods. The first input consists of a mechanically operated microswitch, SW_2, which provides a signal varying from ground to -12 v. As shown in Fig. 18-57, capacitor C_1 is placed across the switch to prevent erroneous triggering caused by transients produced by switch chatter. A second input, SW_1, is provided by a Schmitt circuit driven by an amplified signal from a photocell.

The bistable output circuit is designed to drive the -6 v, 200-ma relay. The output circuit is triggered into one state by the output of the last decade, and is returned to the other state by means of an externally operated switch, SW_3. Alternatively, the output stage may be a monostable circuit triggered by the last decade and then returned to its original state after some predetermined time.

Sufficient power to drive the output stage was obtained by

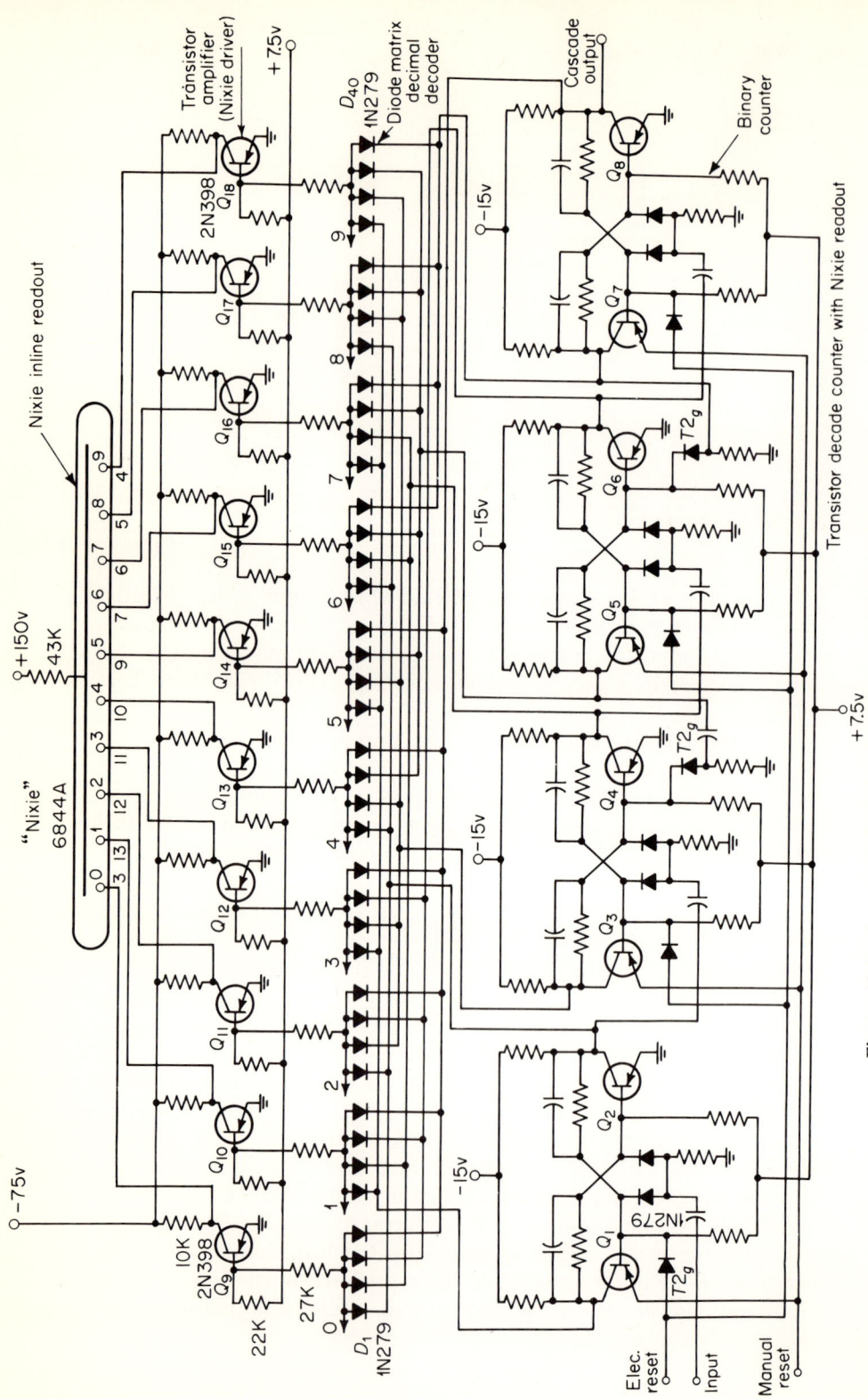

Figure 18-58 A transistorized decade counter with Nixie inline readout.

redesign of the fourth flip-flop in the second decade. The counter uses RCA type 2N406 and 2N270 transistors and type 1N38A diodes, and operates satisfactorily at temperatures up to 55°C with power supply variations of ±50 per cent.

A transistor decade with a Nixie readout, manufactured by the Burroughs Corporation, is shown in Fig. 18-58. This counter uses two feedback loops, one from the collector of Q_7 through a diode to the base of Q_6, and the other from the collector of Q_5 to the base of Q_4, also through a diode. Resetting is accomplished by applying a positive pulse to the base of the input stage of each binary, driving each to cutoff. The input pulses are applied through steering diodes to the transistor bases. Transistors Q_9 through Q_{18} are required to drive the Nixie tube since the potential available in the binary stages is not sufficient for direct operation. This has been a common difficulty in transistor counters.

The drivers are controlled by a diode matrix (a special form of switching circuit) connected to the binary stages. Notice that there are ten groups of four diodes each in the matrix. Each of these groups corresponds to a unique set of binary conditions, and each group is connected to the base of a driver. When a number of input pulses corresponding to the diode group number (0, 1, 2, etc.) occurs, the diodes in that group become nonconductive and turn off the base current of their associated driver. When the driver switches off, its collector potential swings negative. This negative potential is applied to a cathode of the Nixie and causes ionization. The ionization glow surrounding the cathode then provides visual readout.

The need for readout drivers just discussed has been eliminated by the use of a new gas-discharge decade indicator tube, the Z550M, developed by Amperex[9]. This tube is purely an indicator that requires less than 5 v at less than 50 μa to produce the discharge. The numerical count is shown by a red neon glow viewed through the top.

The extremely low triggering voltage and current of this tube are due largely to the proprietary molybdenum sputtering technique used in the manufacture of the tube, and to the geometry of the electrodes. The sputtering technique is a method by which molybdenum is sprayed on the cathode and on a large area of the glass envelope. This technique improves cathode stability and helps maintain the purity of the neon-argon gas within the tube.

The unique geometry of the tube is shown in Fig. 18-59(a). The cathode is ring-shaped and has ten evenly spaced holes into which the trigger electrodes are placed. Clearance between the triggers and

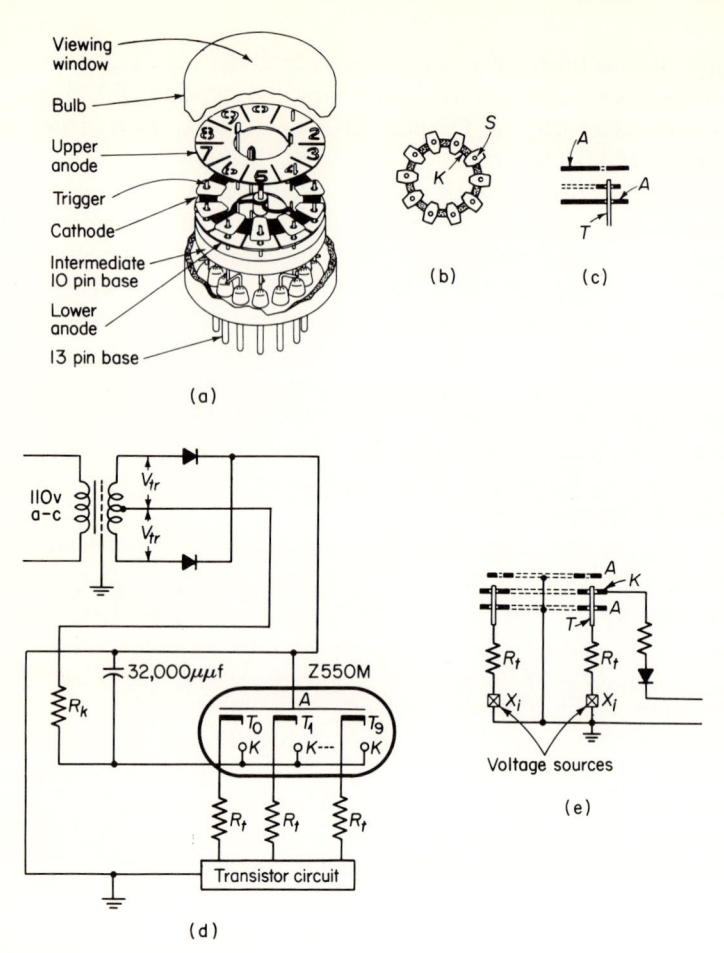

Figure 18-59 Amperex Z550M.

the cathode sectors is about 0.3 mm. Because of this very small gap, a correspondingly low voltage can initiate the discharge.

As shown in Fig. 18-59(b), the face of the molybdenum ring, K, is coated with ten sectors of a material with a higher work function (shaded in the diagram). The ten sectors, S, between them act as a cathode. About 3 mm above and below the ring, Fig. 18-59(c), two other rings are mounted, acting as anodes. The upper anodes are provided with cut-out figures 0 to 9 that can be read by the glow behind them. Through a hole in each of the ten cathode sectors, a wire electrode, T, the trigger, initiates discharge at the desired place. The clearance between the triggers and the cathode sectors is about 3 mm.

The tube is filled with neon gas to which a small percentage of argon has been added. This choice of gas filling helps keep down the relative difference in breakdown voltage between the various trigger-cathode spaces. To obtain a clean cathode surface, the cathode is sputtered during manufacture. The sputtered material on the glass wall helps to keep the gas uncontaminated.

The tube is fed with a rectified alternating voltage, Fig. 18-59(d), which is not smoothed. A discharge is initiated when the amplitude of this voltage is sufficiently high. For a half-wave rectified supply, the supply voltage rises to a maximum and drops to zero once in every power-supply cycle. However, a full-wave rectified voltage may also be used.

As seen in Fig. 18-50(c), the triggers are at the same potential as the anode so long as there is no discharge. A discharge between the cathode and one of the triggers has a lower ignition potential than a discharge between the cathode and the anode. When the voltage begins to rise from zero, therefore, a discharge first occurs between the anode and one of the triggers. If the current produced by this auxiliary discharge is high enough, the anode takes over the discharge almost immediately. The potential difference between the cathode and the anode then drops to the burning potential of the main discharge now occurring between these electrodes (a glow discharge), so that for the rest of that particular half-cycle none of the other triggers can reach their breakdown potential.

The place where the auxiliary discharge occurs can be selected by making the potential of the relevant trigger higher than that of the other triggers (and of the anode) by a small amount, X. As a result, this trigger reaches the breakdown potential earlier than the others and the discharge always recurs at the same position. If the voltage X is transferred to another trigger, the re-ignition in the next power-supply cycle will take place at that trigger, and so on.

The periodic extinction of the discharge is thus essential to make it possible to displace the discharge from one position to another. It follows that the tube can be driven with a signal whose amplitude (always less than 5 v) is much smaller than the breakdown voltage itself.

This small signal can be supplied by a transistor circuit. If the circuit is so designed that a signal X is applied to the trigger T_1 for a count of 1, to the adjacent trigger T_2 for a count of 2, and so on, one can read from the tube the total result of the count.

It is immaterial whether or not the tube can follow a rapid counting operation, because upon the next re-ignitition after the completion of the counting operation, the tube always glows at a position corresponding to the final result of the count. Since the power for the main discharge is not drawn from the transistor circuit, this discharge is bright enough to provide a clear visual indication.

The anode in the circuit is grounded. This makes it possible to ground one of the two terminals of the voltage sources that supply the control signal. In practical applications, these sources are part of the transistor circuit.

REFERENCES

1. Weissman, R., "High-Speed Counter Uses Ternary Notation," *Electronics*, Oct. 1952.
2. Chance, B., et al., *Waveforms*. New York: McGraw-Hill Book Company, Inc., 1949.
3. Grossdoff, I. E., "Electronic Counter," *RCA Review*, Sept. 1946.
4. Potter, J. T., "A Four-Tube Decade Counter," *Electronics*, June, 1944.
5. Partridge, G. R., *Principles of Electronic Instruments*. Englewood Cliffs, N. J.: Prentice-Hall, Inc., 1958.
6. *Operating and Service Manual*, Hewlett-Packard Model 524C and Model 524D Electronic Counters, Hewlett-Packard Co., Palo Alto, Calif., 1959.
7. Brochure BX535, *Beam-X, Switch*, Burroughs Corporation, Plainfield, N. J.
8. Painter, R. R., and R. A. Christensen, "A Low-Cost Pre-Set Transistorized Counter," *Semiconductor Products*, May 1961.
9. Staff, "Decade Indicator Operates Off Transistors," *Electronics*, Nov. 3, 1961.

EXERCISES

18-1 Explain how simultaneous feedback is used to convert a scale-of-sixteen counter to a decade counter.

18-2 Explain how two separate feedback paths may be used to convert a scale-of-sixteen counter to a binary counter.

18-3 Describe the basic operation of the Beam-X Switch.

18-4 Explain the operation of the tri-stable counter shown in Fig. 18-6.

18-5 Why are speed-up capacitors used in some of the coupling networks of a ring counter and not in others? Also, what are the major disadvantages of ring counters?

18-6 How may the Beam-X Switch be cleared?

18-7 Explain the operation of the diode matrix used in connection with the transistor counter of Fig. 18-58.

Exercises

18-8 How can the initial condition be set in a decade counter? Draw a representative circuit.

18-9 Draw a circuit showing how binaries may be used to form a ternary counter.

18-10 Explain the operation of the Nixie and Pixie readout indicators.

19

PULSE MODULATION

Pulse modulation techniques are used exclusively in mobile radio telemetry to obtain data from such vehicles as weather balloons, manned or unmanned aircraft and spacecraft, missiles, and satellite probes. As part of such systems, various types of *transducers* (devices that convert various measurable quantities into electrical signals) are used to measure such quantities as mechanical stress, cosmic-ray intensity, acceleration and deceleration, heart beat, respiration, etc. The transducer-produced electrical signal, which is proportional to the variable quantity, is then used to vary (modulate) one or more parameters of a series of pulses. The modulated pulses in turn are used to modulate the carrier of one or more transmitters before the composite signal is radiated.

The process at the ground receiving station is essentially the reverse of that which occurs in the remote vehicle. The modulated pulses are first detected from the carrier, and the desired intelligence is then detected from the modulated pulses. The output signal is essentially a reproduction of the original transducer signal and is, therefore, a reliable indication of the magnitude of the measured variable.

The purpose of this chapter is not to discuss mobile radio

Sec. 19.1 Pulse-amplitude modulation (PAM)

telemetry, as such; several excellent texts devoted to this subject are already available. Rather, we will be concerned only with various methods of pulse modulation. The reference to mobile radio telemetry is used only to illustrate the application of such techniques to a relatively new and rapidly expanding area.

19.1 Pulse-amplitude modulation (PAM)[1]

In pulse-amplitude modulation, a pulse train is amplitude-modulated (AM) by a modulating signal. The r-f carrier shown in Fig. 19-1(b)

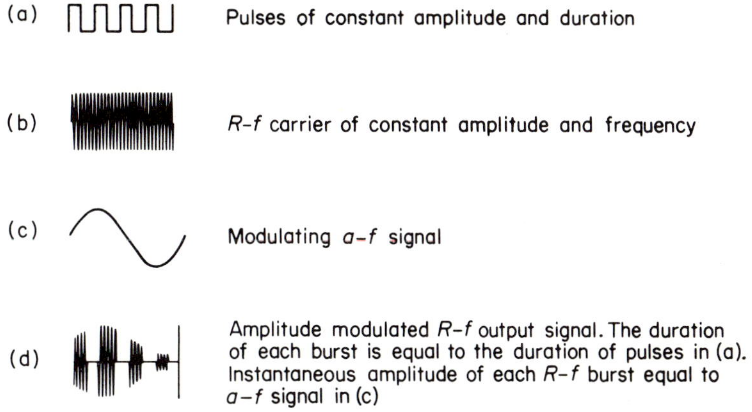

(a) Pulses of constant amplitude and duration

(b) R-f carrier of constant amplitude and frequency

(c) Modulating a-f signal

(d) Amplitude modulated R-f output signal. The duration of each burst is equal to the duration of pulses in (a). Instantaneous amplitude of each R-f burst equal to a-f signal in (c)

Figure 19-1 Illustrating PAM.

is usually in the microwave region, and the duration of the r-f pulses is of the order of one to three microseconds. The pulse-repetition rate should be at least three times the highest modulating frequency.

The major advantages of pulse-amplitude modulation (abbreviated PAM) are that PAM makes it possible to utilize a peak power much greater than that obtainable by AM, and that PAM achieves considerable improvement in the signal-to-noise ratio (SNR) of the system.

Suppose a transmitter produces r-f pulses of 1 μsec duration at a repetition rate of 20,000 pulses per second (pps). This means that one pulse is produced every 50 μsec, and the transmitter is in operation one-fiftieth of the total transmitting time. Under these conditions, the average power required is only one-fiftieth of the peak power, and the transmitter can be driven much harder than is possible for continuous operation.

The improvement in the signal-to-noise ratio results from the fact that the transmitter is in operation for only a small portion of the total time. The receiver can be silenced for comparatively long periods between pulses. Internally generated and externally introduced noises normally present during these periods are, therefore, effectively removed. The SNR is increased by a reduction of the noise output, with no loss in signal strength.

The detector circuits in the receivers of pulsed systems are somewhat different from those used in CW systems. The PAM detector maintains the output voltage at the peak value of the pulse until the next pulse arrives. This permits the output to fluctuate at the audio-frequency rate but not at the pulse-repetition rate. A PAM detector is shown in Fig. 19-2. The first input pulse charges capacitor

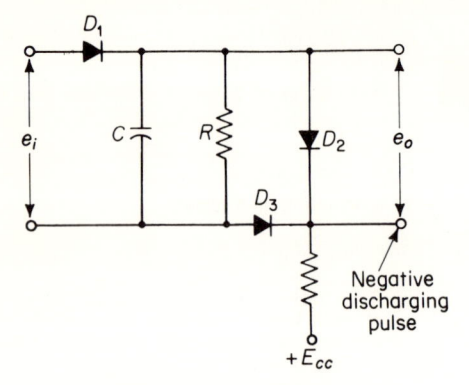

Figure 19-2 PAM Detector.

C through diode D_1 to a voltage (e_o) equal to its peak value. The discharge time constant (RC) is made at least ten times the period between pulses, so that the change in e_o between pulses is quite small. Just before the next pulse is applied to the input, a negative pulse is applied to the cathodes of D_2 and D_3, discharging capacitor C through D_2 to zero. Diode D_3, therefore, clamps e_o at or near 0 v during the discharge period.

19.2 Pulse-duration modulation (PDM)[1]

Other names used in referring to the process of pulse-duration modulation (PDM) are pulse-width modulation (PWM) and pulse-length modulation (PLM). In PDM, the width of each pulse in a train is made proportional to the instantaneous value of the modulating signal at the instant of the pulse. The leading or trailing edge of the pulse, or both, can be modulated to produce the variations in pulse width.

Sec. 19.2   *Pulse-duration modulation (PDM)*   415

Pulse-duration modulation can be obtained by a number of methods, one of which is illustrated in Fig. 19-3. A sine-wave oscillator, shown in (a) of the figure, is used to trigger a sawtooth generator (b) during each negative alternation, and thereby control the pulse-repetition frequency. The positive-going sawtooth waveform is applied

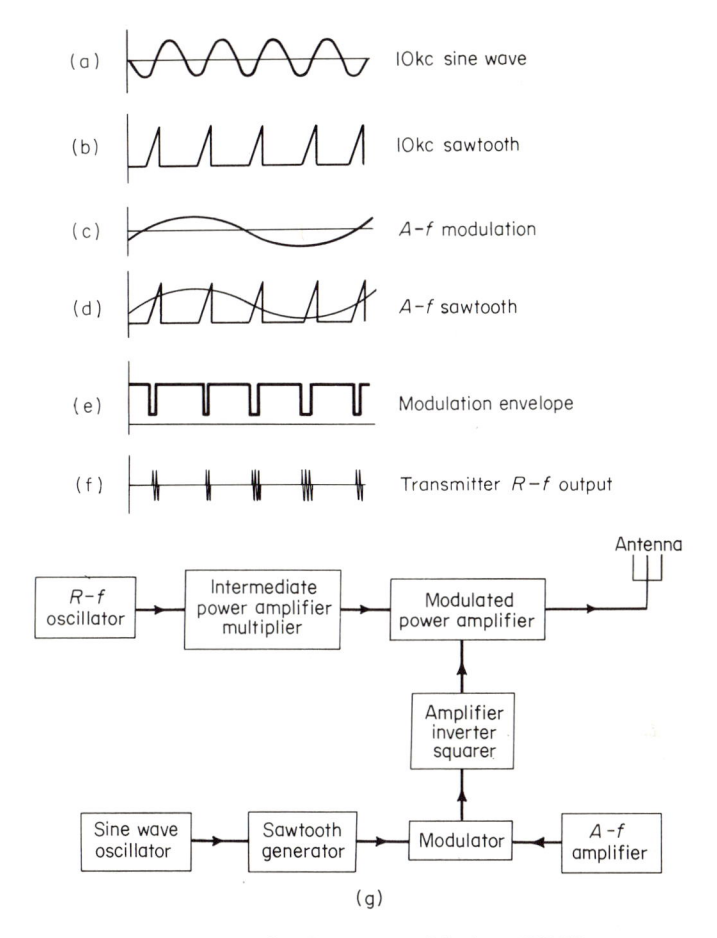

Figure 19-3 Pulse duration modulation (PDM).

to the grid of the modulator, and an audio-frequency signal (c) is applied to the cathode. When the grid of the modulator is positive with respect to its cathode, conduction occurs and an output pulse is produced. This condition is indicated in (d) as the time when the sawtooth waveform is of greater amplitude than the sinusoidal waveform. The width of the output pulse depends on the relative amplitudes of the two modulator input waveforms. Following amplification, inversion, and squaring (e), the pulses are used to modulate the r-f carrier, as shown in (f). A block diagram that serves to clarify the sequence and operation is shown in Fig. 19-3(g).

Pulse-width modulation is readily adaptable to a multi-channel form of operation known as *multiplexing*.[2] The basis of multiplexing is shown in Fig. 19-4. Information from each of three channels is multiplexed in this illustration, but more channels can be handled, the number of channels depending on the width of the modulated pulses from each channel. Synchronizing pulses, Fig. 19-4(b), are transmitted at the beginning of each modulated pulse group so that the receiver can separate the channels. Suppose that the sync pulses have a duration of 20 μsec and are transmitted at a constant rate of 10,000 per second. Under these conditions, a sync pulse occurs every 100 μsec and is separated from the succeeding pulse by 80 μsec. If the average duration of the modulated pulses is 35 μsec, there is sufficient time in this interval to transmit information from eight separate channels. The order in which the pulses from each channel are carried is shown in Figs. 19-4(c), (d), and (e), respectively. Pulses 1, 4, 7, 10, and 13 represent information from channel 1; pulses 2, 5, 8, 11, and 14, information from channel 2; and pulses 3, 6, 9, 12, and 15, information from channel 3.

In the transmitter, a sine-wave oscillator is used to trigger a sawtooth waveform generator. The output of this circuit is shown as waveform (a) in Fig. 19-4; this output is used to trigger a sqaure-wave generator that produces the sync signal, waveform (b), at the beginning of each sawtooth cycle. The sawtooth waveform also triggers a series of separate sawtooth generators, one in each information channel. These generators are all cut off in the absence of a triggering pulse, but the bias voltage on each is progressively greater than that applied to the preceding one in the series. As a result, each is triggered in turn as the amplitude of the input signal becomes adequate to overcome the bias, and the sawtooth pulses of waveforms (c), (d), and (e) occur in the order and at the instant shown. Each series of pulses may be modulated by a different audio frequency in the manner previously described.

The outputs of the individual modulators are finally combined to form a series of square pulses, and this composite waveform (not shown) modulates the transmitter r-f carrier.

In the receiver, the PWM signal is applied to separate channels and also to an integrator for separation of the sync pulses from the information channel pulses.

Sec. 19.3   Pulse-position modulation (PPM)   417

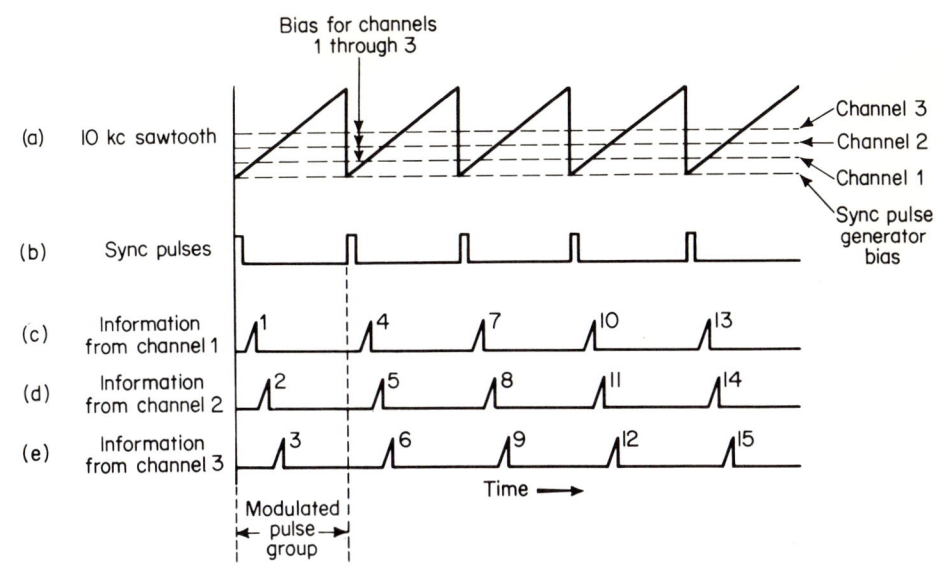

Figure 19-4 Multiplexing with PPM.

The integrator triggers a sawtooth generator, and the output of this circuit controls the action of a gate generator in each channel. The bias on each gate generator is progressively greater than on the preceding one, so that successive positive gates are produced by each during the interval between sync pulses. The amplifiers in each channel do not operate until they receive their gating pulses. Since the gating pulse in each channel occurs only when information for that channel is being received, no channel interaction occurs.

In each channel, the signal pulses pass through low-pass filters with a selected cut-off frequency, to recover the modulation frequencies. Following amplification, the recovered intelligence is then used to actuate an output device.

19.3 Pulse-position modulation (PPM)[1]

Pulse-position modulation (PPM) is also referred to as pulse-time modulation (PTM). In PPM, a modulating signal causes the position in time of a pulse to vary relative to its unmodulated time of occurrence. Refer to Fig. 19-5. The frequency of the pulse is determined by the frequency of the modulating voltage.

PPM has numerous advantages over other types of transmission that make it particularly useful in operations of a multiplex type.

Some of these advantages are: (1) The signals are very simple, consisting of short pulses of constant shape and duration with variable timing. (2) The SNR is superior to that of AM or FM under comparable conditions. (3) The cumulative amplifier distortion encountered in AM and FM systems is eliminated. (4) The total bandwidth is independent of the number of channels used. In other multiplex systems based on frequency selection, the total bandwidth required for satisfactory results increases as the number of channels employed increases.

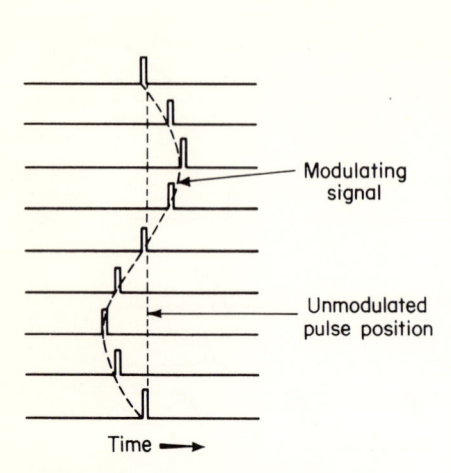

Figure 19-5 Illustrating pulse-position modulation (PPM).

Pulse-position modulation used in a typical system is illustrated in Fig. 19-6. An audio signal, shown in (b), is applied to a synchronized blocking oscillator (a) which produces a series of very narrow pulses in each of its positive and negative half-cycles. In the absence of an audio signal, the pulse-repetition frequency of the blocking oscillator (a) is constant. When audio is present, the position of the

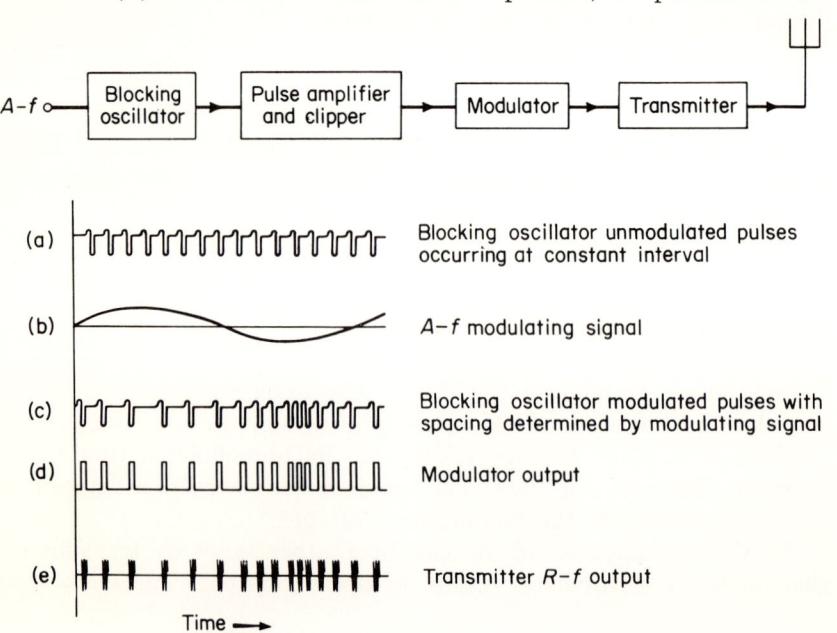

Figure 19-6 Pulse-position modulation system and wave forms.

Sec. 19.3 Pulse-position modulation (PPM)

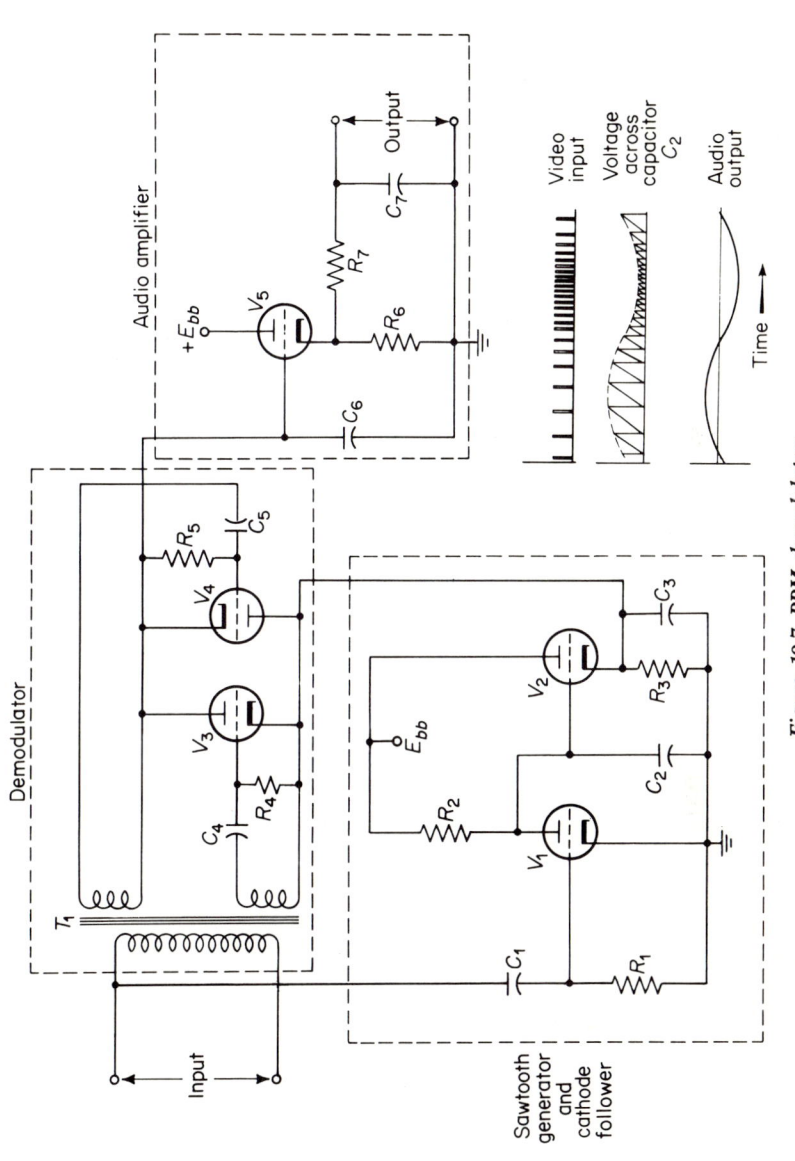

Figure 19-7 PPM demodulator.

pulses is shifted in the manner shown in Fig. 19-6(c). The position-modulated pulses are then amplified, the negative portions are clipped, and the positive portions are squared. The resulting waveform (d) modulates the transmitter to produce the output (e).

A receiver of a superheterodyne type is used for reception of the PPM signal. At the output of the i-f amplifier, the video signals [waveform (d) in Fig. 19-6] are recovered. Following several stages of video amplification, a second detector is used to recover the original audio.

The video signal is also applied to a clipper which selects a portion of the pulses occurring at the time of steepest rise and fall. This voltage is fed back to the first video amplifier to gate the receiver and to remove all noise between pulses.

A diagram of the circuit used to recover the audio intelligence is shown in 19-7. The video-input pulses are applied simultaneously to the grid of V_1 through capacitor C_1, and to V_3 and V_4 across transformer T_1. Triode V_1 is a sawtooth generator in which the the R_2C_2 time constant is the same as that of the blocking oscillator used in the transmitter. Capacitor C_2 charges exponentially through R_2 and discharges through V_1 when a video pulse is applied to the grid of that tube. The variation in the period between pulses produces a corresponding variation in the amplitude of the sawtooth voltage across C_2. This sawtooth voltage is coupled to cathode-follower V_2 and is developed across the cathode resistor and capacitor, R_3C_3.

From here, the sawtooth signal is applied through V_3 and V_4 to capacitor C_6. The video input pulses are also applied to the control grids of V_3 and V_4 as positive pulses. The resulting grid currents develop, across R_4 R_5, bias voltages of sufficient amplitude to keep both tubes cut off except for the duration of each video pulse.

At the instant that the sawtooth voltage developed across capacitor C_2 reaches its peak value, the video pulses applied to V_3 and V_4 permit either triode to conduct. Capacitor C_6 can then discharge to the same voltage as C_3. If C_3 is charged to a higher value than C_6 at this instant, the plate of V_4 becomes positive with respect to its cathode, conduction occurs, and C_6 charges to the same voltage as C_3. If the voltage on C_6 is lower than that on C_3, the cathode of V_3 is made negative with respect to its plate. Triode V_3 then conducts until both capacitors are equally charged. The two triodes cannot conduct simultaneously. Thus, capacitor C_6 sees the peak sweep voltage for a brief period during the reception of each video pulse and maintains its voltage between pulses, since V_3 and V_4 are cut off during this interval. The voltage across C_2 and C_6 are shown in the diagram. When the voltage across C_6 is filtered, the output of the detector

Sec. 19.4 Pulse-code modulation (PCM)

is a reproduction of the original audio-frequency modulating waveform.

It is interesting to note that the resistor-capacitor cathode-follower output combination, R_3C_3, also acts as a delay network. If C_3 were removed, the cathode voltage of V_2 would drop rapidly when C_2 discharged. This would occur at the same time as the video pulses applied to V_3 and V_4. As a result, C_6 would see the discharge of C_2 instead of its peak value. The insertion of C_3 in the circuit prevents this. When C_2 discharges, the grid voltage of V_2 drops swiftly. The cathode voltage cannot follow this change as rapidly, however, because C_3 is discharging through R_3. As a result, V_2 is cut off when C_2 discharges, and the peak voltage across C_3 is maintained for the required time. The charge of C_3 is not delayed, since it occurs through the lower resistance of V_2.

19.4 Pulse-code modulation (PCM)[2]

In pulse-code modulation (PCM), the audio signal is sampled and a coded group of pulses that represent the amplitude of the sample is transmitted at regular intervals. To approximate the modulating signal closely, a sufficient number of samples must be taken at frequent intervals.

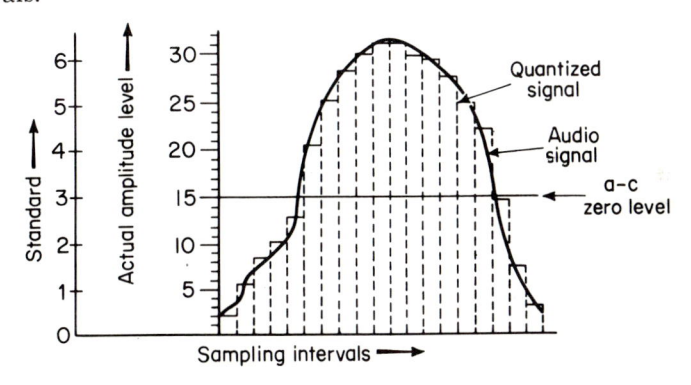

Figure 19-8 PCM-sampling and quantizing.

The principal concepts involved in PCM are illustrated in Fig. 19-8. To avoid the use of a prohibitively large number of code pulses to indicate the sample amplitudes, the signal is quantized into standard amplitudes. Actual amplitudes are then transmitted as the nearest standard amplitude. Examining the quantized signal, it is apparent that some distortion, called quantizing noise, is introduced. The percentage of distortion is minimized by increasing the number of standard amplitudes, but this necessitates the use of more pulses in each code group. In the binary system to be described below, any actual discrete amplitude level between units of 0 to 31 is transmitted by a different combination of pulses in a five-place code, in which a pulse may be present or absent in each place. If a six-place code is used, any actual amplitude between 0 and 63 can be transmitted. As the number of places in the code is increased, fidelity is improved, but the bandwidth requirements are also increased.

A PCM system in block-diagram form is shown in Fig. 19-9. At the transmitter, an audio signal is applied to the audio-frequency

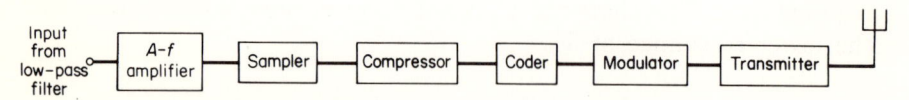

Figure 19-9 Block diagram of an PCM system.

amplifier through a low-pass filter with a cut-off frequency of about 3000 cps. The amplifier output is maintained at a level where the sampler operates most efficiently. Samples are taken at the rate of 8000 per second and quantized into 32 discrete amplitude levels. The compressor reduces the amplitude of each pulse by an amount proportional to its amplitude, so that the number of quantizing levels is reduced without increasing over-all distortion. The coder produces a group of five pulses which, by their presence or absence only, indicate the amplitude of the sample. The code groups are then combined with coded signals from as many as twelve other audio channels to produce a complex pulse train. Sync pulses are also included to permit proper operation of the receiver. The pulse train then modulates the transmitter.

The pulses, upon entering the receiver, are first sorted into the channels used to modulate the transmitter. A decoder in each channel converts the code groups into pulses of varying amplitude, which, when expanded, are used to reconstruct the audio signal from the transmitter. The high-frequency components of the pulses produced by the decoding units are removed by low-pass filters. The recovered audio then passes through the audio-frequency section of the receiver.

It has been mentioned that the arrangement described here uses the binary number system. Intelligence is conveyed only by the presence or absence of pulses at specific regular instants of time and not by variation of any specific parameter such as amplitude or position.

A comparison between the decimal and binary number systems is shown in Fig. 19-10. In the five-digit decimal group, the columns

Sec. 19.4   *Pulse-code modulation (PCM)*   423

reading from left to right are ten-thousands, thousands, hundreds, tens, and units. Reading in the same direction, therefore, represents the exponent to which the base 2 has been raised. Thus, in the decimal system, the five-digit group 11111 is 10,000 + 1,000 + 100 + 10 + 1 = 11,111, while in the binary system the same five-digit group is 1 + 2 + 4 + 8 + 16 = 31. As another example, the five-digit group 10101 in the decimal system is 10,000 + 0 + 100 + 0 + 1 = 10,101, and in the binary system is 1 + 0 + 4 + 0 + 16 = 21.

If the binary number 1 is represented by a pulse, and the binary number 0 is represented by the absence of a pulse, amplitude levels of 10, 20, and 30 in the five-pulse binary code would be represented as shown in Fig. 19-11.

A sampling circuit is shown in Fig. 19-12. Its output consists of a series of equally-spaced constant-duration pulses, whose individual amplitudes are equal to the instantaneous amplitude of the input voltage.

The input audio-frequency signal is applied across parallel triode circuits in series with load resistor R_3. The tubes are arranged to conduct on alternate half-cycles of input signal. Both are normally biased beyond cutoff as a result of grid-leak bias developed by the R_1C_1 and R_2C_2 networks. Both tubes conduct briefly 8000 times per second when a positive trigger pulse is applied through T_1. If the audio-frequency input is positive, V_1 conducts and a positive output pulse appears across R_3. If the audio-frequency input is negative, V_2 conducts and a negative output pulse appears across R_3. Regardless of its polarity, the amplitude of e_0 is nearly equal to e_i at that instant because the resistance of the tube when conducting is very low compared to the resistance of R_3.

The output pulses from the sampler are passed to a coding circuit. This circuit translates each sample pulse into a coded group of five pulses which, in the binary number system, represent the amplitude of the sample. One form of coding device is shown in Fig. 19-13.[3] It consists of a vacuum tube similar in construction to a CRT with an electron gun and deflection plates. Mounted on the face of the

Decimal	Binary	Code group
00001	10000	I
00002	01000	I
00003	11000	I I
00004	00100	I
00005	10100	I I
00006	01100	I I
00007	11100	I I I
00008	00010	I
00009	10010	I I
00010	01010	I I
00011	11010	I I I
00012	00110	I I
00013	10110	I I I
00014	01110	I I I
00015	11110	I I I I
00016	00001	I
00017	10001	I I
00018	01001	I I
00019	11001	I I I
00020	00101	I I
00021	10101	I I I
00022	01101	I I I
00023	11101	I I I I
00024	00011	I I
00025	10011	I I I
00026	01011	I I I
00027	11011	I I I I
00028	00111	I I I
00029	10111	I I I I
00030	01111	I I I I
00031	11111	I I I I I

Figure 19-10 Comparison of decimal and binary number systems.

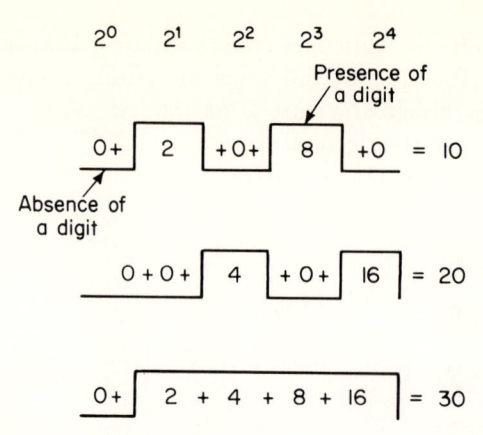

Figure 19-11 Five digit electrical code groups representing magnitudes of 10, 20, and 30.

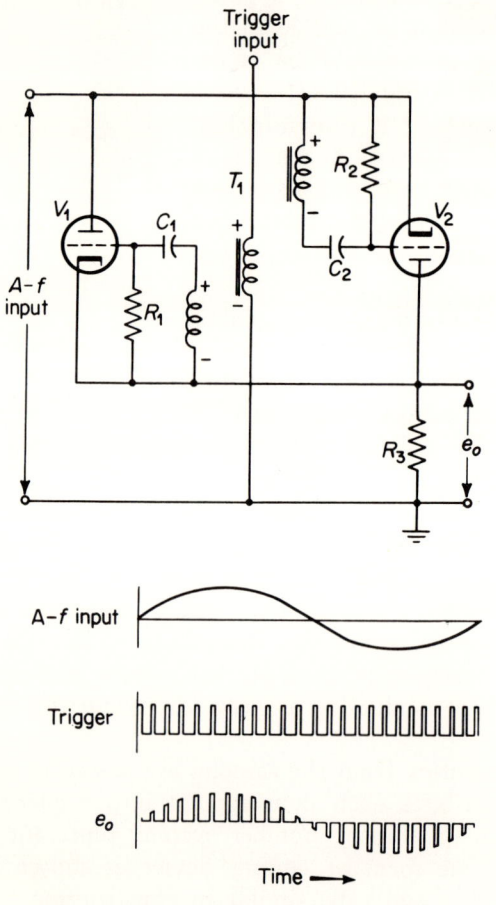

Figure 19-12 Sampling circuit for PCM.

Sec. 19.4 *Pulse-code modulation (PCM)*

tube is an impulse plate that receives the electron stream from the gun. Between the impulse plate and the electron gun is an aperture plate containing coded slots through which the electrons must pass to reach the impulse plate. Between the aperture plate and the gun is a quantizing grid framed in a rectangular metal collector.

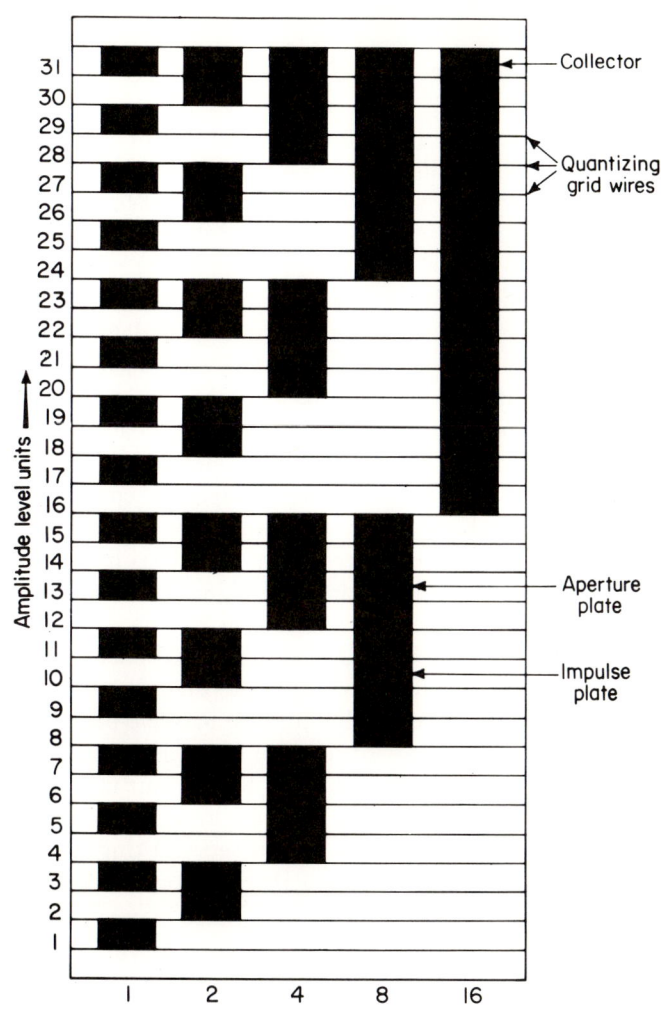

Figure 19-13 Coding device as seen from gun assembly.

The sampler output pulses are applied to the vertical deflection plates. The level of vertical deflection is held constant by the pulse during the horizontal sweep of the electron beam across the quadrature plate assembly. The beam is then cut off until the next pulse deflects it again. As the sweep occurs, electrons pass through the aperture corresponding to the vertical level and are prevented from reaching the impulse plate where it is covered by the solid portions of the aperture plate. An electrical impulse is received from the impulse plate for each aperture that occurs in the path of the electron beam as the sweep occurs. Suppose, for example, the amplitude level of the sample pulse is 20. The binary representation of this value is 00101. The pulse causes the electron beam to be deflected vertically, in Fig. 19-13, to amplitude level 20.

As the horizontal sweep takes place, from left to right, electrons pass through the apertures in columns 3 and 5, but are obtained by the aperture plate in columns 1, 2, and 4. The impulse plate, therefore, delivers a code group of space, space, pulse, space, pulse (00101). These pulses are amplified and used to modulate the transmitter in the same sequence.

The horizontal quantizing grid wires and the collector ensure that the electron beam is swept across the plates at a definite quantizing level. If the beam falls on a quantizing-grid wire, secondary emission occurs. These secondary electrons are attracted to the collector and produce a bias which moves the beam into the space between the wires. The correcting bias is provided by a feedback amplifier.

At the receiver, the five-place code groups are converted to amplitude pulses from which the audio signal is reconstructed. The device that accomplishes this is called a decoder; it consists of a simple parallel R-C network to which the signal is applied. The final output voltage is taken across the capacitor at the end of each code group, and the voltage depends upon the number of pulses in the group, the time at which each is received, and the amount of voltage lost by discharge through the resistor. Each pulse in a five-place code group contributes an equal amount of charge (in this case, 32 units) to the capacitor. The time constant of the R-C circuit is chosen so that the capacitor voltage drops 50 per cent during the interval between pulses. Thus, when the binary code group representing 20 is applied, zero units of charge are applied during the first two time divisions, t_1 t_2. At the third time division, however, a pulse is present and applies 32 units of charge to the capacitor. At time interval t_4, no pulse is present to add charge to the capacitor which has now lost 16 units of charge. The decrease in capacitor charge continues until time interval t_5, when a new pulse adds 32 units to the 8 remaining. This makes a total of 40. The final output occurs at time interval t_6 and is $40/2 = 20$.

In describing the quantizing process, quantizing noise was mentioned. If the number of quantizing levels is increased, distortion is decreased, but only at the cost of requiring more pulses in each code group. Because the response of the ear to sound intensity is logarithmic, the quantizing voltage steps can be increased for signals of large amplitude and decreased for signals of small amplitude without loss of fidelity. A compressor performs this operation. It increases the quantizing (or absolute) error for signals of large amplitude, but maintains the fractional quantitizing error (ratio of intrinsic) constant for a wide range of amplitude variations.

The response of the ear to sound is logarithmic rather than linear. Thus, if there are three sound intensities of relative amplitudes 10 and 100 and 1000, the ear will perceive the same difference between 10 and 100 as it does between 100 and 1000. On an absolute or linear intensity scale, the difference between 10 and 100 is much smaller than the difference between 100 and 1000, but the ear is unable to distinguish this difference and treats each increase in intensity in the ratio of ten to one.

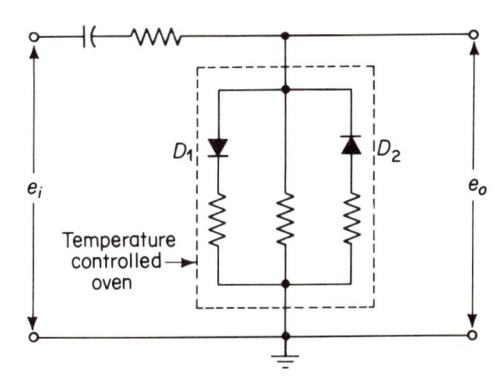

Figure 19-14 A Compressor.

An instantaneous compressor is shown in Fig. 19-14. Diodes D_1 and D_2 provide nonlinear paths in parallel with the output. They are arranged to conduct for opposite polarities of input signal. The diodes have a short-circuiting effect on the signal, which increases as the amplitude of the signal increases. Thus, steps of equal size after the compressor correspond to steps of varying size before the compressor.

An expander, located in the receiver, reverses the effect of the compressor. The expander is simply an amplifier with a compressor circuit connected in a negative-feedback path.

REFERENCES

1. Moskowitz, Sidney, and Joseph Racker, *Pulse Techniques*. Englewood Cliffs, N. J.: Prentice-Hall, Inc., 1951.

2. Rideout, Vincent C., *Active Networks*. Englewood Cliffs, N. J.: Prentice-Hall, Inc., 1951.
3. Sears, R. W., "Beam Deflection Tube for Coding in PCM," *Bell Labs Record*, Oct. 1948.

EXERCISES

19-1 What are the advantages of PAM compared to AM? Explain.

19-2 What is meant by PAM? PWM? PCM? PPM?

19-3 What advantages of PPM make it particularly useful for multiplexing?

19-4 What is meant by the quantizing process, used in PCM?

19-5 What is quantizing noise?

19-6 How would you represent 36 in the binary numbering system?

19-7 What is the function of a decoder in a PCM system?

19-8 What is the purpose of a compressor in a PCM system? An expander?

19-9 Draw the schematic diagram of a typical sampling circuit.

19-10 Describe a PAM detector.

20

ELECTROMAGNETIC DELAY LINES

An electromagnetic delay line is essentially a transmission line that, through circuitry, has been compressed. Its function is to delay information for a specific length of time, usually in the microsecond range. For longer delays, *acoustic delay* lines and *storage devices* are used.

Among the common applications for electromagnetic delay lines are air navigation (avigation) systems, such as DME (distance-measuring equipment), using electrical information related to time. A cathode-ray oscilloscope for observing fast waveforms may also use an electromagnetic delay line to delay the input signal to the vertical plates until the sweep is started. If the delay line were not included, the initial portion of the waveform might not be visible on the scope face.

Electromagnetic delay lines may be classified as either *distributed constant* or *lumped constant* delay lines. The distributed-constant line more closely approaches a transmission line, whereas the lumped-constant line more closely approaches a filter. Thus, it should prove helpful to begin with a brief review of transmission lines.

20.1 Transmission lines—general

A transmission line may be a conductor which serves to guide energy from one point (input) to another (output). The line may consist of a straight length of wire. Any linear conductor possesses both inductance and resistance. In a two-wire transmission line, the two wires are separated by a dielectric, and the property of capacitance is also present. Because no capacitor is perfect, a dielectric leakage resistance must also be considered. The equivalent circuit for a short length of line, therefore, must include all of these elements, and it is represented as shown in Fig. 20-1, with resistance and inductive reactance in series, and with resistance and capacitive reactance in parallel. A long line may be considered as made up of a number of similar sections distributed along the length of the line.

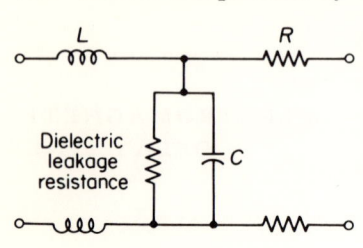

Figure 20-1 Equivalent circuit for a short length of line.

Because a transmission line contains both resistive and reactive elements, it offers an impedance to any signal it carries. This impedance is called the *characteristic impedance* (also the surge impedance), and is designated Z_o. Mathematically

$$Z_o = \sqrt{\frac{L}{C}} \qquad (20\text{-}1)$$

where Z_o = characteristic impedance in ohms

L = inductance in henrys

C = capacitance in farads

The velocity, v, of an electromagnetic wave changes whenever there is a change in the electric or magnetic properties of the medium in which it travels, but the frequency, f, does not change. The wavelength, λ, therefore, increases or decreases in accordance with

$$v = f\lambda \qquad (20\text{-}2)$$

Thus, the velocity, v, of a wavefront moving along a transmission line is dependent on $1/(LC)^{1/2}$ only to the extent that the characteristic wavelength of the transmission line changes.

A wavefront, moving along an infinite line, is illustrated in Fig. 20-2. Beginning at the generator, the wavefront requires a definite length of time, T, to reach any designated point along the line. For a given characteristic impedance and time delay, we can first rewrite Eq. 20-1 either as

Sec. 20.1   Transmission lines—general

$$Z_o = \frac{\sqrt{L}}{\sqrt{C}} \quad (20\text{-}3)$$

or as

$$\sqrt{L} = Z_o\sqrt{C} \quad (20\text{-}4)$$

The equation for T is

$$T = \sqrt{LC} \quad (20\text{-}5)$$

which may be written

$$T = \sqrt{L}\sqrt{C} \quad (20\text{-}6)$$

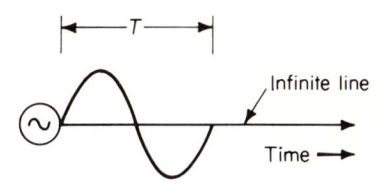

Figure 20-2 *A wavefront moving along an infinite line requires a finite period of time, T, to travel from the generator to a specified point on the line.*

and transposed to

$$\sqrt{L} = \frac{T}{\sqrt{C}} \quad (20\text{-}7)$$

Now, if we substitute $T/\sqrt{C}$ from Eq. 20-7 into Eq. 20-4, we obtain

$$\frac{T}{\sqrt{C}} = Z_o\sqrt{C} \quad (20\text{-}8)$$

and

$$T = Z_o\sqrt{C}\sqrt{C} \quad (20\text{-}9)$$

Thus

$$T = Z_o C \quad (20\text{-}10)$$

and

$$C = \frac{T}{Z_o} \quad (20\text{-}11)$$

Similarly, it can be shown that

$$\sqrt{C} = \frac{\sqrt{L}}{Z_o} \quad (20\text{-}12)$$

and

$$\sqrt{C} = \frac{T}{\sqrt{L}} \quad (20\text{-}13)$$

Then

$$\frac{\sqrt{L}}{Z_o} = \frac{T}{\sqrt{L}} \quad (20\text{-}14)$$

and

$$T = \frac{L}{Z_o} \quad (20\text{-}15)$$

or

$$L = TZ_o \quad (20\text{-}16)$$

In all of the equations above, T is in seconds, L is in henrys, and C is in farads.

If an impedance equal to Z_o is used to terminate a transmission line, or if the line is infinitely long, all energy traveling down the line is absorbed by the load and no energy is reflected back to the source. If the terminating impedance is not equal to Z_o, part of the energy is reflected. If the far end of the line is an open or short circuit, there is effectively no termination, and all energy traveling down the line is reflected toward the source. The amplitude of the signal on the transmission line is then the algebraic sum of the reflected and incident waves. This produces standing waves of voltage and current on the line.

By taking advantage of the standing wave patterns, a section of transmission line can be made to act as a circuit element. When the line is not terminated in an impedance equal to Z_o, the input impedance is subject to wide variation along its length. Thus a transmission line can be made to act as an inductor, a capacitor, a resistor, or a combination of these components.

Figure 20-3 illustrates the relationship between voltage, current, and impedance for various lengths of open-ended resonant lines. The impedance which the generator sees is shown at the top of the figure. The impedance curve indicates the relative values of impedance presented to the generator as it is moved from right to left along the line. The impedance values are derived from the very simple principle that the impedance at any point is equal to the voltage at that point divided by the current at the same point. Thus, if the standing waves of voltage and current indicate that, at a point, the voltage is high and the current low, the line at that point has a high impedance.

The reactance curve shows the reactance relationship along the sections and indicates whether the reactance at any point is purely resistive. A comparison of the reactance curve with the impedance curve also shows the points where the impedance of the line is purely resistive. This is because the reactances either have cancelled themselves out or are zero. The circuit symbols indicate the conventional circuit that is equivalent to a resonant line of that particular length.

The one-quarter wavelength section exhibits the characteristics of a conventional series-resonant circuit with high current, low voltage, and an impedance minimum. Thus, the generator looking into a quarter-wave open-ended section sees a very low impedance or, in effect, a short circuit. Similarly, the three-quarters wavelength section presents the same load to the generator. In general, an open-ended resonant-line section that is an odd number of quarter-wavelengths long exhibits the characteristics of a conventional series-resonant circuit and therefore looks to the generator like a short circuit.

The open-ended resonant section may also be used to act as nearly pure capacitance or inductance. Fig. 20-3 shows that an open-ended line section less than one-quarter wavelength long acts as a capacitance; also, a section from one-quarter to one-half wavelength long acts as an inductance; and so on. In addition, a one-eighth wave-

Sec. 20.1   Transmission lines—general   433

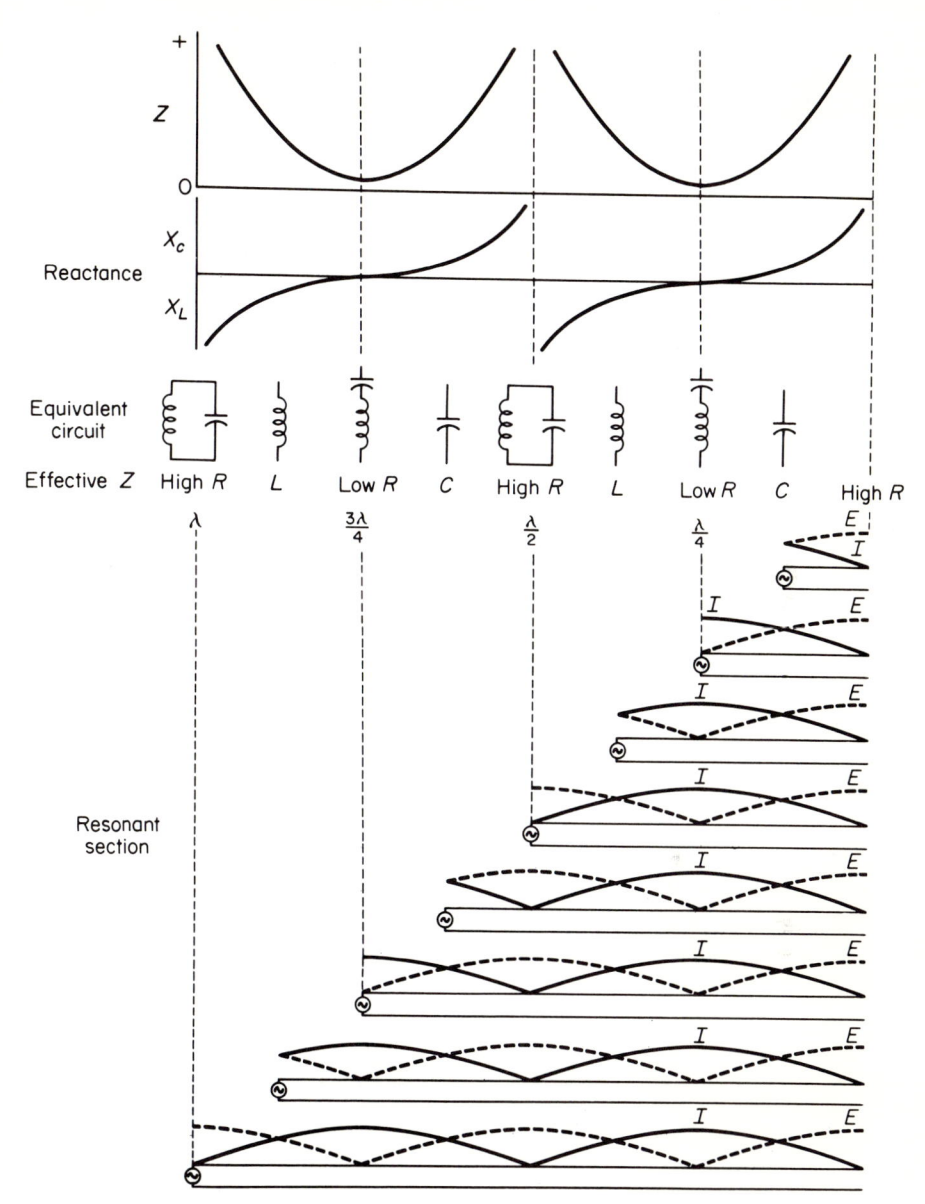

Figure 20-3 Relationships between E, I, and Z for various lengths of an open resonant line.

length open-ended section acts as a capacitive reactance numerically equal to the characteristic impedance Z_o, and a three-eighths wavelength section acts as an inductive reactance numerically equal to Z_o.

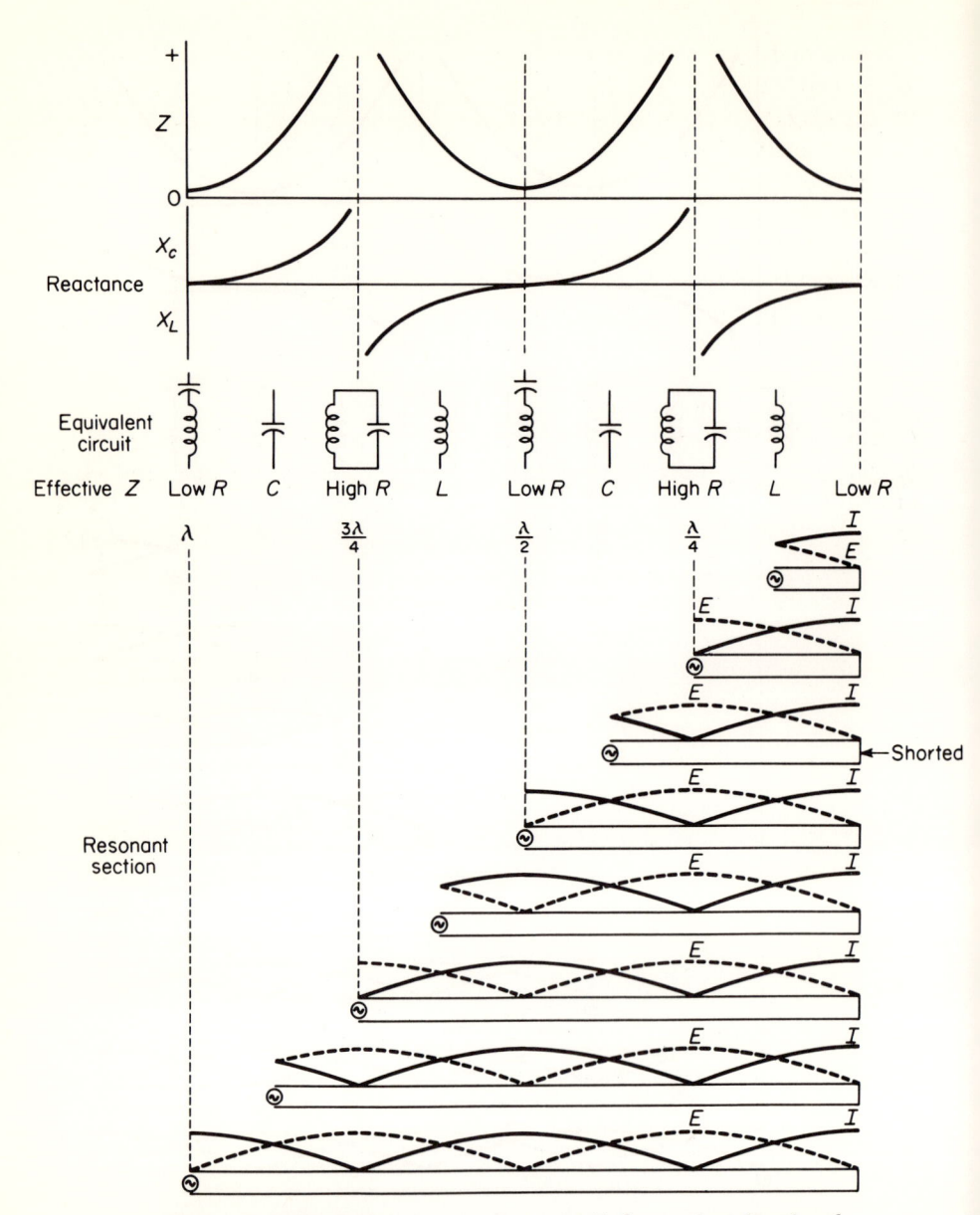

Figure 20-4 Relationship between E, I, and Z for various lengths of a short-circuited line.

Figure 20-4 illustrates the relationship between voltage, current, and impedance for various sections of a short-circuited resonant line.

Sec. 20.1   *Transmission lines—general*   435

The one-quarter wavelength section exhibits the characteristics of a conventional parallel-resonant circuit with low current, high voltage, and an impedance maximum. Thus, the generator looking into a quarter-wave short-circuit section sees a very high impedance or, in effect, an open circuit. Similarly, the three-quarters wavelength section presents the same load to the generator. In general, a short-circuited reasonant-line section that is an odd number of quarter-wavelengths long exhibits characteristics of a parallel-resonant circuit and therefore looks to the generator like an open circuit.

The one-half wavelength section exhibits the characteristics of a conventional series-resonant circuit with high current, low voltage, and a minimum impedance. Thus, the generator, looking at a one-half wavelength short-circuited section, sees a very low impedance or, in effect, a short circuit. Similarly, the full-wavelength section presents the same load to the generator. In general, a short-circuited resonant-line section which is an even number of quarter-wavelengths

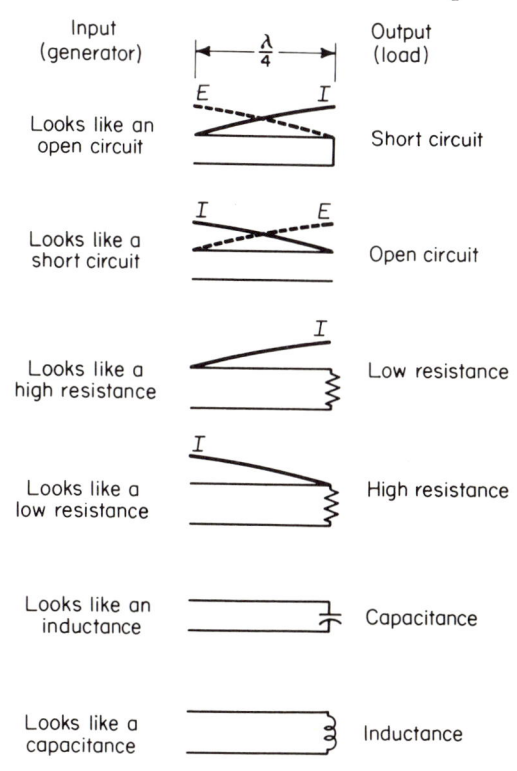

Figure 20-5 Impedance presented to a generator by a quarter-wave section with different terminations.

long exhibits the characteristics of a conventional series-resonant circuit and therefore looks to the generator like a short circuit.

In addition to acting as conventional resonant circuits, resonant short-circuited lines may also be used to act as nearly pure capacitance and inductance. For example, a short-circuit line less than one-quarter wavelength long acts as an inductance, from one-quarter to one-half wavelength as a capacitance, and so on, as indicated in Fig. 20-4. A one-eighth wavelength short-circuited line section acts as an inductive reactance equal to the characteristic impedance Z_o.

It has been seen from Figs. 20-3 and 20-4 that in the case of the quarter-wave section, different terminations place different loads on the generator. The impedance presented to a generator by a

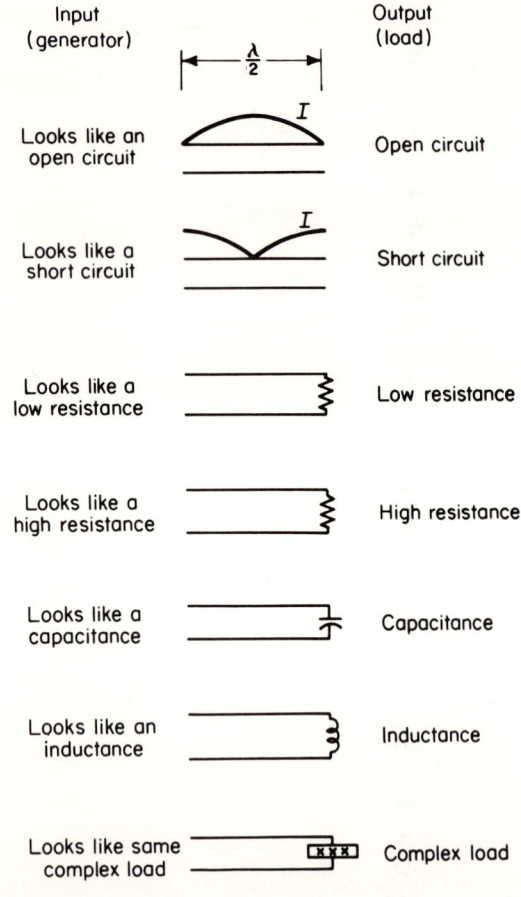

Figure 20-6 Impedance presented to a generator by a half-wave section with different terminations.

quarter-wave section with different terminations is shown in Fig. 20-5. In each case it is important to note that the quarter-wave section always inverts its load: that is, a short circuit appears as an open circuit to the generator; a capacitance appears as an inductance; a low resistance looks like a high resistance. Further, any line which is an odd number of quarter-wavelengths long ($\lambda/3$, $3\lambda/4$, etc.) acts like a quarter-wave section.

It has also been shown that an open-ended half-wave section presents a high impedance to the generator, and a short-circuited half-wave line presents a low impedance to the generator. In effect, the important fact is that a half-wave section always presents to its generator a load equal to its termination. In other words, a half-wave section repeats its load. The impedance presented to a generator by half-wave sections with different terminations is shown in Fig. 20-6. It follows that a line any multiple of a half-wavelength long ($\lambda/2$, λ, $3\lambda/2$, etc.) also repeats its load.

20.2 Distributed-parameter delay lines[1]

From what has been said, it may appear that we would only have to use a length of uniform lossless transmission line terminated in its characteristic impedance to obtain a required delay. Unfortunately, the solution is not this simple, for in order to achieve a delay of one microsecond, we would have to use a line about 200 miles long. To increase the delay time, T, certain constructional modifications are required.

A simple form of distributed-parameter delay line is derived from a coaxial cable or a parallel line by changing one conductor into a long thin coil; the resulting increase in inductance increases both the impedance and the delay. Closer spacing of the two conductors increases capacitance; this also increases delay but reduces the impedance.

The inductance of a long cylindrical coil at low frequencies is

$$L_o = \frac{10^{-9}\pi^2 d^2}{w}, \text{ henrys per centimeter} \qquad (20\text{-}17)$$

where w = pitch in centimeters

d = average diameter in centimeters

At higher frequencies, currents in different turns along a delay line, although still magnetically linked, are less and less in phase with each other, and less coupling occurs between them. The separation of two turns having a given phase difference decreases in proportion to their frequency; their mutual inductance ($M = K\sqrt{L_iL_2}$) thus drops and later reverses. This effect manifests itself as a steady drop in the effective inductance, L, of the winding, as shown in Fig. 20-7. The ordinate is L/L_o, that is, the effective inductance L compared with the low-frequency inductance L_o. The abscissa is

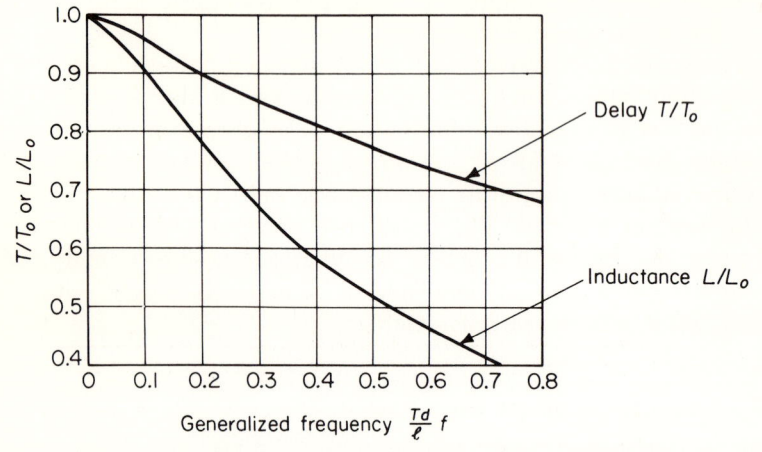

Figure 20-7 Effective inductance and time delay vs frequency.

proportional to frequency. The drop in effective inductance at a given frequency will increase as the phase difference between any two points increases; that is, in proportion to the delay T per unit length of line. The drop will also increase as the coupling between two given turns becomes stronger; that is, proportionally to the diameter d of the turn and inversely proportional to the separation l. Thus, the generalized frequency scale of Fig. 20-7 is presented in units of $(Td/l)f$.

Because the distributed capacitance, C, of a delay line varies widely with the geometry of the design and the dielectric properties of the spacer, it is not easily computed. The capacitance between conductors does not vary significantly with frequency.

Because inductance decreases as frequency increases, the delay, T, of a simple delay line is a function of frequency, decreasing steadily as plotted in Fig. 20-7. This decrease in T shows up as distortion, the phases of the high-frequency components in a signal being advanced relative to the phase of the low-frequency components. Phase distortion severely affects the shape of pulses, but it can be held within acceptable limits by conservative design.

Signals are also distorted by the attenuation in the line. If the attenuation at all frequencies is equal, it is not considered distortion and is harmful only in applications such as pulse-forming circuits where the absolute rather than the relative steepness of the echo front (front of the returning wave) is important. Because of increased attenuation of the high-frequency components, pulses that were originally square are rounded as they are in a system with too narrow a pass band.

Attenuation in delay lines is a result of the resistance of the conductor and of the dielectric losses between conductors. Attenuation of high-frequency components is also caused by mismatch at the ends of a terminated line. The impedance of a line does not stay exactly constant over its whole useful range, but, because of the drop in effective inductance and the unavoidable load capacitance, it usually changes slightly at the higher frequencies. At lower frequencies for which the delay line is accurately matched, all power is delivered to the load. At the higher frequencies at which the line is mismatched, a fraction of the power is reflected at the termination, partly reflected again at the input terminal, and then appears at the output terminal as a short pip after each transition, followed by twice the delay time of the line. This is illustrated in Fig. 20-8.

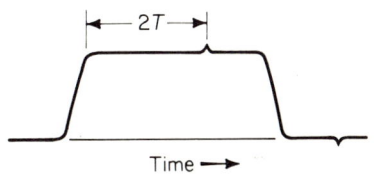

Figure 20-8 Effect of mismatch at high frequencies.

Another appreciable mismatch occurs within the line just before the end. In this region, the inductance per unit length of the line changes because each turn is coupled to fewer and fewer other turns. Flaring of the ends, or insertion of conical pieces of iron-dust core, may be used as a means of compensation for manufactured lines. Another means of compensation is the simple expedient of cutting the line into small pieces. Thus the line may be wound in many equal sections, each of which produces an end echo; the sections, however, are so short that the echoes form a ripple of invisibly high frequency.

Since delay lines are normally used for step functions and square waves rather than sine waves, it is convenient to express the line in terms of the rise time (at the output) of a perfect step function introduced at the input. The relationship between F (the frequency response at the 3 db points) and rise time (t_r) is expressed by the equation

$$F = \frac{0.36}{t_r} \tag{20-18}$$

Modern distributed-constant delay lines are made by winding a continuous coil (usually multi-layer) on a glass or ceramic rod that has been coated with a conductive material, such as silver, to provide a ground conductor pattern. This rod is covered with a thin layer of dielectric such as Polystyrene, Teflon, or Mylar. The coil provides a continuous inductance along the rod, and the capacitance is obtained through the coil wire and the ground plane. See Fig. 20-9.

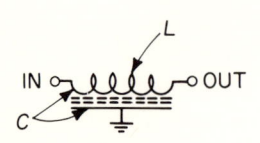

Figure 20-9 Distributed-constant delay line produces a continuous inductance along the rod and the capacitance, C, is obtained through the coil wire and ground plane.

Distributed-parameter lines are limited in their delay-to-rise time ratio and generally do not exceed a delay-to-rise time ratio of 10 to 1. Delay lines also have a temperature coefficient, usually expressed as a percentage change in delay per degree Centigrade or as parts per million per degree Centigrade (PPM/°C). The distributed-delay lines are poor in respect to temperature, having a coefficient of delay of 150 PPM/°C normally, and 100 PPM/°C at best.

To attain a 10-to-1 delay-to-rise time ratio, a distributed-parameter delay line winding must be at least six inches long. Delays of up to five microseconds can be achieved with one six-inch winding. Longer delays in shorter windings can be attained, but at the sacrifice of the delay-to-rise time ratio. Distributed-parameter delay lines may be cascaded to obtain longer delays. Not all impedances are possible for every delay because of the limitation of the amount of capacitance that can be attained from the area of the winding.

The vast majority of delay lines presently manufactured are built to meet military specifications and are generally epoxy-encapsulated.

20.3 Lumped-parameter delay lines

For very low impedances or for very high voltages, it is more convenient to lump the ground capacitances than to distribute them along the winding. Also, when suitable capacitors are chosen, the attenuation, due largely to dielectric losses, is appreciably lower in lumped-parameter than in distributed-parameter delay lines, and the temperature coefficient can be made as low as about 5 PPM/°C.

The lumped-parameter delay line in its simplest form consists of a number of inductors and capacitors all similar in value. See Fig. 20-10. The inductors are connected in series, and the capacitors are connected through the junction between the inductors to the ground lead. The number of inductor-capacitor sections required to obtain a particular frequency response or delay-to-rise time ratio is represented by the empirical equation

$$n = \left(\frac{T}{t_r}\right)^{2/3} \quad (20\text{-}19)$$

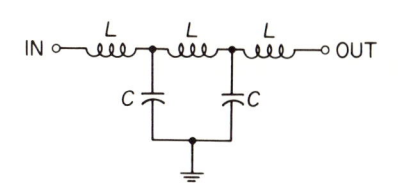

Figure 20-10 Lumped-parameter delay line.

This equation holds fairly well for ratios up to 50 to 1. Higher ratios are obtainable in certain cases, but this is difficult because of the required very high "Q" for the coils. Complex sections, special iron cores, and novel construction are used to reduce to a basic minimum the number of sections required.

Lumped-parameter delay lines can be tapped at several points to give a series of delays, and this feature is very useful in many applications.

Ripple is a function of impedance-matching within the delay line, the number of sections of the line, and the external effects of the casing and construction on the delay line. Ripple as little as 0.1 per cent can be achieved by careful construction.

Normally, lumped-parameter delay lines are made in small packages consistent with the use of existing components which are available at low cost. In some cases, however, sub-miniature components are used. In practically all cases, they are resin-encapsulated with a foam material and hermetically sealed in metal containers.

Lumped-parameter delay lines typically have delays from about 0.01 μsec to 100,000 μsec, delay-to-rise time ratios up to 170 to 1, and impedances from 20 ohms to 20,000 ohms.

20.4 Delay-line applications[2,3]

Before describing several typical applications of electromagnetic delay lines, reference is made to Fig. 20-11, which shows both pictorial and schematic drawings of the distributed-parameter type in (a) and the lumped-parameter type in (b).

A common application of lumped-parameter delay lines is in the driver stages of radar systems. A typical arrangement will now be discussed.

Multivibrators are used extensively in radar to generate pulses at low power levels. For microwave radar systems, the biased multi-

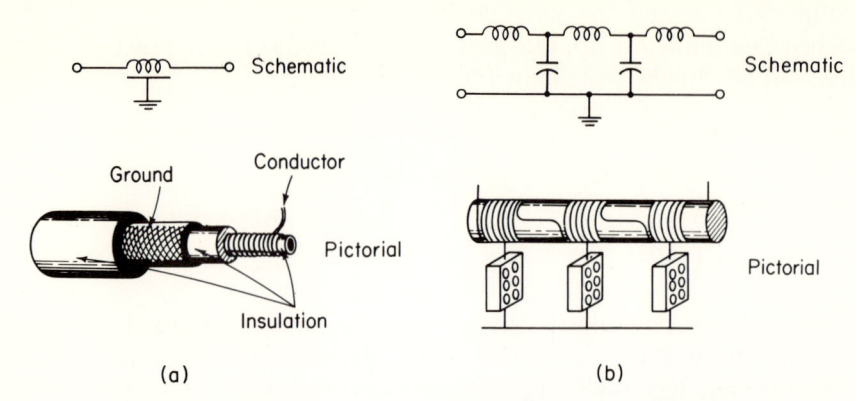

Figure 20-11 Schematic and pictorial drawings of (a) distributed-parameter delay line and (b) lumped-parameter delay line.

vibrator has been adopted for use as the pulse generator in the driver of a hard-tube pulser. This circuit is suitable for driver applications because it has a single stable state. It is therefore possible to obtain an output pulse from a biased multivibrator only when the proper triggering impulse is applied to one of the tubes, and the length of the interpulse intervals can be determined by a timing circuit that is independent of the pulser.

It was noted, in the chapter dealing with biased multivibrators, that the shape and duration of the output pulse depends on the load and on the value of the circuit elements and tube voltages. In practice, some variations occur in these parameters, and allowances must be made to accommodate a reasonable range of values. The most serious aspect of this characteristic of the multivibrator in the driver application is the possible variation of pulse duration. This difficulty may be avoided by connecting a delay line between the output of the driver and the input of the amplifier following the multivibrator, to control the duration of the pulse.

The block diagram of the driver, shown in Fig. 20-12, indicates

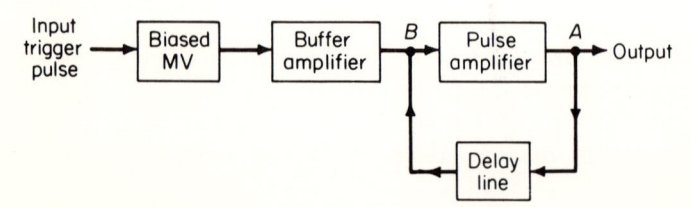

Figure 20-12 Block diagram of driver for hard-tube pulse using an MV pulse generator.

the way in which the pulse duration is determined. The multivibrator pulse generator is constructed so that the pulse fed into the amplifier at B is of longer duration than the pulse desired at the output of the driver. When the leading edge of the pulse reaches A, a voltage wave starts to move from A to B through the delay line. This voltage wave reaches B after a time determined by the constants of the delay line. If the pulse fed into the amplifier at B is positive and the pulse voltage at A is negative, it is possible to neutralize the voltage at the amplifier input, and thereby to terminate the pulse appearing at A. The pulse duration at the driver output is thus fixed by the time it takes a voltage wave to travel the length of the delay line, independent of the multivibrator output. This arrangement has been expressively called the *tail-biting* circuit.

To ensure that only one pulse appears at the driver output for each trigger pulse, the output pulse of the multivibrator must not last longer than twice the transit time of the delay line. For pulse durations of about a microsecond, this latitude for the multivibrator output is ample. When the desired pulse duration is very short, however, more care must be exercised in the design of the circuit. The buffer amplifier is introduced to minimize the effect of loading on the output pulse shape and pulse duration obtained from the multivibrator. If the amplifier is carefully designed, the shape of the pulse at the driver output can be considerably better than that fed into the amplifier, and the dependence of driver output on multivibrator output is further reduced.

A simplified schematic diagram for this driver circuit is shown in Fig. 20-13. The biased multivibrator delivers a negative pulse to the grid of the normally *on* buffer amplifier V_3. A positive pulse is then obtained at the grid of pulse amplifier V_4. The negative pulse that appears at the plate of V_4 is inverted by means of a pulse transformer, and the resulting positive pulse is applied to the grids of V_5 and V_6 in parallel. The negative pulse that appears at the plates of these tubes is also inverted by a pulse transformer to give a positive pulse at the grids of the parallel-connected pulse tubes. A part of the negative pulse at A is impressed across the end of the delay line by means of voltage divider consisting of R_1 and R_2. After traversing the delay line, this negative pulse appears at B with an amplitude sufficient to neutralize the positive pulse output from V_3. This negative pulse lasts for a time corresponding to the pulse duration at A; and, if this time plus the delay time is greater than the duration of the pulse from the multivibrator, the bias voltage maintains V_4 nonconducting when the pulse is over.

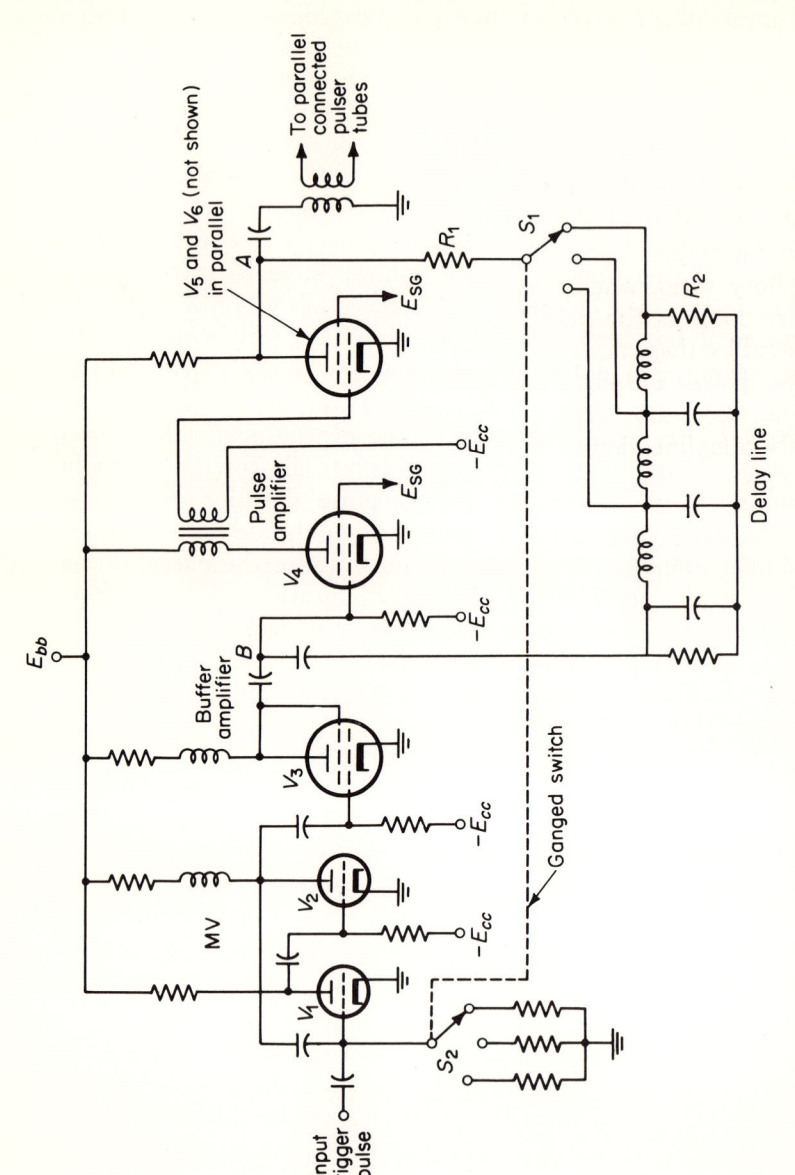

Figure 20-13 Simplified schematic diagram of driver for hard-tube pulser using an MV pulse generator and a delay line to determine pulse duration.

Sec. 20.4 Delay-line applications

The pulse duration from this driver may be varied by changing the length of the delay line between points A and B. Switch S_1 in Fig. 20-13 is provided to facilitate this change. Switch S_2 is mechanically coupled with S_1 to change the pulse duration at the multivibrator output at the same time that the length of the delay line is changed. The constants in the multivibrator circuit are chosen so that the pulse delivered to the amplifier is about 25 to 40 per cent longer than the pulse duration desired from the driver.

The operation of this driver requires V_1 and V_3 to be conducting during the interpulse interval. When the pulse is being operated at a low-duty ratio, where the interpulse interval is about a thousand times as long as the pulse interval, it is desirable to have the power dissipated by these normally *on* tubes as small as possible. The multivibrator power output also should be kept small for the same reason, and the necessary output power from the driver must be provided by pulse amplification.

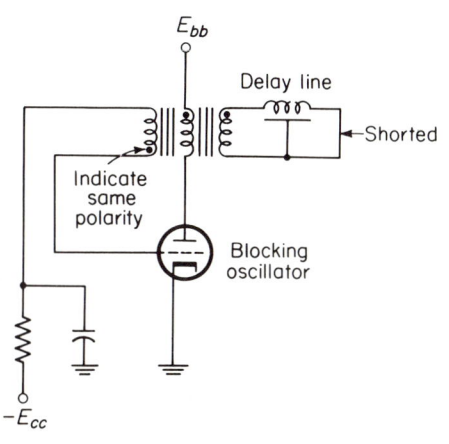

Figure 20-14 *The shorted delay line controls the pulse width of the blocking oscillator.*

An application of distributed-parameter delay line is illustrated in Fig. 20-14. In this circuit, the short-circuited delay line controls the pulse width of the blocking oscillator. When a triggering pulse is applied to the blocking oscillator, a positive step is produced at the input to the delay line. This step travels down the line and, upon reaching the short-circuited end, is reversed in polarity. The reflected wave arriving back at the input to the line produces a positive step at the plate of the tube. This initiates the regenerative switching action and, very quickly, the blocking oscillator is driven *off*. The

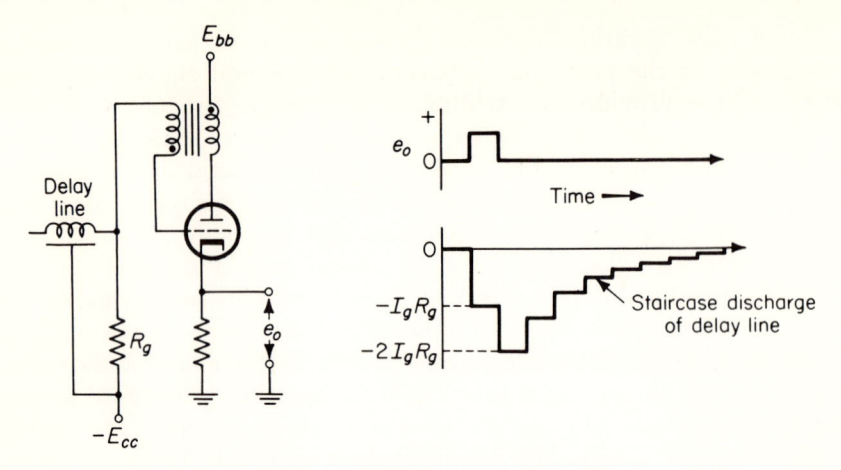

Figure 20-15 Alternate method of using delay line to control pulse width of blocking oscillator.

width of the output pulse will be $2T$, provided that the natural width is greater than $2T$.

Alternatively, an open-circuited delay line can be connected in the grid circuit in place of capacitor C_g. This arrangement is shown in Fig. 20-15. When the blocking oscillator is triggered, a negative voltage step, $I_g R_g$, is produced by current through resistor R_g. This step travels down the line, is reflected without change in polarity, and reaches the input at time $2T$. Neglecting attenuation, when the reflected step reaches the input, the line voltage becomes $-2I_g R_g$ and initiates the switching action which drives the blocking oscillator *off*. The line discharges in a staircase manner to zero. The waveforms shown in Fig. 20-15 will help to clarify the operation.

REFERENCES

1. Blackburn, J.F. (ed.), *Components Handbook.* New York: McGraw-Hill Book Co., Inc., 1949.
2. Glasoe G. N., and J. V. Lebacqz, *Pulse Generators.* New York: McGraw-Hill Book Co., Inc., 1948.
3. Benjamin, R., "Blocking Oscillators," *J. I. Elec. Engrs* (*London*), v. 93, pt. IIIA. No. 7, 1946.

EXERCISES

20-1 The type RG-65U cable has the following parameters: $Z_o = 1100$ ohms and $T = 0.14$ μ sec/meter. Determine L and C.

Exercises 447

20-2 What impedance does the generator see looking into a quarter-wave open-end section? Explain.

20-3 What impedance does the generator see looking into a quarter-wave shorted-end section? Explain.

20-4 Explain three causes of distortion in transmission lines.

20-5 Describe the construction of a modern distributed-parameter delay line.

20-6 What is meant by the temperature coefficient of a delay line?

20-7 Explain how a distributed-parameter delay line may be used to control the pulse width of a blocking oscillator.

20-8 Show and explain an alternate circuit for Prob. 20-7.

20-9 What factors determine ripple on a lumped-parameter line?

20-10 Describe the "tail-biting" circuit.

21

CONCLUSION

Having completed the study of individual pulse circuits, the reader should be interested in examining a system or device that utilizes pulse circuits of all of the general types. With this thought in mind, the final chapter is devoted to a discussion of the Hewlett-Packard Model 170A Oscilloscope, a general-purpose, high-speed, laboratory-type instrument, which has a vertical-channel passband capability of d-c to 30 megacycles, and its associated plug-in units.

21.1 The oscilloscope—general

Figure 21-1 is an over-all block diagram of the oscilloscope and, as shown, the main functional sections are: (1) a plug-in vertical amplifier, (2) a main vertical amplifier, (3) a sweep generator, (4) a horizontal amplifier, (5) a high-voltage power supply, (6) a calibrator, and (7) low-voltage power supplies.

The plug-in vertical amplifier receives the vertical input signals, amplifies them, and applies them to the main vertical amplifier.

The main vertical amplifier amplifies the signals received

Sec. 21.1   *The oscilloscope—general*   449

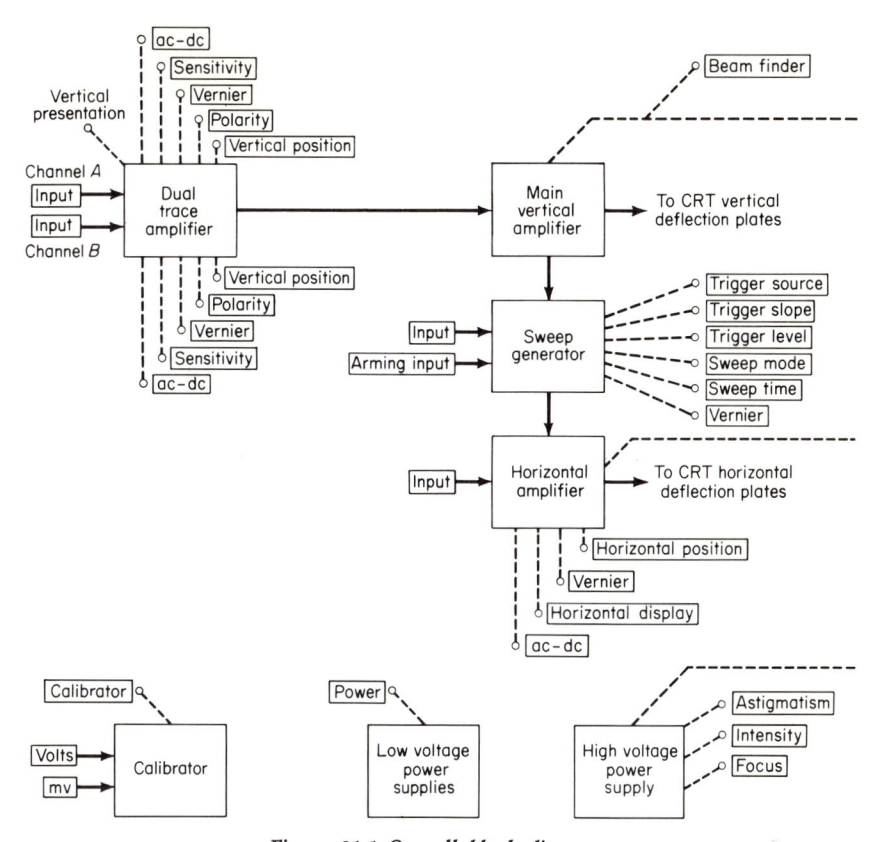

Figure 21-1 Overall block diagram.

from the dual-trace amplifier, and applies these signals to the vertical deflection plates of the CRT. The main vertical amplifier also applies part of the vertical signal to the sweep generator for internal triggering of the sweep.

The sweep generator generates a linearly rising voltage (sweep) which is used to drive the CRT beam horizontally across the CRT screen for internal sweep operation. The sweep generator thus provides a linear time base on which to measure the vertical signals. The generator can be operated as a triggered or free-running circuit.

The horizontal amplifier amplifies the sweep signal or external horizontal signals, and applies them to the horizontal deflection plates of the CRT.

The calibrator generates a square wave of about one kilocycle for checking the calibration of the vertical and horizontal amplifiers and for matching the test probes to particular input channels.

The high-voltage power supply generates the high voltages required by the CRT circuits for proper operation.

The low-voltage power supplies provide the operating voltages required by the various circuits throughout the oscilloscope and its plug-in units. All d-c voltages provided by the supplies (except to the fan) are regulated.

The time-axis plug-in units provide a variety of functions. One of the units, the Model 166A, provides the necessary switching for single-sweep operation and external intensity modulation.

21.2 Main vertical amplifier

The main vertical amplifier receives a balanced input signal from the plug-in vertical amplifier, amplifies the signal, and drives the vertical deflection plates of the CRT. The main vertical amplifier also feeds trigger signals to the horizontal plug-in unit and the sweep generator. A block diagram of the main vertical amplifier is shown in Fig. 21-2 and a shematic diagram in Fig. 21-3 (see back-flap envelope).

A balanced input signal is applied to the input cathode-followers V1A and V1B. The input cathode-followers provide a low-impedance driving source for the cascode amplifiers (described below). Cascode

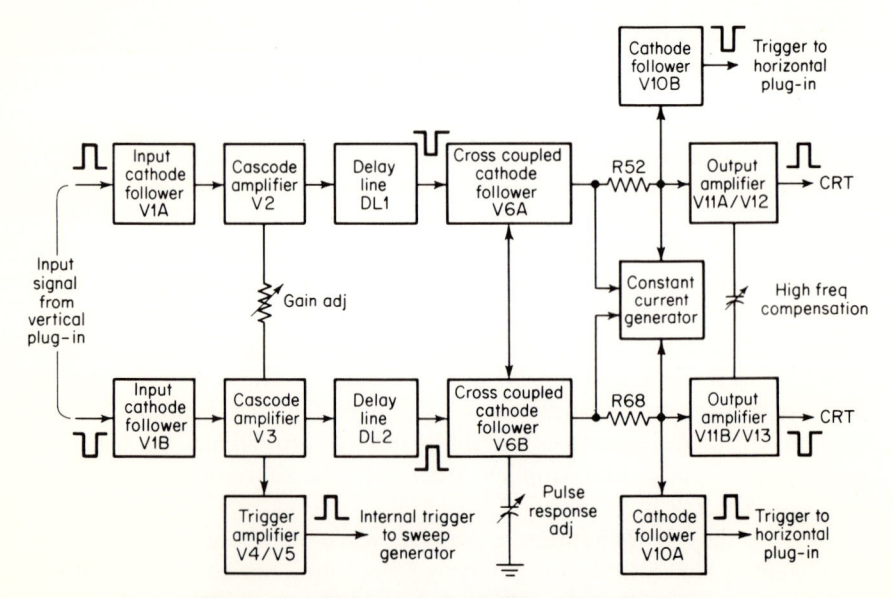

Figure 21-2 Main vertical amplifier block diagram.

Sec. 21.2   *Main vertical amplifier*   451

amplifiers V2 and V3 amplify the signal and drive the delay lines. The GAIN ADJUST potentiometer between the balanced cascode amplifiers varies the output amplitude of the cascode amplifiers and thereby sets the over-all gain of the main vertical amplifier. A trigger signal is picked off cascode amplifier V3 and fed to the sweep generator for internal sweep. This trigger signal is undelayed so that the sweep will start before the vertical signal reaches the deflection plates. Delay lines DL1 and DL2 delay the signal approximately 0.2 μsec (the time necessary for the sweep to start) before it is supplied to the cross-coupled cathode-followers (described later). The cross-coupled cathode-followers V6A and V6B isolate the delay lines and provide

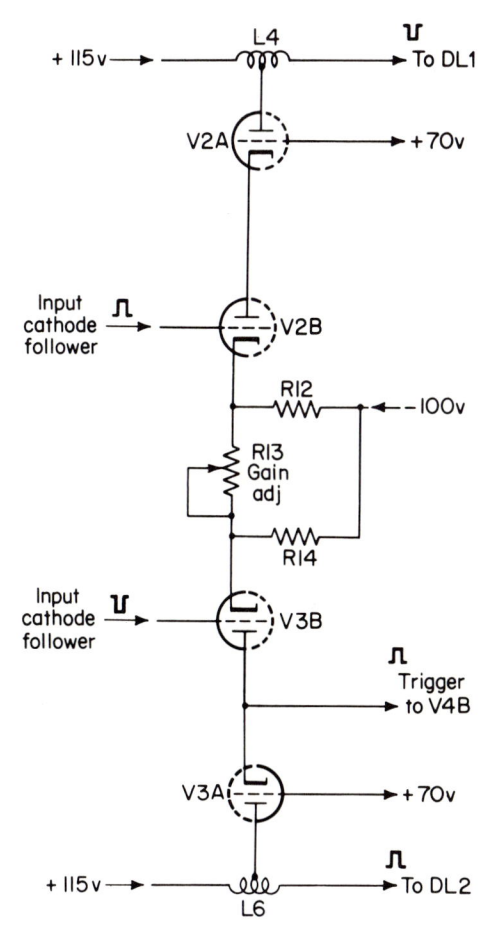

Figure 21-4 Cascode amplifier—simplified schematic.

a low-impedance driving source for cathode-followers V10A and V10B and the output amplifier.

The d-c voltage at the cathodes of the cross-coupled cathod-followers is +50 v. If this voltage were allowed to reach the control grids of the output amplifier, they would be overdriven. The constant-current generator (described later) dissipates this d-c voltage by providing a constant-voltage drop across resistors R52 and R68 equal to the cathode voltages at V6. Changes in the d-c level at the cathodes of V6A and V6B are sensed by the constant-current generator. The constant-current generator compensates for this voltage shift by drawing more current when the voltage rises above +150 v, and by drawing less when the voltage level drops below +150 v. The output amplifier (a modified cascode amplifier) supplies the driving voltages to the vertical deflection plates. The HIGH FREQ. COMPENSATION capacitor varies the effective feedback at high frequencies, and therefore controls the high-frequency gain in the output amplifier.

A cascode amplifier is a two-stage amplifier with a conventional grid-coupled input and a grounded-grid output used where the high gain of a pentode and the low noise of a triode are required. Figure 21-4 shows a simplified schematic of the V2/V3 balanced cascode amplifier circuit. A positive input signal from the input cathode-follower is fed to the control grid of V2B. The signal is amplified and the phase is shifted 180°, but because the plate current of V2B flows through V2A, there is no additional phase shift in the grounded-grid section. Resistor R13 GAIN ADJ in the cathode circuit adjusts the gain of the cascode amplifier, and sets the over-all gain of the main vertical amplifier. An output is taken from the plate of V3B and fed to amplifier V4B for internal triggering of the sweep generator. The output signals are coupled through inductors L4 and L6 to the delay lines. The output amplifier (V11/V12/V13) is a modified cascode amplifier. The modification consists of adding an extra tube in parallel with the conventional grid-coupled input section to double the gain of the stage. To prevent excessive current in the output sections (V12A and V13A), half of the current is shunted around these tubes through resistors R91 and R94.

Two sections of the BEAM FINDER are included in the output cascode amplifier. One section increases the common-cathode resistance to decrease the gain; the second section limits the CRT beam defocusing to a reasonable limit by: (1) disconnecting resistors R91 and R94 from the +110 v supply, and (2) adding R99 between the +110 v supply and the plates of V12A and V13A to prevent an exces-

Sec. 21.2   Main vertical amplifier   453

sive voltage change at these plates because of the reduced plate current.

The cross-coupled cathode-follower circuit provides a low-impedance driving source for cathode-follower V10 and the output amplifier. Figure 21-5 shows a simplified schematic of the cross-coupled cathode-follower circuit. The delayed input signal is applied to the control

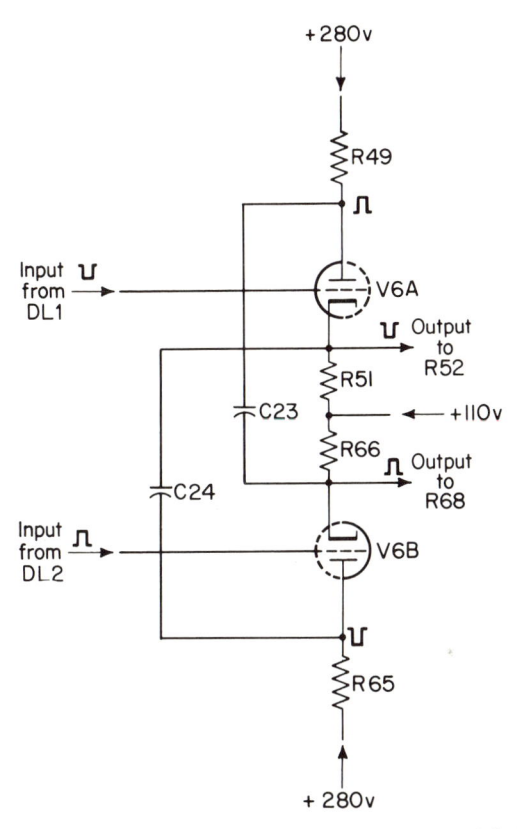

Figure 21-5 Cross coupled cathode follower—simplified schematic.

grids of V6A and V6B. At low frequencies, the stage acts as a simple cathode-follower. High-frequency signals appearing at the plate circuits readily pass through coupling capacitors C23 and C24. Since these high-frequency signals are in phase with the signals at the opposite cathode, they add to the load current. The decrease in load impedance at high frequencies is partially compensated by the increase in high-frequency load current offered by the cross-coupled cathode-follower. Resistors R49, R51, R66, and R65 are all of the same value; therefore, the plates and cathode voltages are equal in amplitude. Since the plate and cathode signals are added, at high frequencies, the effective output impedance is reduced. The output signals are then fed through resistors R52 and R68 to the constant-current generator circuit.

The constant-current generator reduces the voltage level from about +150 v d-c to about 0 v at the output stage. The constant-current generator is employed to: (1) reduce the +150 v d-c appearing at the cathodes of V6 to essentially 0 v at the grids of the output amplifier, and (2) pass low-frequency signals without attenuation. Figure 21-6 is a simplified schematic diagram of the constant-current generator. High-frequency signals are bypassed through C22 and C25.

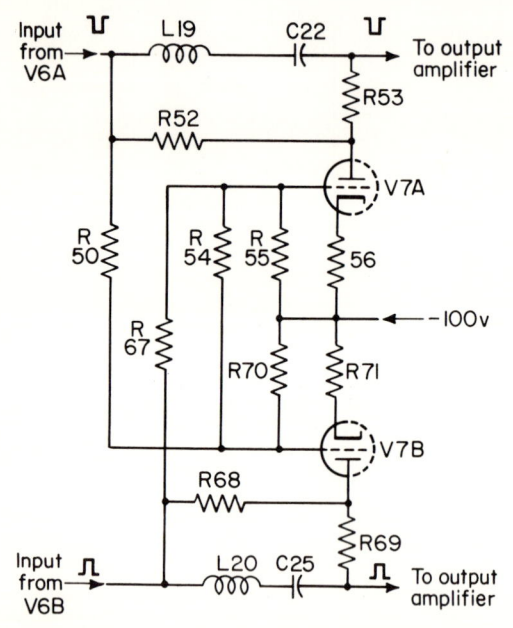

Figure 21-6 Constant current generator-simplified schematic.

Since small plate voltage changes (due to the signal) will affect the plate current slightly, the stage is cross-coupled to compensate for this effect. Therefore, a constant current is maintained through R52 and R68, and any voltage at the cathodes of V6 will be coupled to the grids of the output stage without being attenuated.

21.3 Sweep generator

The sweep generator consists of sweep-generating circuits plus an amplifier and a trigger generator which actuate the sweep-generating circuits. A block diagram of the sweep generator is shown in Fig. 21-7 and a schematic diagram is shown in Fig. 21-8 (see back-flap envelope). Schmitt trigger circuits (cathode-coupled bistable multivibrators) are used in the sweep generators.

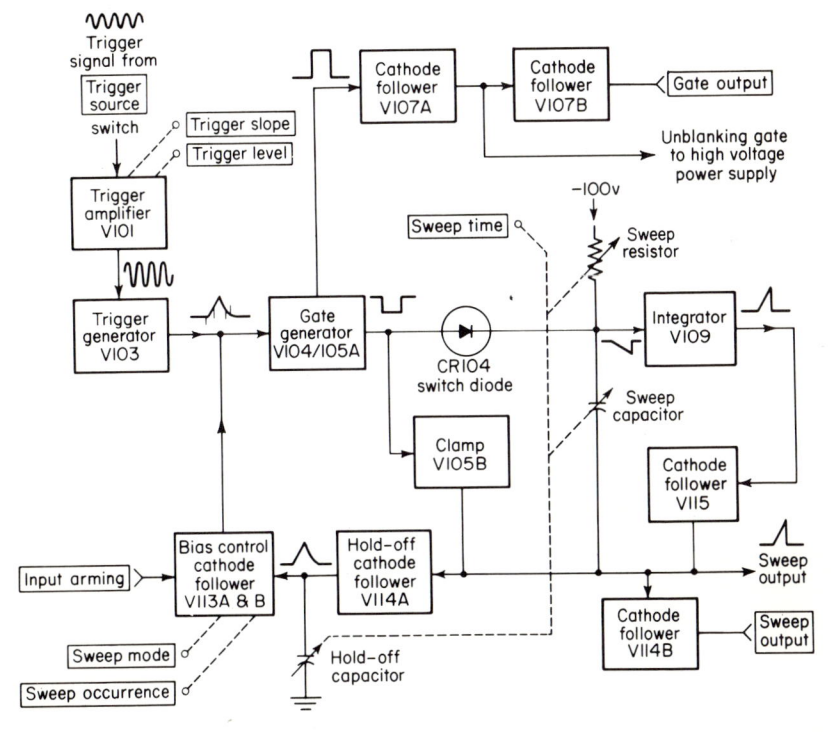

Figure 21-7 Sweep generator block diagram.

The TRIGGER SOURCE switch selects the trigger signal and applies it to trigger amplifier V101. The amplified trigger signal is applied to trigger generator V103, a Schmitt circuit with narrow hysteresis limits. Provided that the amplified trigger signal crosses both hysteresis limits, the trigger generator applies sharp pulses or triggers to the gate generator V104/V105A.

The TRIGGER LEVEL control sets the zero-signal output level of trigger amplifier V101. Since the trigger amplifier is d-c coupled to trigger generator V103, the control determines the voltage levels that the trigger signal must cross if the amplified trigger signal is to cross the hysteresis limits of V103. The TRIGGER SLOPE switch determines whether or not the trigger amplifier inverts the trigger signal, and thereby determines whether the sweep starts on the positive-going or negative-going portion of the trigger signal.

Gate generator V104/V105A is a Schmitt circuit with wide hysteresis limits. Between sweeps, the A section of bias-control cathode-follower V113 holds the bias at the input of the gate generator close to the lower hysteresis limit. Trigger generator V103 applies both positive and negative triggers. The positive triggers are reduced in amplitude and have no effect, but a negative pulse drives the input to the gate generator below the lower hysteresis limit and causes the gate generator to switch.

When it switches, gate generator V104/V105A provides a positive and a negative gate. The positive gate is applied to the high-voltage power supply to turn on the CRT beam and to the front-panel GATE OUTPUT connector for external use. The negative gate applies reverse bias to switch diode CR104. Prior to the gate, the switch diode had been forward-biased and had been holding the input to integrator V109 at about zero volts. The negative gate opens the diode switch and frees the input to the integrator.

Once freed, the input to integrator V109 starts going negative because it is connected to -100 v through the sweep resistor. The integrator amplifies and inverts its input to produce a large, positive-going output which is applied back to the input through cathode-follower V115 and the sweep capacitor. As a result, the voltage at the input to integrator V109 changes by about one volt during sweep time. The voltage across the sweep resistor, then, changes by about one per cent, and the current through the resistor changes by the same amount. The current through the sweep resistor is the charging current for the sweep capacitor; therefore the voltage across the sweep capacitor changes quite linearly with time, and the sweep signal is a nearly linear voltage ramp. The SWEEP TIME switch changes the value of the sweep resistor or capacitor to change the sweep time. The sweep output is applied to the horizontal amplifier and to the front-panel SWEEP OUTPUT connector.

An attenuated sweep signal is applied to the input of gate generator V104/V105A through hold-off cathode-follower V114A and the B section of bias-control cathode-follower V113. This signal drives the input of the gate generator up to the upper hysteresis limit and causes the gate generator to switch back to its pre-sweep state. The gate generator then ends the gates, blanking the CRT and forward-biasing switch diode CR104. The switch diode returns the input to integrator V109 to its pre-sweep level, discharging the sweep capacitor.

During a sweep time, hold-off cathode-follower V114A charges a hold-off capacitor. After the sweep ends, this capacitor lets the input to gate generator V104/V105A down slowly enough to prevent that circuit from being triggered again until the remaining sweep circuits have recovered completely. The SWEEP TIME switch changes the size of the hold-off capacitor with sweep time.

Clamp V105B ensures that each sweep starts from the same voltage level, about −50 v.

The SWEEP MODE control determines the no-signal bias at the input to gate generator V104/V105A by setting the bias on the A section of bias-control cathode-follower V113. With the control set to PRE-SET or in the TRIGGER portion of its adjustable range, the gate generator bias cannot drop below its lower hysteresis limit unless the trigger generator provides a trigger. With the control set in the FREE RUN portion of its adjustable range, the gate bias is allowed to drop below its lower hysteresis limit. Thus, as the hold-off capacitor discharges, it lets the gate generator bias all the way down to the lower hysteresis limit, and another sweep starts automatically.

The SWEEP OCCURRENCE switch selects normal or single-sweep operation. Normal operation is discussed above. For single-sweep operation, the SWEEP OCCURRENCE switch converts V113 into a Schmitt circuit. As the sweep signal from hold-off cathode-follower V114A rises to end the gate from the gate generator, the sweep signal also switches the Schmitt circuit of V113 so that V113B conducts and V113A is cut off. The B section of V113 then holds the input to gate-generator V104/V105A high enough so that triggers from the trigger generator cannot actuate the gate generator, and the sweep-generating circuits are effectively disabled. A positive signal applied to V113A switches the Schmitt circuit of V113 so that V113A conducts and V113B is cut off. The A section of V113 then sets the input to the gate generator according to the setting of the SWEEP MODE control, and the sweep-generating circuits are effectively armed. The switching signal for V113A can be an external signal applied to the ARMING INPUT connector, or an internal signal obtained from the SWEEP MODE control when the control is rotated just out of PRE-SET.

The Schmitt trigger circuit is a form of bistable multivibrator used where fast-rising signals are required. Figure 21-9 shows a simplified Schmitt trigger circuit, and input and output waveforms.

If initially the input voltage is such that V1 is cut off, V2 conducts. As the input voltage becomes more positive, it will eventually reach a predetermined level (a) at which the circuit changes state; then

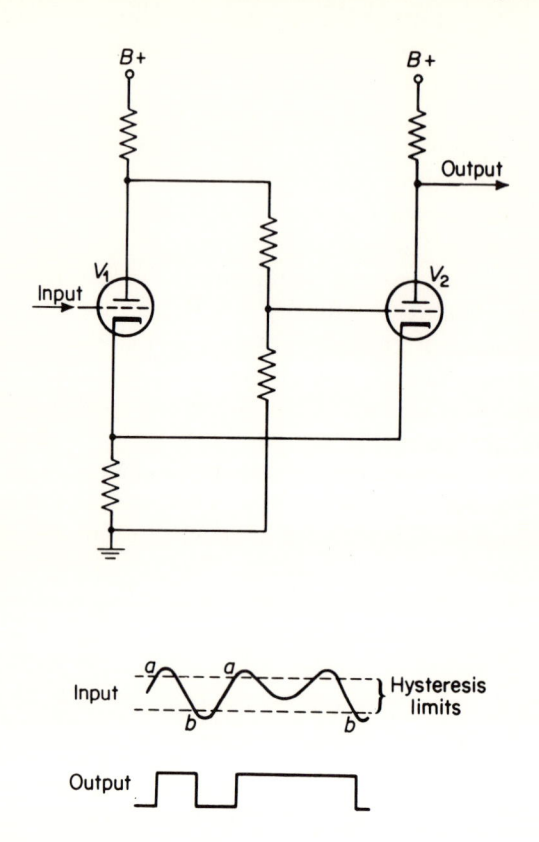

Figure 21-9 *Simplified Schmitt Trigger circuit and waveforms.*

V1 conducts and V2 is cut off. If the input voltage then goes negative, the common-cathode potential decreases and the V2 grid goes positive. When the input reaches a second predetermined level (b), V2 conducts and the circuit switches back to its initial state. The output of the circuit is a voltage step, either positive or negative depending upon the slope of the input. In the case of trigger-generator V103, a differentiating network differentiates the voltage steps in short pulses.

The input voltage levels at which a Schmitt trigger circuit switches are the hysteresis limits. Note that the circuit does not switch unless the input crosses both limits.

Sec. 21.4   Horizontal amplifier

21.4 Horizontal amplifier

The horizontal amplifier amplifies the internal sweep or an external signal applied to the horizontal INPUT connector, and drives the horizontal deflection plates of the CRT. Figure 21-10 is a simplified

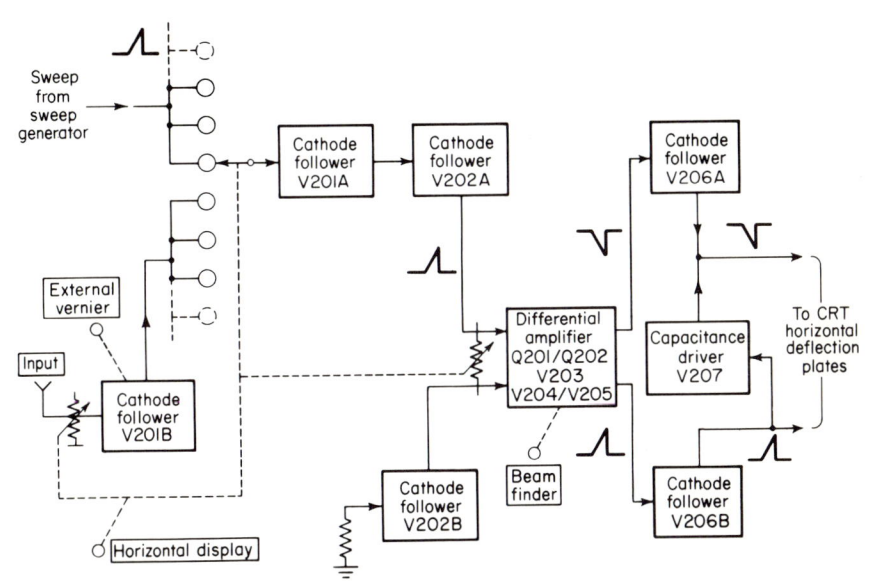

Figure 21-10 Horizontal amplifier block diagram.

block diagram of the horizontal amplifier, and a schematic diagram is shown in Fig. 21-11 (see back-flap envelope). The sweep signal is applied through the HORIZONTAL DISPLAY switch to cathode-follower V201A. From V201A, the signal passes through cathode-follower V202A and one side of a balanced attenuator to one input of a differential amplifier. Cathode-follower V202B controls the second input to the differential amplifier through the other side of the balanced attenuator. The input to cathode-follower V202B is grounded through a resistance. The differential amplifier amplifies the difference between its two input signals and provides a balanced output signal, which is applied to cathode-followers V206A and V206B. These cathode-followers drive the CRT deflection plates. Cathode-follower V206B also drives capacitance driver V207, which acts as the cathode resistance for cathode-follower V206A. When high-speed signals such as the faster sweeps drive the CRT beam from left to right, the capacitance driver discharges the capacitance of the deflection plate associated with V206A. The capacitance driver thereby prevents cutoff in V206A and the resulting degradation of the signal.

External horizontal signals applied to the INPUT connector pass through an attenuator, cathode-follower V201B, and the HORIZONTAL DISPLAY switch to cathode-follower V201A. Otherwise the operation is the same as described above for the sweep signal.

The HORIZONTAL DISPLAY switch selects the signal to be applied to the horizontal deflection plates. The switch also controls the input attenuator and the balanced attenuator. The balanced attenuator provides a means of sweep expansion and, in combination with the input attenuator, provides steps of external horizontal sensitivity. The EXTERNAL VERNIER control varies the series resistance in the output of cathode-follower V201B and thereby varies the output of V201B. The range of the EXTERNAL VERNIER is sufficient to provide continuous adjustment of the external horizontal sensitivity between the calibrated settings of the HORIZONTAL DISPLAY switch. A section of the BEAM FINDER switch is in the common-cathode circuit of V204 and V205. When pressed, the switch reduces the gain so that an unbalance prior to V204 and V205 cannot deflect the CRT beam off the screen.

21.5 High-voltage power supply

The high-voltage power supply provides the operating voltages for the CRT. A block diagram of the high-voltage power supply is shown in Fig. 21-12. The 50-kc output of r-f oscillator V304 is applied to high-voltage transformer T301. The transformer steps up the oscillator output to high a-c voltages, which rectifiers V310 and V311, and doubler stage V308 and V309, convert to d-c. The d-c voltages are then supplied to the CRT.

Control for the high-voltage supply is accomplished by comparing the -1500 v applied to the CRT cathode with $+370$ v from the regulated low-voltage power supply. Changes in the -1500 v supply produce an error voltage, which is amplified by the d-c control amplifier V301 and is applied to the oscillator as a control voltage. The control voltage changes the output amplitude of the oscillator, and hence corrects for the change in the -1500 v supply.

The CRT is normally biased off. The positive gate pulse from the sweep generator, applied to the CRT control grid, overrides the bias and unblanks the CRT. The gating pulse time is identical to the sweep time so that the CRT remains on during sweep time and external horizontal operation.

Sec. 21.6 Calibrator

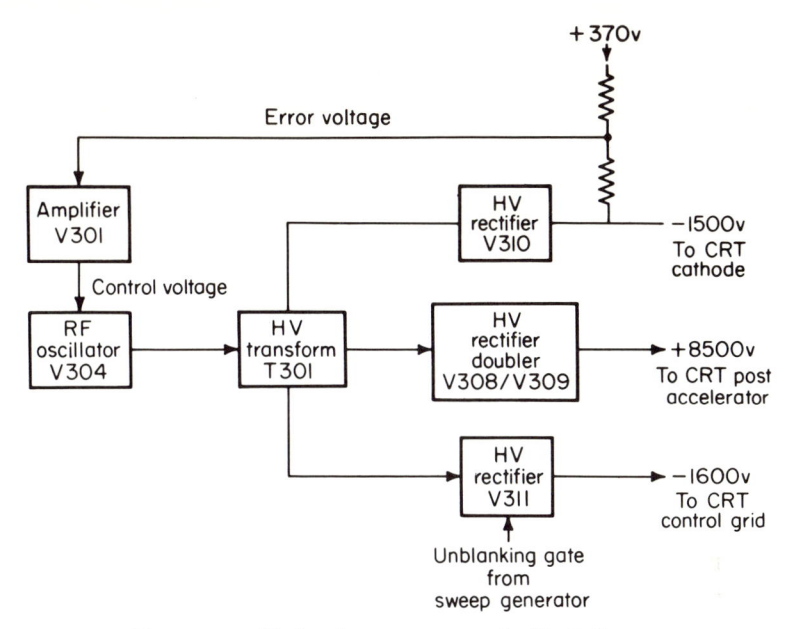

Figure 21-12 High voltage power supply block diagram.

The INTENSITY control varies the -1600 v supplied to the CRT control grid. The FOCUS control varies the voltage supplied to the CRT grid to produce a sharply defined trace. The ASTIGMATISM control, part of the voltage divider from the $+370$ v supply to ground, varies the voltage on the CRT grid to compensate for the electron beam defocusing when the beam is being deflected by the vertical or horizontal signals. A section of the BEAM FINDER is in the CRT cathode supply. When pressed, the switch returns the cathode supply to -100 v instead of ground. The bias voltage to amplifier V301 decreases, resulting in a decreased oscillator output. This decrease in oscillator output holds the CRT cathode constant at -1500 v, but the grid supply decreases enough to turn the CRT on.

21.6 Calibrator

The calibrator signal is generated by a plate-coupled multivibrator, V306 (see the schematic diagram, Fig. 21-13). A clamp diode, V307A, limits the positive output of the multibrator to $+110$ v. A disconnect diode, V307B, disconnects the multivibrator from the output when the multivibrator signal goes below 0 v. The output signal, then, is a square wave which varies between 0 v and $+110$ v. The square

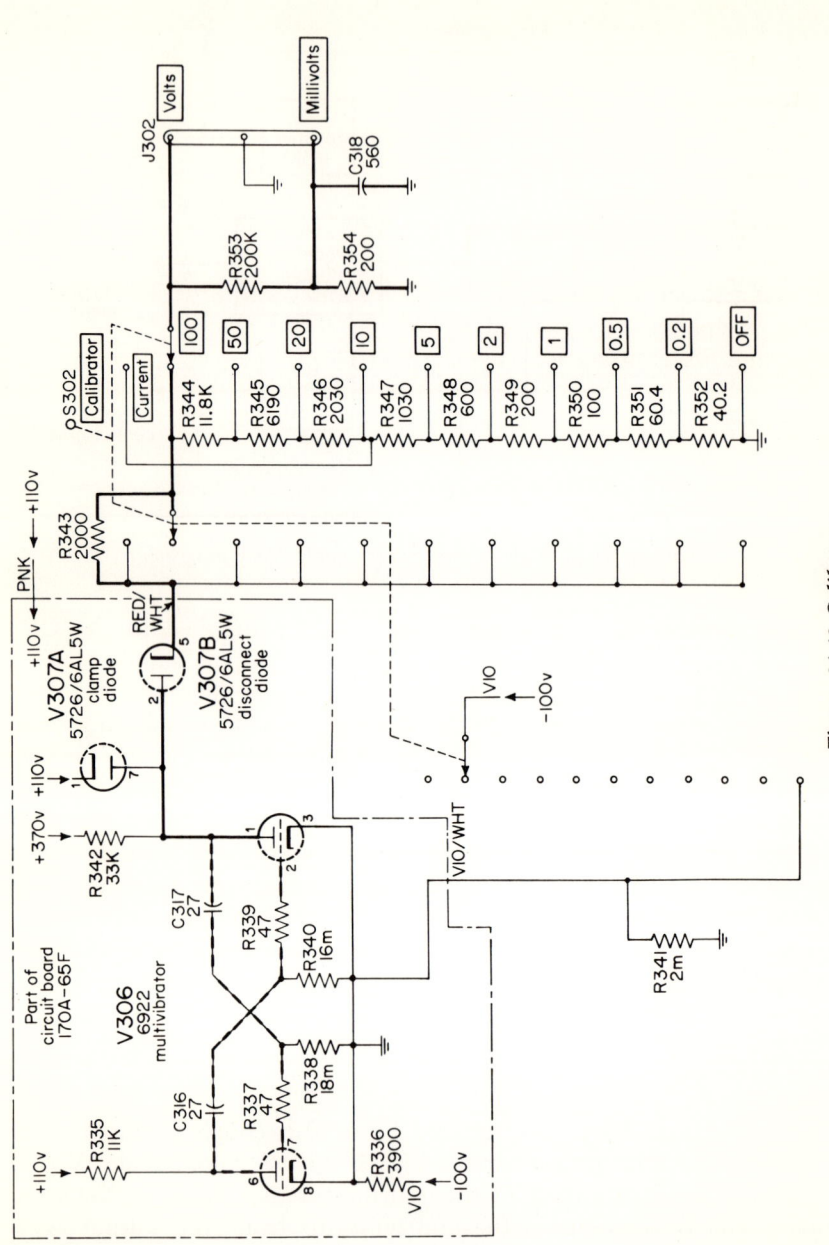

Figure 21-13 Calibrator.

Sec. 21.7   Dual-trace plug-in amplifier Model 162A   463

wave is applied to a step attenuator and then to the VOLTS terminal of the calibrator output on the front panel. A second, fixed attenuator reduces the VOLTS signal 1000:1 and connects the smaller signal to the MV terminal of the calibrator output. The CALIBRATOR switch controls the step attenuator and selects the voltage of the calibrator output, which is measured in millivolts.

21.7 Dual-trace plug-in amplifier Model 162A

Referring to the block diagram in Fig. 21-14, the Model 162A Dual-Trace Amplifier contains two identical but independent low-gain wide-band differential amplifiers. The two channels, A and B, are

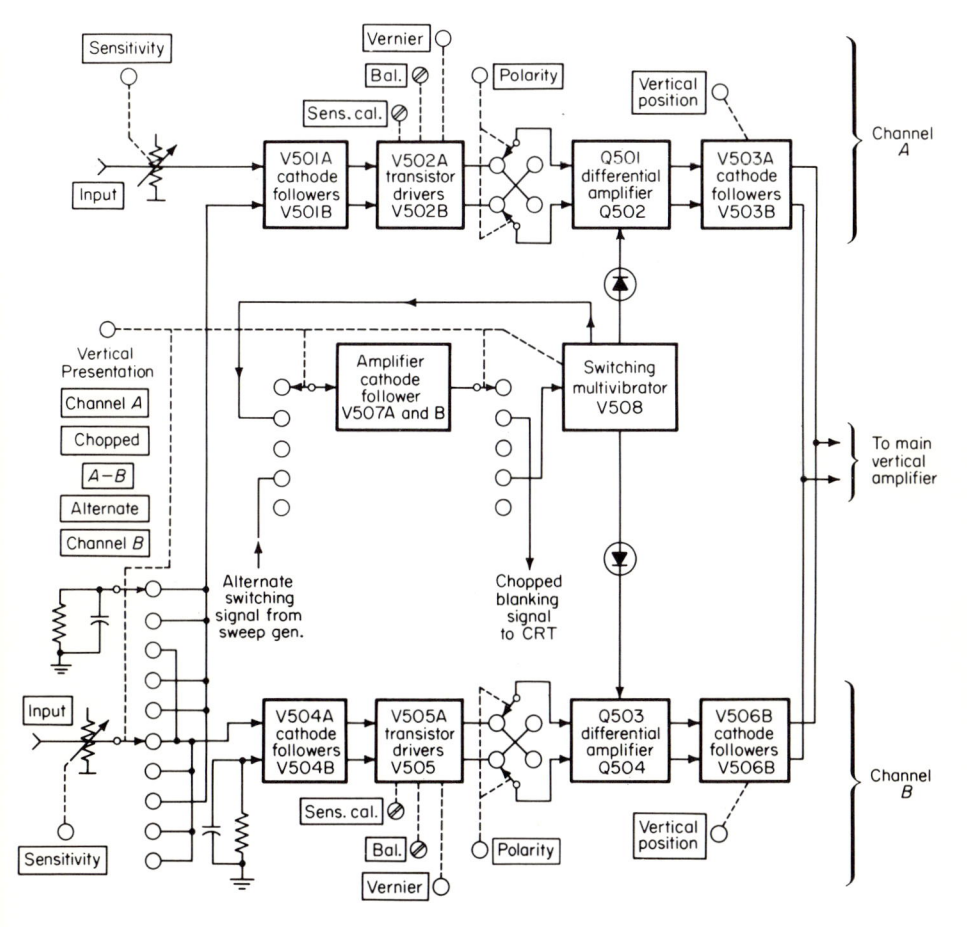

Figure 21-14 Dual-trace amplifier block diagram.

connected to a common output, the input of the main vertical amplifier of the oscilloscope. A multivibrator turns on one channel and turns off the other, preventing their simultaneous operation. Signals may be applied to either channel or to both channels at the same time. A VERTICAL PRESENTATION switch selects the mode of presentation. In the CHANNEL A presentation, the multivibrator switches on channel A only, so that only the signal applied to the CHANNEL A INPUT is presented. The same is true for CHANNEL B operation. With the switch positioned on CHOPPED, the multivibrator free-runs, switching the two channels on and off alternately at a one-megacycle rate, so that both signals are presented during the same sweep. On ALTERNATE, the horizontal sweep signal from the oscilloscope triggers the multivibrator, and the two signals appear on alternate sweeps. On the A-B presentation, channel B is switched off while A is on, and the two signals are connected to the two inputs of the channel A differential amplifier; the difference between the two signals is displayed on the screen.

The SENSITIVITY selector switch is in reality a step attenuator, attenuating the input signal by a selected degree to bring to the oscilloscope screen a convenient figure. All together there are four different kinds of attenuation circuitry, represented by the four switch positions (a), (b), (c), and (d) in Fig. 21-15. The low resist-

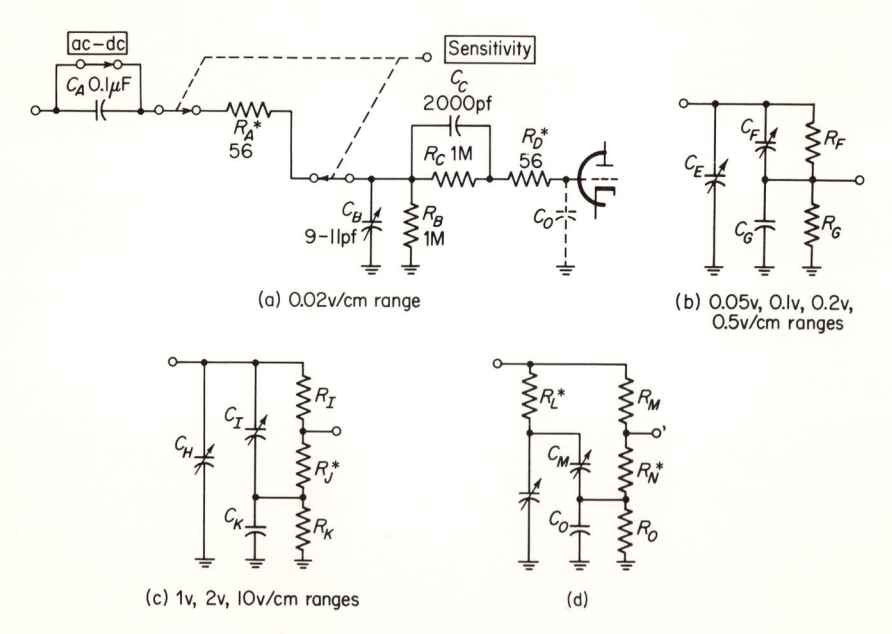

Figure 21-15 Sensitivity selector switch.

Sec. 21.7 Dual-trace plug-in amplifier Model 162A

ances R_A, R_D, R_J, R_L, and R_N, marked out by asterisks in the figure, are used for the suppression of high-frequency parasitic oscillations, and may be regarded as short circuits, so far as their function in attenuation is concerned. In the 0.02 v/cm range, the signal is fed straight through, and there is practically no attenuation. C_A may be switched into the circuit to block the d-c component from the input signal, or it may be short-circuited out of the circuit, permitting the transmission of all the components. C_B and R_B provide a high input impedance, one megohm shunted by a small capacitance. C_B and R_B furnish high-frequency compensation for input capacitor C_D of the following stage.

In switch position (b), R_F and R_G provide the desired voltage step-down, and present a total of 1 megohm impedance to the input. When the product $C_F R_F$ equals $C_G R_G$, the relative phase and amplitude of all frequency components in the input signal are preserved after the attenuation. This is important in the preservation of the waveform of the signal, so that the figure presented on the screen may represent point-by-point on the time base the signal input to the Model 162A. The remaining ranges have the same configuration as circuit (b) when all the parasitic suppressing resistors in the circuits are considered short circuits. With an AC-21 voltage-divider probe plugged into the input, the product of the probe r-c network equals the product $C_B R_B$, and the input waveform is faithfully preserved.

Two stages of cascaded cathode-followers (Fig. 21-16) are required

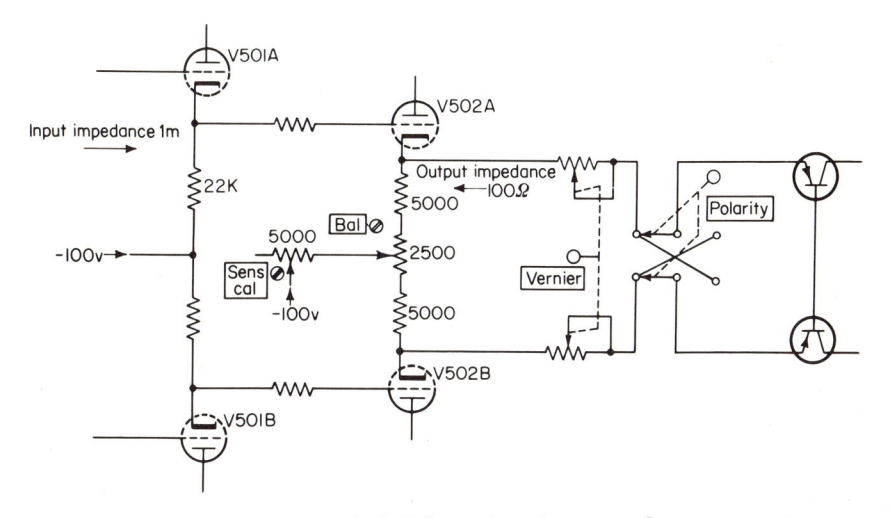

Figure 21-16 Cathode-follower impedance transformer.

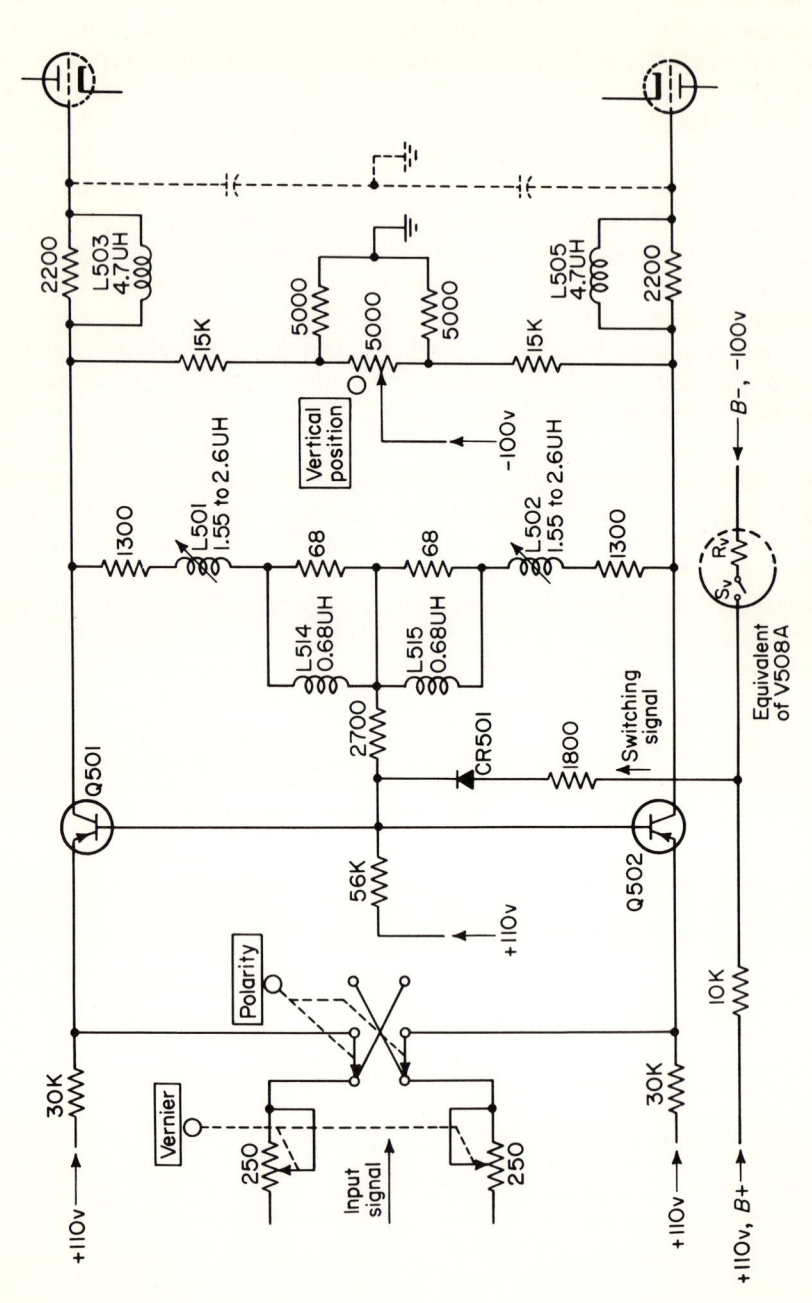

Figure 21-17 Output cathode-follower.

Sec. 21.7   *Dual-trace plug-in amplifier Model 162A*   467

to transform an input impedance of 1 megohm to an output impedance of 100 ohms. The first stage is a true cathode-follower of constant gain, about 0.97, functioning as a buffer to isolate the input from the rest of the channel. The second stage, though connected in the same configuration as the first, does not behave as a true cathode-follower. By varying its bias through the SENS. CAL potentiometer, its gain is adjustable between 0.4 and 0.7. As cathode-followers are inherently broad in bandwith, and since only comparatively low impedances are involved, no frequency compensation is necessary.

The total gain of each half of the differential amplifier of the entire channel may be equalized by the BAL potentiometer so that at the zero input signal level, switching of the POLARITY switch produces no d-c changes in the amplifier and therefore no vertical shift on the CRT.

The transistor switching differential amplifier consists of two transistors in a back-to-back common-base configuration (Fig. 21-17), which offers very low impedance to the input, but relatively high

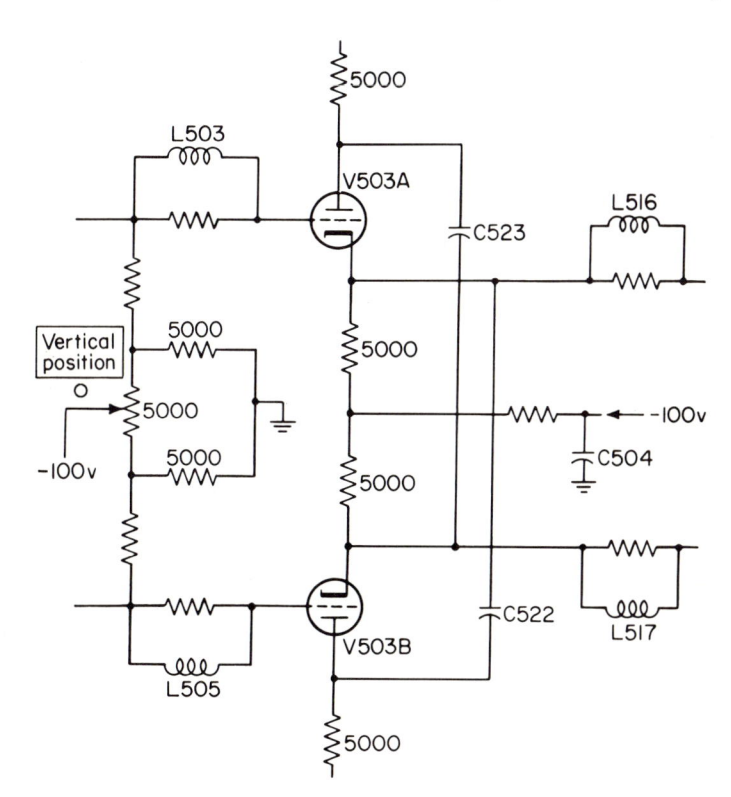

Figure 21-18 Output cathode follower.

impedance to the output. It is the only stage that provides a signal voltage gain. The VERNIER control, by adjusting the emitter resistance, varies the gain of the stage between 5 and 10. The POLARITY switch reverses the signal to the emitters of the transistors. The inductances in the collector circuit (L501, L502, L503, L505, L514, and L515) compensate for losses in the high-frequency response resulting from interelectrode and stray capacitances between this stage and the next, and the transistor effects.

Switching is accomplished through a silicon diode CR501 and half of the multivibrator triode V508A, which functions as a switch, S_V, in series with a resistor, R_V. When the triode is cut off, which is equivalent to opening S_V, positive voltage is applied to the bases of transistors Q501 and Q502 through CR501 to switch off the transistors, blocking the incoming signal. When V508A conducts; that is, when S_V closes, negative voltage from B– switches off CR501, leaving Q501 and Q502 free to function as amplifiers.

The output cathode-follower, Fig. 21-18, isolates the entire channel from the output load, and it is connected in parallel with the output cathode-follower of the other channel of the unit (not shown). As the switching multivibrator switches on only one channel at a time, the signals from the two channels do not mix. Capacitor C504 in the cathode bias is a high-frequency decoupling capacitor. C522 and C523 are additional output coupling capacitors for improvement of the high-frequency response of this stage. Pairs of inductors L503, L505 and L516, L517 provide high-frequency compensation for the input and output respectively. The VERTICAL POSITION potentiometer adjusts the relative biases to the two halves of the output differential cathode-follower, varying the relative d-c voltages on the input grids of the main vertical amplifier of the oscilloscope.

The switching control circuit, Fig. 21-19, consists of a multivibrator, an amplifier, and a selector switch which is represented as a seven-section, five-position switch. Sections (1) and (2) connect the input signal from channel B to the upper half of differential amplifier B or to the lower half of differential amplifier A (V501A/B). Sections (3) and (4) select respectively the input signal to and the output signal from the switch signal amplifier (V507A/B). Sections (6) and (7) apply negative switch-off bias to the grids of the multivibrator (V508A/B).

With the VERTICAL PRESENTATION switch at position A, both the input and output of the switch signal amplifier are disconnected, and

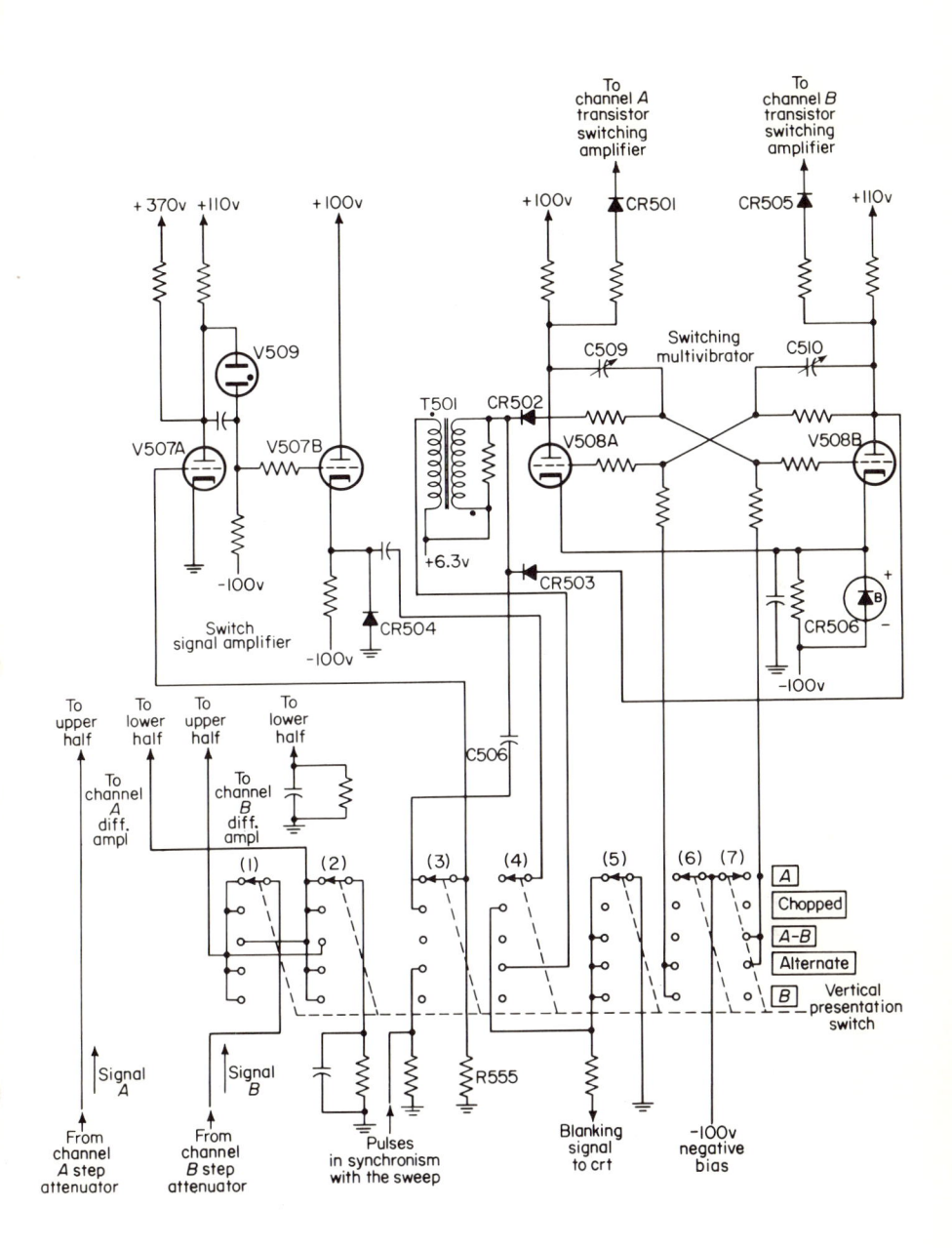

Sec. 21.7   *Dual-trace plug-in amplifier Model 162A*   469

Figure 21-19 Switching control circuitry.

the signal from input A is connected to the A channel amplifier (V501A/B), and then input B to the B channel amplifier (V504A/B). Negative bias is applied to switch off the B half (V508B) of the multivibrator. CR505 conducts, applying a positive voltage to the bases of transistors Q503, Q504 (not shown in Fig. 21-19), thereby switching off channel B. With the A half (V508A) conducting, diode CR501 cuts off, and with Q501, Q502 connecting, channel A is free to amplify the signal from input A. The reverse is true with the VERTICAL PRESENTATION switch on position B.

On CHOPPED, the negative bias is eliminated, and the switching multivibrator free-runs at a frequency set by C509 and C510, switching on and off channels A and B. Breakdown diode CR506 limits the saturation to assure the multivibrator self-starting.

(a) Dual signals on alternate

(b) Chopped signals without blanking

(c) Chopped signals with blanking

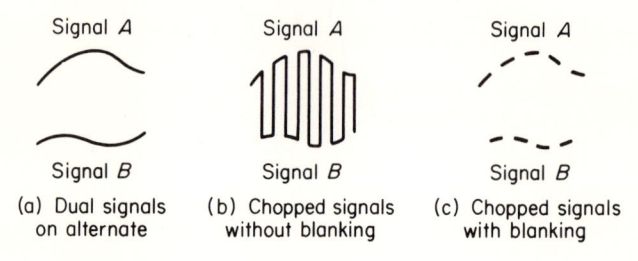

Figure 21-20 Formation of two chopped signal traces.

During chopping, illustrated in Fig. 21-20(b), the transient at the beginning of each chop is high. Blanking out the oscilloscope trace at such times erases the transients from observation, and the smooth chopped traces shown in Fig. 21-20(c) result. The blanking signal to the CRT is derived from the switching multivibrator, transferred through CR502 and CR503, amplified by V507A, and isolated by cathode-follower V507B. Diode CR504 clips off all the negative pulses; the positive pulses are applied to the cathode of the CRT to blank the switching transient.

On A-B, the switching multivibrator switches channel A on and channel B off. The signal input to B is applied to the lower half of differential amplifier A, and the differential input, channel A minus channel B, is amplified and presented.

On ALTERNATE, the switching multivibrator is rendered bistable by applying negative bias on both halves. Pulses, from the oscilloscope, and synchronized with the horizontal sweep signal, are amplified by the switch signal amplifier (V507A/B) to trigger the switching multivibrator, presenting the signals A and B on the screen on alternate sweeps.

21.8 High-gain vertical plug-in amplifiers Model 162D

The 162D High-Gain Vertical Amplifier is a d-c coupled circuit that converts either a single-ended or differential signal to a balanced signal for application to the vertical amplifier in the Model 170A Oscilloscope. Single-ended input signals are applied to INPUT A or INPUT B; differential operation requires application of signals to both INPUT A and INPUT B.

Refer to the block diagram, Fig. 21-21. With the VERTICAL DISPLAY control, S1, in the INPUT A or INPUT B position, input signals are a-c–d-c coupled to a frequency-compensated step attenuator to the grid of the cathode-follower V1A. The single-ended signal is taken at the cathode of V1A and applied to the base of transistor Q1 of differential amplifier Q1/Q2, where the signal is converted to a balanced signal. The Q1/Q2 output is in turn applied to the differential amplifier Q3/Q4, where it is amplified and applied to the cathode-follower V2A/B. The signal is also coupled back to the emitters of Q1/Q2, as degenerative feedback. The balanced output at the cathodes of V2A/B is applied to the vertical amplifier in the 170A Oscilloscope.

With the VERTICAL DISPLAY control S1 in the DIFFERENTIAL (A-B) position, two signals are simultaneously applied to INPUT A and INPUT B. The INPUT A signal is applied to the grid of V1A; the INPUT B signal is applied to the grid of V1B. The two input signals produce a difference voltage at the cathode of V1A/B, which in turn is applied to the bases of the differential amplifier Q1/Q2. From this point, the operation is the same as for the single-ended input.

Refer to Figs. 21-22 and 21-23 (see back-flap envelope) throughout the following circuit description.

The step attenuator provides attenuation of the input signal in twelve range steps to produce sensitivities from 0.005 v/cm to 20 v/cm. Each range is selected by the SENSITIVITY control S2. Attenuation for the four most sensitive ranges (0.005 to 0.05 v/cm) is obtained by degenerative feedback; attenuation for the eight less sensitive ranges (0.01 to 20 v/cm) is obtained by degenerative feedback plus resistive dividers. The feedback network is connected from the collectors of Q3/Q4 to the emitters of Q1/Q2. It includes feedback capacitors C49 through C56, coils L1 through L4, and resistors R57 through R62. The feedback capacitors provide the feedback coupling at high frequencies; the coils compensate for excessive phase shift produced by Q1 through Q4, at high frequencies.

Feedback attenuation is provided, equally, to both halves of the balanced circuit on all sensitivity ranges; resistive divider attenua-

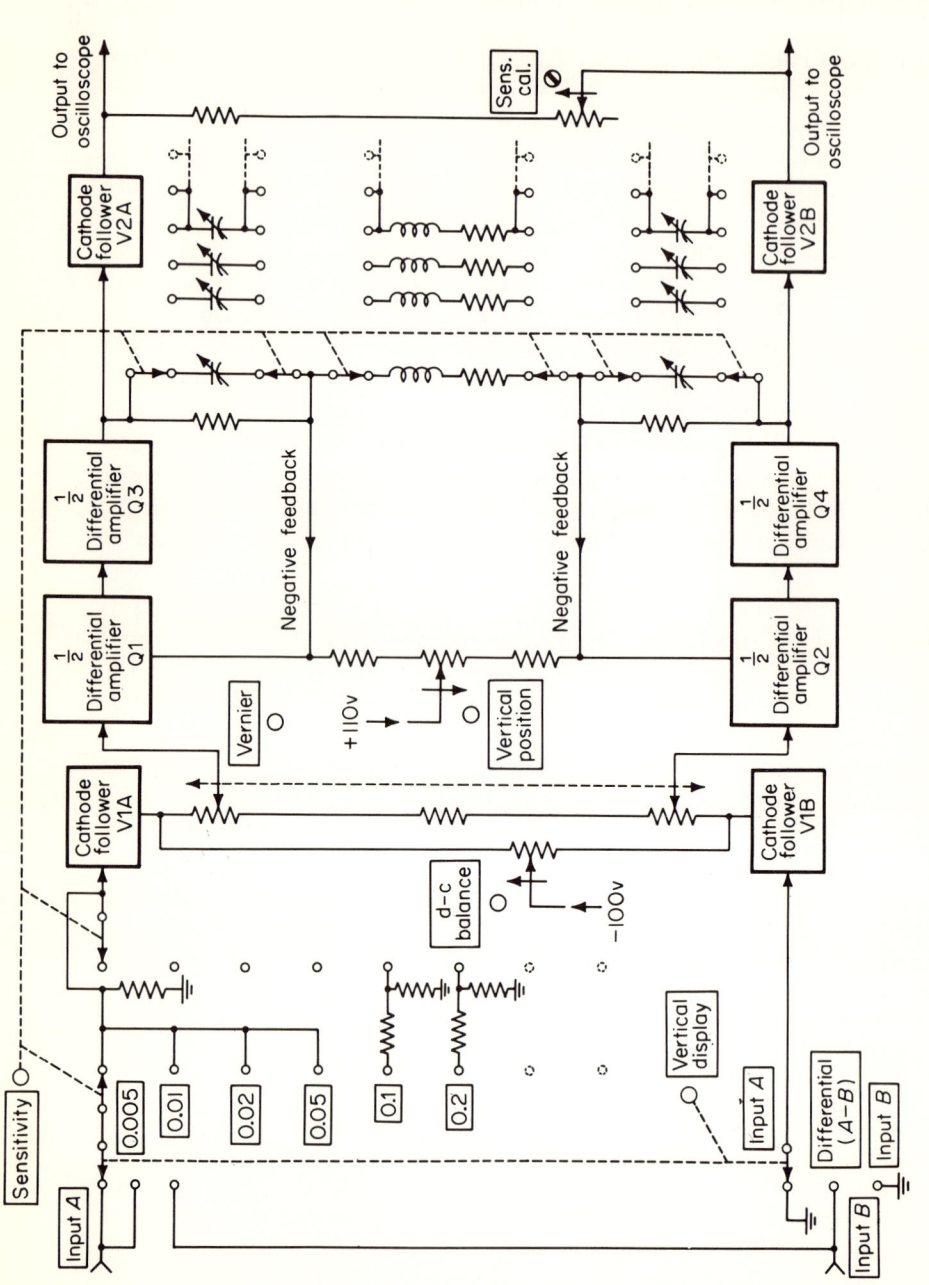

Figure 21-21 Block diagram of Model 162D.

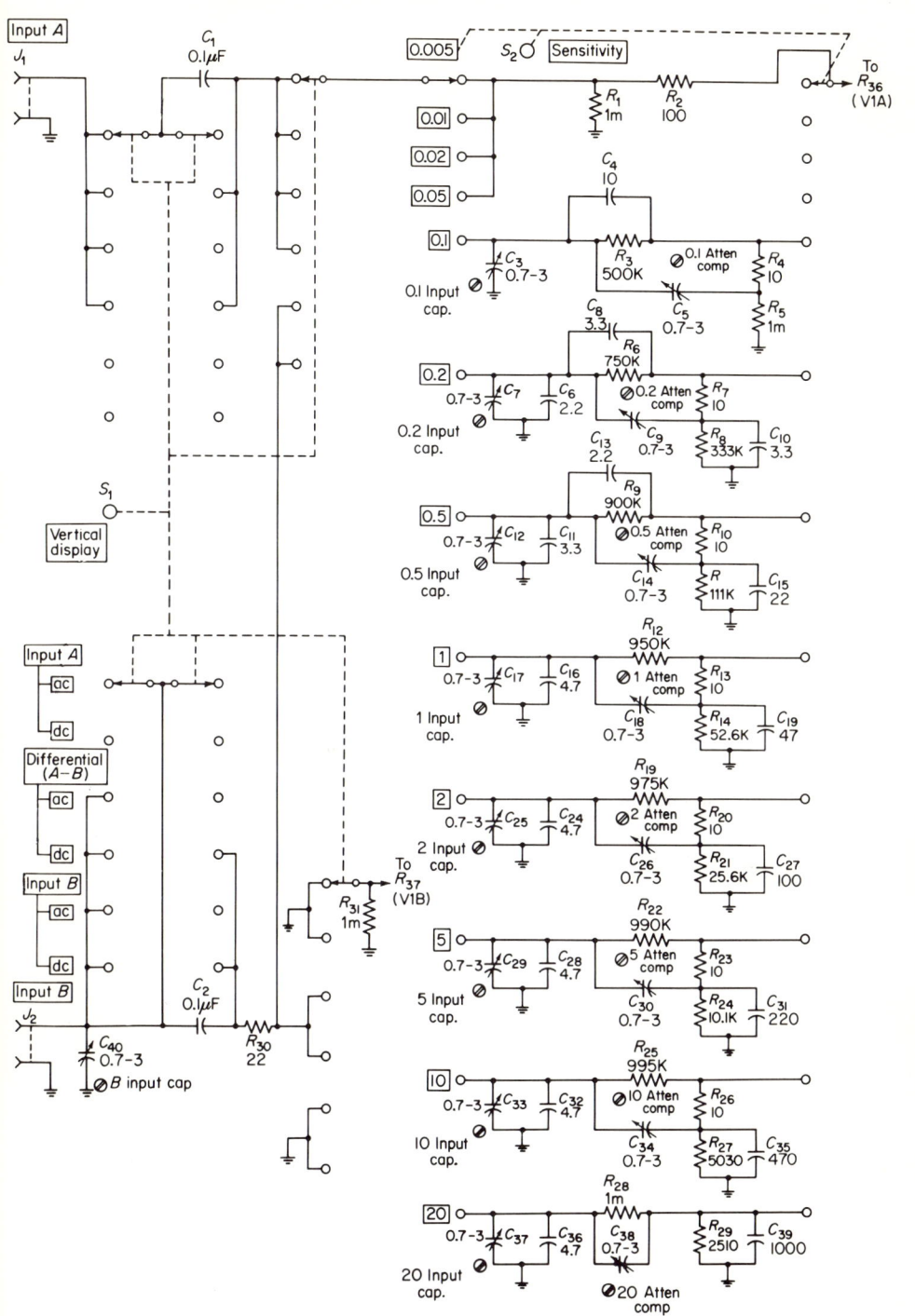

Sec. 21.8 High-gain vertical plug-in amplifiers Model 162D

Figure 21-22 Input and step attenuator.

tion is provided at one input only. Consequently the entire range of attenuation is provided for single-ended input signals, but only on the four most sensitive ranges for DIFFERENTIAL (A-B) input signals.

The resistive component of the input impedance is 1 megohm for all positions of the sensitivity switch. Input capacitors for the ranges 0.01 v/cm to 20 v/cm are adjusted to agree with the input capacitance of the 0.005 v/cm to 0.05 v/cm sensitivity ranges. The attenuator compensation capacitors are adjusted for equal attenuation of low- and high-frequency signals.

Cathode-follower V1A/B serves two functions: it couples the single-ended input signals to the base of Q1 (in this condition the grid of V1B is grounded) and couples a differential signal to the bases of Q1/Q2.

The DC BALANCE control, R43, controls the bias voltage of V1A/B. The VERNIER control, R42A/B, decreases the amplitude of the signal applied to the differential amplifier Q1/Q2, thus decreasing vertical sensitivity of the range used.

Differential amplifier Q1/Q2 converts the single-ended signal at the cathode of V1A to a balanced signal. Its output is applied to differential amplifier Q3/Q4. Differential amplifier Q3/Q4 amplifies the signal and couples it to the cathode-follower V2A/B. The signal is also coupled back to the emitters of differential amplifier Q1/Q2 as degenerative feedback. The VERTICAL POSITION control R49 controls the d-c current in the feedback network.

Cathode-follower V2A/B provides a low impedance to drive the vertical amplifier in the 170 A Oscilloscope. The SENS. CAL. control R71, in the cathode circuit, permits limited adjustment of the gain of the 162D Amplifier to compensate for variations in gain of the main vertical amplifier of the 170 A Oscilloscope.

21.9 Wide-band vertical plug-in amplifiers Model 162F

The Model 162F Wide-Band Vertical Amplifier is a d-c coupled amplifier which converts a single-ended input signal into a balanced signal for application to the main vertical amplifier of either the 160B or 170A Oscilloscope. The amplifier section is preceded by a step attenuator and an input coupling (a-c–d-c) switch.

A schematic diagram of the Model 162F is shown in Fig. 21-24 (see back-flap envelope). The input signal is a-c or d-c coupled to a frequency-compensated attenuator controlled by the SENSITIVITY switch. The attenuator determines the calibrated sensitivity of the vertical channel oscilloscope.

Sec. 21.10   *Time mark generator Model 166B*   475

The output of the attenuator is coupled through cathode-follower V1 to one input of differential amplifier V2. The second input to V2 is provided by the DC BALANCE control, which permits compensation for unbalance in the amplifier. The output of differential amplifier V2 is a balanced signal that is applied to cathode-followers V4A and V4B, which drive the input of the main vertical amplifier in the oscilloscope. The VERNIER and SENS. CAL. controls both affect the gain of differential amplifier V2. The VERNIER provides continuous adjustment of sensitivity between calibrated steps selected by the SENSITIVITY switch. The SENS. CAL. control provides a means of setting the calibrated sensitivity. (Sensitivity is calibrated when the amplifier section has a gain of 4.)

Constant-current generator V3 maintains constant the sum of the currents through resistors R50 and R56. As a result, the d-c level drops from about +100 v at the output of differential amplifier V2 to about −5 v at the input to cathode-followers V4A and V4B, with essentially no attenuation of the signal. The drop is across resistors R50 and R56; resistors R51 and R57 isolate the constant-current generator from the output cathode-followers. Compensating capacitors provide signal paths for high-frequency components.

The VERTICAL POSITION control determines the division of current through resistors R50 and R56. The control thereby determines the vertical position of the display on CRT.

21.10 Time mark generator Model 166B.

The time markers provided by the Model 166B are clipped sine waves generated by a Hartley oscillator. The frequency of the markers is determined by the MARKER INTERVAL switch, which connects one of three tuned circuits into the oscillator circuit. The markers are applied either to the CRT cathode as display markers or to the front-panel MARKER OUTPUT connector for external use.

Refer to Fig. 21-25 (see back-flap envelope). Oscillator V2 is controlled by amplifier Q1, which acts as a switch between −6.3 v and the junction of R10 and CR2. When the amplifier is off, it is an open switch. Diode CR2 is then forward-biased, and R10 is effectively in parallel with the selected tuned circuit (the +6.3 v supply is a-c ground). Resistor R10 lowers the Q of the tuned circuit and thereby prevents oscillations. In addition, current flows from the +6.3 v supply through R10, CR2, and the coil of the selected tuned circuit, storing energy in the field of the coil.

When amplifier Q1 is turned on, it becomes a closed switch and applies -6.3 v to the anode of CR2. Diode CR2 is then reverse-biased, disconnecting R10 from the tuned circuit, and oscillator V2 oscillates at the selected frequency. However, because of the energy stored in the field of the coil prior to oscillation, the first half-cycle of oscillation is always the same polarity and full amplitude.

With the INTENSITY MODULATION switch set to INTERNAL, and the TIME MARKER switch set to DISPLAY, the unblanking gate from the oscilloscope switches amplifier Q1 on and off. Between gates, the amplifier is held off, and no oscillations occur. During the unblanking gates, the positive gate signal turns on the amplifier through cathode-followers V1A and V1B, and the oscillator oscillates at the selected frequency.

The oscillator output is applied to cathode-follower V3A. The MARKER AMPLITUDE control adjusts the bias on the cathode-follower and thereby determines how much of the oscillator signal is above the cut-off level. In any case, V3A conducts for less than half of each cycle of the oscillator output. The cathode-follower output is applied to amplifier V3B, and the plate signal of V3B is applied to the CRT cathode as an intensity-modulation signal.

With the TIME MARKER switch set to OUTPUT, the switch holds amplifier Q1 on, and oscillator V2 oscillates continuously. The signal is applied to V3B, but in this case the output signal is taken from the cathode of V3B. Thus the signal at the MARKER OUTPUT connector is a continuous train of positive, clipped sine waves.

The INTENSITY MODULATION switch permits the use of external signals in place of the intensity markers. With the switch set to EXTERNAL, the front-panel INPUT connector is capacitively coupled to the CRT cathode, and amplifier Q1 is held off, Q1 disabling the oscillator. With the INTENSITY MODULATION switch set to INTERNAL, the markers are generated as described above.

21.11 Delay generator Model 166D

The Model 166D Delay Generator provides the Model 170A Oscilloscope with delayed sweep operation. The unit inserts a known amount of delay, which can be selected at the front panel, between a reference trigger and the start of the main sweep generated by the oscilloscope. The Model 166D itself consists of a sweep generator and a delayed-trigger generator, as shown in Fig. 21-26. The sweep generator generates a linear voltage ramp, the delaying sweep, which is applied to the delayed-trigger generator. The delayed-trigger generator generates a trigger at the end of the delay period selected

Sec. 21.11   *Delay generator Model 166D*   477

at the front panel, and delivers the trigger to the main sweep generator of the oscilloscope, which then provides the delayed sweep.

A schematic diagram of the delaying sweep generator is shown in Fig. 21-27 (see back-flap envelope). In addition to the actual

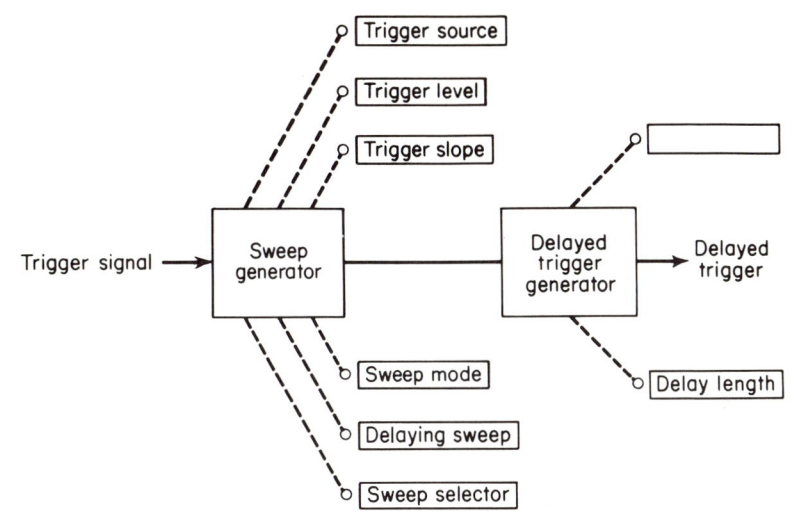

Figure 21-26 Overall block diagram of model 166D.

sweep generating circuits (V5, V6, V9, V10, and V13) the sweep generator contains amplifying and shaping circuits (V1 and V4) and gating circuits (V8). The amplifying and shaping circuits provide adequate triggering of the sweep circuits, and the gating circuits provide unblanking to the CRT.

The input or reference trigger is applied to one grid of amplifier V1, a differential amplifier, and a d-c signal from the TRIGGER LEVEL control is applied to the other grid. The output of the amplifier is a signal-ended signal which is proportional to the instantaneous difference between the trigger and the d-c signals. As shown in the figure, the TRIGGER SLOPE switch reverses the trigger and the d-c signals when switched from one polarity position to the other. The switch thereby determines the phase between the trigger signal and the output of the amplifier. With the switch set to +, the output of amplifier V1 is 180° out of phase with the trigger signal; with the switch set to −, the output of the amplifier is in phase with the trigger signal.

The output of amplifier V1 must be negative-going and must cross the +110 v level to start a sweep. Since the output of the amplifier is proportional to the difference between the trigger signal and the d-c value selected by the TRIGGER LEVEL control, the point on the trigger signal at which the amplifier output is $+100$ v depends upon the setting of the control. Thus the TRIGGER LEVEL control permits selection of the voltage level which the trigger signal must cross to start a sweep.

The signal from amplifier V1 is applied to trigger generator V4, a Schmitt trigger with narrow hysteresis limits. Provided that the signal crosses both hysteresis limits, the trigger generator switches back and forth between its two stable states, generating positive-going and negative-going voltage steps at its output. These steps are differentiated to form short pulses to be applied as triggers to gate generators V5/V7A. Only the negative triggers are used, and CR1 reduces the amplitude of the positive triggers.

Gate generator V5/V7A is a Schmitt trigger with wide hysteresis limits. Between sweeps, the A section of bias-control cathode-follower V13 holds the bias at the input of the gate generator close to the lower hysteresis limit. A positive trigger from trigger generator V4 has no effect, but a negative trigger drives the input to the gate generator below the lower hysteresis limit and causes the gate generator to switch.

When it switches, gate generator V5/V7A provides a positive and a negative gate. The positive gate is applied to the high-voltage power supply in the oscilloscope to unblank the CRT beam. The negative gate is applied to diode CR3 to start the sweep. Prior to the gate, CR3 has been forward-biased and has been holding the input to integrator V9 at about -2 v. The negative gate reverse-biases the diode and frees the integrator output.

Once freed, the input to the integrator starts going more negative, for it is connected to -100 v through the sweep resistor. Integrator V9 amplifies and inverts its input and produces a large, positive-going output which is applied back to the input through the sweep output cathode-follower V10A and the sweep capacitor. As a result, the input to the integrator changes by about 0.5 v during the sweep time. The voltage across the sweep resistor therefore changes about 0.5 per cent during the sweep time, and the current through the resistor changes by the same percentage. Since the current through the sweep resistor is the charging current for the sweep capacitor, the voltage across the sweep capacitor changes quite linearly with time, and the sweep signal is a nearly linear voltage ramp. The DELAYING SWEEP switch changes the value of the sweep resistor or capacitor to change the sweep time. The sweep output is applied to the delayed-trigger generator and to the SWEEP SELECTOR switch.

An attenuated sweep signal is applied to the input of gate generator

Sec. 21.11   *Delay generator Model 166D*   479

V5/V7A through cathode-follower V10B and B section of bias-control cathode-follower V13. This signal drives the input of the gate generator to the upper hysteresis limit and causes the gate generator to switch back to its pre-sweep state. The gate generator then ends the gates, removing its unblanking signal from the CRT and forward-biasing CR3. The diode then returns the input to integrator V9 to its pre-sweep level, resetting the sweep.

During the sweep time, cathode-follower V10B charges a hold-off capacitor. After the sweep ends, this capacitor lets the input to gate generator V5/V7A down slowly enough to prevent that circuit from being triggered again until the remaining sweep circuits have recovered. The DELAYING SWEEP TIME switch changes the size of the hold-off capacitor with sweep time.

Clamp V7B ensures that each sweep starts from the same voltage level, about −50 v.

The SWEEP MODE control determines the pre-sweep bias at the input to gate generator V5/V7A by setting the bias on the A section of bias-control cathode-follower V13. With the control set to PRE-SET or in the TRIGGER portion of its adjustable range, the gate generator bias cannot drop below its lower hysteresis limit unless trigger generator V4 provides a trigger. However, with the SWEEP MODE control set in the FREE RUN part of its range, the gate generator bias can drop below its lower hysteresis limit. Thus, as the hold-off capacitor discharges, it lets the gate generator bias fall to the lower hysteresis limit, and another sweep starts automatically.

The SWEEP SELECTOR switch determines the way the delaying and main sweeps appear on the CRT. With the switch set to MAIN SWEEP, the delaying sweep generator is disabled by the fixed bias applied to gate generator V5/V7A, the main sweep from the oscilloscope is routed back to the horizontal amplifier of the oscilloscope, and the main unblanking gate from the oscilloscope is applied back to the oscilloscope through gate-out cathode-follower V8B.

With the SWEEP SELECTOR set to DELAYING SWEEP, the delaying sweep generator operates normally, the delaying sweep is applied to the horizontal amplifier in the oscilloscope, and the main sweep is disconnected. The unblanking gates from the delaying sweep generator and the main sweep generator in the oscilloscope are mixed in the common-cathode circuits of V8. The delaying sweep unblanking gate is reduced in amplitude by R110; and, as a result, the main unblanking gate appears as a pedestal on top of the delaying sweep unblanking gate. The pedestal brightens the trace on the CRT during the time of the delayed main sweep.

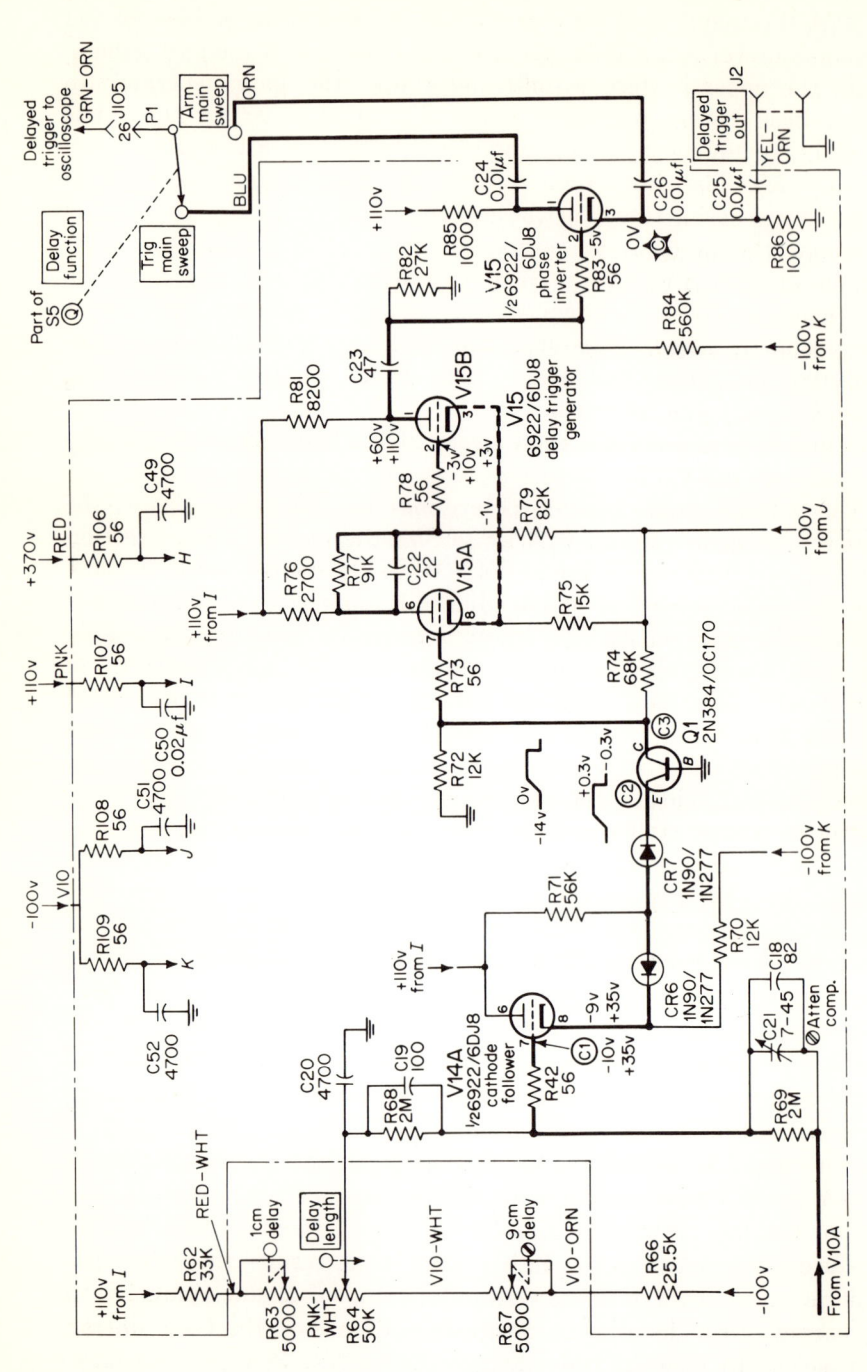

Figure 21-28 Sweep generator.

Sec. 21.11   *Delay generator Model 166D*

With the SWEEP SELECTOR set to MAIN SWEEP DELAYED, the delaying sweep generator operates normally, but the delayed main sweep is applied to the horizontal amplifier in the oscilloscope. The main unblanking gate is applied to the oscilloscope, and the delaying sweep unblanking gate is disconnected.

With the SWEEP SELECTOR set to MIXED SWEEP, the delaying sweep is applied to the horizontal amplifier in the oscilloscope, and the delayed main sweep is applied to the anode of diode CR5. The cathode of CR5 is connected to the delaying sweep output. Therefore, the delaying sweep signal is applied to the oscilloscope as long as the delaying sweep is more positive than the delayed main sweep. When the main sweep becomes the more positive signal, CR5 becomes forward-biased, and the main sweep is applied both to the oscilloscope and to gate generator V5/V7A. Thus the main sweep completes the trace on the CRT and terminates the delaying sweep as well.

The delayed-trigger generator is shown in Fig. 21-28. The signal applied to cathode-follower V14A is the algebraic sum of the delaying sweep signal and a d-c voltage selected by the DELAY LENGTH control. The delay period is the time required for the delaying sweep to make the sum equal to about 0 v. As long as the sum is negative, the cathode of V14A is negative. Diode CR6 is therefore forward-biased and holds the junction of CR6 and CR7 negative. Diode CR7 and transistor Q1 are then cut off. As the sweep progresses, the algebraic sum of the sweep and delay voltage approaches 0 v, the cathode of V14A goes positive, and the junction of CR6 and CR7 also goes positive. Diode CR7 then becomes forward-biased, and current flows into the emitter of Q1. As the delaying sweep continues, CR6 becomes reverse-biased and disconnects CR7 from the output of cathode-follower V14A.

As Q1 starts to conduct, it produces a positive-going signal at its collector. The positive-going signal is applied to delayed-trigger generator V15, a Schmitt trigger, which then switches states and produces a positive voltage step at its output. The step is differentiated into a short pulse and applied to phase inverter V14B. The phase inverter provides both positive and negative pulses. The positive pulse is applied to the DELAY FUNCTION switch and to the front-panel DEL. TRIG. OUTPUT connector. The negative pulse is applied only to the DELAY FUNCTION switch.

The DELAY FUNCTION switch selects either the positive or the negative pulse from phase inverter V14B and applies the pulse to bias-control cathode-follower V113A in the sweep generator of the oscilloscope. In addition, the DELAY FUNCTION switch determines the type of operation of the main sweep generator. With the DELAY FUNCTION switch set to TRIGGER MAIN SWEEP, the main sweep generator operates normally, and the negative pulse from phase inverter V14B starts the main sweep.

When set to ARM MAIN SWEEP, the DELAY FUNCTION switch converts V113 in the oscilloscope sweep generator to a Schmitt trigger, thereby setting the main sweep generator for single-sweep operation. The positive pulse from phase inverter V14B sets the Schmitt trigger circuit of V113 to arm the main sweep generator, which then produces a sweep when triggered through the triggering circuits of the oscilloscope itself.

INDEX

ACCELERATION, transducers used to measure, 412
Active region, 169
  in common-emitter amplifier, 170
Air navigation systems, electromagnetic delay line used in, 429
Alpha, current amplification factor, 57, 97
AM (see Amplitude modulation)
Amplification systems, conditionally stable, 125
Amplifier:
  buffer, 444
  cascade, 450–451, 452
  common-base, 74–75
    h parameter equivalent circuit for, 91
    high-frequency equivalent circuit for, 97–98
    leakage current in, 60–61
    small-signal graphical analysis, 74–76
    voltage source T-equivalent circuit, 83
  common-collector, 56, 144–145 (see also Emitter-follower circuit)
  common-emitter, 58, 76–77
    a-c equivalent-circuit for, 86–87
    frequency current gain in, 98
    ideal collector characteristics for, 170–171
    leakage current in, 60–61

Amplifier (cont.):
    small-signal graphical analysis, 76–77
    using emitter bias, 69–70
    using fixed bias, 67
    using self bias, 70–71
    using single-battery bias, 68–69
  d-c, 117
    simple two-stage, 117
  dual-trace plug-in, 463–471
  electron-tube, conventional, 149
  feedback, 120–138
    block diagram, 121
    defined, 120
    operation of, 121–122
    three-stage, 130–131
    two-stage, 128–129
  grounded-cathode, 140
  grounded-plate (see Cathode-follower circuit)
  high-gain vertical, Hewlett-Packard Model 162D, 471–474
  horizontal, 448, 449, 459–460
  linear and thermal runaway, 73–74
  low-gain wide-band, in dual-trace amplifier, 463–464
  main vertical, in Hewlett-Packard 170A oscilloscope, 448–449, 450–454
  NPN junction transistor, 56, 58

Amplifier (cont.):
  paraphase, 149
    modified, 151–152
    single-tube, 150–151
  pentode, 106
  power and thermal runaway, 73–74
  pulse, 104
    wide band width in, 106
  R-C:
    equivalent circuit for low- and mid-frequency ranges, 99
    frequency compensated, 104
    need for compensation, 109
    operated under dynamic conditions, 43
  R-C coupled, 47, 48
    dynamic operation, 42–43
    push-pull, exciting arrangement based on, 140
    quiescent operation, 42
    static operation, 42
    two-stage, 128–129
    universal response curve for, 51–52
  R-C single-stage, Nyquist diagram for, 126
  R-C transistor, 99–102
    methods of high-frequency compensation, 109–112
  R-C two-stage, Nyquist diagram for, 127
  R-C vacuum-tube, 41–54
  shunt-compensated, 110
  shunt-series compensated, 111, 112
  transistor:
    feedback in, 136–137
    direct coupled, 117–118
    gain-bandwidth product, 107
    reducing nonlinear distortion in, 125
    sag in, 113
    single-stage:
      using current feedback, 136
      using voltage feedback, 136
  vacuum-tube:
    direct-coupled, 115–117
    feedback concepts of, applicable to transistor amplifiers, 136
  vertical plug-in, in Hewlett-Packard 170A oscilloscope, 448–449, 450
  video, 104
    sag in, 114–115
  wideband, 104–119
  wideband vertical, 474–475

Amplifier circuits:
  common-base transistor, 57–58
  positive feedback used in, 120–121
Amplifier signal-to-noise ratio, 125
Amplifiers:
  amount of sag permissible, 114–115
  bandwidth of, 104
    increase, by negative feedback, 135
  considered as black box, 80
  frequency response of, 47
  frequency-response characteristics, 52
  low-frequency response, 112–115
  nonlinear distortion in, 123–124
  overdriven, 188, 189
  signal-to-noise ratio, 125
  stability, effect of negative feedback on, 122
  using negative current feedback, 133–134
Amplitude distortion, 52–53
Amplitude modulation, compared with pulse-position modulation, 418
Amplitude-modulated carrier, 2
Amplitude selector, 176 (*see also* Clipper)
Antenna, in radar system, 2, 3
Astable multivibrators (*see* Multivibrators, astable)
Astigmatism control, 461
Attenuation, in delay lines, 438–439
Attenuation circuitry, in dual-trace amplifier, 464–465
Attenuator, frequency-compensated, 474–475

BACKSWING, 291
Base circuits, resistance in, 73
Base spreading resistance, 109
Baseline stabilizer, 196
Beam, resetting in Beam-X switch, 394–395
Beam finder, 461
  included in cascode amplifier, 452
Beam-setting, operation in Beam-X switch, 394
Beam-X switch, 385–403
  application of, 400–403
  method of driving, 400
  method of zero-setting, 395–396
  ten-position circuit for, 398–399
  theory of operation, 385–400
  used with flip-flop driver circuit, 395

Index

Bearing, in radar, 3
Beta, current amplification factor, 58
Bias, single-battery, 68–69
Bias battery, 179ff
Bias circuits, 67–71
Binaries, 348–354
   in 3-stage ring counter, 360–361
   in transistor counter, flip-flop circuit, 404
Binary:
   cascaded, 349–351
      preset counting techniques applied to, 365–367
   cathode-coupled, 262–264
   as scale-of-two counter, 348–349
   single, used to construct decade counter, 361
Binary addition, performed by binary counter, 352
Binary chain, 351, 365
Binary circuits:
   transistor, 254–255
   vacuum-tube, 253–255
Binary counter, 252, 259
Binary number system:
   compared with decimal number system, 422–423
   used in pulse-code modulation, 421–422
Bistable multivibrators (*see* Multivibrators, bistable)
Black box concept, 89–90
Blanking pulse:
   horizontal, 5
   vertical, 6
Blocking oscillator (*see* Oscillator, blocking)
Bootstrap sweep circuit, 321–324
Bottomed region, 63, 169 (*see also* Saturation region)
Breakdown diodes, 163, 164, 167, 173, 255
   use, for nonsaturated transistor, 260–261
Breakpoint, in voltage-current plot of ideal diode, 161
Burroughs Corporation:
   Beam-X switch manufactured by, 385
   readout indicators manufactured by, 384, 385

CALIBRATOR, in Hewlett-Packard oscilloscope, 448, 449, 461–463

Capacitance:
   in cathode-follower circuits, 142–143
   in distributed-parameter lines, 437, 438
   junction, 165, 166
   low-effective input, 139
   resonant short-circuited lines as, 436
Capacitor:
   cathode-bypass, 149
   high-frequency compensation, 452
   in linear wave-shaping circuits, 33, 34
   rate of charge, 15
   transmission line as, 432
Capacitor discharge period, 17–19
Capacitors:
   commutating, in soft-biased binary, 253
   speed-up, 253, 259
Carrier lifetime, 166
Cascaded decades, 368
Cathode-bias circuits, 114
Cathode-bias network, 113
Cathode-bias resistor, 212
Cathode follower, 56, 108
   cascaded, in dual-trace amplifier, 465, 467
   cross-coupled, 451ff
   output, in dual-trace amplifier, 466ff
Cathode-follower circuits, 139–159
   advantages, 139
   circuit arrangement of, 140–141
   defined, 139
   disadvantages, 139
   frequency response, 144
   input capacitance, 142–143
   input resistance, 142–143
   output impedance, 143
   power gain, 142
   voltage gain, 141–142
Cathode-ray oscilloscope, 37
   astable multivibrator used in, 244–245
   electromagnetic delay line used in, 429
Cathode-ray tube circuits:
   balanced, 206–207
   unbalanced, 206
Cathode-ray tubes:
   diode clamps used in sweep circuits of, 204–206
   electromagnetically-deflected, 331, 332
   time-based generator used with, 314
Cathodes, oxide-coated, tubes using, 185
Chemical processing, counters used in, 347

Circuit:
　scale-of-two, 265–266
　time constant of, 15, 31
Clamp:
　biased diode, 207–209
　　simple, 214–215
　grid, 210–212
　keyed:
　　one-way, 215
　　two-way, 216–217
　unbiased, 207
Common-emitter amplifier (*see* Amplifier, common-emitter)
Common-emitter collector characteristics, 64
Common-emitter configuration:
　h parameter equivalent circuit for, 95
　in transistor physics, 56
Common-emitter h parameter, 148
Common-emitter stages, R-C coupling used between, 99
Complementary symmetry, 117–118
　properties of transistors, in monostable MV circuit, 281–282
Compressor, in PCM system, 427
Cosmic-ray intensity, 412
Counter:
　binary, 252, 259
　blocking oscillator, 310–311
　diode, 309
　preset decade, 367–368
　preset/reset, 402–403
　ring, 358
　　three-stage, using binaries, 360–361
　　vacuum-tube, 358–360
　scale-of-2, 348–359
　scale-of-3, 354–361
　scale-of-4, 349
　scale-of-9, 357
　scale-of-16, 351–354
　　converting to decade counter, 361–365
　scale-of-32, 354
　step, 288
　step-by-step, 309
　ternary, 356
Counters, 347–411 (*see also* Hewlett-Packard electronic counters)
　application of gating circuit, 346
　to base 2, 354
　to base other than 2, 354

Counters (*cont.*):
　decade, 361–365
　　bidirectional, 402
　　defined, 361
　　preset counting techniques applied to, 365, 367–368
　　transistor, 406, 407
　preset, 365–368
　　application, 365
　　Beam-X switch for applications, 402–403
　　defined, 365
　pulse, frequency divider circuits used in, 309
　transistor, 403–410
　used with gating circuit, 368–384
　vacuum-tube, 347, 402
Counting:
　to base 2, 348–354
　decade, 400–403
Counting section, of Hewlett-Packard electronic counter, 379, 380–382
CRO (*see* Cathode-ray oscilloscope)
CRT (*see* Cathode-ray tube)
Current source T-equivalent circuit, 84
Cutoff limiting, 186–188
Cutoff region, 64, 168
　in common-emitter amplifier, 170
　of a pentode, 173
Clamper, 199
　diode, 199–207, 212
　triode-grid, 210, 211
Clamping circuits, 196–220, 315
　classifications, 196
　defined, 196
　diodes, 196
　disadvantages, 212–214
　grid, 196
　keyed, 196, 214–219
　semiconductor diodes used in, 209–210
　unilateral characteristics, 212, 213
Clamping diode, 260–261
Clipper:
　biased-diode, 179–183
　cathode-coupled, 188, 190
　defined, 176
　nonlinear wave-shaping circuit, 176
　parallel-diode, 178–179, 181
　series-diode, 177–178
Clipper circuit, transistor, 190, 191
Clipping, by collector saturation, 191

Index 487

Collector current, 60
Common-base amplifier (*see* Amplifier, common-base)
Common-base configuration:
  a-c equivalent circuit for, 84–86
  h parameters for, 90
  in transistor physics, 56
  two-generator equivalent circuit, 82
Common-collector amplifier (*see* Amplifier, common-collector)
Common-collector h parameter, 147–148

DAMPED ringing effect, 295
Damping, 36ff
  in blocking oscillator, 297, 298–299
db (*see* Decibels)
D-C amplifier (*see* Amplifier, d-c)
D-C circuit analysis, 66
D-C restorer, 196
Decade counters (*see* Counters, decade)
Decay time, 294, 295
Decibels, 122–123
Decimal number system, 422–423
Decoder, 426
Degeneration, 114, 120
Degenerative effect, in common-emitter amplifier, using self-bias, 71
Deionizing potential, of neon tube, 317
Delay function switch, 481–482
Delay lines, acoustic, 429
Delay lines:
  distributed-constant, 429, 440
  distributed-parameter, 437–440
    capacitance, 437–438
    cascaded, 440
    inductance, 437–438
    impedance, 437
    use in blocking oscillators, 445–446
  electromagnetic, 429–447
    applications, 429, 441–446
    attenuation, 438–439
    classifications, 429
    defined, 429
    delay-to-rise time ratio, 440
    in Hewlett-Packard oscilloscope, 451
    temperature coefficient, 440
  lumped-constant, 429
  lumped-parameter, 440–441
    in radar systems, driver stages, 441–445

Delay time:
  cascade, in decade counting, 401, 402
  storage, 172
Depletion region, 166
Differentials, 25–26
Differentiating circuit, 27
Differentiator output, 27–28
Detector circuits, in pulsed systems, 414
Digital computers:
  counters used in, 347
  nickel iron used in circuits, 290
Diode matrix, 407
Diode rectifier, 161
Diodes (*see also* Breakdown diodes, Zener diodes):
  back resistance, 209, 210
  crystal, 176, 256
  damping, 299, 300
  as electronic switch, 167
  gas, 164
    use in neon sawtooth generator, 318
  germanium, 209–210
    back resistance in, 210
    saturation current, 163
  high-vacuum, 167
  junction:
    increase in reverse current, 209
    transient response, 164–167
  plate-catching, 276, 330
  plate-catching action, 256
  pulse-steering, 258
  series-connected, as triggering device, 256
  semiconductor, 208
    in clamping circuits, 209–210
    switching delay, 171
  semiconductor junction, 162–164
  silicon:
    back resistance, 210
    breakdown in, 164
    saturation current, 163
  silicon junction, limitations of, 210
  switching characteristics, 208
  switching speed, 166–167
  thermionic, 176, 208, 256
    compared with semiconductor junction diode, 162–163
  thermionic high-vacuum, 162, 209
Distortion (*see also* Amplifier distortion, Frequency distortion, Harmonic

Distortion (cont.):
  distortion, Nonlinear distortion, Phase distortion):
  from pulse passage through linear network, 22
  in low-pass filter, 31
  in R-C amplifiers, 52–53
Driver, for hard-tube pulse, 442–445
Driver circuits (see Flip-flop driver circuits)
Driving pulse, in Bistable MV, 251
Droop, 112 (see also Sag)
Duration (pulse), defined, 8
Dynamic operation:
  in transistors, 57
  in vacuum tubes, 42–43
Dynamic plate resistance, 114
  of vacuum tube, 44, 46
Dynamic screen resistance, 113–114

EAR, response to sound, 426–427
Eccles-Jordan trigger circuit, 251–252
  modified, 354, 355–357
Echo, in radar, 3
Electric fields, 386–387
  utilized by Beam-X switch, 386, 388
Electrical signal, transducer-produced, 412
Electromagnetic deflection, of electron beam, 314, 331
Electromagnetic deflection system, 314
Electromagnetic delay lines (see Delay lines)
Electromotive force, 288
Electron, behavior of, 386–387
Electron beam, 314, 331 (see also Beam-X switch)
Electron scanning beam, in TV system, 5
Electron transit time, 162
Electronic counters (see Counters, electronic)
Electronic data processing systems, 347
Electronics, importance of semiconductors in, 55
Electrostatic deflection system, 314
Electrostatic horizontal deflection, of electron beam, 314, 331
emf (see Electromotive force)
Emission saturation, 185
Emitter biasing, 69–70
Emitter-follower circuit, 140
Equalizing pulse, 6

Equivalent circuit:
  common-emitter, 145
  constant-current, 50–51
  constant-voltage, 45–47, 51
    of amplifier with feedback, 131–132
    for analyzing tetrodes and pentodes, 50
    effects of frequency, 47–49
    practical use of results of using, 46
  high-frequency:
    for common-base amplifier, 97–98
    for R-C coupled amplifier, 49
    for R-C transistor amplifier, 101
    simplified, 101
  low frequency, of R-C coupled amplifier, 48–49
  middle frequency, of R-C coupled amplifier, 49
  transformer, 292
  two-generator, 82
Equivalent circuits, 41, 430
  in astable MV, cathode-coupled, 235, 236–237
  transistor, 226
  vacuum-tube, 225–226
  modified by increase in transistor frequency response, 97
  in monostable MV, transistor, 279
  transistor, 79–103
Exponential curve, 14
Exponential waveform, defined, 10

FALL time, defined, 8
Feedback:
  current, 128, 133
    in cathode-follower circuit arrangement, 140
  defined, 120
  degenerative, 107, 120
  difference in, at high frequencies, 151
  inverse, 120
  loss in signal gain owing to, 71
  negative, 120, 121, 139
    effects of use, 122
    effects on transistor amplifiers, 136
    increase of amplifier bandwidth, 135
    reduce nonlinear distortion, 123–125
  negative current, 133–135
  positive, 120
  regenerative, 120
  simultaneous, 361–362

Index

Feedback (*cont.*):
  in transistor amplifiers, 136–137
  in transistor astable MV, 224
  in vacuum-tube astable MV, 223
  voltage, 128
    effect on input impedance, 132–133
    effect on output impedance, 131–132
Feedback amplifier (*see* Amplifier, feedback)
Feedback attenuation, 471, 474
Feedback factor, 122
Feedback network, 471
Feedback oscillator (*see* Oscillator, feedback)
Feedback transformers, 287
Feedback voltage ratio (*see* Reverse voltage ratio)
Ferrites:
  core loss in, 291
  magnetic, saturation level of, 290
  permeability of, 289
Figure-of-merit:
  transistor, 108–109
  vacuum tube, 107–108
Firing potential:
  of neon tube, 316, 317
  of thyratron, 318, 319
Flip-flop circuit, 251
  application of, 265–267
  electronic equivalent of SPDT switch, 392
  method of driving Beam-X switch, 400
  in transistor counter, 404
  triggering, 258, 259
Flip-flop driver circuits, 395
Flip-flop transistor:
  in direct-coupled bistable MV, 264
  nonsaturated, 260–261
  saturated, 260
Flop-over circuit, 251
Flux density, magnetic, 289
  saturation, 290
Flyback time, 316
  defined, 315
FM (*see* Frequency modulation)
Focus control, 461
Forward current transfer ratio, 90
Forward transient, 164–165
Four-terminal networks, 80
  typical arrangements, 112

Frequency:
  of astable MV, 238–240
  effects on constant-current equivalent circuits, 51
  effects on constant-voltage equivalent circuits, 47–49
Frequency compensation, 106
Frequency converters, Hewlett-Packard, 374
Frequency distortion, 52
  reduction of, 104, 120
Frequency dividers, 252
  blocking oscillator used as, 288, 308–309
Frequency division, in monostable MV circuit, 282
Frequency division and multiplication, transistor astable MV used for, 247
Frequency measurements, 368, 372–373, 374–376
Frequency modulation, 418
  limiter circuit, 191
  PAM signal used to modulate transmitter, 4
Frequency range, wide:
  maintaining stability over, 125–127
  requirements for uniform response, 139
Frequency response:
  of cathode-follower circuit, 144
  improvement of, 135–136
Frequency stability, of astable MV, 240–244
Frequency unit switch, 373
Function, defined, 81

Gain:
  maximum available, 94
  variation in transistor replacement, 67
Gain bandwidth product:
  transistor, 107
  vacuum tube, 106–107
Gas discharge decade indicator tube, 407, 408–409
Gate:
  bidirectional, 338–340
  coincidence, 334, 336–337
  defined, 334
  four-diode, 340–341
  linear, 334
  six-diode, 341–342
  threshold, 335–336
  time selector, 334

Gate (*cont.*):
  transmission (*see* Transmission gates)
  triode, simple, 342–343
  two-triode, 343–344
  unidirectional (*see* Gate, coincidence)
  unidirectional diode, 334–336
Gate generator, 456–457
Gate section, of Hewlett-Packard electronic counter, 378
Gate switch, manual, in Hewlett-Packard electronic counter, 379
Gating circuits:
  astable MV used in, 247
  used with counters, 368–384
Gating pulse, 335ff
General magnetic theory, 288–291
Generation:
  square-wave, 247
  time-base, 247
Generator:
  constant current, in Hewlett-Packard oscilloscope, 451, 452, 454
  delay, used in Hewlett-Packard oscilloscope, 476–482
  delayed trigger, 476–477, 481–482
  gate (*see* Gate generator)
  pulse (*see* Pulse generator)
  sawtooth, 415–417
    neon, 316–318
  sawtooth wave form:
    negative, 316
    positive, 215–216
  sinusoidal-signal, 2
  sweep (*see* Sweep generator)
  time mark, 475–476
Generators, time-base, 314–333
  current, 315, 316, 330–332
  thyratron, 319–321
  voltage, 314ff, 330–331
  neon lamp as, 316–318
Grid-cathode circuit, 183
  clamping in, 210–212
Grid conduction, 167–168
Grid-limiter circuit, 183–184
  combined with cutoff limiting, 188
Grid-to-cathode circuit, as a diode, 167

h parameters, 79, 89–96 (*see also* Common-collector h parameter, Common-emitter h parameter)
  black box concept, 89

h parameters (*cont.*):
  related to r parameters, 96–97, 148
Hard-tube pulse, 442
Harmonic distortion, 53, 120
Hartley oscillator, 475
Heart beat, transducer used to measure, 412
Hewlett-Packard electronic counters:
  Model 524C, 368, 369
  Model 524D, 347, 368ff
    gate section, 378
    as high speed totalizer, 377
    time base section, 337–378
    used for measurements, 372–377
Hewlett-Packard frequency converters, 374
Hewlett-Packard High-Gain Vertical Amplifier, Model 162D, 471–474
Hewlett-Packard oscilloscope, Model 170A, 448–482
Hewlett-Packard transfer oscillator, 372
High-frequency circuit, and pulse transformers, 292–293
High-frequency discriminator, 28 (*see also* R-C filter, low-pass)
High-frequency half-power point, 52
High-frequency response, methods of improving, 109–112
High-speed totalizer, 377
Horizontal centering controls, 206
Horizontal display switch, 459, 460
Hybrid parameters (*see* h parameters)
Hysteresis, 263
Hysteresis loop, 290

IMPEDANCE:
  characteristic, 430, 437
  in distributed parameter delay lines, 437
  forward transfer, 81, 82
  input (*see* Input impedance)
  load (*see* Load impedance)
  output (*see* Output impedance)
  surge, 430
  reverse transfer, 82
  terminating, 431–432
  in transmission lines, 430–432, 436–437
Impedance-matching devices, 94
Impedance parameters, open-circuit, 82
Indicator, in radar system, 2, 3
Indicator circuit, in radar system, 2

Index

Inductance:
  in distributed-parameter delay lines, 437–438
  short-circuited resonant lines as, 436
Inductors:
  equivalent circuit of, 330, 331–332
  in linear wave-shaping circuits, 33–34
  part of R-L-C circuit, 34
  transmission line as, 432
Industrial control circuits, relays in, 266–267
Input, differentiation of, 28
Input capacitance, low-effective, 139
Input circuit equation, 81
Input impedance, open circuit, 81, 82
Input resistance, 90
  in cathode-follower circuits, 142–143
Input waveform, 176, 177
Instrumentation systems, flip-flops in, 266
Interelectrode capacitances, 143
  in a pentode, 174
  reduction in, 169–170
  vacuum-tube, 325–326
Interlacing, 6
Intermodulation distortion, 53
Internal resistance parameters (*see r* parameters)
Internal short-circuit forward-current gain, 83–84
Interval measurements, 377–378
Inventory control, 354
Ionizing potential, of neon tube, 316

JITTER, 239, 272
Junction capacitance, 165, 166
Junction transistor, 66, 170–173

KIRCHHOFF voltage law, 85, 87
  applied to input, output loops, 145

LEAKAGE current, 60–61, 64, 67
Leakage resistance, dielectric, 430
Letter subscripts, to specify transistor characteristics, 82
Limiter, 176 (*see also* Clipper)
Limiter circuits, 191–194
Line pulse, horizontal, 5 (*see also* Synchronizing pulse)
Linear amplifier (*see* Amplifier, linear)
Linear conductors, transmission lines as, 43

Linear network, defined, 12
Linear wave shaping, 12–40
  defined, 12
Linear wave-shaping networks, 176
Load impedance, 47–48
  low-effective, 139
Load resistor, 56
Load resistor, spade, in Beam-X switch, 389
Locking circuit, 252
Locus, 126
Low-frequency amplification, 115
Low-frequency compensation, 113
Low-frequency discriminator (*see* R-C filter, high-pass)
Low-frequency half-power point, 52
Low-frequency response, 112–115

MAGNETIC circuit, 289
Magnetic fields, 386–387
Magnetic flux density, unit of, 289
Magnetic forces, 288
Magnetic materials, characteristics of, 289–291
Magnetic theory, general, 288–291
Magnetomotive force, 288
Measurements, counters used for, 347 (*see also* Frequency measurement, Interval measurement, Period measurement, Ratio measurement, Time-interval measurement)
Mechanical stress, transducers used to measure, 412
Medical research, counters used in, 347
Miller effect, 108, 254, 326
Miller run-down, 329
Miller sweep circuit, 325–327
Miller sweep generator, 328
Minority-carrier injection, 164
Minority carrier storage, 165, 166, 210
mmf (*see* Magneto motive force)
Modulation, in radar system, 2
Molybdenum sputtering technique, of tube manufacture, 407, 409
Monostable multivibrators (*see* Multivibrators, monostable)
Multivibrators:
  astable, 221–250
    applications, 244–249
    compared with astable blocking oscillator, 287

Multivibrators (cont.):
  defined, 222
  frequency, 238–240
  synchronization, 240–244
  waveforms, 227–232
  astable cathode-coupled, 232–237
    circuit, 233
    equivalent circuits, 235, 236–237
    waveforms:
      asymmetrical operations of, 237
      symmetrical operations of, 234, 237
  astable electron-coupled, 237
  astable transistor, 224–225
    equivalent circuits, 226
    waveforms:
      asymmetrical operation of, 232
      symmetrical operation of, 229–232
  astable vacuum-tube, 222–224
    duration, 238
    equivalent circuits, 225–226
    feedback action, 223
    waveforms:
      asymmetrical operation of, 232
      symmetrical operation of, 227–229
  biased, 441–445
  bistable, 251–268
    applications of, 265–267
    defined, 222
    Eccles-Jordan trigger circuit, 251–252
    operation of, 252–253
    self-biasing in, 253–255
    Schmitt trigger circuit a form of, 457–458
  bistable direct-coupled, 264–265
  bistable transistor:
    operation of, 253
    triggering devices, 257–260
  bistable vacuum-tube:
    operation of, 252–253
    triggering devices, 255–257
  compared with blocking oscillator, 287, 300
  definitions of terms, 221–222
  driven, 269
  monostable, 269–285
    applications, 283–284
    compared with monostable blocking oscillator, 288
    defined, 222
    described, 269
    nonsaturated, 283

Multivibrators (cont.):
  monostable cathode-coupled, 273–275
  monostable transistor, 278–279
    circuit modifications, 279–283
  monostable vacuum-tube, 269–273
    circuit modifications, 275–278
    triggering diode used in, 276–277
  one-shot, 269
  in radar, 441–442
  switching, 470
  transistor symmetrical, 231
  triggered, 269
  type of relaxation oscillator, 221
  vacuum-tube symmetrical, 229
Multiplex system, single, 5
Multiplexing, 416–417, 418
  time-division, 4
MV (see Multivibrators)

NEON lamp, as time-base generator, 316–318
Network response, 19–22
Nickel iron, permeability of, 289
  saturation level, 290
Nixie readout indicator, 384–385, 402
Nonlinear distortion, 52–53
  reducing, by negative feedback, 123–125
Nonlinear waveshaping, 176–195
Nonsinusoidal waveform:
  applied to linear networks, 12
  faithfully reproduced, 104
NPN junction transistor amplifier (see Amplifier, NPN junction transistor)
NPN transistor, 61, 62, 117
  negative base triggering for, 259
Nucleonics, counters used in, 347
Number systems (see Binary number system, Decimal number system)
Numerical subscripts, use for general circuit analysis, 82
Nyquist diagram, 126, 128 129, 130

OERSTED, 289
Oscillating circuit, 237
Oscillation:
  in time-base generators, 332
  recovery time, 330
  time required for, 300
Oscillators:
  blocking:
    applications, 308–312

Index

Oscillators (cont.):
  astable, 287, 300
  bistable, 288
  compared with multivibrators, 287, 288, 300
  defined, 286
  distributed-parameter delay-line used in, 445–446
  frequency stability, 300
  monostable, 287–288
  output, frequency and duration, 300
  resonant period of transformer, 287
  synchronized by sinc signal, 308
  transistor, 302–303
    triggering methods, 306–308
  uses, 258, 287, 288
    in pulse-position modulation system, 418–420
  vacuum-tube, 295–298, 303
    triggering methods, 304–306
  feedback, 286, 287
  Hartley, 475
  and positive feedback, 121
  relaxation, 221, 316 (see also Generator, neon sawtooth)
  ringing, 37, 39
    compared with R-L-C peaker circuit, 38
  sine-wave, 221, 415, 416
  transfer, Hewlett-Packard, 372
  triggered blocking, 288
Oscilloscope:
  cathode-ray (see Cathode-ray oscilloscope)
  electronic-switching, 245–246
  Hewlett-Packard Model 170 A, 448–482
    amplifiers used:
      dual-trace plug in, 463–471
      high-gain vertical, 471–474
      horizontal, 459–460
      main vertical, 450–454
      wide-band vertical, 474, 475
    block diagram, 449
    calibrator, 461–463
    functional sections, 448–450
    generators used:
      delay, 476–482
      sweep, 455–458
      time-mark, 475–476
    high-voltage power supply, 460–461

Output circuit equation, 81
Output conductance, 90
Output impedance, 143
  open-circuit, 82
Output pulse:
  in bistable MV, 251
  shape of, obtained from blocking oscillator, 298–300
Output voltage amplitude, 26–27
Output waveform, 31
  critically damped, 36
  effect of rise time on amplitude, 24
Overdamping, 36
Overshoot, 105, 293, 300
  defined, 8
  result of high-frequency deficiencies, 106

PAM (see Pulse-amplitude modulation)
PAM signal, 5
Parameters, 10 (see also h parameters, r parameters)
  transistor, 79–103
  vacuum-tube, 43–45
Paraphase amplifier (see Amplifier, paraphase)
Parallel-resonant circuit, 435
PCM (see Pulse-code modulation)
PDM (see Pulse-duration modulation)
Peak-to-peak amplitude, 393
Peaker circuit, 191–194
Pedestal, 336, 341
  removal of, 344–345
Pentode amplifier (see Amplifier, pentode)
Pentode amplifier design, 51
Pentode gates, 345
  multicoincidence, 345
Pentodes, 173–174
  grid-cathode circuit, 183
  operating regions, 173–174
  region, in switching operations, 168–169
  special, for wideband application, 107–108
Period measurement, 373–375, 377
Petroleum processing, counters used in, 347
Phantastron, 328
  monostable screen-coupled, 328–330
Phantastron-frequency divider, in Hewlett-Packard electronic counter, 382–383

Phase distortion, 53
 in delay lines, 438
 negative feedback to reduce, 120
Phase inverters, 140, 149
 cathode-coupled, 154–155
 common-emitter, 158–159
  two-stage, 157–158
 single-tube, 149–152
 transistor, 149, 156–158
  split load, 155–156
 two-tube, 152–155
Phase shift, 113 (*see also* Phase distortion)
 in two-tube phase inverter, 152
Phase-splitter circuit, simple, 151
Phase-splitters, 149
Photoelectric cell, application of binary counter, 354
Pips, in radar, 3
Plate-current saturation, 185
PNP transistor, 61, 64, 117, 258
Polarity inverter, 149
Power amplifier (*see* Amplifier, power)
Power gain:
 available, 94
 in cathode-follower circuit, 142
Power gain calculations, for common-base and common-emitter circuits, 60
Power supply:
 high-voltage, in Hewlett-Packard Model 170A oscilloscope, 448, 450, 460–461
 low-voltage, in Hewlett-Packard Model 170A oscilloscope, 448, 450
 in radar systems, 2
PPM (*see* Pulse-position modulation)
Preamplifiers, and thermal runaway, 73
Preset counter (*see* Counter, preset)
Preset relay driver, 366
prf (*see* Pulse repetition frequency)
prr (*see* Pulse repetition rate)
Production records, binary counter used in, 354
prt (*see* Pulse repetition time)
Pulse, amplitude value of, 212
Pulse amplifier (*see* Amplifier, pulse)
Pulse-amplitude modulated signal, 4
Pulse-amplitude modulation, 413–414
Pulse applications:
 amount of sag in, 115
 ferrites used in, 290
 silicon steel used in, 290
 wave forms useful in, 176

Pulse circuits, defined, 1
Pulse-code modulation, 421–427
 block diagram of system, 422
Pulse-duration modulation, 414–417
 methods of obtaining, 415–416
Pulse equalization, monostable MV used for, 283
Pulse-forming, by differentiating circuit, 27
Pulse-forming circuits, 439
Pulse-generation circuits, classifications of, 221
Pulse generators:
 active, 221
 gas-tube, 287
 low-impedance, 287
 multivibrator, 444
 passive, 221
Pulse input, response of low-pass R-C filter, 29–31
Pulse-length modulation (PLM) (*see* Pulse-duration modulation)
Pulse modulation, 412–428
 use of techniques, 412
Pulse-position modulation, 417–421
 demodulator, 419, 420
 in multiplex operations, 417–418
Pulse repetition frequency, 8
Pulse repetition rate, 8
Pulse repetition time, 8
Pulse-shaping circuit, 221
Pulse synchronization, 242–244
Pulse-time modulation (PTM) (*see* Pulse-position modulation)
Pulse train, 8
Pulse transformers, 288–295
 compared with conventional types, 291
 high-frequency response, 292–293
Pulse-type circuits:
 narrow pulses required, 24
 semiconductor arrangements in, 55
 use of controlled amplitude distortion in, 53
 vacuum tube arrangements in, 55
Pulse waveform, 221
Pulse width, 292
Pulse-width modulation (PWM) (*see* Pulse-duration modulation)
Pulses:
 defined, 7
 historical development, 1–2

Index

Pulses (cont.):
  typical applications, 2–7
  use in radar, 2
Push-pull transformer method, of driving Beam-X switch, 393

Q point stability, 67, 69
Quantized pulse, 393
Quantizing noise, in PCM, 421, 426–427
Quantizing process, in PCM, 421–422
Quiescent operation:
  in transistors, 57
  in vacuum tubes, 42

r parameters:
  common-collector equivalent circuit, 145–147
  related to h parameters, 96–97, 148
Radar, 2–3
  counters used in, 347
  defined, 2
  equipment, 1
  use of multivibrators in, 441–442
Radar Plan Position Indicator, 217–218
Radar pulse-forming networks, 191–194
Radar sets, and clamping circuits, 206
Radar systems:
  block diagram, 2
  lumped-parameter delay lines in driver stages, 441–445
  microwave, 441–442
Radio communication, 2
  equipment, compared with pulse circuits, 1
Radio telemetering, 5
  use of pulses in, 3–5
Radio telemetry, mobile, 412
Ramp input, 31–32
Ramp voltage, 24–26
Ramp waveform, 9
Range, in radar, 2
Ratio measurements, 376
R-C amplifier (*see* Amplifier, R-C)
R-C circuit, compared with R-L circuit, 33–34
R-C coupling, 99
R-C coupling network, action of, 196–199
R-C differentiator, 26–28
R-C filter:
  high-pass:
    related to low-pass, 28

R-C filter (cont.)
  response to ramp voltage, 24–26
  low-pass, 28–29
    as an integrator, 32
    response:
      to pulse input, 29–31
      to ramp input, 31–32
      to square wave input, 31
      to step-voltage input, 28
  pass-pass, 13
R-C filter circuit, high-pass, 12–14 (*see also* R-C differentiator)
R-C integrator, 32–33
R-C time constant, 15–16, 20
Readout indicators, 384–385
Readout section, of Hewlett-Packard electronic counter, 379, 382
Receiver, in radar system, 2, 3
Recovery time, 166
Rectangular pulse, 19
  ideal, defined, 7–8
Regeneration, 120
Regenerator, 252
Reset circuit:
  electronic:
    requiring negative pulse, 396–397
    requiring positive pulse, 397–398
  in transistor counter, 403–405
Reset switch, 351
  in Hewlett-Packard electronic counter, 379
  in vacuum-tube ring counter, 360
Resetting, operation in Beam-X switch, 394
Resistance-capacitance amplifier (*see* Amplifier, R-C)
Resistance, input (*see* Input resistance)
Resistor, transmission line as, 432
Resolution time, in self-biased binary counter, 254
Resonant lines:
  open-ended, 432–433
  short-circuited, 434–436
Resonant period, of transformer, 287
Respiration, transducer used to measure, 412
Response characteristics, 10
Reverse current (*see* Leakage current)
Reverse transient, 165
Reverse voltage ratio, 90
Ringing oscillator (*see* Oscillator, ringing)

Rise time, 105, 106
  defined, 8
  effect on output waveform amplitude, 24, 26–27
  of output waveform, reduction in, 110
R-L circuits, 33–34
R-L-C circuits, 34–36
R-L-C peaker, 38
Run-down, 327, 329

Sag, 112
  amount permissible in given amplifier, 114–115
  in common-emitter transistor amplifier, 113
  in pulse applications, 115
  in vacuum-tube circuits, 113
Sampling, in PCM, 421, 422
Sampling circuit, for PCM, 423–424
Saturation flux density, 290
Saturation limiting, 185–186
Saturation region, 63, 169
  in common-omitter amplifier, 170–171
  of a pentode, 173–174
Saturation resistance, 63
Sawtooth waveform, 314–316
Sawtooth time-base waveform, 314–315
Schmitt trigger circuit, 262, 457, 481 (see also Cathode-coupled binary)
  in Hewlett-Packard delay generator, 478, 482
  in Hewlett-Packard electronic counter, 378, 379
  in Hewlett-Packard 170A oscilloscope, 457–458
  in sweep generator, 455, 456
  in transistor counter, 404, 405
  transistor version, 263–264
Screen network, 113
Self-bias, 70–71, 184
  in bistable MV, 253–255
Self-oscillating circuit, 221
Semiconductor switching, 160
Semiconductors, importance in electronics, 55
Separator circuit, 334
Series compensation, 111
Series peaking, 111
Series-resonant circuit, 432, 435
Short-circuit current-amplification factors, 60

Short-time-constant circuit, 31–32
Shunt compensation, 109, 111
Shunt peaking, 109
Shunt-series compensated amplifier (see Amplifier, shunt-series compensated)
Shunting resistor, 299
Signal inversion, 58
Signal-to-noise ratio, 125, 413–414
Silicon steel:
  core loss in, 291
  permeability of, 289
  saturation level, 290
Sine wave, 12
Sine-wave oscillator (see Oscillator, sine-wave)
Sinusoidal waveform, 4
  in clipping circuits, 176
Sinusoidal waves, 1
Small-signal graphical analysis, 96–97, 148
  of common-base amplifier, 74–76
  of common-emitter amplifier, 76–77
SNR (see signal-to-noise ratio)
Spade bus voltage, in Beam-X switch, 389, 390, 396
Spade load resistor, in Beam-X switch, 389
SPDT switch, 392–393
Spikes, 23
Square wave, 9
Square wave input:
  response to, 23–24
  of low-pass R-C filter, 31
Squarer circuit, 191–194
Squaring circuit, Schmitt trigger circuit used as, 263
Stable state, 221
Standard Frequency Output Connector, 378
State, in multivibrators, defined, 222
Static operation:
  in transistors, 56–57
  in vacuum tubes, 42–44
Step attenuator, 471
Step voltage, 14, 19
Step voltage waveform, 8–9
Storage delay time, 172
Storage devices, 429
Subscripts, numerical and letter, 82
Superheterodyne receiver, 419
Sweep generator:
  boot strap, 324

Index

Sweep generator (*cont.*):
  constant-current pentode, 319–320
  in Hewlett-Packard delay generator, 476–477, 480
  in Hewlett-Packard Model 170A oscilloscope, 448–450, 455–458
  Miller, 328
  thyratron, 318–319
  triggered, 322
Sweep mode control, 457, 479
Sweep speed, 37
Sweep time, 460
  defined, 315
Sweep time switch, in sweep generator, 456, 457
Sweep waveform, 315
  integrator circuit to improve linearity, 324
  time-base generator used to generate, 314
Switch:
  ideal, 160–161
  low impedance, blocking oscillator used as, 288
  vacuum-tube, 160
Switch output, 4
Switching circuits, 72, 160
Switching control circuit, in dual-trace amplifier, 468–470
Switching devices, electronic, practical forms of, 161–162
Switching multivibrator (*see* Multivibrator, switching)
Switching operations, regions in, 168–169
Switching speeds in pentode, 174
Symmetrical waveforms, 23
  in astable multivibrator:
    cathode-coupled, 234
    transistor, 229–232
    vacuum-tube, 227–229
Sync, 240
Sync pulse:
  negative, 243–244
  positive, 242
  positive-going, applied to thyratron control grid, 318, 319
  serrated vertical, 32–33
Sync signal, 240–242
Synchronizing pulse, 6, 196 (*see also* Horizontal line pulse)
  in multiplexing, 416

Tail-biting circuit, 444
Telegraph systems, change of pulse characteristics in, 10
Telemetering (*see* Radio telemetering)
Television, 1
  use of pulses, 5–6
Television receivers:
  change in pulse characteristics in, 10
  sag in video amplifier, 114–115
  use of R-C integrator, 32–33
Temperature, effect on recovery time, 166
Tetrode, grid-cathode circuit, 183
Tetrode amplifier design, 51
Thermal excitation, 163
Thermal runaway, 71–74
Thyratron sweep circuit, 317, 318, 320–321
Thyratron sweep generator, 318–319
Thyratron tubes, use of, 318–319
Thyratrons, 267
Time-axis plug-in units, in Hewlett-Packard Model 170A oscilloscope, 450
Time-base section, of Hewlett-Packard electronic counter, 377–378
  phantastron frequency divider, 382–383
Time-bases, free-running, 288
Time constant, 19–21
  circuit, charging, 214
  defined, 15
Time-constant chart, universal, 16–17, 18
Time delay (*see* Phase distortion)
Time-interval measurements, 368, 373, 376–377
Timer, in radar system, 2
Toroidal sample, 290
T-R switch (*see* Transmit-receive switch)
Transconductance, of vacuum tube, 44
Transducer gain, 94
Transducers, 3–4, 412
Transformer coupling, 99
Transformer inverter, 149
Transformer output pulse, 291
  frequency components, 292
Transformers, 140
  simplest type of phase inverter, 149
Transient region, 166
Transient response, 105
Transient reverse current, 166
Transistor amplifiers (*see* Amplifiers, transistor)

498 *Index*

Transistor circuitry, electronic reset circuit with positive pulse adaptable to, 398
Transistor design, constant-current equivalent circuit in, 51
Transistor equivalent circuits, 79–103
Transistor limiters, 190–191
Transistor parameters, 79–103
Transistor phase inverters, 149
   two-stage, 156–158
Transistor storage time, elimination of, 173
Transistor sweep circuits, 321
Transistor switching differential amplifier, in dual-trace amplifier, 467–468
Transistor transient response, 171
Transistors, 55–78
   alloy junction, 279, 281
   basic circuits, 56–60
   basic current gains, equations showing relationships, 59
   characteristics, 79
   circuit notation, 56–60
   compared with vacuum tubes, 55
   determining junction temperature, 71–72
   dynamic operation, 57
   effects on internal capacitances, with frequency increase, 97
   equivalent circuits to replace, 79
   establishing correct operating point in, 64–66
   figure-of-merit, 108–109
   frequency response, 97–98
   germanium, 63, 66–67
   and h parameters, 79, 80
   high-frequency, 173
   high-frequency performance of, 97–98
   junction, 66, 170–173
   leakage currents, 60–61
   power dissipation in, 74
   quiescent operation, 57
   and r parameters, 79, 80–89
   replacement, important variations in, 66–67
   silicon, 63, 67
   static operation, 56–57
   and switching circuits, 72
   switching operations, region in, 168–169
   use, at high-junction temperatures, 71
   variations in given type, 64

Transistors (*cont.*):
   volt-ampere plots used in studying, 61–64
   for wideband application, 109
Transmission, defined, 430
Transmission gates, 334–346
   applications, 346
   ideal, 334
Transmission lines, 429, 430–437
   as circuit element, 431
   impedance in, 430–432, 435, 436–437
   two-wire, 430
Transmit-receive switch, in radar system, 2, 3
Transmitter, in radar system, 2–3
Transmitter circuits, in radar system, 2
Trigger circuit:
   Eccles-Jordan (*see* Eccles-Jordan trigger circuit)
   Schmitt (*see* Schmitt trigger circuit)
Trigger level control, 477, 478
   in sweep generator, 455–456
Trigger slope switch, 477
   in sweep generator, 456
Trigger source switch, in sweep generator, 455
Triggered blocking oscillator (*see* Oscillator, blocking, triggered)
Triggered sweep circuit, 321
Triggered sweep generator, 322
Triggering, parallel:
   of transistor blocking oscillators, 307
   of vacuum-tube blocking oscillators, 304–305, 306
Triggering, series:
   of transistor blocking oscillator, 307
   of vacuum-tube blocking oscillator, 304–306
Triggering devices:
   in transistor bistable MV, 257–260
   in vacuum-like bistable MV, 255–257
Triggering methods, in blocking oscillators, 303–308
Triggering pulse, 222
   in blocking oscillators:
     tarnsistor, 303
     vacuum-tube monostable, 301–302
   in multivibrators, bistable, 251ff
     transistor, 257–260
     vacuum-tube, 255–257
   in multivibrators, monostable, 269–271

Index

Triggering pulse (*cont.*):
    vacuum-tube cathode-coupled, 273–275
  negative, in bistable multivibrators, 253
Triode, high-vacuum thermionic, 167–170
Triodes, 2, 108, 176
  as electronic switch, compared to diode, 167
  grid-cathode circuit, 183
  region, in switching operations, 168–169
  switching speed, factors acting to reduce, 169
  as triggering device, 257
  twin, 12B27, plate characteristics, 44–45
Two-terminal pair network, 80 (*see also* Four-terminal network)

UNDERSHOOT, 300
  defined, 8
Universal time-constant chart, 16–18
Unsymmetrical waveforms, 24

VACUUM-TUBE amplifier (*see* Amplifier, vacuum-tube)
Vacuum-tube circuits, sag in, 113
Vacuum-tube sweep circuits, 320, 321
  improving linearity of, 321–325

Vacuum tubes, 79, 107, 108
  amplification factor, 43, 44
    in constant-voltage equivalent circuit, 46
  compared with transistor, 55
  figure-of-merit used in selecting, 107–108
  gain-bandwidth product, 106
  methods of high-frequency compensation, 109–112
  reducing nonlinear distortion in, 124–125
  use, for bandwidth application, 108
Vernier control, 468, 475
  external, in horizontal amplifier, 460
Video amplifier (*see* Amplifier, video)
Video pulses, in radar, 3
Visual readout device, for scale-of-16 counter, 353–354
Volt-ampere plots, used in studying transistors, 61–64
Voltage gain, in cathode-follower circuit, 141–142
Voltage selector, 176 (*see also* Clipper)
Voltage source T-equivalent circuit, 82–83

WAVEFORM, 1
  reproduced, defects in, 105–106

ZENER diodes, 164
Zener region, 209